Progress in Probability

Volume 43

Series Editors
Thomas Liggett
Charles Newman
Loren Pitt

High Dimensional Probability

Ernst Eberlein
Marjorie Hahn
Michel Talagrand
Editors

1998

Birkhäuser Verlag
Basel · Boston · Berlin

Editors' addresses:

Ernst Eberlein
Institut für Mathematische Stochastik
Universität Freiburg
Eckerstraße 1
79104 Freiburg
Germany

Marjorie Hahn
Department of Mathematics
Tufts University
Medford, MA 02155
USA

Michel Talagrand
Equipe d'analyse, Tour 46
Université Paris VI
4, place Jussieu
75230 Paris Cedex 05
France

1991 Mathematics Subject Classification: 60B11, 60B12, 60G15

A CIP catalogue record for this book is available from the Library of Congress, Washington D.C., USA

Deutsche Bibliothek Cataloging-in-Publication Data

High dimensional probability / Ernst Eberlein ... ed. – Basel ;
Boston ; Berlin : Birkhäuser, 1998
  (Progress in probability ; Vol. 43)
  ISBN 3-7643-5867-X (Basel ...)
  ISBN 0-8176-5867-X (Boston)

© 1998 Birkhäuser Verlag Basel, P.O. Box 133, CH-4010 Basel
Printed on acid-free paper produced from chlorine-free pulp. TCF ∞
ISBN 3-7643-5867-X
ISBN 0-8176-5867-X

9 8 7 6 5 4 3 2 1

Contents

Introduction

What is high dimensional probability? Under this broad name we collect topics with a common philosophy, where the idea of high dimension plays a key role, either in the problem or in the methods by which it is approached. Let us give a specific example that can be immediately understood, that of Gaussian processes.

Roughly speaking, before 1970, the Gaussian processes that were studied were indexed by a subset of Euclidean space, mostly with dimension at most three. Assuming some regularity on the covariance, one tried to take advantage of the structure of the index set. Around 1970 it was understood, in particular by Dudley, Feldman, Gross, and Segal that a more abstract and intrinsic point of view was much more fruitful. The index set was no longer considered as a subset of Euclidean space, but simply as a metric space with the metric canonically induced by the process. This shift in perspective subsequently lead to a considerable clarification of many aspects of Gaussian process theory, and also to its applications in other settings.

While the theory of "abstract" Gaussian processes is an important part of high dimensional probability, there are many other topics currently studied. The authors whose articles appear in this volume participated in a conference entitled "High Dimensional Probability" which was held at the Mathematical Research Institute at Oberwolfach in August 1996. That conference followed a long series of meetings that were called "Probability in Banach Spaces". The historical reason for this name was the long-time challenge of extending the classical results of probability to the vector valued case either by proving infinite dimensional results, or by proving finite dimensional results with estimates independent of the dimension.

Particular attention was focused on the Central Limit Theorem, the Law of Large Numbers, and the Law of the Iterated Logarithm. This goal has largely been achieved through the efforts of many, and hence the name of "Probability in Banach Spaces", that carried the idea of the study of these limit laws, is no longer appropriate to describe the full range of activities in this area.

One of the most remarkable features of the study of Probability in Banach spaces is that this study has given impetus to a number of methods whose importance goes far beyond the original goal of extending limit laws to the vector valued case. For example, the theory of empirical processes over "general" classes of functions is a topic that is close, both in spirit and by the techniques it uses, to the study of limit laws in Banach spaces. The theory of empirical processes is very much alive and well represented in this volume. The method of majorizing measures, a tool to study regularity of stochastic processes, and the theory of concentration of measure are examples which demonstrate that the search for proper generality and abstraction can be an essential motor of progress.

The papers in this volume reflect the vitality and diversity of the newly evolving field of "High Dimensional Probability". Yet, most of the papers exhibit the influence of the past through two common ingredients – high dimensions and the use of abstract methods.

The Editors

Ernst Eberlein
Marjorie Hahn
Michel Talagrand
June, 1997

Progress in Probability, Vol. 43
© 1998 Birkhäuser Verlag Basel/Switzerland

Weak Convergence of the Row Sums of a Triangular Array of Empirical Processes

Miguel A. Arcones[*]

ABSTRACT. We study the weak convergence of the row sums of a general triangular array of empirical processes converging to an arbitrary limit. In particular, our results apply to infinitesimal arrays and random series processes. We give some sufficient finite dimensional approximation conditions for the weak convergence of these processes. These conditions are necessary under quite minimal regularity assumptions.

1. Introduction

We consider the weak convergence of the row sums of a triangular array of empirical processes. The set–up that we consider is as follows. Let $(\Omega_n, \mathcal{A}_n, Q_n)$ be a sequence of probability spaces. Let $(S_{n,j}, \mathcal{S}_{n,j})$ be measurable spaces for $1 \leq j \leq k_n$, where $\{k_n\}_{n=1}^{\infty}$ is a sequence of positive integers converging to infinity. Let $\{X_{n,j} : 1 \leq j \leq k_n\}$ be $S_{n,j}$–valued independent r.v.'s defined on Ω_n. To avoid measurability problems, we assume that $\Omega_n = \prod_{j=1}^{k_n} S_{n,j}$, $\mathcal{A}_n = \prod_{j=1}^{k_n} \mathcal{S}_{n,j}$ and $Q_n = \prod_{j=1}^{k_n} \mathcal{L}(X_{n,j})$. Let $f_{n,j}(\cdot, t) : S_{n,j} \to \mathbb{R}$ be a measurable function for each $1 \leq j \leq k_n$, each $n \geq 1$ and each $t \in T$. Let $c_n(t)$ be a real number for each $t \in T$ and each $n \geq 1$. Let

$$Z_n(t) := \left(\sum_{j=1}^{k_n} f_{n,j}(X_{n,j}, t) \right) - c_n(t). \tag{1.1}$$

In this paper we study the weak convergence of the sequence of stochastic processes $\{Z_n(t) : t \in T\}$. As a particular case, we consider sums of i.i.d. stochastic processes. Let $\{X_j\}_{j=1}^{\infty}$ be a sequence of i.i.d.r.v.'s with values in a measurable space $(S, \mathcal{S})$, let X be a copy of X_1, let $f(\cdot, t) : S \to \mathbb{R}$ be a measurable function for each $t \in T$, let $\{a_n\}_{n=1}^{\infty}$ be a sequence of positive numbers converging to infinity and let $c_n(t)$ be a real number. The sequence of processes

$$\left\{ Z_n(t) := \left(a_n^{-1} \sum_{j=1}^{n} f(X_j, t) \right) - c_n(t) : t \in T \right\}, \quad n \geq 1, \tag{1.2}$$

is a particular case of the sequence of processes in (1.1). It is well known that the limit set of (1.2) is either a Gaussian or a stable process. It is also known that if

*)   Research partially supported by NSF Grant DMS–93–02583 and carried out at the Department of Mathematics of the University of Utah.

(1.2) converges to a nondegenerate limit, then $\{a_n\}_{n=1}^{\infty}$ is regularly varying with exponent α^{-1}, with $0 < \alpha \leq 2$. We refer for more on this topic to Gnedenko and Kolmogorov (1968). The processes $\{Z_n(t) : t \in T\}$ in (1.2) are a generalization of the processes called empirical processes ($a_n = n^{1/2}$ and $c_n(t) = n^{1/2}E[f(X,t)]$). We refer to Dudley (1984), Giné and Zinn (1986), Pollard (1990) and Talagrand (1987) for more on this area. Another interesting sequence of stochastic processes that is a particular case of (1.1) is

$$\{Z_n(t) := \sum_{j=1}^{n} f_j(X_j, t) : t \in T\}, \tag{1.3}$$

i.e. a random series process. The set–up in this case is as follows. $\{X_j\}$ are independent r.v.'s. X_j takes values in $(S_j, \mathcal{S}_j)$. $f_j(\cdot, t) : S_j \to \mathbb{R}$ is a measurable function for each $1 \leq j$ and each $t \in T$. Particular types of random series processes have been considered by Kahane (1968), Marcus and Pisier (1981) and Kwapień and Woyczyński (1992). We may have processes that are a mixture of the processes in (1.2) and (1.3). For example, a triangular array that is a mixture of the processes in (1.2) and (1.3) is the following: let $\{\xi_j\}_{j=1}^{\infty}$ be a sequence of i.i.d. symmetric r.v.'s with finite second moment. Let $\{a_j\}_{j=1}^{\infty}$ be a sequence of real numbers such that $\sum_{j=1}^{n} a_j^2 < \infty$. Let $X_{n,j} = n^{-1/2}\xi_j$, for $1 \leq j \leq n$, and let $X_{n,j} = a_{j-n}\xi_{j-n}$, for $n+1 \leq j \leq 2n$. Then, $\sum_{j=1}^{2n} X_{n,j}$ converges in distribution. The limit distribution is the mixture of a normal distribution and $\sum_{j=1}^{\infty} a_j\xi_j$.

There are several applications of the weak convergence of the type of stochastic processes in (1.1) to statistics (see for example Le Cam, 1986; Kim and Pollard, 1990; Pollard, 1990; and Arcones, 1994a, 1994b). The processes called partial–sum processes, which have been studied by several authors, are a particular case of the processes in (1.1) (see Alexander, 1987, Section IV; Arcones, Gaenssler, and Ziegler, 1992; and Gaenssler and Ziegler, 1994). The study of the weak convergence of sequences of processes similar to the one in (1.1) has been considered by several authors. Alexander (1987) and Pollard (1990) considered the weak convergence of the sequence of processes in (1.1) to a Gaussian process. Andersen, Giné, Ossiander and Zinn (1988) and Andersen, Giné and Zinn (1988) considered the weak convergence of the process in (1.1), under restricted bracketing conditions, when $f_{n,j}(\cdot, t)$ does not change neither with n nor with j to particular types of limit processes. Here, we give minimal sufficient conditions for the weak convergence of an arbitrary sequence of stochastic processes as in (1.1).

We will use the following definition of weak convergence:

DEFINITION 1.1. *(Hoffmann–Jørgensen, 1991). Let $\{Z_n(t) : t \in T\}$, $n \geq 1$, be a sequence of stochastic processes, and let $\{Z(t) : t \in T\}$ be another stochastic process. We say that the sequence of stochastic processes $\{Z_n(t) : t \in T\}$, $n \geq 1$, converges weakly to $\{Z(t) : t \in T\}$ in $l_{\infty}(T)$ if:*

(i) *For each n large enough, $\sup_{t \in T} |Z_n(t)| < \infty$ a.s.*
(ii) *There exists a separable set A of $l_{\infty}(T)$ such that $\Pr^*\{Z \in A\} = 1$.*
(iii) *For each bounded, continuous function H in $l_{\infty}(T)$, $E^*[H(Z_n)] \to E[H(Z)]$, where E^* means outer expectation.*

We will denote this by $\{Z_n(t) : t \in T\} \xrightarrow{w} \{Z(t) : t \in T\}$. It is well known that the sequence of stochastic processes $\{Z_n(t) : t \in T\}$, $n \geq 1$, converges weakly to $\{Z(t) : t \in T\}$ in $l_\infty(T)$ if and only if:

(i) The finite dimensional distributions of $\{Z_n(t) : t \in T\}$ converge to those of $\{Z(t) : t \in T\}$.

(ii) For each $\eta > 0$, there exists a map $\pi : T \to T$ such that $\pi(\pi(t)) = \pi(t)$, for each $t \in T$, the cardinality of $\{\pi(t) : t \in T\}$ is finite and

$$\limsup_{n \to \infty} \Pr^* \{ \sup_{t \in T} |Z_n(t) - Z_n(\pi(t))| \geq \eta \} \leq \eta. \tag{1.4}$$

Condition (1.4) is called a finite dimensional approximation condition, because $\{\pi(t) : t \in T\}$ takes only finitely many values. We will denote to a finite partition π of T to a map $\pi : T \to T$ such that $\pi(\pi(t)) = \pi(t)$, for each $t \in T$, and the cardinality of $\{\pi(t) : t \in T\}$ is finite. In Section 2, we will give sufficient finite dimensional approximation conditions in different truncations of the process to guarantee the weak convergence of the empirical processes in (1.1). Under minimal regularity conditions, previous conditions are also necessary. We will see how these conditions specialize in different cases.

Weak convergence of the row sums of a triangular array of empirical processes in specific cases, such as VC and bracketing cases, will be discussed in future publications.

2. Finite dimensional approximation conditions for the weak convergence of row sums of a triangular array of empirical processes

First, we give some sufficient conditions for the weak convergence of processes in (1.1). These conditions are finite dimensional approximation conditions in different truncations of the processes.

THEOREM 2.1. *Let $b > 0$ and suppose that:*

(i) *The finite dimensional distributions of*

$$\left\{ Z_n(t) := \left(\sum_{j=1}^{k_n} f_{n,j}(X_{n,j}, t) \right) - c_n(t) : t \in T \right\}$$

converge to those of $\{Z(t) : t \in T\}$.

(ii) *For each $t \in T$, $\sup_{n \geq 1} \sum_{j=1}^{k_n} \Pr\{|f_{n,j}(X_{n,j}, t)| \geq 2^{-1}b\} < \infty$.*

(iii) *For each $\eta > 0$, there exists a finite partition π of T such that*

$$\limsup_{n \to \infty} \sum_{j=1}^{k_n} \Pr^* \{ \sup_{t \in T} |f_{n,j}(X_{n,j}, t) - f_{n,j}(X_{n,j}, \pi(t))| \geq \eta \} \leq \eta.$$

(iv) *For each $\eta > 0$, there exists a finite partition π of T such that*

$$\limsup_{n \to \infty} E^* [\sup_{t \in T} |S_n(t, b) - S_n(\pi(t), b) - E[S_n(t, b) - E[S_n(\pi(t), b)]|] \leq \eta,$$

where

$$S_n(t,b) = \sum_{j=1}^{k_n} f_{n,j}(X_{n,j},t) I_{F_{n,j}(X_{n,j}) \leq b}$$

and.

$$F_{n,j}(x) = \sup_{t \in T} |f_{n,j}(x,t)|.$$

(v) *For each $\eta > 0$, there exists a finite partition π of T such that*

$$\limsup_{n \to \infty} \sup_{t \in T} |E[S_n(t,b) - S_n(\pi(t),b)] - c_n(t) + c_n(\pi(t))| \leq \eta.$$

Then, the sequence of stochastic processes $\{Z_n(t) : t \in T\}$, $n \geq 1$, converges weakly to $\{Z(t) : t \in T\}$.

Proof. It suffices to prove that for each $\eta > 0$, there exists a partition π of T such that

$$\limsup_{n \to \infty} \Pr{}^* \{\sup_{t \in T} |Z_n(t) - Z_n(\pi(t))| \geq 4\eta\} \leq 4\eta.$$

Observe that conditions (ii) and (iii) imply that

$$\sup_{n \geq 1} \sum_{j=1}^{k_n} \Pr\{F_{n,j}(X_{n,j}) \geq b\} < \infty.$$

Given $1/10 > \eta > 0$, take $a > 0$ such that

$$a \sup_{n \geq 1} \sum_{j=1}^{k_n} \Pr\{F_{n,j}(X_{n,j}) \geq b\} \leq \eta^2.$$

Take a finite partition π of T such that

$$\limsup_{n \to \infty} \sum_{j=1}^{k_n} \Pr{}^* \{\sup_{t \in T} |f_{n,j}(X_{n,j},t) - f_{n,j}(X_{n,j},\pi(t))| \geq a\} < \eta,$$

$$\limsup_{n \to \infty} E^* [\sup_{t \in T} |S_n(t,b) - S_n(\pi(t),b) - E[S_n(t,b) - S_n(\pi(t),b)]|] < \eta^2,$$

and

$$\limsup_{n \to \infty} \sup_{t \in T} |E[S_n(t,b) - S_n(\pi(t),b)] - c_n(t) + c_n(\pi(t))| < \eta.$$

Let $Y_{n,j} = \sup_{t \in T} |f_{n,j}(X_{n,j}, t) - f_{n,j}(X_{n,j}, \pi(t))|$. Then,

$$\sum_{j=1}^{k_n} (f_{n,j}(X_{n,j}, t) - f_{n,j}(X_{n,j}, \pi(t))) - (c_n(t) - c_n(\pi(t)))$$

$$= S_n(t, b) - S_n(\pi(t), b) - E[S_n(t, b) - S_n(\pi t, b)]$$

$$+ \sum_{j=1}^{k_n} (f_{n,j}(X_{n,j}, t) - f_{n,j}(X_{n,j}, \pi(t))) I_{F_{n,j}(X_{n,j}) > b, \, Y_{n,j} \leq a}$$

$$+ \sum_{j=1}^{k_n} (f_{n,j}(X_{n,j}, t) - f_{n,j}(X_{n,j}, \pi(t))) I_{F_{n,j}(X_{n,j}) > b, \, Y_{n,j} > a} \qquad (2.1)$$

$$+ E[S_n(t, b) - S_n(\pi(t), b)] - c_n(t) + c_n(\pi(t))$$

$$=: I(t) + II(t) + III(t) + IV(t).$$

We have that

$$\Pr{}^* \{ \sup_{t \in T} |I(t)| \geq \eta \} \leq \eta,$$

$$\Pr{}^* \{ \sup_{t \in T} |II(t)| \geq \eta \} \leq a\eta^{-1} \sum_{j=1}^{k_n} \Pr\{ F_{n,j}(X_{n,j}) \geq b \} \leq \eta,$$

$$\Pr{}^* \{ \sup_{t \in T} |III(t)| > 0 \} \leq \sum_{j=1}^{k_n} \Pr{}^* \{ Y_{n,j} > a \} \leq \eta,$$

and

$$\sup_{t \in T} |IV(t)| \leq \eta.$$

Therefore, the claim follows. $\qquad\square$

Conditions (i)–(iii) and (v) in the previous theorem are easy to check. Only condition (iv) is a difficult to check condition. There are different techniques for checking this condition in the empirical process literature.

Observe that condition (i) in Theorem 2.1 is a necessary condition. Condition (ii) is not necessary, but it is necessary under symmetry. For example, if $X_{n,j} = a_{n,j}$, for each $1 \leq j \leq k_n$, for some constant $a_{n,j}$. We may have that $\sum_{j=1}^{k_n} a_{n,j}$ converge due to cancellations, but $\sum_{j=1}^{k_n} I_{|a_{n,j}| \geq \eta} \to \infty$, for each $\eta > 0$. Suppose that $\{X_{n,j}, \, 1 \leq j \leq k_n\}$ is a triangular array of symmetric r.v.'s, such that $\sum_{j=1}^{k_n} X_{n,j}$ converges in distribution to L, then if $\Pr\{|L| \geq M\} < 1/2$, then

$$\limsup_{n \to \infty} \sum_{j=1}^{k_n} \Pr\{|X_{n,j}| \geq M\} \leq (1 - 2\Pr\{|L| \geq M\})^{-1} \Pr\{|L| \geq M\}$$

(this follows from Lévy's inequality (see e.g. Proposition 2.3 in Ledoux and Talagrand, 1991) and Lemma 2.6 in Ledoux and Talagrand, 1991). Next, we see that under some regularity conditions, (iii)–(v) are necessary in Theorem 2.1.

THEOREM 2.2. *Let $b > 0$ and suppose that:*

(i) $\left\{ \left(\sum_{j=1}^{k_n} f_{n,j}(X_{n,j}, t) \right) - c_n(t) : t \in T \right\}$ *converge weakly to* $\{Z(t) : t \in T\}$.

(ii) *For each* $t \in T$, $\sup_{n \geq 1} \sum_{j=1}^{k_n} \Pr\{|f_{n,j}(X_{n,j}, t)| \geq 2^{-1}b\} < \infty$.

(iii) *For each* $\eta > 0$, *there exists a finite partition* π *of* T *such that*

$$\limsup_{n \to \infty} \max_{1 \leq j \leq k_n} \sup_{t \in T} \Pr\{|f_{n,j}(X_{n,j}, t) - f_{n,j}(X_{n,j}, \pi(t))| \geq \eta\} \leq \eta.$$

Then,

(b.1) *For each* $\eta > 0$, *there exists a finite partition* π *of* T *such that*

$$\limsup_{n \to \infty} \sum_{j=1}^{k_n} \Pr{}^{*}\{\sup_{t \in T} |f_{n,j}(X_{n,j}, t) - f_{n,j}(X_{n,j}, \pi(t))| \geq \eta\} \leq \eta.$$

(b.2) *For each* $\eta > 0$, *there exists a finite partition* π *of* T *such that*

$$\limsup_{n \to \infty} E^{*}[\sup_{t \in T} |S_n(t, b) - S_n(\pi(t), b) - E[S_n(t, b) - S_n(\pi(t), b)]|] \leq \eta,$$

where $S_n(t, b) = \sum_{j=1}^{k_n} f_{n,j}(X_{n,j}, t) I_{F_{n,j}(X_{n,j}) \leq b}$.

(b.3) *For each* $\eta > 0$, *there exists a finite partition* π *of* T *such that*

$$\limsup_{n \to \infty} \sup_{t \in T} |E[S_n(t, b) - S_n(\pi(t), b)] - c_n(t) + c_n(\pi(t))| \leq \eta.$$

Proof. Let $\{X'_{n,j} : 1 \leq j \leq k_n, 1 \leq n\}$ be an independent copy of $\{X_{n,j} : 1 \leq j \leq k_n, 1 \leq n\}$. Let $\{\epsilon_j\}_{j=1}^{\infty}$ be a Rademacher sequence independent of $\{X_{n,j} : 1 \leq j \leq k_n, 1 \leq n\}$ and $\{X'_{n,j} : 1 \leq j \leq k_n, 1 \leq n\}$. Then, (i) implies that

$$\{\sum_{j=1}^{k_n} \epsilon_j(f_{n,j}(X_{n,j}, t) - f_{n,j}(X'_{n,j}, t)) : t \in T\}$$

converges weakly. So, given $1/20 > \eta > 0$, there exists a finite partition π of T such that

$$\Pr{}^{*}\{\sup_{t \in T} | \sum_{j=1}^{k_n} (f_{n,j}(X_{n,j}, t) - f_{n,j}(X_{n,j}, \pi(t)) \tag{2.2}$$
$$- f_{n,j}(X'_{n,j}, t) + f_{n,j}(X'_{n,j}, \pi(t)))| \geq \eta\} \leq \eta$$

and

$$\sup_{t \in T} \sup_{1 \leq j \leq k_n} \Pr\{|f_{n,j}(X_{n,j}, t) - f_{n,j}(X_{n,j}, \pi(t))| \geq \eta\} \leq 1/2. \tag{2.3}$$

(2.2) and Lemma 2.6 and Proposition 2.3 in Ledoux and Talagrand (1991) give
that

$$
\begin{aligned}
\sum_{j=1}^{k_n} & \Pr{}^*\{\sup_{t\in T} |f_{n,j}(X_{n,j},t) - f_{n,j}(X_{n,j},\pi(t)) \\
& - f_{n,j}(X'_{n,j},t) + f_{n,j}(X'_{n,j},\pi(t))| \geq \eta\} \\
& \leq 4\Pr{}^*\{\sup_{t\in T} |\sum_{j=1}^{k_n} \epsilon_j (f_{n,j}(X_{n,j},t) - f_{n,j}(X_{n,j},\pi(t)) \\
& - f_{n,j}(X'_{n,j},t) + f_{n,j}(X'_{n,j},\pi(t)))| \geq \eta\} \leq 4\eta.
\end{aligned}
\tag{2.4}
$$

(2.3) and Lemma 1.2.1 in Giné and Zinn (1986) imply that

$$
\begin{aligned}
\Pr{}^*\{&\sup_{t\in T} |f_{n,j}(X_{n,j},t) - f_{n,j}(X_{n,j},\pi(t))| \geq 2\eta\} \\
& \leq 2\Pr{}^*\{\sup_{t\in T} |f_{n,j}(X_{n,j},t) - f_{n,j}(X_{n,j},\pi(t)) \\
& - f_{n,j}(X'_{n,j},t) + f_{n,j}(X'_{n,j},\pi(t))| \geq \eta\},
\end{aligned}
$$

for each $1 \leq j \leq k_n$ and n large enough. From this and (2.4)

$$
\sum_{j=1}^{k_n} \Pr{}^*\{\sup_{t\in T} |f_{n,j}(X_{n,j},t) - f_{n,j}(X_{n,j},\pi(t))| \geq 2\eta\} \leq 8\eta,
$$

for n large enough. Therefore, (b.1) holds.

Next, we prove (b.2). Given $1/10 > \eta > 0$, choose $a > 0$ such that

$$
a \sup_{n\geq 1} \sum_{j=1}^{k_n} \Pr\{F_{n,j}(X_{n,j}) \geq b\} \leq \eta^2.
$$

Choose a finite partition π of T such that

$$
\begin{aligned}
\Pr{}^*\{&\sup_{t\in T} |\sum_{j=1}^{k_n} (f_{n,j}(X_{n,j},t) - f_{n,j}(X_{n,j},\pi(t)) \\
& - f_{n,j}(X'_{n,j},t) + f_{n,j}(X'_{n,j},\pi(t)))| \geq \eta\} \leq \eta,
\end{aligned}
$$

and

$$
\sum_{j=1}^{k_n} \Pr{}^*\{\sup_{t\in T} |f_{n,j}(X_{n,j},t) - f_{n,j}(X_{n,j},\pi(t))| \geq a\} \leq \eta.
$$

Let $Y_{n,j} = \sup_{t\in T} |f_{n,j}(X_{n,j},t) - f_{n,j}(X_{n,j},\pi(t))|$, let $Y'_{n,j} = \sup_{t\in T} |f_{n,j}(X'_{n,j},t) - f_{n,j}(X'_{n,j},\pi(t))|$, let $S_n(t,b) = \sum_{j=1}^{k_n} f_{n,j}(X_{n,j},t) I_{F_{n,j}(X_{n,j})\leq b}$ and let $S'_n(t,b) =$

$\sum_{j=1}^{k_n} f_{n,j}(X'_{n,j},t)I_{F_{n,j}(X'_{n,j})\leq b}$. Then,

$$S_n(t,b) - S_n(\pi(t),b) - S'_n(t,b) + S'_n(\pi(t),b)$$

$$= \sum_{j=1}^{k_n}(f_{n,j}(X_{n,j},t) - f_{n,j}(X_{n,j},\pi(t)) - f_{n,j}(X'_{n,j},t) + f_{n,j}(X'_{n,j},\pi(t)))$$

$$- \sum_{j=1}^{k_n}(f_{n,j}(X_{n,j},t) - f_{n,j}(X_{n,j},\pi(t)))I_{F_{n,j}(X_{n,j})>b,\ Y_{n,j}\leq a}$$

$$+ \sum_{j=1}^{k_n}(f_{n,j}(X'_{n,j},t) - f_{n,j}(X'_{n,j},\pi(t)))I_{F_{n,j}(X'_{n,j})>b,\ Y'_{n,j}\leq a}$$

$$- \sum_{j=1}^{k_n}(f_{n,j}(X_{n,j},t) - f_{n,j}(X_{n,j},\pi(t)))I_{F_{n,j}(X_{n,j})>b,\ Y_{n,j}>a}$$

$$+ \sum_{j=1}^{k_n}(f_{n,j}(X'_{n,j},t) - f_{n,j}(X'_{n,j},\pi(t)))I_{F_{n,j}(X'_{n,j})>b,\ Y'_{n,j}>a}.$$

It is easy to see from this decomposition that

$$\Pr{}^*\{\sup_{t\in T}|S_n(t,b) - S_n(\pi(t),b) - S'_n(t,b) + S'_n(\pi(t),b)| \geq 5\eta\} \leq 5\eta,$$

for n large enough. So, by the Hoffmann Jørgensen inequality (see for example Proposition 6.8 in Ledoux and Talagrand (1991)

$$E^*[\sup_{t\in T}|S_n(t,b) - S_n(\pi(t),b) - S'_n(t,b) + S'_n(\pi(t),b)|]$$

$$\leq 6E^*\Big[\sup_{1\leq j\leq k_n}\sup_{t\in T}|(f_{n,j}(X_{n,j},t) - f_{n,j}(X_{n,j},\pi(t)))I_{F_{n,j}(X_{n,j})\leq b}$$

$$- (f_{n,j}(X'_{n,j},t) - f_{n,j}(X'_{n,j},\pi(t)))I_{F_{n,j}(X'_{n,j})\leq b}|\Big] + 30\eta$$

$$\leq 24b\Pr{}^*\Big\{\sup_{1\leq j\leq k_n}\sup_{t\in T}|(f_{n,j}(X_{n,j},t) - f_{n,j}(X_{n,j},\pi(t)))I_{F_{n,j}(X_{n,j})\leq b}$$

$$- (f_{n,j}(X'_{n,j},t) - f_{n,j}(X'_{n,j},\pi(t)))I_{F_{n,j}(X'_{n,j})\leq b}| \geq 5\eta\Big\} + 60\eta$$

$$\leq 48b\Pr{}^*\{\sup_{t\in T}|S_n(t,b) - S_n(\pi(t),b) - S'_n(t,b) + S'_n(\pi(t),b)| \geq 5\eta\} + 60\eta$$

$$\leq 240b\eta + 60\eta,$$

for n large enough. Hence, (b.2) follows.

(b.3) follows from (b.1) and (b.2) and the decomposition (2.1). □

Condition (ii) and (iii) in Theorem 2.2 are weak regularity conditions. Previous arguments give that the weak convergence of $\{\sum_{j=1}^{k_n}\epsilon_j f_{n,j}(X_{n,j},t) : t \in T\}$ implies (ii) and (iii).

In the case of random series processes, we have the following theorem:

THEOREM 2.3. *With the notation in (1.3), let $b > 0$. Suppose that for each $\eta > 0$ and each $j \geq 1$, there exists a finite partition π of T such that*

$$\sup_{t \in T} \Pr\{|f_j(X_j, t) - f_j(X_j, \pi(t))| \geq \eta\} \leq \eta. \tag{2.5}$$

Then, the following sets of conditions ((a) and (b)) are equivalent:

(a) *The sequence of stochastic processes $\{\sum_{j=1}^n f_j(X_j, t) : t \in T\}$, $n \geq 1$, converges weakly.*

(b.1) *For each $t \in T$, $\sum_{j=1}^n f_j(X_j, t)$ converges in distribution.*

(b.2) *For each $\eta > 0$, there exists a finite partition π of T such that*

$$\sum_{j=1}^{\infty} \Pr{}^*\{\sup_{t \in T} |f_j(X_j, t) - f_j(X_j, \pi(t))| \geq \eta\} \leq \eta.$$

(b.3) *For each $\eta > 0$, there exists a finite partition π of T such that*

$$\limsup_{n \to \infty} E^*[\sup_{t \in T} |S_n(t, b) - S_n(\pi(t), b) - E[S_n(t, b) - S_n(\pi(t), b)]|] \leq \eta,$$

where $S_n(t, b) = \sum_{j=1}^n f_j(X_j, t) I_{F_j(X_j) \leq b}$ and $F_j(x) = \sup_{t \in T} |f_j(x, t)|$.

(b.4) *For each $\eta > 0$, there exists a finite partition π of T such that*

$$\limsup_{n \to \infty} \sup_{t \in T} |E[S_n(t, b) - S_n(\pi(t), b)]| \leq \eta.$$

Proof. Part (a) implies (b) follows from Theorem 2.2. Observe that the weak convergence of $\{\sum_{j=1}^n f(X_j, t) : t \in T\}$, $n \geq 1$, implies that $\{\sum_{j=1}^n f(X_j, t) : t \in T\}$, $n \geq 1$, converges in distribution for each $t \in T$. This fact, by the three series theorem, implies that

$$\sum_{j=1}^{\infty} \Pr\{|f_j(X_j, t)| \geq 2^{-1}b\} < \infty,$$

for each $b > 0$ and each $t \in T$. So, condition (ii) in Theorem 2.2 holds. In this situation, the part of Ito–Nisio theorem that says convergence in distribution implies convergence almost surely is true. This follows just doing minor variation in the regular proof of this theorem (see for example the proof of Theorem 6.1 in Ledoux and Talagrand, 1991). So,

$$\Pr\{\sup_{t \in T} |f_j(X_j, t)| \geq \eta\} \to 0,$$

for each $\eta > 0$. This and (2.5) imply condition (iii) in Theorem 2.2. Part (b) implies (a) follows from Theorem 2.1. Condition (ii) in Theorem 2.1 follows from the three series theorem. $\qquad\square$

In the considered situation, it is not true that $\sup_{t\in T}|\sum_{j=1}^{n} f_j(X_j,t)|$ converges almost surely implies that $\{\sum_{j=1}^{n} f_j(X_j,t) : t \in T\}$, $n \geq 1$, converges weakly. In other words the whole Ito–Nisio theorem is not true. Let $\{f_1(X_1,t) : t \in T\}$ be a stochastic process such $\sup_{t\in T}|f_1(X_1,t)| < \infty$ a.s., but there exists no separable set A of $l_\infty(T)$ such that $\Pr^*\{\{f_1(X_1,t) : t \in T\} \in A\} = 1$. Let $f_j(X_j,t) = 0$, for $j \geq 2$. Here, we have that $\sup_{t\in T}|\sum_{j=1}^{n} f_j(X_j,t)|$ converges a.s. But, $\{\sum_{j=1}^{n} f_j(X_j,t) : t \in T\}$, $n \geq 1$, does not converge weakly. An example of process $\{f_1(X_1,t) : t \in T\}$ as above is the following. Let $\{\xi_k\}_{k=1}^{\infty}$ be a sequence of i.i.d.r.v.'s with standard normal distribution. Let $T = \{1,2,\ldots\}$, let $X_1 = (\xi_1,\xi_2,\ldots)$ and let $f(x,k) = (\log(k+1))^{-1/2}x_k$. Then, $\{f(X_1,k) : k \in T\}$ satisfies $\sup_{k\in T}|f(X_1,k)| < \infty$, but $\{f(X_1,k) : k \in T\}$ does not have a tight law in $l_\infty(T)$.

Condition (2.5) is a very weak condition. In the symmetric case, (a) implies this condition.

We also have that Theorems 2.1 and 2.2 simplify for infinitesimal arrays:

THEOREM 2.4. *With the above notation, suppose that*

$$\max_{1\leq j\leq k_n} \sup_{t\in T} \Pr\{|f_{n,j}(X_{n,j},t)| \geq \eta\} \to 0, \tag{2.6}$$

for each $\eta > 0$. Let $b > 0$. Then, the following sets of conditions ((a) and (b)) are equivalent:

(a) *The sequence of stochastic processes $\{Z_n(t) : t \in T\}$, $n \geq 1$, converges weakly to $\{Z(t) : t \in T\}$.*

(b.1) *The finite dimensional distributions of $\{Z_n(t) : t \in T\}$ converge to those of $\{Z(t) : t \in T\}$.*

(b.2) *For each $\eta > 0$, there exists a finite partition π of T such that*

$$\limsup_{n\to\infty} \sum_{j=1}^{k_n} \Pr^*\{\sup_{t\in T}|f_{n,j}(X_{n,j},t) - f_{n,j}(X_{n,j},\pi(t))| \geq \eta\} \leq \eta.$$

(b.3) *For each $\eta > 0$, there exists a finite partition π of T such that*

$$\limsup_{n\to\infty} E^*[\sup_{t\in T}|S_n(t,b) - S_n(\pi t, b) - E[S_n(t,b) - S_n(\pi(t),b)]|] \leq \eta,$$

where $S_n(t,b) = \sum_{j=1}^{k_n} f_{n,j}(X_{n,j},t)I_{F_{n,j}(X_{n,j})\leq b}$.

(b.4) *For each $\eta > 0$, there exists a finite partition π of T such that*

$$\limsup_{n\to\infty} \sup_{t\in T}|E[S_n(t,b) - S_n(\pi(t),b)] - c_n(t) + c_n(\pi(t))| \leq \eta.$$

Proof. Part (a) implies (b) follows from Theorem 2.2. (ii) in Theorem 2.2 follows from Theorem 1.2. The convergence of the finite dimensional distributions imply (ii) in Theorem 2.2. (2.6) implies (iii) in Theorem 2.2. Part (b) implies (a) follows from Theorem 2.1. Observe that (ii) in Theorem 2.1 follows from Theorem 1.2.   □

We must observe that we need condition (2.6) in Theorem 2.4. It is not sufficient to have that

$$\max_{1 \le j \le k_n} \Pr\{|f_{n,j}(X_{n,j}, t)| \ge \eta\} \to 0,$$

for each $\eta > 0$ and each $t \in T$. Let $T = \mathbb{R}$, $f_{n,j}(X_{n,j}, t) = n^{-2}t$, $k_n = n$, $c_n = n^{-1}t$. Then, $Z_n(t) = 0$, for each $t \in T$ and each $n \ge 1$. So, the sequence $\{Z_n(t) : t \in T\}$ converges weakly. However, condition (b.2) in Theorem 2.1 does not hold. Instead of (2.6), previous authors Alexander (1987), Andersen, Giné, Ossiander and Zinn (1988) and Andersen, Giné and Zinn (1988) imposed the stronger condition

$$\max_{1 \le j \le k_n} \Pr\{\sup_{t \in T} |f_{n,j}(X_{n,j}, t)| \ge \eta\} \to 0,$$

for each $\eta > 0$. A triangular array satisfying (2.6) is called an infinitesimal array.

Next, we consider infinitesimal arrays with a Gaussian limit.

THEOREM 2.5. *With the above notation, suppose that:*

$$\max_{1 \le j \le k_n} \sup_{t \in T} \Pr\{|f_{n,j}(X_{n,j}, t)| \ge \eta\} \to 0,$$

for each $\eta > 0$. Let $b > 0$. Then, the following sets of conditions ((a) and (b)) are equivalent:

(a) *The sequence of stochastic processes $\{Z_n(t) : t \in T\}$, $n \ge 1$, converges weakly to a Gaussian process $\{Z(t) : t \in T\}$.*

(b.1) *For each $\eta > 0$, $\sum_{j=1}^{k_n} \Pr\{F_{n,j}(X_{n,j}) \ge \eta\} \to 0$.*

(b.2) *For each $s, t \in T$, the following limit exists*

$$\lim_{n \to \infty} \operatorname{Cov}(S_n(s, b), S_n(t, b)).$$

(b.3) *For each $t \in T$, the following limit exists*

$$\lim_{n \to \infty} (E[S_n(t, b)] - c_n(t))$$

(b.4) *For each $\eta > 0$, there exists a finite partition π of T such that*

$$\limsup_{n \to \infty} E^*[\sup_{t \in T} |S_n(t, b) - S_n(\pi(t), b) - E[S_n(t, b) - S_n(\pi(t), b)]|] \le \eta.$$

(b.5) *For each $\eta > 0$, there exists a finite partition π of T such that*

$$\limsup_{n \to \infty} \sup_{t \in T} |E[S_n(t, b) - S_n(\pi(t), b)] - c_n(t) + c_n(\pi(t))| \le \eta.$$

Moreover, if either (a) or (b) holds:

$$E[Z(t)] = \lim_{n \to \infty} (E[S_n(t, b)] - c_n(t)),$$

for each $t \in T$, and

$$\text{Cov}(Z(s), Z(t)) = \lim_{n \to \infty} \text{Cov}(S_n(s, b), S_n(t, b))$$

for each $s, t \in T$.

Proof. First, we prove that (b) implies (a). By Theorem 2.4, we only need to prove convergence of the finite dimensional distributions. Observe that (b.2) is equivalent to: for each $\lambda_1, \ldots, \lambda_m \in \mathbb{R}$ and each $t_1, \ldots, t_m \in T$, the following limit exists $\lim_{n \to \infty} \text{Var}(\sum_{l=1}^{m} \lambda_l S_n(t_l, b))$. By (b.1) and by Theorem 25.1 in Gnedenko and Komogorov (1968) it suffices to show that:

(b.2)$'$ For each $\lambda_1, \ldots, \lambda_m \in \mathbb{R}$ and each $t_1, \ldots, t_m \in T$, the following limit exists

$$\lim_{n \to \infty} \sum_{j=1}^{k_n} \text{Var}\left(\sum_{l=1}^{m} \lambda_l f_{n,j}(X_{n,j}, t_l) I_{|\sum_{l=1}^{m} \lambda_l f_{n,j}(X_{n,j}, t_l)| \leq b} \right).$$

(b.3)$'$ For each $\lambda_1, \ldots, \lambda_m \in \mathbb{R}$ and each $t_1, \ldots, t_m \in T$, the following limit exists

$$\lim_{n \to \infty} E[\sum_{j=1}^{k_n} \sum_{l=1}^{m} \lambda_l f_{n,j}(X_{n,j}, t_l) I_{|\sum_{l=1}^{m} \lambda_l f_{n,j}(X_{n,j}, t_l)| \leq b}] - \sum_{l=1}^{m} \lambda_l c_n(t_l).$$

We will show that under (b.1), (b.2) is equivalent to (b.2)$'$, and (b.3) is equivalent to (b.3)$'$. We may assume that $\lambda_l \neq 0$, for each $1 \leq l \leq m$. Since

$$|\sum_{l=1}^{m} \lambda_l f_{n,j}(X_{n,j}, t_l)| \leq \sum_{l=1}^{m} |\lambda_l| F_{n,j}(X_{n,j}),$$

$$0 \leq \sum_{j=1}^{k_n} E[(\sum_{l=1}^{m} \lambda_l f_{n,j}(X_{n,j}, t_l))^2$$

$$(I_{|\sum_{l=1}^{m} \lambda_l f_{n,j}(X_{n,j}, t_l)| \leq b} - I_{\sum_{l=1}^{m} |\lambda_l| F_{n,j}(X_{n,j}) \leq b})]$$

$$= \sum_{j=1}^{k_n} E[(\sum_{l=1}^{m} \lambda_l f_{n,j}(X_{n,j}, t_l))^2 I_{|\sum_{l=1}^{m} \lambda_l f_{n,j}(X_{n,j}, t_l)| \leq b, \ \sum_{l=1}^{m} |\lambda_l| F_{n,j}(X_{n,j}) > b}]$$

$$\leq b^2 \sum_{j=1}^{k_n} \Pr\{\sum_{l=1}^{m} |\lambda_l| F_{n,j}(X_{n,j}) > b\} \to 0.$$

$$(2.7)$$

We also have that

$$
\left| \sum_{j=1}^{k_n} \left((E[\sum_{l=1}^{m} \lambda_l f_{n,j}(X_{n,j}, t_l) I_{|\sum_{l=1}^{m} \lambda_l f_{n,j}(X_{n,j},t_l)| \le b}])^2 \right. \right.
$$

$$
\left. \left. - (E[\sum_{l=1}^{m} \lambda_l f_{n,j}(X_{n,j}, t_l) I_{\sum_{l=1}^{m} |\lambda_l| F_{n,j}(X_{n,j}) \le b}])^2 \right) \right|
$$

$$
\le \sum_{j=1}^{k_n} E[|\sum_{l=1}^{m} \lambda_l f_{n,j}(X_{n,j}, t_l)| (I_{|\sum_{l=1}^{m} \lambda_l f_{n,j}(X_{n,j},t_l)| \le b} + I_{\sum_{l=1}^{m} |\lambda_l| F_{n,j}(X_{n,j}) \le b})]
$$

$$
\times E[|\sum_{l=1}^{m} \lambda_l f_{n,j}(X_{n,j}, t_l)| I_{|\sum_{l=1}^{m} \lambda_l f_{n,j}(X_{n,j},t_l)| \le b, \ \sum_{l=1}^{m} |\lambda_l| F_{n,j}(X_{n,j}) > b}]
$$

(2.8)

$$
\le 2b^2 \sum_{j=1}^{k_n} \Pr\{\sum_{l=1}^{m} |\lambda_l| F_{n,j}(X_{n,j}) > b\} \to 0.
$$

From (2.7) and (2.8), we get that (b.2) is equivalent to (b.2)$'$. Similarly, under (b.1),

$$
|\sum_{j=1}^{k_n} E[\sum_{l=1}^{m} \lambda_l f_{n,j}(X_{n,j}, t_l)(I_{|\sum_{l=1}^{m} \lambda_l f_{n,j}(X_{n,j},t_l)| \le b} - I_{\sum_{l=1}^{m} |\lambda_l| F_{n,j}(X_{n,j}) \le b})]| \to 0
$$

and

$$
|\sum_{j=1}^{k_n} E[\sum_{l=1}^{m} \lambda_l f_{n,j}(X_{n,j}, t_l)(I_{\sum_{l=1}^{m} |\lambda_l| F_{n,j}(X_{n,j}) \le b} - I_{F_{n,j}(X_{n,j}) \le b})]| \to 0
$$

So, (b.3) is equivalent to (b.3)$'$. Therefore, (a) implies (b).

Assume (a). Convergence to a normal limit implies that

$$
\sum_{j=1}^{k_n} \Pr\{|f_{n,j}(X_{n,j}, t)| \ge \eta\} \to 0,
$$

for each $\eta > 0$ and each $t \in T$. This and condition (b.2) in Theorem 2.4 imply (b.1). From the convergence of finite dimensional distributions, we have (b.2)$'$ and (b.3)$'$, which imply (b.2) and (b.3). By Theorem 2.4, conditions (b) in this theorem hold. So, (b.4) and (b.5) hold. $\qquad \square$

It follows from Theorem 2.5 that, under condition (2.6), if

$$
\left\{ \left(\sum_{j=1}^{k_n} f_{n,j}(X_{n,j}, t) \right) - c_n(t) : t \in T \right\}
$$

converges weakly to a Gaussian process, then

$$\left\{\sum_{j=1}^{k_n} \left(f_{n,j}(X_{n,j},t) - E[f_{n,j}(X_{n,j},t)I_{F_{n,j}(X_{n,j})\leq b}]\right) : t \in T\right\}$$

also converges weakly to another Gaussian process. We prefer to truncate on $f_{n,j}(X_{n,j},t)$, since $F_{n,j}(X_{n,j})$ may be difficult to find. In this situation, we have:

COROLLARY 2.6. *With the above notation, suppose that:*

$$\max_{1\leq j\leq k_n} \sup_{t\in T} \Pr\{|f_{n,j}(X_{n,j},t)| \geq \eta\} \to 0,$$

for each $\eta > 0$. Let $b > 0$. Then, the following sets of conditions ((a) and (b)) are equivalent:

(a) *The sequence of stochastic processes*

$$\{\sum_{j=1}^{k_n} \left(f_{n,j}(X_{n,j},t) - E[f_{n,j}(X_{n,j},t)I_{|f_{n,j}(X_{n,j},t)|\leq b}]\right) : t \in T\},$$

$n \geq 1$, *converges weakly to a Gaussian process $\{Z(t) : t \in T\}$.*
(b.1) *For each $\eta > 0$, $\sum_{j=1}^{k_n} \Pr\{F_{n,j}(X_{n,j}) \geq \eta\} \to 0$.*
(b.2) *For each $s,t \in T$, the following limit exists*

$$\lim_{n\to\infty} \sum_{j=1}^{k_n} \mathrm{Cov}(f_{n,j}(X_{n,j},s)I_{|f_{n,j}(X_{n,j},s)|\leq b}, f_{n,j}(X_{n,j},t)I_{|f_{n,j}(X_{n,j},t)|\leq b}).$$

(b.3) *For each $\eta > 0$, there exists a finite partition π of T such that*

$$\limsup_{n\to\infty} E^*[\sup_{t\in T} |S_n(t,b) - S_n(\pi(t),b) - E[S_n(t,b) - S_n(\pi(t),b)]|] \leq \eta.$$

Moreover, if either (a) or (b) holds: $\{Z(t) : t \in T\}$ is a mean zero Gaussian process with covariance given by

$$\mathrm{Cov}(Z(s),Z(t)) = \lim_{n\to\infty} \sum_{j=1}^{k_n} \mathrm{Cov}(f_{n,j}(X_{n,j},s)I_{|f_{n,j}(X_{n,j},s)|\leq b},$$

$$f_{n,j}(X_{n,j},t)I_{|f_{n,j}(X_{n,j},t)|\leq b}),$$

for each $s,t \in T$.

Proof. By Theorem 2.5, it suffices to show that condition (b.1) implies that

$$\sum_{j=1}^{k_n} E[f_{n,j}(X_{n,j},s)I_{|f_{n,j}(X_{n,j},s)|\leq b}f_{n,j}(X_{n,j},t)I_{|f_{n,j}(X_{n,j},t)|\leq b}]$$

$$- \sum_{j=1}^{k_n} E[f_{n,j}(X_{n,j},s)f_{n,j}(X_{n,j},t)I_{F_{n,j}(X_{n,j})\leq b}]\to 0,$$

for each $s, t \in T$;

$$\sum_{j=1}^{k_n} E[f_{n,j}(X_{n,j}, s)I_{|f_{n,j}(X_{n,j},s)|\leq b}]E[f_{n,j}(X_{n,j}, t)I_{|f_{n,j}(X_{n,j},t)|\leq b}] \tag{2.9}$$

$$-\sum_{j=1}^{k_n} E[f_{n,j}(X_{n,j}, s)I_{F_{n,j}(X_{n,j})\leq b}]E[f_{n,j}(X_{n,j}, t)I_{F_{n,j}(X_{n,j})\leq b}] \to 0$$

for each $s, t \in T$; and

$$\sup_{t \in T}|\sum_{j=1}^{k_n} E[f_{n,j}(X_{n,j}, t)I_{|f_{n,j}(X_{n,j},t)|\leq b}] - E[f_{n,j}(X_{n,j}, t)I_{F_{n,j}(X_{n,j})\leq b}]| \to 0, \tag{2.10}$$

for each $s, t \in T$.

We have that

$$|\sum_{j=1}^{k_n} E[f_{n,j}(X_{n,j}, s)I_{|f_{n,j}(X_{n,j},s)|\leq b}f_{n,j}(X_{n,j}, t)I_{|f_{n,j}(X_{n,j},t)|\leq b}]$$

$$-\sum_{j=1}^{k_n} E[f_{n,j}(X_{n,j}, s)f_{n,j}(X_{n,j}, t)I_{F_{n,j}(X_{n,j})\leq b})]|$$

$$\leq \sum_{j=1}^{k_n} E[|f_{n,j}(X_{n,j}, s)f_{n,j}(X_{n,j}, t)|I_{|f_{n,j}(X_{n,j},s)|,|f_{n,j}(X_{n,j},t)|\leq b<F_{n,j}(X_{n,j})}]$$

$$\leq b^2 \sum_{j=1}^{k_n} \Pr\{F_{n,j}(X_{n,j}) > b\} \to 0.$$

(2.9) and (2.10) follow by similar arguments. $\qquad\square$

We will need the following lemma:

LEMMA 2.7. Let $\{a_n\}$ be a sequence of real numbers regularly varying of order α^{-1} such that $a_n \nearrow \infty$.

(i) If $1 < \alpha \leq 2$, then,

$$\limsup_{n\to\infty} na_n^{-1}E[F(X)I_{F(X)\geq\delta a_n}] \leq \alpha(\alpha - 1)^{-1}\delta^{1-\alpha}\limsup_{n\to\infty} n\Pr\{F(X) \geq a_n\}.$$

(ii) If $0 < \alpha < 2$, then,

$$\limsup_{n\to\infty} na_n^{-2}E[F^2(X)I_{F(X)\leq\delta a_n}] \leq 2(2-\alpha)^{-1}\delta^{2-\alpha}\limsup_{n\to\infty} n\Pr\{F(X) \geq a_n\}.$$

(iii) If $0 < \alpha < 1$, then,

$$\limsup_{n\to\infty} na_n^{-1}E[F(X)I_{F(X)\leq\delta a_n}] \leq (1-\alpha)^{-1}\delta^{1-\alpha}\limsup_{n\to\infty} n\Pr\{F(X) \geq a_n\}.$$

Proof. First, we consider (i). Assume $\delta > 1$. The case $1 \geq \delta > 0$ is similar. Let $c := \limsup_{n \to \infty} n \Pr\{F(X) \geq a_n\}$ and let $0 < \tau < \alpha - 1$. By regular variation (see for example Lemma 1.9.6 and Theorem 1.9.7 in Bingham, Goldie and Teugels, 1987) $a_n/a_{n+1} \to 1$ and

$$a_n = b(n)n^{\alpha^{-1}} \exp\left(\int_1^n u^{-1}\epsilon(u)\,du\right),$$

where $b(n) \to b(\infty)$, for some finite positive constant $b(\infty)$, and $\epsilon(u) \to 0$. Therefore, there exists n_0 such that

$$n\Pr\{F(X) \geq a_n\} \leq c(1+\tau),$$
$$(a_{k+1}/a_k)^{\alpha/(\tau+1)} \leq (1+\tau), \quad \text{and} \quad ((1+\tau)k/n)^{\tau+1} \geq (a_k/a_n)^\alpha, \tag{2.11}$$

for $k > n \geq n_0$. Then,

$$na_n^{-1}E[F(X)I_{F(X)\geq\delta a_n}] = n\delta\Pr\{F(X) \geq \delta a_n\} + \int_\delta^\infty n\Pr\{F(X) \geq a_n u\}\,du.$$

Take n such that $\delta a_n > a_{n_0}$. If $u > \delta$, there exists a $k \geq n_0$ such $a_k < ua_n \leq a_{k+1}$. So,

$$n\Pr\{F(X) \geq a_n u\} \leq n\Pr\{F(X) \geq a_k\} \leq c(1+\tau)(n/k)$$
$$\leq c(1+\tau)^2(a_n/a_{k+1})^{\alpha/(\tau+1)}(a_{k+1}/a_k)^{\alpha/(\tau+1)} \leq c(1+\tau)^3 u^{-\alpha/(\tau+1)}. \tag{2.12}$$

Hence,

$$\int_\delta^\infty n\Pr\{F(X) \geq a_n u\}\,du \leq \int_\delta^\infty c(1+\tau)^3 u^{-\alpha/(\tau+1)}\,du$$
$$= c(1+\tau)^4(\alpha - 1 - \tau)^{-1}\delta^{1-\alpha/(\tau+1)},$$

for $n \geq n_0$. Take $k_n = [\delta^\alpha n(1+\tau)^{-\alpha}]$. Then $a_{k_n}a_n^{-1} \to \delta(1+\tau)^{-1}$. So, $a_{k_n} \leq \delta a_n$, for n large enough. Hence

$$n\delta\Pr\{F(X) \geq \delta a_n\} \leq n\delta\Pr\{F(X) \geq a_{k_n}\} \leq c\delta(1+\tau)nk_n^{-1}$$
$$\leq c\delta(1+\tau)^2(a_n/a_{k_n})^{\alpha/(\tau+1)} \to c(1+\tau)^{2+\alpha/(\tau+1)}\delta^{1-\alpha/(\tau+1)}.$$

Therefore, (i) follows. (ii) and (iii) follow similarly, using (2.11) and (2.12). $\square$

Theorem 2.5 gives the following in the the case of the sequence of processes in (1.2) with a normal limit:

COROLLARY 2.8. *Let $\{X_j\}_{j=1}^\infty$ are i.i.d.r.v.'s with values in a measurable space $(S,\mathcal{S})$, let $f(\cdot,t) : S \to \mathbb{R}$ be a measurable function for each $t \in T$, let $\{a_n\}_{n=1}^\infty$ is a sequence of positive numbers and let $b > 0$. Suppose that:*

(i) $\sup_{t\in T}\Pr\{|f(X,t)| \geq a_n\eta\} \to 0$, *for each $\eta > 0$.*
(ii) $\{a_n\}$ *is regularly varying of order $1/2$.*
(iii) $n^{1/2}a_n^{-1} = O(1)$.

Then, the following sets of conditions ((a) and (b)) are equivalent:

(a) *The sequence of stochastic processes*

$$\left\{ Z_n(t) := a_n^{-1} \sum_{j=1}^{n} (f(X_j,t) - E[f(X,t)]) : t \in T \right\}, \quad n \geq 1,$$

converges weakly to a stochastic process $\{Z(t) : t \in T\}$.

(b.1) *For each* $\eta > 0$, $n \Pr\{F(X) \geq a_n\eta\} \to 0$.

(b.2) *For each* $s,t \in T$, *the following limit exists*

$$\lim_{n \to \infty} na_n^{-2}\text{Cov}(f(X,s)I_{|f(X,s)| \leq ba_n}, f(X,t)I_{|f(X,t)| \leq ba_n}).$$

(b.3) *For each* $\eta > 0$, *there exists a finite partition* π *of* T *such that*

$$\limsup_{n \to \infty} a_n^{-1} E^*[\sup_{t \in T} | \sum_{j=1}^{n} \epsilon_j(f(X_j,t) - f(X_j,\pi(t)))I_{F(X) \leq ba_n}|] \leq \eta.$$

Moreover, if either (a) or (b) holds: $\{Z(t) : t \in T\}$ *is a mean–zero Gaussian process with covariance given by*

$$E[Z(s)Z(t)] = \lim_{n \to \infty} na_n^{-2}\text{Cov}(f(X,s)I_{|f(X,s)| \leq ba_n}, f(X,t)I_{|f(X,t)| \leq ba_n}),$$

for each $s,t \in T$.

Proof. Assume (a). (b.1) and (b.2) follow from Theorem 2.5. (b.1) implies $E[F(X)] < \infty$. So, for each $\eta > 0$, there exists a finite partition π of T such that

$$\limsup_{n \to \infty} a_n^{-1} E[\sup_{t \in T} | \sum_{j=1}^{n} ((f(X_j,t) - f(X_j,\pi(t)))I_{F(X_j) \leq ba_n}$$

$$- E[(f(X_j,t) - f(X_j,\pi(t)))I_{F(X_j) \leq ba_n}])|] \leq \eta$$

and

$$\sup_{t \in T} E[|f(X,t) - f(X,\pi(t))|] \leq \eta.$$

From this and symmetrization (see for example Lemma 6.3 in Ledoux and Talagrand, 1991),

$$\limsup_{n \to \infty} a_n^{-1} E[\sup_{t \in T} | \sum_{j=1}^{n} \epsilon_j((f(X_j,t) - f(X_j,\pi(t)))I_{F(X_j) \leq ba_n}$$

$$- E[(f(X_j,t) - f(X_j,\pi(t)))I_{F(X_j) \leq ba_n}])|] \leq 2\eta.$$

We also have that

$$a_n^{-1} E[\sup_{t \in T} | \sum_{j=1}^{n} \epsilon_j E[(f(X_j,t) - f(X_j,\pi(t)))I_{F(X_j) \leq ba_n}]|]$$

$$\leq n^{1/2}a_n^{-1} \sup_{t \in T} E[|f(X,t) - f(X,\pi(t))|].$$

Therefore, condition (b.3) holds.

Assume (b). We apply Theorem 2.5. Obviously, conditions (b) imply (b.1) and (b.2) in Theorem 2.5. By Lemma 2.7,

$$\sup_{t \in T} |E[S_n(t,b)] - na_n^{-1}E[f(X,t)]| \leq na_n^{-1}E[F(X)I_{F(X) \geq ba_n}] \to 0.$$

So, (b.3) and (b.5) follows. Condition (b.4) in Theorem 2.5 follows from (b.3) via symmetrization.   □

Next, we consider the case when the limit process does not have a Gaussian part.

THEOREM 2.9. *With the above notation, suppose that:*

$$\max_{1 \leq j \leq k_n} \sup_{t \in T} \Pr\{|f_{n,j}(X_{n,j},t)| \geq \eta\} \to 0,$$

for each $\eta > 0$. Let $b > 0$. Then, the following sets of conditions ((a) and (b)) are equivalent:

(a) *The sequence of stochastic processes $\{Z_n(t) : t \in T\}$, $n \geq 1$, converges weakly to an infinitely divisible process $\{Z(t) : t \in T\}$ without a Gaussian part.*

(b.1) *The finite dimensional distributions of $\{Z_n(t) : t \in T\}$ converge to those of an infinitely divisible process $\{Z(t) : t \in T\}$ without a Gaussian part.*

(b.2) *For each $\eta > 0$, there exists a finite partition π of T such that*

$$\limsup_{n \to \infty} \sum_{j=1}^{k_n} \Pr^*\{\sup_{t \in T} |f_{n,j}(X_{n,j},t) - f_{n,j}(X_{n,j}, \pi(t))| \geq \eta\} \leq \eta.$$

(b.3) $\lim_{\delta \to 0} \limsup_{n \to \infty} E^*[\sup_{t \in T} |S_n(t,\delta) - E[S_n(t,\delta)]|] = 0.$

(b.4) *For each $\eta > 0$, there exists a finite partition π of T, such that*

$$\limsup_{n \to \infty} \sup_{t \in T} |E[S_n(t,b) - S_n(\pi(t),b)] - c_n(t) + c_n(\pi(t))| \leq \eta.$$

Proof. The part (b) implies (a) follows similarly to the proof of Theorem 2.1.

Assume (a). (b.1), (b.2) and (b.4) follow from Theorem 2.4. As for (b.3), given $1/10 > \eta > 0$, there exists a finite partition π of T such that

$$\limsup_{n \to \infty} E^* \left[\sup_{t \in T} \left| \sum_{j=1}^{k_n} \epsilon_j \left((f_{n,j}(X_{n,j},t) - f_{n,j}(X_{n,j}, \pi(t))) \right) I_{F_{n,j}(X_{n,j}) \leq 1} - \right. \right.$$

$$\left. \left. E[(f_{n,j}(X_{n,j},t) - f_{n,j}(X_{n,j}, \pi(t))) I_{F_{n,j}(X_{n,j}) \leq 1}]) \right| \right] \leq \eta.$$

For $1 > \delta > 0$,

$$2 \sum_{j=1}^{k_n} \epsilon_j (f_{n,j}(X_{n,j},t) I_{F_{n,j}(X_{n,j}) \leq \delta} - E[f_{n,j}(X_{n,j},t) I_{F_{n,j}(X_{n,j}) \leq \delta}])$$

$$= \sum_{j=1}^{k_n} \epsilon_j \left(f_{n,j}(X_{n,j},t) I_{F_{n,j}(X_{n,j}) \leq \delta} + f_{n,j}(X_{n,j},t) I_{\delta < F_{n,j}(X_{n,j}) \leq 1} \right.$$

$$-E[f_{n,j}(X_{n,j},t)I_{F_{n,j}(X_{n,j})\leq\delta} + f_{n,j}(X_{n,j},t)I_{\delta<F_{n,j}(X_{n,j})\leq1}])$$

$$+\sum_{j=1}^{k_n}\epsilon_j\left(f_{n,j}(X_{n,j},t)I_{F_{n,j}(X_{n,j})\leq\delta} - f_{n,j}(X_{n,j},t)I_{\delta<F_{n,j}(X_{n,j})\leq1}\right.$$

$$\left.-E[f_{n,j}(X_{n,j},t)I_{F_{n,j}(X_{n,j})\leq\delta} - f_{n,j}(X_{n,j},t)I_{\delta<F_{n,j}(X_{n,j})\leq1}]\right),$$

where the two summands have the same distribution. So, for any $1 > \delta > 0$,

$$\limsup_{n\to\infty} E^*[\sup_{t\in T}|\sum_{j=1}^{k_n}\epsilon_j((f_{n,j}(X_{n,j},t) - f_{n,j}(X_{n,j},\pi(t)))I_{F_{n,j}(X_{n,j})\leq\delta}$$

$$- E[(f_{n,j}(X_{n,j},t) - f_{n,j}(X_{n,j},\pi(t)))I_{F_{n,j}(X_{n,j})\leq\delta}])|] \leq 2\eta. \tag{2.13}$$

By Theorem 25.1 in Gnedenko and Komogorov (1968),

$$\lim_{\delta\to0}\limsup_{n\to\infty}\text{Var}\left(\sum_{j=1}^{k_n}f_{n,j}(X_{n,j},t)I_{|f_{n,j}(X_{n,j},t)|\leq\delta}\right) = 0,$$

for each $t \in T$. Since $|f_{n,j}(x,t)| \leq F_{n,j}(x)$, for each $x \in S_{n,j}$ and each $t \in T$,

$$\sum_{j=1}^{k_n}\left(E[f_{n,j}^2(X_{n,j},t)I_{F_{n,j}(X_{n,j})\leq\delta}] - (E[f_{n,j}(X_{n,j},t)I_{F_{n,j}(X_{n,j})\leq\delta}])^2\right)$$

$$\leq \sum_{j=1}^{k_n}\left(E[f_{n,j}^2(X_{n,j},t)I_{|f_{n,j}(X_{n,j},t)|\leq\delta}] - (E[f_{n,j}(X_{n,j},t)I_{|f_{n,j}(X_{n,j},t)|\leq\delta}])^2\right)$$

$$+ \sum_{j=1}^{k_n}\left((E[f_{n,j}(X_{n,j},t)I_{|f_{n,j}(X_{n,j},t)|\leq\delta}])^2 - (E[f_{n,j}(X_{n,j},t)I_{F_{n,j}(X_{n,j})\leq\delta}])^2\right).$$

We also have that

$$|\sum_{j=1}^{k_n}(E[f_{n,j}(X_{n,j},t)I_{|f_{n,j}(X_{n,j},t)|\leq\delta}])^2 - (E[f_{n,j}(X_{n,j},t)I_{F_{n,j}(X_{n,j})\leq\delta}])^2|$$

$$= \left|\sum_{j=1}^{k_n}\left(E[f_{n,j}(X_{n,j},t)(I_{|f_{n,j}(X_{n,j},t)|\leq\delta} + I_{F_{n,j}(X_{n,j})\leq\delta})]\right.\right.$$

$$\left.\left.\times E[f_{n,j}(X_{n,j},t)(I_{|f_{n,j}(X_{n,j},t)|\leq\delta} - I_{F_{n,j}(X_{n,j})\leq\delta})])\right|$$

$$\leq 2\delta \max_{1\leq j\leq k_n} E[|f_{n,j}(X_{n,j},t)|I_{|f_{n,j}(X_{n,j},t)|\leq\delta}]\sum_{j=1}^{k_n}\text{Pr}\{F_{n,j}(X_{n,j}) \geq \delta\} \to 0,$$

for each $\delta > 0$ and each $t \in T$. Hence,

$$\lim_{\delta \to 0} \limsup_{n \to \infty} E\left[\left|\sum_{j=1}^{k_n} \epsilon_j(f_{n,j}(X_{n,j},t)I_{F_{n,j}(X_{n,j})\leq\delta}\right.\right.$$

$$\left.\left.-E[f_{n,j}(X_{n,j},t)I_{F_{n,j}(X_{n,j})\leq\delta}])\right|\right] = 0,$$

for each $t \in T$. From this and (2.13),

$$\lim_{\delta \to 0} \limsup_{n \to \infty} E\left[\sup_{t \in T}\left|\sum_{j=1}^{k_n} \epsilon_j(f_{n,j}(X_{n,j},t)I_{F_{n,j}(X_{n,j})\leq\delta}\right.\right.$$

$$\left.\left.-E[f_{n,j}(X_{n,j},t)I_{F_{n,j}(X_{n,j})\leq\delta}])\right|\right] \leq \eta,$$

for each $\eta > 0$. Therefore, condition (b.3) follows. □

The last theorem generalizes Proposition 2.2 in Andersen, Giné and Zinn (1988). This proposition asserts a weaker result than (b) implies (a) in Theorem 2.9 under strong conditions than the ones above. They assumed (b.1) and (b.2) in Theorem 2.9 plus the extra condition

$$\lim_{\delta \to 0} \limsup_{n \to \infty} \Pr\{\sup_{t \in T}|S_n(t,\delta)| \geq \tau\} = 0,$$

for each $\tau > 0$. It is easy to see that this condition implies (b.3) and (b.4) with $c_n(t) = 0$, for each $t \in T$. In Proposition 2.2 in Andersen, Giné and Zinn (1988) there is the restriction that $f_{n,j}(x,t) = f(x,t)$ for some fixed function $f(x,t)$.

Next, we consider stable limits for the process (1.2). According to the order of the stability of the limit, the shift $c_n(t)$ and other conditions are different. So, we make three cases:

COROLLARY 2.10. *Let $\{X_j\}_{j=1}^{\infty}$ be a sequence of i.i.d.r.v.'s with values in a measurable space $(S, \mathcal{S})$, let $f(\cdot, t) : S \to \mathbb{R}$ be a measurable function for each $t \in T$, let $\{a_n\}_{n=1}^{\infty}$ is a sequence of positive numbers and let $1 < \alpha < 2$. Suppose that:*

(i) $\sup_{t \in T} \Pr\{|f(X,t)| \geq a_n\eta\} \to 0$, *for each $\eta > 0$.*
(ii) $a_n \nearrow \infty$ *and a_n is regularly varying of order α^{-1}.*

Then, the following sets of conditions ((a) and (b)) are equivalent:

(a) *The sequence of stochastic processes*

$$\{a_n^{-1}\sum_{j=1}^{n}(f(X_j,t) - E[f(X,t)]) : t \in T\}, \ n \geq 1,$$

converges weakly.

(b.1) *For each $\lambda_1, \ldots, \lambda_m \in \mathbb{R}$ and each $t_1, \ldots, t_m \in T$, there exists a finite constant $N(\lambda_1, \ldots, \lambda_m, t_1, \ldots, t_m)$ such that*

$$\lim_{n \to \infty} n\Pr\{\sum_{l=1}^{m} \lambda_l f(X,t_l) \geq ua_n\} = \alpha^{-1}u^{-\alpha}N(\lambda_1, \ldots, \lambda_m, t_1, \ldots, t_m),$$

for each $u > 0$.

(b.2) *For each $\eta > 0$, there exists a finite partition π of T such that*

$$\limsup_{n \to \infty} n \Pr\{\sup_{t \in T} |f(X,t) - f(X, \pi(t))| \geq a_n \eta\} \leq \eta.$$

(b.3) $\lim_{\delta \to 0} \limsup_{n \to \infty} E[\sup_{t \in T} a_n^{-1} | \sum_{j=1}^n \epsilon_j f(X_j, t) I_{F(X_j) \leq \delta a_n} |] = 0.$

Proof. Assume (a), convergence of the finite dimensional distributions imply (b.1). By Theorem 2.9, (b.2) follows and

$$\lim_{\delta \to 0} \limsup_{n \to \infty} E[\sup_{t \in T} a_n^{-1} | \sum_{j=1}^n \epsilon_j (f(X_j, t) I_{F(X_j) \leq \delta a_n}$$

$$- E[f(X_j, t) I_{F(X_j) \leq \delta a_n}]) |] = 0. \tag{2.14}$$

Now, (b.1) and (b.2) imply that $E[F(X)] < \infty$. So,

$$E[\sup_{t \in T} a_n^{-1} | \sum_{j=1}^n \epsilon_j E[f(X_j, t) I_{F(X_j) \leq \delta a_n}]) |] \leq n^{1/2} a_n^{-1} E[F(X)] \to 0.$$

Therefore, (b.3) follows.

Assume (b). We apply Theorem 2.9. By Theorem 25.1 in Gnedenko and Komogorov (1968) and (b.1), we have convergence of the finite dimensional distributions, i.e. (b.1) in Theorem 2.9 holds. Obviously, (b.2) in Theorem 2.9 is satisfied. (b.2) and a previous argument implies (2.13), i.e. (b.3) in Theorem 2.9 holds. As to condition (b.4), given $\eta > 0$, take a partition π of T such that

$$\limsup_{n \to \infty} n \Pr\{G(X, \pi) \geq \eta c^{-1} a_n\} < \eta,$$

where $c = \sup_{n \geq 1} n \Pr\{F(X) \geq a_n\}$ and $G(X, \pi) := \sup_{t \in T} |f(X,t) - f(X, \pi(t))|$. We have that

$$n a_n^{-1} E[G(X, \pi) I_{F(X) > a_n}]$$

$$\leq \eta c^{-1} n \Pr\{F(X) > a_n\} + n a_n^{-1} E[G(X, \pi) I_{G(X, \pi) > \eta c^{-1} a_n}].$$

By Lemma 2.7,

$$\limsup_{n \to \infty} n a_n^{-1} E[G(X, \pi) I_{F(X) > a_n}] \leq \eta + \alpha(\alpha - 1)^{-1} \eta, \tag{2.15}$$

and condition (b.4) in Theorem 2.9 holds. $\qquad\square$

Observe that we need condition (i) in Corollary 2.10. Suppose that $T = \mathbb{R}$ and $f(X,t) = t$, then (a) holds, but neither (b.2) nor (b.3) hold.

COROLLARY 2.11. *With the above notation, suppose that:*

(i) $\sup_{t \in T} \Pr\{|f(X,t)| \geq a_n \eta\} \to 0$, *for each $\eta > 0$.*
(ii) $a_n \nearrow \infty$ *and a_n is regularly varying of order 1.*

Then, the following sets of conditions ((a) and (b)) are equivalent:

(a) *The sequence of stochastic processes*

$$\{(a_n^{-1} \sum_{j=1}^{n} f(X_j, t)) - c_n(t) : t \in T\}, \ n \geq 1,$$

converges weakly.

(b.1) *For each $\lambda_1, \ldots, \lambda_m \in \mathbb{R}$ and each $t_1, \ldots, t_m \in T$, there exists a finite constant $N(\lambda_1, \ldots, \lambda_m, t_1, \ldots, t_m)$ such that*

$$\lim_{n \to \infty} n \Pr\{\sum_{l=1}^{m} \lambda_l f(X, t_l) \geq u a_n\} = u^{-1} N(\lambda_1, \ldots, \lambda_m, t_1, \ldots, t_m),$$

for each $u > 0$.

(b.2) *For each $\lambda_1, \ldots, \lambda_m \in \mathbb{R}$ and each $t_1, \ldots, t_m \in T$, the following limit exists*

$$\lim_{n \to \infty} n a_n^{-1} E[\sum_{l=1}^{m} \lambda_l f(X, t_l) I_{|\sum_{l=1}^{m} \lambda_l f(X, t_l)| \leq a_n}] - \sum_{l=1}^{m} \lambda_l c_n(t_l).$$

(b.3) *For each $\eta > 0$, there exists a finite partition π of T such that*

$$\limsup_{n \to \infty} n \Pr\{\sup_{t \in T} |f(X, t) - f(X, \pi(t))| \geq a_n \eta\} \leq \eta.$$

(b.4) $\lim_{\delta \to 0} \limsup_{n \to \infty} E[\sup_{t \in T} a_n^{-1} |\sum_{j=1}^{n} \epsilon_j f(X_j, t) I_{F(X_j) \leq \delta a_n}|] \leq \eta.$

(b.5) *For each $\eta > 0$, there exists a finite partition π of T such that*

$$\lim_{n \to \infty} \sup_{t \in T} |n a_n^{-1} E[(f(X, t) - f(X, \pi(t))) I_{F(X) \leq a_n}] - c_n(t) + c_n(\pi(t))| \leq \eta.$$

Proof. It follows directly from Theorem 25.1 in Gnedenko and Komogorov (1968) and Theorem 2.9. Observe that (b.4) is equivalent to

$$\lim_{\delta \to 0} \limsup_{n \to \infty} E[\sup_{t \in T} a_n^{-1} |\sum_{j=1}^{n} (f(X_j, t) I_{F(X_j) \leq \delta a_n} - E[f(X_j, t) I_{F(X_j) \leq \delta a_n}])|] \leq \eta,$$

because $\lim_{\delta \to 0} \limsup_{n \to \infty} n^{1/2} a_n^{-1} E[F(X) I_{F(X) \leq \delta a_n}] = 0$. □

COROLLARY 2.12. *With the above notation, let $0 < \alpha < 1$, suppose that:*

(i) $a_n \nearrow \infty$ *and a_n is regularly varying of order α^{-1}.*

Then, the following sets of conditions ((a) and (b)) are equivalent:

(a) *The sequence of stochastic processes*

$$\{Z_n(t) := a_n^{-1} \sum_{j=1}^{n} f(X_j, t) : t \in T\}, \ n \geq 1,$$

converges weakly.

(b.1) *For each $\lambda_1, \ldots, \lambda_m \in \mathbb{R}$ and each $t_1, \ldots, t_m \in T$, there exists a finite constant $N(\lambda_1, \ldots, \lambda_m, t_1, \ldots, t_m)$ such that*

$$\lim_{n \to \infty} n \Pr\{\sum_{l=1}^{m} \lambda_l f(X, t_l) \geq u a_n\} = \alpha^{-1} u^{-\alpha} N(\lambda_1, \ldots, \lambda_m, t_1, \ldots, t_m),$$

for each $u > 0$.

(b.2) *For each $\eta > 0$, there exists a finite partition π of T such that*

$$\limsup_{n \to \infty} n \Pr\{\sup_{t \in T} |f(X, t) - f(X, \pi(t))| \geq a_n \eta\} \leq \eta.$$

Proof. Assume (a). Let $\eta > 0$. We have that $\{a_n^{-1} \sum_{j=1}^{n} f(X_j, t) : t \in T\}$ and $\{a_n^{-1} \sum_{j=1}^{n-1} f(X_j, t) : t \in T\}$ converge weakly. So, there is a partition π of T such that

$$\limsup_{n \to \infty} \Pr\{\sup_{t \in T} |\sum_{j=1}^{n} (f(X_j, t) - f(X_j, \pi(t)))| \geq \eta a_n\} \leq \eta$$

and

$$\limsup_{n \to \infty} \Pr\{\sup_{t \in T} |\sum_{j=1}^{n-1} (f(X_j, t) - f(X_j, \pi(t)))| \geq \eta a_n\} \leq \eta.$$

Hence,

$$\limsup_{n \to \infty} \Pr\{\sup_{t \in T} |f(X, t) - f(X, \pi(t))| \geq 2\eta a_n\} \leq 2\eta.$$

Since $\lim_{n \to \infty} \Pr\{|f(X, t)| \geq \eta a_n\} = 0$, for each $\eta > 0$ and each $t \in T$,

$$\lim_{n \to \infty} \Pr\{F(X) \geq 3\eta a_n\} \leq 3\eta.$$

Therefore, condition (2.6) holds. Now by Theorem 2.9, (b.1) and (b.2) hold.

Assume (b), in order to prove (a), it suffices to show that (b.1) and (b.2) imply that for each $\eta > 0$ there exists a finite partition π of T, such that

$$\limsup_{n \to \infty} n a_n^{-1} E[\sup_{t \in T} |f(X, t) - f(X, \pi(t))| I_{F(X) \leq a_n}] \leq \eta. \tag{2.16}$$

and

$$\lim_{\delta \to 0} \limsup_{n \to \infty} E^*[\sup_{t \in T} a_n^{-1} |\sum_{j=1}^{n} (f(X_j, t) I_{F(X_j) \leq \delta a_n}$$
$$- E[f(X_j, t) I_{F(X_j) \leq \delta a_n}])|] = 0. \tag{2.17}$$

Given $\eta > 0$, take a finite partition π of T such that

$$\limsup_{n \to \infty} n \Pr\{G(X, \pi) \geq \eta a_n\} \leq \eta,$$

where $G(X, \pi) = \sup_{t \in T} |f(X, t) - f(X, \pi(t))|$. Then,

$$n a_n^{-1} E[G(X, \pi) I_{F(X) \leq a_n}]$$

$$\leq 2n \Pr\{G(X, \pi) \geq a_n \eta\} + n a_n^{-1} E[G(X, \pi) I_{G(X, \pi) \leq a_n \eta}].$$

So, by Lemma 2.7,

$$\limsup_{n\to\infty} na_n^{-1} E[G(X,\pi)I_{F(X)\le a_n}] \le 2\eta + (1-\alpha)^{-1}\eta^2.$$

Since η is arbitrary, (2.16) follows. As to (2.17),

$$E^*[\sup_{t\in T} a_n^{-1} | \sum_{j=1}^{n} (f(X_j,t)I_{F(X_j)\le\delta a_n} - E[f(X_j,t)I_{F(X_j)\le\delta a_n}])|]$$

$$\le 2a_n^{-1} nE[F(X)I_{F(X)\le\delta a_n}].$$

So, (2.17) follows from Lemma 2.7. □

References

Alexander, K. S. (1987). Central limit theorems for stochastic processes under random entropy conditions. *Probab. Theor. Rel. Fields* **75** 351–378.

Andersen, N. T., Giné, E.; Ossiander, M. and Zinn, J. (1988). The central limit theorem and the law of iterated logarithm for empirical processes under local conditions. *Probab. Theor. Rel. Fields* **77** 271–305.

Andersen, N. T.; Giné, E. and Zinn, J. (1988). The central limit theorem for empirical processes under local conditions: the case of Radon infinitely divisible limits without Gaussian components. *Transact. Amer. Mathem. Soc.* **308** 603–635.

Arcones, M. A. (1994a). Distributional convergence of M–estimators under unusual rates. *Statist. Probab. Lett.* **21** 271–280.

Arcones, M. A. (1994b). On the weak Bahadur–Kiefer representation for M–estimators. *Probability in Banach Spaces, 9 (Sandjberg, 1993).* 357–372. Edts. J. Hoffmann–Jørgensen, J. Kuelbs and M. B. Marcus. Birkhäuser, Boston.

Arcones, M. A., Gaenssler, P. and Ziegler, K. (1992). Partial-sum processes with random locations and indexed by Vapnik–Červonenkis classes of sets in arbitrary sample space. *Probability in Banach Spaces, 8 (Brunswick, ME, 1991).* 379–389. Birkhäuser, Boston.

Bingham, N. H.; Goldie, C. M. and Teugels, J. L. (1987). *Regular Variation.* Cambridge University Press, Cambridge, United Kingdom.

Dudley, R. M. (1984). A course on empirical processes. *Lect. Notes in Math.* **1097** 1–142. Springer–Verlag, New York,

Gaenssler, P. and Ziegler, K. (1994). A uniform law of large numbers for set-indexed processes with applications to empirical and partial-sum processes. *Probability in Banach Spaces, 9 (Sandjberg, 1993).* 385–400. Birkhäuser, Boston.

Giné, E. and Zinn, J. (1986). Lectures on the central limit theorem for empirical processes. *Lect. Notes in Math.* **1221** 50–112. Springer–Verlag, New York.

Gnedenko, B. V. and Kolmogorov, A. N. (1968). *Limit Distributions for Sums of Independent Random Variables.* Addison–Wesley Publishing Company. Reading, Massachusetts.

Hoffmann–Jørgensen, J. (1991). *Stochastic Processes on Polish Spaces.* Various Publications Series, 39. Aarhus University, Matematisk Institut, Aarhus, Denmark.

Kahane, J. P. (1968). *Some Random Series of Functions.* D. C. Heath, Lexington, Massachusetts.

Kim, J. and Pollard, D. (1990). Cube root asymptotics. *Ann. Statist.* **18**, 191–219.

Kwapień, S. and Woyczyński, W. A. (1992). *Random Series and Stochastic Integrals: Single and Multiple*. Birkhäuser, Boston.

Le Cam, L. (1986). *Asymptotic Methods in Statistical Decision Theory*. Springer–Verlag, New York.

Ledoux, M. and Talagrand, M. (1991). *Probability in Banach Spaces*. Springer–Verlag, New York.

Marcus, M. B. and Pisier, G. (1981). *Random Fourier Series with Applications to Harmonic Analysis. Ann. Math. Studies* **101**. Princeton University Press, Princeton, New Jersey.

Pollard, D. (1990). *Empirical Processes: Theory and Applications*. NSF–CBMS Regional Conference Series in Probab. and Statist., Vol. 2. I.M.S., Hayward, California.

Talagrand, M. (1987). Donsker classes and random entropy. *Ann. Probab.* **15** 1327–1338.

Miguel A. Arcones
Department of Mathematics
University of Texas
Austin, TX 78712–1082
arcones@math.utexas.edu

Progress in Probability, Vol. 43
© 1998 Birkhäuser Verlag Basel/Switzerland

Self-Normalized Large Deviations in Vector Spaces

Amir Dembo[1] and Qi-Man Shao[2]

ABSTRACT. In this short note we define and study properties of Partial Large Deviation Principles (PLDP), using them to extend Cramér's theorem to self-normalized partial sums of i.i.d. random vectors obeying no moment assumption.

Let $\mathcal{Z}$ be a topological space equipped with a σ-field $\mathcal{F}$, $\mathcal{K} = \{K : K \in \mathcal{F}$ is pre-compact$\}$, $\mathcal{S}$ be a subset of $\mathcal{F}$, $I : \mathcal{Z} \to [0, \infty]$ be a lower semicontinuous function, and $a_n \to \infty$. We say that the $\mathcal{Z}$-valued random variables $\mathbf{Z}_n$ satisfy a Partial Large Deviation Principle (PLDP) of speed a_n and rate function $I(\mathbf{z})$ with respect to $\mathcal{S}$ if the lower bound

$$- \inf_{\mathbf{z} \in A^o} I(\mathbf{z}) \leq \liminf_{n \to \infty} a_n^{-1} \log P(\mathbf{Z}_n \in A) \,, \qquad (1.1)$$

holds for every $A \in \mathcal{F}$ and the upper bound

$$\limsup_{n \to \infty} a_n^{-1} \log P(\mathbf{Z}_n \in A) \leq - \inf_{\mathbf{z} \in \bar{A}} I(\mathbf{z}) \,, \qquad (1.2)$$

holds for every $A \in \mathcal{S}$. The *full large deviation principle (LDP)* is an extreme case of PLDP in which $\mathcal{S} = \mathcal{F}$. Another common case of PLDP is the *weak LDP*, in which $\mathcal{S} = \mathcal{K}$. Using [1, Lemma 1.2.15], it may be assumed without loss of generality that

$$A, B \in \mathcal{S} \implies A \cup B \in \mathcal{S} \,. \qquad (1.3)$$

Exponential tightness, that is, the existence of $K_r \in \mathcal{K}$ such that

$$\lim_{r \to \infty} \limsup_{n \to \infty} a_n^{-1} \log P(\mathbf{Z}_n \in K_r^c) = -\infty \,,$$

suffices for upgrading the weak LDP to a full LDP (c.f. [1, page 8]). Indeed, a typical route to the full LDP consists of establishing the weak LDP and exponential tightness. In case $I^{-1}[0, b]$ is compact for all $b > 0$, exponential tightness is also necessary for the full LDP to hold (c.f. [1, Exercise 4.1.10]).

In establishing PLDP the exponential tightness property is replaced by *partial exponential tightness*, that is, the existence for each $A \in \mathcal{S}$ of $K_r \in \mathcal{K}$ such that

$$\lim_{r \to \infty} \limsup_{n \to \infty} a_n^{-1} \log P(\mathbf{Z}_n \in A \cap K_r^c) = -\infty \,. \qquad (1.4)$$

---

1) Research partially supported by NSF DMS-9403553 grant and by a U.S.-Israel BSF grant while at the Department of Electrical Engineering, Technion, Haifa, Israel.

2) Research partially supported by a National University of Singapore Research Project when teaching in Singapore.

Indeed, following the proof of [1, Lemma 1.2.18] we see that weak LDP and partial exponential tightness imply the PLDP. Moreover, in this case (1.4) holding for $A \in \mathcal{S}$ implies

$$B \in \mathcal{F}, \ B \subseteq A \in \mathcal{S} \quad \Longrightarrow \quad B \in \mathcal{S}. \tag{1.5}$$

Adapting [1, Exercise 4.1.10] we see that for $\mathbf{Z}_n$ Borel on a complete separable metric space $\mathcal{Z}$, in case (1.5) holds and $I^{-1}[0, b] \cap \bar{A}$ is compact for all $b > 0$, $A \in \mathcal{S}$, the partial exponential tightness is also necessary for the PLDP.

While a further study of the abstract properties of the PLDP is of independent interest, we do not pursue it here, using instead the PLDP as a convenient framework for statements about self-normalized partial sums of (heavy-tailed) independent and identically distributed (i.i.d.) random variables.

Hereafter, $\mathcal{Z} = \mathcal{B}$ is a separable Banach space with $\mathcal{F}$ its Borel σ-field, in which case the rate function for the PLDP is unique as soon as $\mathcal{K} \subseteq \mathcal{S}$ (c.f. [3, Proposition 1.2 and Theorem 2.1]). Let $\mathbf{S}_n = n^{-1} \sum_{i=1}^{n} \mathbf{Y}_i$ be the empirical means of i.i.d. $\mathcal{B}$-valued random variables $\{\mathbf{Y}, \mathbf{Y}_i\}$ and consider the self-normalizing sequence $V_n = n^{-1} \sum_{i=1}^{n} g(\mathbf{Y}_i)$ for $g \in \mathcal{G}_{\mathbf{Y}}$ defined as follows.

DEFINITION 1.1 *Let $\mathcal{B}$ be a separable Banach space and $\mathbf{Y}$ a $\mathcal{B}$-valued random variable. A function $g = h \circ f : \mathcal{B} \to [0, \infty]$ is said to belong to $\mathcal{G}_{\mathbf{Y}}$ if $f : \mathcal{B} \to [0, \infty]$ is a convex function of compact level sets (i.e., $f^{-1}[0, a]$ is compact for each $0 < a < \infty$) such that $f(\mathbf{Y}) < \infty$ almost surely and $h : [0, \infty] \to [0, \infty]$ is convex, monotone nondecreasing, such that $h(z) < \infty$ for $z < \infty$ and*

$$\liminf_{t, z \to \infty} \frac{h(tz)}{th(z)} = \infty \tag{1.6}$$

(excluding h which is identically zero).

PROPOSITION 1.1 *For any $\mathbf{Y}$ there exists $f : \mathcal{B} \to [0, \infty]$ convex of compact level sets with $f(\mathbf{Y}) < \infty$ almost surely, hence $\mathcal{G}_{\mathbf{Y}}$ is non-empty.*

Proof. Since $\mathbf{Y}$ is tight, there exists an increasing sequence of compact, symmetric sets K_m such that $P(\mathbf{Y} \notin K_m) < m^{-1}$. Let $r_m = \text{diam}(K_m) < \infty$. Since the set $\cup_m (mr_m)^{-1} K_m$ is totally bounded, its closed convex hull K is compact and symmetric. Let $q_K(\mathbf{y}) = \inf\{\lambda > 0 : \mathbf{y} \in \lambda K\}$ which is convex of compact level sets ($q_K^{-1}[0, b] = bK$ for all $b > 0$). Then,

$$P(q_K(\mathbf{Y}) > mr_m) \leq P(\mathbf{Y} \notin mr_m K) \leq P(\mathbf{Y} \notin K_m) < m^{-1}.$$

For $m \to \infty$ we deduce that $q_K(\mathbf{Y}) < \infty$ a.s. so that $q_K^p \in \mathcal{G}_{\mathbf{Y}}$ for all $p > 1$. $\square$

Cramér's theorem provides the weak LDP for $(\mathbf{S}_n, V_n)$ with the following convex rate function on $\mathcal{B} \times R$ (equipped with the product topology):

$$\Lambda^*(\mathbf{s}, v) = \sup_{\substack{\boldsymbol{\lambda} \in \mathcal{B}^* \\ \theta \in R}} \{\langle \boldsymbol{\lambda}, \mathbf{s} \rangle + \theta v - \log E(\exp(\langle \boldsymbol{\lambda}, \mathbf{Y} \rangle + \theta g(\mathbf{Y})))\} \tag{1.7}$$

(c.f. [1, Theorem 6.1.3]). Making no assumptions about the law of $\mathbf{Y}$, we strengthen this weak LDP to the following PLDP for $(\mathbf{S}_n, V_n)$.

THEOREM 1.1 *Let $\mathcal{B}$ be a separable Banach space with its Borel σ-field $\mathcal{F}$ and $g \in \mathcal{G}_{\boldsymbol{Y}}$. Then $(\boldsymbol{S}_n, V_n)$ satisfy a PLDP of speed $a_n = n$ and rate function $\Lambda^*(\boldsymbol{s}, v)$ with respect to finite unions of sets $A \subset \mathcal{B} \times \mathbb{R}$ that are either open and convex or such that*

$$\liminf_{\substack{v \to \infty \\ (\boldsymbol{s}, v) \in A}} \{g(\boldsymbol{s})/v\} > 0 \tag{1.8}$$

(in particular (1.8) holds when $\sup\{v : (\boldsymbol{s}, v) \in A \text{ for some } \boldsymbol{s}\} < \infty$).

Note that a set satisfying (1.8) is typically neither convex nor compact (for which the large deviations upper bound (1.2) is known, for example see [1, Exercise 6.1.16(a)]).

We next prove Theorem 1.1. To this end, since Cramér's weak LDP of speed $a_n = n$ holds for every law of $\boldsymbol{Y}$, in view of [1, Exercise 6.1.16(a)] and (1.3), it suffices to establish the following proposition.

PROPOSITION 1.2 *The random variables $(\boldsymbol{S}_n, V_n)$ are partially exponentially tight in $\mathcal{B} \times \mathbb{R}$ for the collection $\mathcal{S}$ of sets satisfying (1.8).*

Proof. Since f has compact level sets, so has g (by the monotonicity and continuity of h). Hence, $K_r = g^{-1}[0, r] \times [0, r]$ is compact. By Jensen's inequality and convexity of $g = h \circ f$, it follows that $V_n \geq g(\boldsymbol{S}_n)$ and hence

$$\left\{(\boldsymbol{S}_n, V_n) \in A \cap K_r^c\right\} \subseteq \left\{(\boldsymbol{S}_n, V_n) \in A, V_n > r\right\}.$$

By the assumption (1.8), there exist $\varepsilon > 0$ and $r_0 \geq 1$ such that for all $r \geq r_0$,

$$\left\{(\boldsymbol{S}_n, V_n) \in A, V_n > r\right\} \subseteq \{g(\boldsymbol{S}_n) \geq \varepsilon V_n \geq \varepsilon r\}.$$

The proof of Proposition 1.2 is thus completed by the following lemma.

LEMMA 1.1 *Condition (1.6) implies that for each fixed $\varepsilon > 0$,*

$$\lim_{r \to \infty} \limsup_{n \to \infty} n^{-1} \log P\left(g(\boldsymbol{S}_n) \geq \varepsilon V_n \geq \varepsilon r\right) = -\infty. \tag{1.9}$$

Proof. Fix $\varepsilon > 0$ and $r \geq 1$. Let

$$W_{n,r} = n^{-1} \sum_{i=1}^{n} I_{\{f(\boldsymbol{Y}_i) \geq m_r\}} \quad \text{with } m_r = \inf_{h(z) \geq \varepsilon r} z/4.$$

Setting $z = n^{-1} \sum_{i=1}^{n} f(\boldsymbol{Y}_i)$ and noting that $g = h \circ f$, by convexity of f and monotonicity of h we have that

$$\{g(\boldsymbol{S}_n) \geq \varepsilon V_n \geq \varepsilon r\} \subseteq \left\{h(n^{-1} \sum_{i=1}^{n} f(\boldsymbol{Y}_i)) \geq n^{-1} \sum_{i=1}^{n} \varepsilon g(\boldsymbol{Y}_i) \geq \varepsilon r\right\}$$

$$\subseteq \bigcup_{z \geq 4m_r} \left\{n^{-1} \sum_{i=1}^{n} [z^{-1} f(\boldsymbol{Y}_i) - \varepsilon g(\boldsymbol{Y}_i)/(2h(z))] \geq 1/2\right\} \tag{1.10}$$

$$\subseteq \{W_{n,r} b_r \geq 1/4\},$$

where

$$b_r = \sup_{\substack{f(y) \geq m_r \\ z \geq 4m_r}} \left\{ \frac{f(y)}{z} - \frac{\varepsilon g(y)}{2h(z)} \right\} \leq t_0 \vee \sup_{\substack{z \geq 4m_r \\ t \geq t_0}} t \left\{ 1 - \frac{\varepsilon}{2} \frac{h(tz)}{th(z)} \right\}$$

for every $t_0 > 0$.

Since h is bounded above on compacts, it follows that $m_r \uparrow \infty$ as $r \to \infty$. By (1.6), there exist $t_0(\varepsilon), r_0(\varepsilon) < \infty$ such that $h(tz) \geq 2th(z)/\varepsilon$ for all $t \geq t_0$ and all $z \geq 4m_{r_0}$. Consequently, $b_r \leq t_0(\varepsilon)$ for all $r \geq r_0(\varepsilon)$.

With $P(f(\boldsymbol{Y}) \geq m_r) \to 0$ as $r \to \infty$, by the tail estimates of the Binomial law it follows that for $\delta = 1/(4t_0(\varepsilon))$, independent of n and r,

$$\lim_{r \to \infty} \limsup_{n \to \infty} n^{-1} \log P(W_{n,r} \geq \delta) = -\infty. \tag{1.11}$$

Since $b_r \leq t_0(\varepsilon)$ for all $r \geq r_0(\varepsilon)$, (1.9) follows by combining (1.10) and (1.11).
$\square$

REMARK 1.1 *Theorem 1.1 applies in any locally convex, Hausdorff, topological real vector space $\mathcal{B}$, for which $\mathcal{B} \times R$ satisfies [1, Assumption 6.1.2].*

Theorem 1.1 also applies for any $g : \mathcal{B} \to [0, \infty]$ convex of compact level sets provided that $g \leq h \circ f$ for f convex with $P(f(\boldsymbol{Y}) = \infty) = 0$ and h monotone nondecreasing such that $h(z) < \infty$ for $z < \infty$ and

$$\liminf_{\substack{z \to \infty \\ z^{-1} f(y) \to \infty}} \frac{zg(y)}{h(z)f(y)} = \infty .$$

Let $\widetilde{\boldsymbol{S}}_n = n^{-1} \sum_{i=1}^n \boldsymbol{X}_i$ for $\{\boldsymbol{X}, \boldsymbol{X}_n, n \geq 1\}$ i.i.d. R^d-valued random variables. Let $\boldsymbol{x}$ denote the point $(x_1, \dots, x_d)$ in R^d, with $\|\boldsymbol{x}\|$ an arbitrary norm corresponding to the usual topology on R^d, while $\|\boldsymbol{x}\|_p = (\sum_{k=1}^d |x_k|^p)^{1/p}$ for $p \geq 1$. Cramér's theorem asserts that $\widetilde{\boldsymbol{S}}_n$ satisfies the full LDP of speed $a_n = n$ if Cramér's condition

$$\inf_{t>0} E(\exp(t\|\boldsymbol{X}\|)) < \infty$$

holds. However, for $d \geq 3$, the full LDP may fail in the absence of Cramér's condition (c.f. [2] for such an example). Thus, this classical theory provides little information on large deviations, in the case of $\boldsymbol{X}$ whose law is heavy-tailed.

In contrast, since $h(z) = z^p$ satisfies (1.6) for any $p > 1$, for *any law* of $\boldsymbol{Y}$ the function $g(\boldsymbol{y}) = \|\boldsymbol{y}\|^p$ is in $\mathcal{G}_{\boldsymbol{Y}}$ when $\mathcal{B} = R^d$ (and conversely, the existence of $f : \mathcal{B} \to [0, \infty)$ convex of compact level sets implies that $\mathcal{B}$ is locally compact). Theorem 1.1 then yields the following PLDP for $(\widetilde{\boldsymbol{S}}_n, \widetilde{V}_{n,p})$ where $\widetilde{V}_{n,p} = n^{-1} \sum_{i=1}^n \|\boldsymbol{X}_i\|^p$.

COROLLARY 1.1 *For $p > 1$ and any i.i.d. R^d-valued $\boldsymbol{X}_i$ the random vectors $(\widetilde{\boldsymbol{S}}_n, \widetilde{V}_{n,p})$ satisfy a PLDP of speed $a_n = n$ and rate function $\Lambda_p^*(\boldsymbol{s}, v)$ corresponding to $g(\boldsymbol{y}) = |\boldsymbol{y}|^p$ in (1.7). This PLDP is with respect to the sets $A \subset R^{d+1}$ such that*

$$\liminf_{\substack{y \to \infty \\ (\boldsymbol{x}, y) \in A}} \{\|\boldsymbol{x}\|^p / y\} > 0. \tag{1.12}$$

REMARK 1.2 *For every $x > 0$, Corollary 1.1 yields the asymptotics of*

$$n^{-1} \log P\left(\|\widetilde{\boldsymbol{S}}_n\| \geq x (\widetilde{V}_{n,p})^{1/p}\right)$$

(with

$$\inf_{\|\boldsymbol{s}\| \geq xv^{1/p} \geq 0} \Lambda_p^*(\boldsymbol{s}, v)$$

in the upper bound and

$$\inf_{\|\boldsymbol{s}\| > xv^{1/p} > 0} \Lambda_p^*(\boldsymbol{s}, v)$$

in the corresponding lower bound). Similarly, it yields the asymptotics of

$$n^{-1} \log P\left(\cap_k\{\widetilde{S}_{n,k} \geq x_k(\widetilde{V}_{n,p})^{1/p}\}\right)$$

for $x_k > 0$ (where now

$$\inf_{\cap_k s_k \geq x_k v^{1/p} \geq 0} \Lambda_p^*(\boldsymbol{s}, v)$$

in the upper bound and

$$\inf_{\cap_k s_k > x_k v^{1/p} > 0} \Lambda_p^*(\boldsymbol{s}, v)$$

in the lower bound).

REMARK 1.3 *In particular, for $d = 1$ one essentially recovers the self-normalized large deviation result of [4, Theorem 1.2]), the proof of which in [4] is direct, with no links to the theory of large deviations. In contrast, the PLDP serves here as the framework for extending the result of [4, Theorem 1.2] to a vast collection of subsets in a general separable Banach space setting as well as for providing a relation with Cramér's classical theorem.*

The next PLDP is another consequence of Theorem 1.1.

COROLLARY 1.2 *Fixing $p > 1$, let $\widehat{V}_{n,k} = n^{-1}\sum_{i=1}^n |X_{i,k}|^p$, $k = 1,\ldots,d$ (where $\boldsymbol{X}_i = (X_{i,1},\ldots,X_{i,d})$, for $i \geq 1$), and $\widehat{\boldsymbol{V}}_n = (\widehat{V}_{n,1},\ldots,\widehat{V}_{n,d})$. The random vectors $(\widetilde{\boldsymbol{S}}_n, \widehat{\boldsymbol{V}}_n)$ satisfy a PLDP of speed $a_n = n$ and rate function*

$$\Gamma_p^*(\boldsymbol{s}, \boldsymbol{v}) = \sup_{(\boldsymbol{\lambda}, \boldsymbol{\theta}) \in R^{2d}} \{\langle \boldsymbol{\lambda}, \boldsymbol{s}\rangle + \langle \boldsymbol{\theta}, \boldsymbol{v}\rangle - \log E(\exp(\langle \boldsymbol{\lambda}, \boldsymbol{X}\rangle + \langle \boldsymbol{\theta}, \boldsymbol{X}^p\rangle))\}$$

(where $\boldsymbol{X}^p = (|X_1|^p,\ldots,|X_d|^p)$), with respect to the sets $A \subset R^{2d}$ such that

$$\liminf_{\substack{\|\boldsymbol{v}\| \to \infty \\ (\boldsymbol{s}, \boldsymbol{v}) \in A}} \{\|\boldsymbol{s}\|^p / \|\boldsymbol{v}\|\} > 0. \tag{1.13}$$

Proof. Since all norms on R^d are equivalent, we may and shall assume that in (1.13) it is $\|\boldsymbol{s}\|_p^p / \|\boldsymbol{v}\|_1$ which is bounded away from zero when $(\boldsymbol{s}, \boldsymbol{v}) \in A$ and $\|\boldsymbol{v}\|_1 \to \infty$. Consider the $\mathcal{B} = R^{2d}$-valued i.i.d. random variables $\boldsymbol{Y}_i = (\boldsymbol{X}_i, |X_{i,1}|^p,\ldots,|X_{i,d}|^p)$. Corollary 1.2 follows by applying Theorem 1.1 with $g =$

$h \circ f$, where $f((x_1,\ldots,x_d,y_1,\ldots,y_d)) = \|x\|_p$ is convex and $h(z) = z^p$ satisfies
(1.6). Indeed, then $S_n = n^{-1}\sum_{i=1}^{n} Y_i = (\widetilde{S}_n, \widehat{V}_n)$. Since $V_n = n^{-1}\sum_{i=1}^{n} g(Y_i) = \|\widehat{V}_n\|_1$ and $g(S_n) = \|\widetilde{S}_n\|_p^p$, we see that (1.13) amounts to (1.8) holding. The sets
$f^{-1}([0,r])$ are not compact and thus $g \notin \mathcal{G}_Y$. However, $\{S_n : g(S_n) \leq r, \widehat{V}_n \leq r\}$
are compact subsets of R^{2d} and the proof of Theorem 1.1 carries through as if
$g \in \mathcal{G}_Y$. $\square$

REMARK 1.4 *Corollary 1.2 yields the asymptotics of* $n^{-1}\log P\left(\|T_n\|_2 \geq x\right)$ *for*
$T_{n,k} = \widetilde{S}_{n,k}(\widehat{V}_{n,k})^{-1/p}$, *and of* $n^{-1}\log P\left(\cap_k\{\widetilde{S}_{n,k} \geq x_k(\widehat{V}_{n,k})^{1/p}\}\right)$ *for every*
$x, x_1,\ldots,x_d > 0$.

Acknowledgement. The authors thank Włodzimierz Bryc and Tiefeng Jiang for
helpful discussions.

References

[1] Dembo, A. and Zeitouni, O. (1993). Large deviations techniques and applications.
    Jones and Bartlett, Boston.
[2] Dinwoodie, I. H. (1991). A note on the upper bound for i.i.d. large deviations. *Ann.
    Probab.*, 19:1732–1736.
[3] O'Brien, G. L. (1996). Sequences of capacities, with connections to large deviation
    theory. *J. Theor. Probab.*, 9:19–35.
[4] Shao, Q. M. (1997). Self-normalized large deviations. *Ann. Probab.*, 25:285–328.

Amir Dembo   Qi-Man Shao
Department of Mathematics   Department of Mathematics
Stanford University   University of Oregon
Stanford, CA, USA   Eugene, OR 97403, USA
amir@math.stanford.edu   qmshao@darkwing.uoregon.edu

Progress in Probability, Vol. 43
© 1998 Birkhäuser Verlag Basel/Switzerland

Consistency of M-Estimators and On€

RICHARD M. DUDLEY[*]

ABSTRACT. Some facts in empirical process theory are based on br... tions, defined by $[f,h] := \{g : f \leq g \leq h\}$. For minimization problems arising in M-estimation it is shown that one can use one-sided brackets $[f,\infty)$. A class $\mathcal{F}$ of functions is called a Glivenko-Cantelli class (GCC) if the law of large numbers for empirical measures $P_n \to P$ holds with respect to uniform convergence over $\mathcal{F}$. A characterization of GCC's by Talagrand (Ann. Probab., 1987) is shown to imply that the GCC property is preserved under the transformation $f \mapsto g \circ f$, where $(g \circ f)(x) = g(f(x))$ for all x, by any monotone Lipschitz function g. A specific convex, decreasing, Lipschitz function g with $-\log(y) \geq g(y) \geq 1 - y$ for all $y > 0$, and equality only at $y = 1$, provides such a transformation which, with one-sided bracketing, applies to log likelihoods. Thus a theorem of van de Geer (Ann. Statist., 1993) is extended.

1. Introduction

This paper is about consistency of (approximate) M-estimators. Here is a sketch of what is done, under stronger assumptions than will actually be needed in some cases. Let $(x,\theta) \mapsto h(x,\theta)$ be a function where $\theta \in \Theta$, a parameter space, and $x \in X$, where $(X,\mathcal{A},P)$ is a probability space. Suppose that $h(\cdot,\theta) \in \mathcal{L}^1(P)$ for all $\theta \in \Theta$ and that $\int h(x,\theta)dP(x)$ is minimized at a unique $\theta = \theta_0(P)$. Replacing h by $h(x,\theta) - h(x,\theta_0)$, we can assume the value of the minimum is 0. Let $X_1, X_2, \cdots$, be i.i.d. (P). Suppose that $P_n h(\cdot,\theta) := \frac{1}{n}\sum_{j=1}^{n} h(X_j,\theta)$ is minimized at $\theta = \hat{\theta}_n \in \Theta$. Then $\hat{\theta}_n = \hat{\theta}_n(X_1, \cdots, X_n)$ is an M-estimator and the problem is of "consistency," i.e. under what conditions will $\hat{\theta}_n$ converge to θ_0.

Brackets $[f,h] := \{g : f(x) \leq g(x) \leq h(x) \text{ for all } x\}$ have been useful in empirical process theory. Specifically, let $\mathcal{F}$ be a class of real-valued measurable functions, let $1 \leq p < \infty$ and suppose that given any $\varepsilon > 0$, $\mathcal{F}$ can be covered by finitely many such brackets $[f_i, h_i]$, $i = 1, \cdots, n(\varepsilon,p)$, where $\int (h_i - f_i)^p dP < \varepsilon^p$ for all i. If $n(\varepsilon,p)$ goes to infinity at not too fast a rate as $\varepsilon \downarrow 0$, then some limit theorems will hold with respect to uniform convergence over $\mathcal{F}$. For example, there is the sharp central limit theorem of Ossiander (1987) for $p = 2$.

For consistency of M-estimates it has turned out that one can use one-sided brackets $h(\cdot,\theta) \in [g_\theta,\infty)$, formalizing methods of Huber (1967). The number of brackets may be infinite; there may be, as the notation suggests, one for each θ. In the simplest case, the uniform law of large numbers $\sup_\theta |\int g_\theta d(P_n - P)| \to 0$ as $n \to \infty$ will be assumed. If for each neighborhood U of θ_0, there is a $\delta > 0$

---

*)   Partially supported by NSF Grants.

$\int g_\theta dP > \delta$ for all $\theta \notin U$, consistency will follow. If, moreover, for each ⟨there⟩ is a neighborhood V of θ_0 such that $\int h(x,\theta)dP > -\varepsilon$ for all $\theta \in V$, ⟨val⟩ues of $P_n h(\cdot, \hat{\theta}_n)$ will also converge to 0 ("value consistency").

The rest of the paper is then organized as follows. The remainder of this ⟨se⟩ction gives more precise and general definitions and assumptions. Sections 2 and 7 are close to the paper of Huber (1967) and show how his work is extended here - the main extension is that Huber's topological assumptions (local compactness) on Θ are replaced by Glivenko-Cantelli properties. By Huber's ingenious method, functions $\rho(x,\theta)$ are replaced by functions $(\rho(x,\theta) - a(x))/b(\theta)$ before taking lower bounds. Section 3 defines one-sided Glivenko-Cantelli and bracketing properties and states some general theorems on how these work to give consistency. Theorem 3.8 is about a curious non-metrizable topology on probability laws defined by the Kullback-Leibler information or "metric."

Then, Section 4 treats the "log likelihood" case where there is a family of laws $\{P_\theta, \ \theta \in \Theta\}$ having densities $f(x,\theta)$ with respect to a fixed σ-finite measure and $\rho(x,\theta) = -\log f(x,\theta)$. Section 5 treats Glivenko-Cantelli classes of functions and recalls Talagrand's beautiful characterization of these classes, using it to show that the Glivenko-Cantelli property is preserved under composition $f \mapsto g \circ f$ with any fixed Lipschitz, monotone function g. Proposition 5.4, extending a theorem of van de Geer (1993), illustrates how the different parts of the paper come together.

For statistics, one would like various properties to hold not only for one probability law P but for families of laws or better yet for all laws. Section 6 gives, as an example, a way of constructing some specific classes of functions on Euclidean spaces for which limit theorems hold uniformly in P. The classes only work so well because of Huber's two devices, subtracting an $a(x)$ and dividing by a $c(\theta)$, so Theorem 6.1 and Corollary 6.5 also bring together different aspects. Appendix A recalls definition of measurable estimators by measurable selection, and Appendix B treats U-statistic analogues of Glivenko-Cantelli classes.

Some precise definitions begin here. Given a probability space $(\Omega, \mathcal{S}, \mu)$ and any set $A \subset \Omega$, the *outer probability* $\mu^*(A)$ is defined by $\mu^*(A) := \inf\{\mu(B) : B \in \mathcal{S}, \ A \subset B\}$. A sequence f_n of not necessarily measurable functions on Ω into a metric space (S,d) is said to converge to f_0 *in outer probability* if for every $\varepsilon > 0$, $\mu^*(d(f_n, f_0) > \varepsilon) \to 0$ as $n \to \infty$. Also, f_n is said to converge to f_0 *almost uniformly* if for every $\varepsilon > 0$,

$$\mu^*(d(f_n, f_0) > \varepsilon \ \text{ for some } \ n > m) \to 0 \ \text{ as } \ m \to \infty.$$

Almost uniform convergence is equivalent to almost sure convergence if $d(f_n, f_0)$ are measurable, e.g. if f_n are measurable and S is separable. If $d(f_n, f_0)$ may be non-measurable then almost sure convergence can be undesirably weak. Almost uniform convergence is stronger and seems to be the appropriate strengthening. For a signed measure ν and function f let $\nu f := \int f d\nu$ (if defined, possibly $\pm\infty$).

Let $(X, \mathcal{A}, P)$ be a probability space and Θ a parameter space (just a set, for the time being). Let $\rho(\cdot, \cdot)$ be a function from $X \times \Theta$ into $[-\infty, \infty]$ such that for each $\theta \in \Theta$, $\rho(\cdot, \theta)$ is measurable on $(X, \mathcal{A})$. Let $X_1, \cdots, X_n, \cdots$ be i.i.d. (P), *strictly*, in the sense that they are coordinates on the countable product $(X^\infty, \mathcal{A}^\infty, P^\infty)$. Let $P_n := \frac{1}{n}\sum_{j=1}^n \delta_{X_j}$, so that $P_n\rho(\cdot, \theta) = \frac{1}{n}\sum_{j=1}^n \rho(X_j, \theta)$.

Classically, an estimator based on n observations would be a function $\hat{\theta}_n = \hat{\theta}_n(X_1, \cdots, X_n)$. Here functions with auxiliary randomization will be allowed.

Let $(X, \mathcal{A})$ and $(\Theta, \mathcal{S})$ be measurable spaces. Let P be a probability measure on $(X, \mathcal{A})$. Take a countable product of copies of $(X, \mathcal{A}, P)$, then the product

$$(\Omega, \mathcal{B}, \mathrm{Pr}) := (X^\infty, \mathcal{A}^\infty, P^\infty) \times (\Omega_0, \mathcal{B}_0, Q)$$

with some probability space $(\Omega_0, \mathcal{B}_0, Q)$. A *sequence of estimators* T_n will mean a sequence of measurable functions $T_n(X_1, \cdots, X_n, \omega)$ into Θ where $\omega \in \Omega_0$. Here ω allows for auxiliary randomization, as in Monte Carlo resampling for bootstrap statistics (e.g. Efron and Tibshirani, 1993; Giné and Zinn, 1990). Some authors, e.g. Huber (1967), Hoffmann-Jørgensen (1994), do not assume measurability of $\hat{\theta}_n$ in proofs of consistency where it is not needed, as it also is not in the present paper. On the other hand, measurability in $(X_1, \cdots, X_n, \omega)$ seems to provide enough generality for estimators at present.

A function $\hat{\theta}_n = \hat{\theta}_n(X_1, \cdots, X_n, \omega)$ is called an *M-estimator* for ρ if

$$P_n \rho(\cdot, \hat{\theta}_n) = \inf_{\theta \in \Theta} P_n \rho(\cdot, \theta).$$

Also, $\hat{\theta}_n = \hat{\theta}_n(X_1, \cdots X_n, \omega)$ will be called a sequence of *weak* (resp. *strong*) *approximate M-estimators* for ρ and P if $\lim_{n \to \infty}[P_n \rho(\cdot, \hat{\theta}_n) - \inf_{\theta \in \Theta} P_n \rho(\cdot, \theta)] = 0$ in outer probability (resp. almost uniformly) as $n \to \infty$ when X_i are strictly i.i.d. (P). (This implies that the latter two expressions involving P_n are each finite except with outer probability $\to 0$ as $n \to \infty$.)

M-estimators may not exist, but under mild measurability conditions, approximate M-estimators always exist and can be chosen by measurable selection so that $\hat{\theta}_n$ is universally measurable in $(X_1, \cdots, X_n)$, see Appendix A.

Let d be a metric on Θ and let $\mathcal{P}$ be a collection of laws (probability measures) on $(X, \mathcal{A})$. Let $\theta_0(\cdot)$ be a function from $\mathcal{P}$ into Θ. Then a sequence of estimators $T_n = T_n(X_1, \cdots, X_n, \omega)$ will be called weakly (resp. strongly) *consistent* for $\theta_0(\cdot)$ and $\mathcal{P}$ if for any $P \in \mathcal{P}$ and $X_1, X_2, \cdots$ strictly i.i.d. (P), as $n \to \infty$, $T_n(X_1, \cdots, X_n, \omega) \to \theta_0(P)$ in outer probability (resp. almost uniformly).

The general notion of (approximate) M-estimation is apparently due to P.J. Huber, e.g. Huber (1967, 1981), who also gave (1967) a very interesting set of sufficient conditions for consistency (listed in Sec. 7 below). M-estimation includes at least three substantial sub-cases:

(I) *The log likelihood ("parametric") case.* Let $\{P_\theta, \theta \in \Theta\}$ be a family of laws on $(X, \mathcal{A})$, where $\theta \mapsto P_\theta$ is 1-1. Suppose that for some σ-finite measure μ on $(X, \mathcal{A})$, each P_θ is absolutely continuous with respect to μ, with a density $f(x, \theta)$. Let $\rho(x, \theta) := -\log f(x, \theta) \leq +\infty$. Assume that $P = P_{\theta_0}$ for some $\theta_0 = \theta_0(P)$. Then an M-estimator is a maximum likelihood estimator, and approximate M-estimators are approximate maximum likelihood estimators.

(II) *Location M-estimators.* Here $X = \mathbb{R}^k$ with Borel σ-algebra and also $\Theta = \mathbb{R}^k$. We will have $\rho(x, \theta) = \rho(x - \theta)$ for some function ρ on $\mathbb{R}^k$. Huber (1967, 1981) treated this case especially. If $\rho(x) = |x|^2$ there is a unique M-estimator $\overline{X} =$

$(X_1 + \cdots + X_n)/n$ which is strongly consistent for any law P in the class $\mathcal{P}$ of all laws P with $P|x| < \infty$, and $\theta_0(P) = Px$.

If $\rho(x) = |x|$, an M-estimator always exists and is a sample median, called a *spatial median* for $k > 1$. It is unique if $X_1, \cdots, X_n$ are not all in any line (Haldane, 1948) or if they are and have a unique sample median there, specifically if n is odd.

Huber defined further functions, where for each $r > 0$, $\rho_r(x) = |x|^2$ for $|x| \le r$ and $2r|x| - r^2$ for $|x| \ge r$. Let $\rho_0(x) := |x|$. Huber (1981, pp. 43-55) treats ρ_r for $k = 1$. For ρ_r, M-estimators are more often unique than medians for $k = 1$, namely if the interval of sample medians has length $\le 2r$. Also, $P\rho_r < \infty$ if and only if $P|x| < \infty$.

(III) *Stochastic programming.* Here $\mathcal{P}$ reduces to a single law P. One wants to find $\theta = \theta_0$ for which $P\rho(\cdot, \theta)$ is minimized. But, while $X_1, X_2, \cdots$ can be generated, suppose that integrals with respect to P are hard enough to compute so that apparently the best available approach is by Monte Carlo, which in this case leads to M-estimation, e.g. Shapiro (1989). Facts on stochastic programming can also yield facts on estimation if the hypotheses hold for all P in some class.

For some P and ρ, $P\rho(\cdot, \theta) = +\infty$ for all θ. For example in Case II, for $\rho(x) = |x|^2$, $P\rho(\cdot, \theta) = +\infty$ for all θ if $P|x|^2 = \infty$, but only $P|x| < \infty$ is needed for consistency. For $\rho(x) \equiv |x|$, $P\rho(\cdot, \theta) = +\infty$ for all θ if $P|x| = \infty$, but for consistency, P need only have a unique median. Huber dealt with such cases by subtracting a suitable function $a(x)$ from $\rho(x, \theta)$. Thus $P(|x - \theta|^2 - |x|^2) < \infty$ for all θ if $P|x| < \infty$, and $P(|x - \theta| - |x|) < \infty$ for all θ and every law P. Such "adjustments" will be treated in Section 2.

For a signed measure ν and class $\mathcal{F}$ of functions such that νf is defined for each $f \in \mathcal{F}$ let $\|\nu\|_{\mathcal{F}} := \sup\{|\nu f| : f \in \mathcal{F}\}$. Now, $d(\mu, \nu) := \|\mu - \nu\|_{\mathcal{F}}$ defines a metric for signed measures with respect to which the functions in $\mathcal{F}$ are integrable. $\mathcal{F}$ is called a *weak* (resp. *strong*) *Glivenko-Cantelli class* for P if, when X_i are strictly i.i.d. (P), we have $\|P_n - P\|_{\mathcal{F}} \to 0$ in outer probability (resp. almost uniformly). Dudley, Giné and Zinn (1991, Proposition 3) give an example of a weak Glivenko-Cantelli class which is not a strong Glivenko-Cantelli class.

One possible method of treating the asymptotics of M-estimators is to apply empirical process theory to the class of functions $\mathcal{F} := \{\rho(\cdot, \theta) : \theta \in \Theta\}$, and try to show that $\mathcal{F}$ is a Glivenko-Cantelli class (for P), so that one has convergence of the empirical P_n to P uniformly over $\mathcal{F}$. We may want to replace $\rho(x, \theta)$ by $h(x, \theta) := \rho(x, \theta) - a(x)$, as done in Sec. 2 below, and then ask whether $\{h(\cdot, \theta) : \theta \in \Theta\}$ is a Glivenko-Cantelli class. Specifically, one can take $a(x) \equiv \rho(x, \theta_0)$. Huber (1967) pointed out that another transformation is helpful, namely to take a function $b(\theta) > 0$ and consider the functions $h(\cdot, \theta)/b(\theta)$. Huber bounded functions $h(\cdot, \theta)/b(\theta)$ below by finite sets of P-integrable functions, which are always Glivenko-Cantelli classes. Thus, cases where $|h(x, \theta)| \to \infty$ as $|\theta| \to \infty$, and the classes $\{h(\cdot, \theta) : \theta \in \Theta\}$ are not themselves Glivenko-Cantelli, can be dealt with. See Sections 2 and 7 below.

While incorporating Huber's adjustment function $a(x)$, and something similar to his function $b(\theta)$, we part company with his local compactness assumption on

Θ. As van de Geer (1993) noted, the theory of Glivenko-Cantelli classes makes local compactness quite unnecessary.

For consistency to make sense, we need at least some minimal topological requirements: something must hold outside of neighborhoods of θ_0, different from what holds inside them, e.g. at θ_0 itself. The main definitions of bracketing conditions in Section 3 below make only such minimal use of the topology. A metric is assumed so that convergence in (outer) probability will make sense. The Hellinger metric, also defined in Sect. 3, is used for some purposes.

On the other hand, in the log likelihood case, there are natural metrics, the Hellinger metric H and the total variation metric, which are uniformly equivalent to each other, and which have long been found useful in dealing with likelihoods. In fact, $H(P_\theta, P_{\theta_0}) \to 0$ if and only if $Ph(\cdot, \theta) \to 0 = Ph(\cdot, \theta_0)$ for $a(x) \equiv \rho(x, \theta_0)$ (see Sec. 4 below).

For $\mathcal{F} \subset \mathcal{L}^1(P)$ let $\mathcal{F}_{0,P} := \{f - Pf : f \in \mathcal{F}\}$. Let $F_{\mathcal{F}} := \sup_{f \in \mathcal{F}} |f(x)|$, called the *envelope function* of $\mathcal{F}$. For any real function f let f^* be the essential infimum of all measurable functions $\geq f$. If $F_{\mathcal{F}}^* \in \mathcal{L}^1(P)$ then $\mathcal{F}$ is called *order bounded* for P. Equivalently, there is some $F \in \mathcal{L}^1(P)$ with $|f| \leq F$ for all f in $\mathcal{F}$.

For uniformly bounded classes $\mathcal{F}$ with a mild measurability condition, Vapnik and Červonenkis (1981) characterized the Glivenko-Cantelli property (the weak and strong properties are equivalent in this case by Theorem A below) in terms of a random metric entropy bound. Giné and Zinn (1984, Sec. 8(a)), for classes $\mathcal{F}$ of functions with some measurability and bounded in $\mathcal{L}^1(P)$ norm, characterized strong Glivenko-Cantelli classes in terms of a random metric entropy condition and order boundedness. Talagrand (1987) gave a characterization of the Glivenko-Cantelli property, without conditions (Sec. 5 below). It turns out that if $\mathcal{F}$ is an order bounded Glivenko-Cantelli class, and g is any monotone function with, for some a, b, $|g(y)| \leq a|y| + b$ for all y, then $g[\mathcal{F}]$ is an order-bounded Glivenko-Cantelli class, via Talagrand's characterization (Propositions 5.2 and 5.3 below). Let's also recall

Theorem A (Talagrand). $\mathcal{F}$ is a strong Glivenko-Cantelli class for P if and only if both

(i) $\mathcal{F}$ is a weak Glivenko-Cantelli class for P and

(ii) $\mathcal{F}_{0,P}$ is order bounded.

Theorem A follows from Talagrand (1987, Theorem 22), as noted in Dudley, Giné and Zinn (1991, Theorem 1).

One of the best known sufficient conditions for the strong Glivenko-Cantelli property is the Blum-DeHardt theorem, whose hypothesis is in terms of *bracketing*, as follows. For any measurable functions $f \leq g$ on $(X, \mathcal{A})$, the *bracket* $[f, g]$ is the set of all measurable functions h such that $f \leq h \leq g$ everywhere. If a family $\mathcal{F} \subset \mathcal{L}^1(P)$ is such that for any $\varepsilon > 0$, $\mathcal{F}$ is included in a finite union of brackets $[f_i, g_i]$ with $\int g_i - f_i dP < \varepsilon$ for all i, then it is not hard to show that $\mathcal{F}$ is a strong Glivenko-Cantelli class for P.

But often, the Glivenko-Cantelli property will not hold even though approximate maximum likelihood estimators do converge. For example, suppose for some $\theta \in \Theta$, $f(x, \theta) = 0$ on a set of positive P-probability. Then $\log f(\cdot, \theta)$ will not be integrable for P, whereas $f(X_j, \theta) = 0$ for some j is actually a help, not a hindrance, to convergence of approximate maximum likelihood estimators. In fact the likelihood will be 0 for all $n \geq j$, and then θ will be not at all in contention to give maximum likelihood.

So it turns out to be useful to consider, not just the Glivenko-Cantelli property of empirical measures or the bracketing property, but *one-sided* properties (Sec. 3 below) which can deal appropriately with densities whose logarithms have integral $-\infty$.

2. Adjustments

Here it will be seen how $\rho(x, \theta)$ can be made more well-behaved by replacing it with $\rho(x, \theta) - a(x)$ for a suitable function $a(\cdot)$.

Definitions. If $(X, \mathcal{A}, P)$ is a probability space, a function $h(\cdot, \cdot) : X \times \Theta \mapsto [-\infty, \infty]$ will be called *adjusted* for P and θ_0 if
  (i) for each $\theta \in \Theta$, $h(\cdot, \theta)$ is measurable on X, $\gamma(\theta) := Ph(\cdot, \theta)$ is well-defined, and
 (ii) $+\infty \geq \gamma(\theta) > \gamma(\theta_0) > -\infty$ for all $\theta \neq \theta_0$.
      A function $\rho(\cdot, \cdot) : X \times \Theta \mapsto (-\infty, \infty]$, measurable in x for each θ, will be called *adjustable* for P and θ_0, and $a(\cdot)$ will be called an *adjustment function* for ρ, P and θ_0, if
(iii) $a(\cdot)$ is measurable from X into $[-\infty, \infty]$,
 (iv) for each θ, $h(x, \theta) := \rho(x, \theta) - a(x)$ is well-defined (not $\infty - \infty$) for P-almost all x, and
  (v) $h(\cdot, \cdot)$ is adjusted for P and θ_0.

The above definitions are a restatement of conditions (A-3) and (A-4) of Huber (1967). We have the following uniqueness of θ_0 and relation between adjustment functions:

2.1 Lemma (Huber). Let $a_1(\cdot)$ be an adjustment function for ρ, P and θ_1. Then $a_2(\cdot)$ is an adjustment function for ρ, P and θ_2 if and only if both $\theta_1 = \theta_2$ and $a_1 - a_2$ is integrable for P.

Proof. "If" is easy to see. To prove "only if," suppose $\theta_1 \neq \theta_2$. Then

$$+\infty \geq P[\rho(\cdot, \theta_2) - a_1(\cdot)] > P[\rho(\cdot, \theta_1) - a_1(\cdot)] > -\infty \quad \text{and}$$

$$+\infty \geq P[\rho(\cdot, \theta_1) - a_2(\cdot)] > P[\rho(\cdot, \theta_2) - a_2(\cdot)] > -\infty, \quad \text{so}$$

$$0 > P[\rho(\cdot, \theta_1) - \rho(\cdot, \theta_2)] > 0,$$

a contradiction, so $\theta_1 = \theta_2$. Then, since $\rho(\cdot, \theta_1) - a_i$ is integrable for $i = 1, 2$, so is $a_1 - a_2$. $\qquad\square$

Thus, it makes sense to say that $\rho(\cdot, \cdot)$ is *adjustable* for P, and that $a(\cdot)$ is an adjustment function for P and ρ. Also, as Huber (1967) noted, if ρ is adjustable for P, the set Θ_1 of θ such that $P|h(\cdot, \theta)| < \infty$ doesn't depend on the choice of adjustment function $a(\cdot)$. If $\rho(\cdot, \cdot)$ is adjustable, then one possible adjustment function is $\rho(\cdot, \theta_1)$ for any fixed $\theta_1 \in \Theta_1$, specifically $\theta_1 = \theta_0$.

It follows from the definitions that for an adjustable function $\rho(x, \theta)$ and an adjustment function $a(x)$ for it, $a(\cdot)$ and $\rho(\cdot, \theta)$ for any $\theta \in \Theta_1$ take finite real values for P-almost all x, while for any $\theta \in \Theta$, $\rho(x, \theta) > -\infty$ for P-almost all x. Clearly, an adjusted function $h(\cdot, \cdot)$ is adjustable.

In case (II), for the three examples of functions ρ mentioned, the adjustment function $a(\cdot)$ can be chosen not depending on P. Specifically:

(i) For $\rho(x, \theta) = |x - \theta|^2$, ρ is adjustable for P if and only if $a(x) = |x|^2$ is an adjustment function, which holds if and only if $P|x| < \infty$.

(ii) For $\rho(x, \theta) = |x - \theta|$, $x, \theta \in \mathbb{R}^k$, ρ is adjustable for P if and only if $a(x) = |x|$ is an adjustment function, which holds if and only if P has a unique median, or spatial median for $k > 1$. The spatial median is unique for any P not concentrated in any line, as shown by Haldane (1948), for finite samples. For the general case see Milasevic and Ducharme (1987).

(iii) Huber's function ρ_r for $x, \theta \in \mathbb{R}$ is adjustable for P if and only if $a(x) = 2r|x|$ is an adjustment function, which holds if and only if the interval of medians of P has length $\leq 2r$.

Adjustment functions can also be helpful in the log likelihood case (I). Suppose for example that μ is Lebesgue measure on $\mathbb{R}$ and P has density $g(x, \theta_0) = 1_{\{x \geq e\}}/(x(\log x)^2)$. Then $P(-\log g(\cdot, \theta_0)) = +\infty$. But, in the log likelihood case, $\rho(x, \theta) := -\log f(x, \theta)$ is *always* adjustable (Theorem 4.1 below), e.g. with adjustment function $a(x) = \rho(x, \theta_0)$. In this case, unlike examples (i), (ii), (iii) of Case II, $a(\cdot)$ may depend on P.

When, as in those examples, $a(\cdot)$ doesn't depend on P and is known, we could replace $\rho(x, \theta)$ by $h(x, \theta) := \rho(x, \theta) - a(x)$ in computing approximate M-estimators, although this may or may not be helpful. But we cannot do this in the log likelihood case (I), since θ_0 is unknown to the statistician.

3. One-sided Glivenko-Cantelli and bracketing properties

Here are some definitions. Let $(X, \mathcal{A}, P)$ be a probability space.

A class $\mathcal{G} \subset \mathcal{L}^1(X, \mathcal{A}, P)$ will be called a *weak* (resp. *strong*) *lower Glivenko-Cantelli class* for P if as $n \to \infty$,

$$\min(0, \inf\{(P_n - P)g : g \in \mathcal{G}\}) \to 0$$

in outer probability (resp. almost uniformly).

For any adjustable function $\rho(\cdot, \cdot)$ and adjustment function $a(\cdot)$ for it, let $h(x, \theta) := h_{\rho, a}(x, \theta) := \rho(x, \theta) - a(x)$ for all x, θ and $H_0 := Ph(\cdot, \theta_0)$. Let $(\Theta, \mathcal{T})$ be a topological space with Borel σ-algebra. As before let $(X, \mathcal{A})$ be a measurable space and let $\mathcal{P}$ be a collection of probability measures (laws) on $(X, \mathcal{A})$. Let $\theta_0(\cdot)$ be a function from $\mathcal{P}$ into Θ.

The sequence $\{T_n\}$ will be called *weakly* (resp. *strongly*) *value-consistent* if $P_n h(\cdot, T_n)$ converges in outer probability (resp. almost uniformly) to H_0, and *weakly* (resp. *strongly*) *doubly consistent* if it is weakly (resp. strongly) both consistent and value-consistent.

Sometimes, as in M-estimation of location (e.g. Huber, 1981, pp. 43-55 for $\Theta = \mathbb{R}$), one is mainly interested in consistency, i.e. approximating θ_0. In other situations, as in model selection, e.g. Akaike (1974), Schwarz (1978), Poskitt (1987), Haughton (1988), one is more interested in estimating the value of H_0 rather than where it occurs.

Often, $Ph(\cdot, \theta)$ is continuous in θ, so that under some conditions, consistent estimators will be value-consistent. Usually, consistency of all approximate M-estimators is only proved when value-consistency also holds. Thus, consistency and value-consistency are closely linked. If, however, $h(x, \theta)$ is not necessarily continuous in θ but only lower semicontinuous, e.g. Huber (1967, (A-2)), value consistency will not follow from consistency. Thus it seems worth while to find conditions for double consistency: H_W and H_S below.

Definitions. Let $(X, \mathcal{A}, P)$ be a probability space, Θ any set, $C \subset \Theta$, $\theta_0 \in \Theta$, $\mathcal{G} \subset \mathcal{L}^1(X, \mathcal{A}, P)$, $\varepsilon > 0$, $\delta > 0$, and $s = \pm 1$. Then let

$$H(\mathcal{G}, \delta, \varepsilon, s, C) \quad := \quad H(\mathcal{G}, \delta, \varepsilon, s, C, P, \theta_0)$$

be the set of adjustable functions ρ on $X \times \Theta$ for P and θ_0 such that for some adjustment function $a(\cdot)$, $h(x, \theta) \equiv \rho(x, \theta) - a(x)$, and $H_0 := Ph(\cdot, \theta_0)$, for each $\theta \in C$, for some $g_\theta \in \mathcal{G}$, $z_\theta \in \mathcal{G}$ and $c(\theta) \geq 1$,

$$Pz_\theta > \delta \quad \text{or} \quad z_\theta \equiv 0, \quad Pg_\theta + Pz_\theta > H_0 + s\varepsilon,$$

and   (3.1)

$$h(x, \theta) \geq g_\theta(x) + c(\theta)z_\theta(x) \qquad \text{for } P\text{-almost all } x.$$

Let $H(\mathcal{E}, W) := H(\mathcal{E}, W, P, \theta_0)$, resp. $H(\mathcal{E}, S) := H(\mathcal{E}, S, P, \theta_0)$, be the set of all adjustable functions ρ for P and θ_0 such that for every neighborhood U of θ_0 there is a weak (resp. strong) lower Glivenko-Cantelli class $\mathcal{G}$ and a $\delta > 0$ such that $\rho \in H(\mathcal{G}, \delta, \delta, +1, U^c)$ where U^c is the complement of U. Here $\mathcal{E}$ connotes "external" and H "Huber" since Huber (1967; 1981, pp. 127-130) considered similar conditions.

Next let $H(\mathcal{N}, W) := H(\mathcal{N}, W, P, \theta_0)$, resp. $H(\mathcal{N}, S) := H(\mathcal{N}, S, P, \theta_0)$, be the set of all adjustable functions ρ for P and θ_0 such that for every $\varepsilon > 0$ there is a $\delta > 0$, a neighborhood V of θ_0 and a weak (resp. strong) Glivenko-Cantelli class $\mathcal{H}$ for P such that $\rho \in H(\mathcal{H}, \delta, \varepsilon, -1, V)$. Here $\mathcal{N}$ connotes "neighborhoods" of θ_0. Let $H_W := H_W(P, \theta_0) := H(\mathcal{E}, W) \cap H(\mathcal{N}, W)$ and likewise for W replaced by S.

The function $c(\theta)$ is often but not always the same as the function $b(\theta)$ of Huber (1967) (see (A-5) in Sec. 7 below). Apparently Huber was the first to use such functions of θ.

In the definition of $H(\mathcal{G}, \cdots)$, the function $a(\cdot)$ and the class $\mathcal{G}$ can depend on each other, but we have:

Proposition 3.2. Each of the definitions of $H(\mathcal{E}, W)$, $H(\mathcal{E}, S)$, $H(\mathcal{N}, W)$, $H(\mathcal{N}, S)$, H_W and H_S, for some P and θ_0 in each case, if it holds for ρ, also holds where $a(\cdot)$ in the definition of $H(\mathcal{G}, \cdots)$ and/or $H(\mathcal{H}, \cdots)$ can be an arbitrary adjustment function for ρ and P, provided that the weak or strong lower Glivenko-Cantelli class $\mathcal{G}$ and/or $\mathcal{H}$ is also chosen suitably.

Proof. Let $\mathcal{G} + f := \{g + f : g \in \mathcal{G}\}$. For any weak (resp. strong) lower Glivenko-Cantelli class $\mathcal{G}$ for P and fixed $f \in \mathcal{L}^1(X, \mathcal{A}, P)$, clearly $\mathcal{G} \cup (\mathcal{G} + f)$ is also a weak (resp. strong) lower Glivenko-Cantelli class. If a_1 and a_2 are two adjustment functions for ρ, θ_0 and P, where (3.1) holds with a_1, apply Lemma 2.1 with $f = a_1 - a_2$ and add f to $h := \rho - a_1$ and g_θ while keeping the same z_θ. The conclusion follows. $\qquad\square$

Definition. An adjustable function ρ on $X \times \Theta$ for P and θ_0, where $(\Theta, \mathcal{T})$ is a topological space, will be called *proper* for P and θ_0 if for an adjustment function $a(\cdot)$ and the corresponding $h(\cdot, \cdot)$, whenever $\theta_k \in \Theta$ are such that $\limsup_{k\to\infty} \gamma(\theta_k) \le \gamma(\theta_0)$, or equivalently $\gamma(\theta_k) \to \gamma(\theta_0)$ (by definition of "adjustable"), then $\theta_k \to \theta_0$.

As in the proof of Lemma 2.1 above one can check that the "proper" condition doesn't depend on the choice of $a(\cdot)$. It is also easily verified that for ρ to be proper is equivalent to the definition of $H(\mathcal{E}, W, P, \theta_0)$ or $H(\mathcal{E}, S, P, \theta_0)$ but for an arbitrary class $\mathcal{G} \subset \mathcal{L}^1(P)$, not necessarily Glivenko-Cantelli. Thus "proper" is a necessary (but far from sufficient) condition for $H(\mathcal{E}, W)$ or $H(\mathcal{E}, S)$.

It will be said that a condition depending on n holds *eventually* if it holds for n except on a set A_n whose outer probability $\to 0$ as $n \to \infty$, and where in the strong case, $A_1 \supset A_2 \supset \cdots$.

3.3 Theorem. Let Θ have a metrizable topology. If $\rho \in H(\mathcal{E}, W, P, \theta_0)$, resp. $\rho \in H(\mathcal{E}, S, P, \theta_0)$, then any sequence $\hat{\theta}_n$ of weak (resp. strong) approximate M-estimators is weakly (resp. strongly) consistent. If $\rho \in H_W(P, \theta_0)$, resp. $\rho \in H_S(P, \theta_0)$, then $\hat{\theta}_n$ are weakly (resp. strongly) doubly consistent.

Proof. If an "$\mathcal{E}$" hypothesis holds, take an adjustment function $a(\cdot)$ as assumed. For any neighborhood U of θ_0 take a corresponding $\delta > 0$ and lower Glivenko-Cantelli class $\mathcal{G}$. Then eventually, $P_n h(\cdot, \theta) > H_0 + \delta/2$ for all $\theta \notin U$, since if $P z_\theta > \delta$ then uniformly in $\theta \notin U$,

$$P_n h(\cdot, \theta) \ge P_n g_\theta + c(\theta) P_n z_\theta \ge P g_\theta - \delta/4 + c(\theta)[P z_\theta - \delta/4]$$

$$\ge P g_\theta + P z_\theta - \delta/2 \ge H_0 + \delta/2,$$

while if $z_\theta \equiv 0$, terms with $c(\theta)$ disappear and the same follows.

On the other hand, eventually

$$P_n h(\cdot, \hat{\theta}_n) < \frac{\delta}{4} + \inf_{\theta \in \Theta} P_n h(\cdot, \theta) \le \frac{\delta}{4} + P_n h(\cdot, \theta_0) < \frac{\delta}{2} + H_0,$$

so $\hat{\theta}_n \in U$ and consistency is proved.

Now suppose $\rho \in H_W$ or H_S. Given $\varepsilon > 0$, take a neighborhood $V = V_\varepsilon$ of θ_0, $\delta > 0$ and lower Glivenko-Cantelli class $\mathcal{H}$ given by the definition. By Proposition 3.2, we can assume that the adjustment function $a(\cdot)$ and function h are the same in both the $\mathcal{E}$ and $\mathcal{N}$ conditions. As in the consistency proof, eventually $P_n h(\cdot, \theta) \ge H_0 - 2\varepsilon$ for all $\theta \in V$. Since we also have eventually $P_n h(\cdot, \theta) > H_0$ for all $\theta \notin V$, it follows that eventually

$$P_n h(\cdot, \hat{\theta}_n) \ge \inf_{\theta \in \Theta} P_n h(\cdot, \theta) \ge H_0 - 2\varepsilon.$$

On the other hand by definition of approximate M-estimator and the law of large numbers, eventually

$$P_n h(\cdot, \hat{\theta}_n) \;\leq\; \inf_{\theta \in \Theta} P_n h(\cdot, \theta) + \varepsilon \;\leq\; P_n h(\cdot, \theta_0) + \varepsilon \;<\; H_0 + 2\varepsilon.$$

So $P_n h(\cdot, \hat{\theta}_n) \to H_0$ in outer probability (resp. almost uniformly) and the value and double consistency hold.   □

Huber (1967, Sec. 2; 1981, Sec. 6.2), gives a proof of consistency of approximate M-estimators under his conditions (A-1) through (A-5) on functions $\rho(x, \theta)$, $a(x)$ and $b(\theta)$, where $b(\cdot)$ is a positive, continuous function on Θ. Huber's conditions, and a proof of the following Theorem, are given in Sec. 7 below.

3.4 Theorem. If conditions (A-1) through (A-5) of Huber (1967) hold, for a function $\rho(\cdot, \cdot)$, P and θ_0, then $\rho \in H_S(P, \theta_0)$.

It follows then from Theorem 3.3 that under Huber's conditions, as he showed, weak (resp. strong) approximate M-estimators converge in outer probability (resp. almost uniformly).

Let $(X, \mathcal{A}, P)$ be a probability space and $(X^k, \mathcal{A}^k, P^k)$ its k-fold Cartesian product. A function f on X^k is called *symmetric* if it is invariant under all permutations of the coordinates. If $f : X^k \mapsto \mathbb{R}$ is symmetric, for any $n \geq k$ and $x_1, \cdots, x_n$ let

$$U_n^k \;:=\; U_n^k(f)(x_1, \cdots, x_n) \;:=\; \binom{n}{k}^{-1} \sum_{\alpha \in I(n,k)} f(x_{\alpha(1)}, \cdots, x_{\alpha(k)}),$$

where $I(n, k)$ is the set of all ordered k-tuples of integers $1 \leq \alpha(1) < \alpha(2) < \cdots < \alpha(k) \leq n$. Then U_n^k are called U-statistics with kernel f, e. g. Dudley (1993, sec. 11.9). A class $\mathcal{F} \subset \mathcal{L}^1(X^k, P^k)$ will be called a *weak* (resp. *strong*) U-*Glivenko-Cantelli class* if as $n \to \infty$, $\sup_{f \in \mathcal{F}} |(U_n^k - P^k)(f)| \to 0$ in outer probability (resp. almost uniformly).

For any real-valued function f on the sample space X and $k = 1, 2, \cdots$, let

$$P_{(k)} f \;:=\; [f(x_1) + \cdots + f(x_k)]/k,$$

viewed as a function on X^k (rather than Ω). Then for any $n \geq k$,

$$P_n f \;=\; U_n^k P_{(k)} f. \tag{3.5}$$

Thus $\{P_n f\}_{n \geq k}$ is a sequence of U-statistics with kernel $P_{(k)} f$. A collection of U-statistics with kernels varying over a class of functions, here $\{P_{(k)} h(\cdot, \theta) : \theta \in \Theta\}$, is called a U-process, e.g. Nolan and Pollard (1987, 1988) for $k = 2$, Arcones and Giné (1993) for general $k > 1$. In the present case the U-process itself might seem uninteresting since it reduces to the sequences $P_n h(\cdot, \theta)$ where $k = 1$, an empirical process. But, bracketing conditions may hold for $P_{(k)} h(\cdot, \theta)$, $\theta \in \Theta$, $k > 1$, when they do not for $k = 1$, as Pitman (1979, Chapter 8) and no doubt others have noted. Perlman (1972) considers cases where a finite class of brackets $[f_i, \infty)$ for

$k = 1$ can be replaced by a single bracket $[g, \infty)$ for some g and $k > 1$. The "admissible Suslin" measurability condition on a class of functions is defined in Appendix A below. If an admissible Suslin class $\mathcal{G}$ of integrable functions on X^k is a strong Glivenko-Cantelli class for P^k, and there is a G with $|g| \leq G$ for all $g \in \mathcal{G}$ and $P^k G < \infty$, then $\mathcal{G}$ is also a strong U-Glivenko-Cantelli class, as E. Giné has shown (Appendix B below). On the other hand a U-Glivenko-Cantelli class is not necessarily a Glivenko-Cantelli class as shown by an example on p. 1511 of Arcones and Giné (1993). But, if $\mathcal{H}$ is a class of functions on X integrable for P and if $\{P_{(k)}h : h \in \mathcal{H}\}$ is a U-Glivenko-Cantelli class then since $P^k P_{(k)} f \equiv P f$ it follows directly from (3.5) that $\mathcal{H}$ is a Glivenko-Cantelli class for P.

Notations $H^{(U)}(\cdots)$ in place of $H(\cdots)$ will mean that instead of a lower Glivenko-Cantelli class $\mathcal{G}$ or $\mathcal{H}$ we have a U-Glivenko-Cantelli class. The proof of Theorem 3.3 then can be adapted straightforwardly to give, with (3.5):

3.6 Theorem. If Θ has a metrizable topology and for some $k \geq 1$, $P_{(k)}h(\cdot, \cdot) \in H^{(U)}(\mathcal{E}, W, P^k, \theta_0)$, respectively $H^{(U)}(\mathcal{E}, S, P^k, \theta_0)$, then any sequence $\hat{\theta}_n$ of weak (resp. strong) approximate M-estimators is weakly (resp. strongly) consistent.

I do not know and would be interested in knowing if there are cases where Theorem 3.6 applies but Theorem 3.3 does not or is harder to apply. Perlman (1972) also gives a condition equivalent to consistency, but naturally somewhat hard to check.

Let $(X, \mathcal{A})$ be any measurable space and P, Q any two laws on $(X, \mathcal{A})$. There exist σ-finite measures v such that P and Q are both absolutely continuous with respect to v, for example $v = P + Q$. Let f and g be Radon-Nikodym derivatives of P and Q respectively with respect to v. Let

$$H(P, Q) := (\textstyle\int (f^{1/2} - g^{1/2})^2 dv)^{1/2}.$$

Then H doesn't depend on v and is called the *Hellinger metric*. It is defined and finite for any two laws P, Q and is indeed a metric. For the metric

$$\|P - Q\|_1 := \textstyle\int |f - g| dv,$$

which also doesn't depend on v, the identity map from the set of all laws on $(X, \mathcal{A})$ to itself is uniformly continuous in both directions between H and $\|\cdot\|_1$, specifically

$$H^2(P, Q) \leq \textstyle\int |f^{1/2} - g^{1/2}|(f^{1/2} + g^{1/2}) dv = \|P - Q\|_1 \leq 2H(P, Q), \quad (3.7)$$

the latter inequality by the Cauchy-Schwarz inequality twice, cf. Pitman (1979, p. 7), or Le Cam (1986, pp. 46-47), except for factors of 2).

Given two laws P and Q on $(X, \mathcal{A})$, by the Lebesgue decomposition take a set $A \in \mathcal{A}$ with maximal $P(A)$ such that $Q(A) = 0$. Then P is absolutely continuous with respect to Q on A^c, with Radon-Nikodym derivative $R_{P/Q}$ there. Set $R_{P/Q} := +\infty$ on A. Then $R_{P/Q}$ is called the likelihood ratio of P to Q, and is defined up to equality $(P+Q)$-almost everywhere. Also, $R_{Q/P} = 1/(R_{P/Q})$ almost everywhere for $P + Q$, where $x \mapsto 1/x$ interchanges 0 and $+\infty$. Let $I(P, Q) :=$

$\int \log(R_{P/Q})dP = -\int \log(R_{Q/P})dP$. Then it is well known that $I(P,Q)$ is well-defined for any P and Q and $0 \leq I(P,Q) \leq +\infty$ with $I(P,Q) = 0$ if and only if $P = Q$: Pitman (1979), p. 63, Kullback (1983), or Bickel and Doksum (1977, pp. 226, 243).

Let $I(\theta, \phi) := I(P_\theta, P_\phi)$. Then $I(P,Q)$ (resp. $I(\theta, \phi)$) is called the *Kullback-Leibler information* for P, Q (resp. θ, ϕ). Kullback and Leibler (1951) note that their information is essentially that of Shannon and Wiener; see also Kullback (1983), Zacks (1971, pp. 212-213).

$I(P,Q)$ is sometimes called a "metric," but it is not symmetric, nor does it satisfy the triangle inequality. In fact we have:

3.8 Theorem. (a) For any measurable space $(X, \mathcal{A})$, there is a topology $\mathcal{T}$ on the set of all probability laws on $(X, \mathcal{A})$ such that $P_k \rightarrow P$ for $\mathcal{T}$ if and only if $I(P, P_k) \rightarrow 0$.
(b) But, no such topology is metrizable for $X = [0,1]$ since the convergence fails the iterated limit property: there exist laws Q_{kj}, P_k, $P = U[0,1]$ on $[0,1]$ such that $P_k \rightarrow P$ for $\mathcal{T}$ as $k \rightarrow \infty$ and for all k, $Q_{kj} \rightarrow P_k$ for $\mathcal{T}$ as $j \rightarrow \infty$, but there are no $j(k)$ such that $Q_{kj(k)} \rightarrow P$ for $\mathcal{T}$ as $k \rightarrow \infty$.
(c) Also for $P = U[0,1]$, if $0 < \varepsilon < 1$, then $\{Q : I(P,Q) < \varepsilon\}$ is not a neighborhood of P for $\mathcal{T}$. For every $\mathcal{T}$-open set U containing P, $\sup\{I(P,Q) : Q \in U\} \geq 1$, and $\sup\{I(P,\mu) : I(P,Q) < \varepsilon, \ I(Q,\mu) < \varepsilon\} \geq 1$.

Proof. (a) Define convergence of sequences by $P_k \rightarrow P$ if and only if $I(P, P_k) \rightarrow 0$ as $k \rightarrow \infty$. It's easily seen that sequences have unique limits, if they converge in this sense. It's also easily seen that a subsequence of a convergent sequence converges to the same limit, and we have: if P_k and P are such that for every subsequence $P_{k(j)}$ there is a subsubsequence $P_{k(j(i))} \rightarrow P$ as $i \rightarrow \infty$, then $P_k \rightarrow P$. Thus the convergence is convergence for a topology $\mathcal{T}$ in which points are closed, e.g. Kisyński (1960, Thme. 3), where

$$\mathcal{T} := \{U : \quad \text{if } P_k \rightarrow P \in U \text{ then } P_k \in U \text{ for } k \text{ large enough}\}.$$

This was apparently first proved by Kantorovich, Vulikh and Pinsker (1950, Theorem 2.42 p. 51; 1951, translation, pp. 61-62). Linearity and partial ordering are not used in the proof.
(b) Let P_k have density $f_k = x$ on $[0, 1/k]$ and $(1 - \frac{1}{2}k^{-2})/(1 - \frac{1}{k})$ for $1/k < x \leq 1$. Then it's easily checked that $I(P, P_k) \rightarrow 0$ where $P = U[0,1]$. For each $k = 1, 2, ...$, and each $j > k$, let P_{kj} have density $f_{kj} := e^{-j}$ on $[0, 1/j]$ and $a_j f_k$ on the rest of the interval, where $a_j = (1 - e^{-j}/j)/(1 - \frac{1}{2}j^{-2})$. Then it can be checked that $P_{kj} \rightarrow P_k$ as $j \rightarrow \infty$ for each k. Let $g_{ki} := f_{k,i+k}$ for all positive integers k and i. Let Q_{ki} have density g_{ki}. For any k and j,

$$-I(P, P_{kj}) = \int_0^1 \log(f_{kj})dx = -1 + (1 - \frac{1}{j})\log(a_j) + \int_{1/j}^1 \log(f_k)dx = -1 + o(1)$$

as $k \rightarrow \infty$ and thus also $j \rightarrow \infty$. It follows that although $Q_{ki} \rightarrow P_k$ as $i \rightarrow \infty$ for $\mathcal{T}$, for each k, there is no subsequence $Q_{ki(k)} \rightarrow P$.

The same example proves all three statements in (c).   □

Parts (b) and (c) show that despite (a), $I(\cdot,\cdot)$ is far from being a metric. The triangle inequality fails very badly. On the topology $\mathcal{T}$ see also Corollary 4.5 below.

For consistency of maximum likelihood estimators it is not sufficient that for any neighborhood U of θ_0, $\inf_{\theta \notin U} I(\theta_0, \theta) > 0$, or even that $I(\theta_0, \theta) = +\infty$ for all $\theta \neq \theta_0$. Without any regularity conditions, for example, let P_{θ_0} be uniform on $[0,1]$ and let P_θ, $\theta \in \Theta$, contain all laws whose densities with respect to P_{θ_0} are either 0 or 2 at each point of $[0,1]$. (This is an infinite-dimensional family.) Then any given sample $X_1, ..., X_n$ i.i.d. (P_{θ_0}) has likelihood 1 for P_{θ_0} but likelihood 2^n for many other values of θ.

4. The log likelihood case

It will be said that the *log likelihood case* applies to $\rho(\cdot,\cdot)$, θ_0 and P if there is a family $\{P_\theta,\ \theta \in \Theta\}$ of laws on $(X, \mathcal{A})$ such that $P = P_{\theta_0}$, $P_\theta \neq P_\phi$ for all $\theta \neq \phi$, there is a σ-finite measure v on $(X, \mathcal{A})$ such that each P_θ is absolutely continuous with respect to v and has density $f(\cdot, \theta) \geq 0$, and for all x and θ, $\rho(x, \theta) = -\log f(x, \theta)$, so that $-\infty < \rho(x, \theta) \leq +\infty$.

If we set $a(x) := \rho(x, \theta_0)$, then from the basic facts mentioned above about Kullback-Leibler information, for $h(x, \theta) := \rho(x, \theta) - a(x)$ and $\gamma(\theta) := Ph(\cdot, \theta)$, we have $\gamma(\theta) \geq 0$ for all θ and $\gamma(\theta) = 0$ if and only if $\theta = \theta_0$. Thus we have the following essentially known fact:

4.1 Theorem. In the log likelihood case, ρ is always adjustable.

If $\rho(x, \theta) \equiv -\log f(x, \theta)$ but P is not in the family $\{P_\theta,\ \theta \in \Theta\}$, then ρ may or may not be adjustable. If it is, then $\theta_0 = \theta_0(P)$ is called the *pseudo-true* value of θ: Sawa (1978), Poskitt (1987). That interesting situation is not treated in this paper except under the general case as in Sec. 3. Returning to the case $P = P_{\theta_0}$, we have:

4.2 Theorem. In the log likelihood case, with the Hellinger metric H on Θ, $\rho(\cdot, \cdot)$ is always proper for P and θ_0.

Proof. Two lemmas will be useful:

4.3 Lemma. Let $r(t) := t - \log(1 + t)$ for $t > -1$ and $r(-1) := +\infty$. Let $q(t) := t^2/8$ for $|t| \leq 1$ and $q(t) := (2t - 1)/8$ for $t \geq 1$. Then $r(t) \geq q(t)$ for all $t \geq -1$.

Proof. Clearly $r(0) = r'(0) = q(0) = q'(0) = 0$ and $r''(t) = (1+t)^{-2} \geq 1/4 = q''(t)$ for $|t| < 1$. Thus $r(t) \geq q(t)$ for $|t| < 1$. Note that q and q' are continuous at $t = 1$. Also $r'(1) = 1/2 > q'(1)$ and $r''(t) > 0 = q''(t)$ for $t > 1$. The conclusion follows.
□

4.4 Lemma. There exists a decreasing, Lipschitz function g from $[0, \infty)$ into $\mathbb{R}$, bounded above, with $g(y) \leq -\log(y)$ for all $y \geq 0$, such that for any $\varepsilon > 0$ there is a $\delta > 0$ such that if P and Q are any laws on $(X, \mathcal{A})$ with $H(P, Q) > \varepsilon$, then for $f := R_{Q/P}$, $Pg(f) > \delta$.

Proof. By the uniform equivalence of Hellinger and L^1 metrics (3.7), for $\zeta = \zeta(\varepsilon) = \varepsilon^2/4 > 0$, $H(P, Q) > \varepsilon$ implies $P(|f - 1| \geq \zeta) > \zeta$, and $0 < \zeta < 1$. Let

$$g(y) = \begin{cases} 1 - y + (y - 1)^2/8, & \text{for } 0 \leq y \leq 2; \\ (5 - 6y)/8, & \text{for } y > 2. \end{cases}$$

Then $g(y) \equiv 1 - y + q(y - 1)$ for $q(\cdot)$ from Lemma 4.3, so for all $y > 0$,

$$g(y) \leq 1 - y + r(y - 1) = -\log y.$$

It is clear by derivatives that g is decreasing and Lipschitz. It is bounded above by $g(0) = 9/8$. We have $P(1 - f) \geq 0$, and $q(t)$ is increasing for $t > 0$, decreasing for $t < 0$. Thus $Pg(f) \geq Pq(f - 1) \geq \zeta \cdot \zeta^2/8 = \zeta^3/8 = \varepsilon^6/2^9 =: \delta$. $\square$

Returning to the proof of Theorem 4.2, we have for each θ,

$$\gamma(\theta) = P(-\log f(\cdot, \theta) + \log f(\cdot, \theta_0)) = P(-\log(R_{P_\theta/P})).$$

Thus Lemma 4.4 gives the conclusion. $\square$

Lemma 4.4 also implies the following:

4.5 Corollary. The topology $\mathcal{T}$ of Theorem 3.8(a) is stronger (has more open sets) than that of the Hellinger or equivalently the total variation metric. Thus $\mathcal{T}$ is Hausdorff.

Proof. By Lemma 4.4, for any $\varepsilon > 0$ and $\delta = \delta(\varepsilon) > 0$, if $I(P, Q) \leq \delta$ then $H(P, Q) \leq \varepsilon$. So if $I(P, P_k) \to 0$ as $k \to \infty$, then $H(P, P_k) \to 0$. The conclusions follow. $\square$

5. Glivenko-Cantelli classes

Let $(X, \mathcal{A}, P)$ be a probability space. A class $\mathcal{F}$ of real-valued functions on X will be called *T-stable (Talagrand-stable)* for P iff for each $A \in \mathcal{A}$ with $P(A) > 0$ and for any $\alpha < \beta$ there is some $n > 0$ such that

$$(P^{2n})^* \{\langle s_1, \cdots, s_n, t_1, \cdots, t_n \rangle \in A^{2n} : \quad \text{for some } f \in \mathcal{F}$$

$$\text{and all } i = 1, \cdots, n, \quad f(s_i) < \alpha < \beta < f(t_i)\} < P(A)^{2n}.$$

Talagrand (1987, Theorems 2 and 22) proved (cf. Theorem A above):

5.1 Theorem (Talagrand). Let $\mathcal{F} \subset \mathcal{L}^1(X, \mathcal{A}, P)$. Then the following are equivalent:
(a) $\mathcal{F}$ is a strong Glivenko-Cantelli class for P and $\{Pf : f \in \mathcal{F}\}$ is bounded;
(b) $\mathcal{F}$ is an order bounded, weak Glivenko-Cantelli class for P;
(c) $\mathcal{F}$ is T-stable and order bounded for P.

Let $(g \circ f)(x) := g(f(x))$ for any functions g and f for which $g(f(x))$ is defined, and $g[\mathcal{F}] := \{g \circ f : f \in \mathcal{F}\}$. Van de Geer (1993) considered such transformations of Glivenko-Cantelli classes. We then have:

5.2 Proposition. If $\mathcal{F}$ is T-stable and g is a continuous, monotone function on a (possibly unbounded) interval J in $\mathbb{R}$ including the ranges of all $f \in \mathcal{F}$, then $g[\mathcal{F}]$ is T-stable.

Proof. We can assume g is non-decreasing (interchanging g with $-g$, α with $-\beta$, β with $-\alpha$, and s_i with t_i). Now $g(J) := \{g(y) : y \in J\}$ is a possibly unbounded interval. If α or β is not in the interior of $g(J)$ then $\{g(f(s_i)) < \alpha\}$ or $\{g(f(t_i)) > \beta\}$ is empty for each i, so assume $\alpha < \beta$ are in the interior. Then for some $u < v$, $g(y) < \alpha$ if and only if $y < u$, and $g(y) > \beta$ if and only if $y > v$. So T-stability of $g[\mathcal{F}]$ follows from that of $\mathcal{F}$. $\square$

It follows easily that:

5.3 Proposition. If $\mathcal{F}$ is an order bounded weak, thus strong Glivenko-Cantelli class for a law P, J is a possibly unbounded interval including the ranges of all $f \in \mathcal{F}$, g is continuous and monotone on J, and for some finite constants c, d, $|g(y)| \leq c|y| + d$ for all $y \in J$, then $g[\mathcal{F}]$ is also an order bounded strong Glivenko-Cantelli class for P.

In the log likelihood case, the Hellinger metric on $\{P_\theta,\ \theta \in \Theta\}$ defines a metric on Θ. Total variation gives a metric uniformly equivalent by (3.7), and so with the same topology and convergence on Θ. Consistency can and will be defined in this metric. Let $R_\theta := R_{P_\theta/P} 1_{\{f(\cdot,\theta_0)>0\}}$. Van de Geer (1993, Lemma 1.1) proved strong consistency of (approximate) maximum likelihood estimators in the log likelihood case whenever $\{R_\theta^{1/2} : \theta \in \Theta\}$ is a strong Glivenko-Cantelli class for P. In fact we have:

5.4 Proposition. If $\{P_\theta,\ \theta \in \Theta\}$ is a family of laws all absolutely continuous with respect to a σ-finite measure v, $0 < \eta \leq 1$ is a constant and $\{R_\theta^\eta : \theta \in \Theta\}$ is a strong Glivenko-Cantelli class for P, then for $\rho(x, \theta) = -\log(dP_\theta/dv)(x)$, $\theta \in \Theta$, we have $\rho \in H_S(P_\theta, \theta)$ for all $\theta \in \Theta$, and so approximate M-estimators (in this case, approximate maximum likelihood estimators) are strongly doubly consistent in the Hellinger metric, for all θ.

Proof. For each θ, $PR_\theta^\eta \leq (PR_\theta)^\eta \leq 1$. Thus by Theorem 5.1, $\{R_\theta^\eta : \theta \in \Theta\}$ is T-stable and order bounded for P. Proposition 5.3 implies that for g as defined in the proof of Lemma 4.4, $\mathcal{G} := \{0\} \cup \{g \circ R_\theta : \theta \in \Theta\}$ is also a strong Glivenko-Cantelli class. Then Lemma 4.4 implies that ρ is in $H(\mathcal{E}, S)$, letting $z_\theta \equiv 0$, so that $c(\theta)$ is irrelevant, say $c(\theta) \equiv 1$, and $g_\theta := g \circ R_\theta$. Since g doesn't depend on ε or δ, we can let $\delta \downarrow 0$ and conclude that $\rho \in H(\mathcal{N}, S)$, where we actually have H_0 in place of $H_0 - \varepsilon$. The last statement follows from Theorem 3.3. $\square$

6. Universal and uniform Glivenko-Cantelli conditions

For statistics, since the law P is unknown, it is desirable that properties of Glivenko-Cantelli type should hold for all P in a rather large class $\mathcal{P}$ of laws and, if possible, uniformly in P.

Let $(X, \mathcal{A})$ be a measurable space and $\mathcal{P}$ a class of laws (probability measures) on $(X, \mathcal{A})$. A class $\mathcal{F}$ of real functions on $(X, \mathcal{A})$ will be called a *weak* (resp. *strong*)

$\mathcal{P}$-*universal Glivenko-Cantelli class* if it is a weak (resp. strong) Glivenko-Cantelli class for all $P \in \mathcal{P}$, and a *weak* (resp. *strong*) $\mathcal{P}$-*uniform* Glivenko-Cantelli class if for all $\varepsilon > 0$, as $n \to \infty$

$$\sup_{P \in \mathcal{P}} \Pr^* \{ \|P_n - P\|_{\mathcal{F}} > \varepsilon \} \to 0, \quad \text{resp.}$$

$$\sup_{P \in \mathcal{P}} \Pr^* \{ \|P_k - P\|_{\mathcal{F}} > \varepsilon \ \text{ for some } \ k \geq n \ \} \to 0.$$

In each case, if $\mathcal{P}$ is the class of all laws on $(X, \mathcal{A})$, "$\mathcal{P}$-" will be omitted. Then, Dudley, Giné and Zinn (1991) gave sufficient conditions for the universal Glivenko-Cantelli property and characterizations of the weak and strong uniform Glivenko-Cantelli properties, under a measurability assumption for the strong case. Sheehy and Wellner (1992) consider $\mathcal{P}$-uniformity in the central limit theorem.

Recall the notion of Vapnik-Červonenkis (VC) class of sets, e. g. Dudley (1984, Chap. 9). Also, a class $\mathcal{F}$ of real-valued functions on a set X is called a *VC major class* if there is a VC class $\mathcal{C}$ of subsets of X such that for all $f \in \mathcal{F}$ and $t \in \mathbb{R}$, the set $\{x \in X : f(x) > t\} \in \mathcal{C}$. Such classes were treated in Dudley (1987) and Quiroz and Dudley (1991). Recall also the notion of uniform Donsker class (Giné and Zinn, 1991). Let $|\cdot|$ be the usual Euclidean norm on $\mathbb{R}^k$. Functions ρ as in the following fact occur in Huber's theory of M-estimators of location, e.g. Huber (1981, Example 4.5.2, (5.22)) which gives an optimality property for $k = 1$ where $\psi = \tau'$.

A function from $\mathbb{R}^k$ into $\mathbb{R}$ will be called a *polar polynomial* if it is a polynomial in $x_1, \cdots, x_k$ and $|x|$. A set $A \subset \mathbb{R}^k$ will be called a *Boolean polynomial set* if it is a finite Boolean combination (allowing complements, intersections and thus unions) of sets $\{x : P_i(x) > 0\}$ where each P_i is a polynomial. A function $\tau : \mathbb{R}^k \mapsto \mathbb{R}$ will be called a *piecewise polar polynomial* if $\mathbb{R}^k$ is a disjoint union of Boolean polynomial sets $A_0, A_1, \cdots, A_m$ such that on each A_i, τ equals some polar polynomial τ_i.

6.1 Theorem. For any $k = 1, 2, \cdots$ and piecewise polar polynomial $\tau : \mathbb{R}^k \mapsto \mathbb{R}$, the classes of functions $\mathcal{H} := \{\tau(\cdot - \theta) : \theta \in \mathbb{R}^k\}$ and $\mathcal{F} := \{\tau(\cdot - \theta) - \tau : \theta \in \mathbb{R}^k\}$ are VC major classes.

Proof. Let $\mathbb{R}^k$ be the disjoint union of Boolean polynomial sets $A_0, A_1, \cdots A_m$, where $\tau = \tau_i$, a polar polynomial, on A_i for each i. Let each A_i be a finite Boolean combination of k_i sets $\{P_{ij} > 0\}$ for polynomials P_{ij}. Any translate $A_i + \theta :=$ $\{x + \theta : x \in A_i\}$ is a Boolean combination of k_i sets $\{Q_{ij} > 0\}$ where each Q_i is a polynomial of the same degree as P_{ij}. Thus by known facts about VC classes (e.g. Dudley, 1984, Proposition 9.1.7, Theorems 9.2.1, 9.2.3), there is a VC class $\mathcal{C}$ containing all sets $A_i + \theta$, $i = 0, 1, \cdots, m$, $\theta \in \mathbb{R}^k$, and also all intersections $(A_i + \theta) \cap A_j$, $j = 0, 1, \cdots, m$.

For each i and θ, we have

$$\tau_i(x - \theta) \equiv \tau_{i1}(x - \theta) + |x - \theta| \tau_{i2}(x - \theta)$$

where τ_{i1} and τ_{i2} are polynomials, since $|x - \theta|^2$ is a polynomial in $x - \theta$ and so in x for any fixed θ.

For any real functions f, g, h on a set X with $f \geq 0$, let $A(f, g, h) := \{f^{1/2}g > h\}$. Then it is straightforward to check that $A(f, g, h) =$

$$[\{g > 0\} \cap \{fg^2 - h^2 > 0\}] \cup [\{g > 0\} \cap \{h < 0\}] \cup [\{h < 0\} \cap \{fg^2 - h^2 < 0\}]. \quad (6.2)$$

For any fixed $\theta \in \mathbb{R}^k$ and $t \in \mathbb{R}$,

$$B_i(\theta, t) := \{x : \tau_i(x - \theta) > t\} = A(f, g, h)$$

where $f(x) \equiv |x - \theta|^2$, $g(x) \equiv \tau_{i2}(x - \theta)$ and $h(x) \equiv t - \tau_{i1}(x - \theta)$. Then f, g and h are polynomials in x, with degrees not depending on θ. Thus the degrees of polynomials appearing in (6.2) are uniformly bounded. By the same general facts about VC classes already mentioned, there is a VC class $\mathcal{D}$ containing all the sets $B_i(\theta, t)$ for $i = 0, 1, \cdots, m$ and $\theta \in \mathbb{R}^k$, $t \in \mathbb{R}$. Then

$$\{x : \tau(x - \theta) > t\} = \bigcup_{i=1}^{m} (A_i + \theta) \cap B_i(\theta, t).$$

It follows that $\mathcal{H}$ as defined in the Theorem is a VC major class.

For $\mathcal{F}$ the following (which may be of independent interest) will be useful. Its proof is similar to some arguments in Quiroz and Dudley (1991, around (7.8)).

6.3 Lemma. If V is a finite-dimensional vector space of real-valued functions on a set X, then

$$\{f_1^{1/2}g_1 + f_2^{1/2}g_2 + \eta : f_1, f_2, g_1, g_2, \eta \in V, \ f_1 \geq 0, f_2 \geq 0\}$$

is a VC major class.

Proof. We can assume that all constants belong to V. Let V^3 be the vector space spanned by all functions $u_1 u_2 u_3$ for $u_i \in V$. Then V^3 is finite-dimensional and includes V. For any function $w \in V^3$, $\{w < 0\} = \{-w > 0\}$ where $-w \in V^3$. For any $t \in \mathbb{R}$,

$$\{f_1^{1/2}g_1 + f_2^{1/2}g_2 + \eta > t\} = A(f_2, g_2, t - \eta - f_1^{1/2}g_1).$$

By (6.2), each such set is of the form $\cup_{r=1}^{3} \cap_{s=1}^{2} B_{rs}$ where each B_{rs} is a set $A(f_1, u, v)$ for some $u, v \in V^3$. Then by (6.2) again, we have for each r and s,

$$B_{rs} = \cup_{i=1}^{3} \cap_{j=1}^{2} B_{rsij}$$

where each B_{rsij} is a set $\{\zeta > 0\}$ for $\zeta \in (V^3)^3$. Any finite-dimensional vector space of real-valued functions is a VC major class, e.g. Dudley (1984, Theorem 9.2.1). A bounded number (in this case, 36) of Boolean combinations of sets preserves the VC property, e.g. Dudley (1984, (9.2.3)). The Lemma follows. $\square$

Now, if τ is a polar polynomial and $\theta \in \mathbb{R}^k$, then

$$\tau(x - \theta) - \tau(x) \equiv a(x - \theta) + |x - \theta|b(x - \theta) + c(x) + |x|d(x)$$

where $a(\cdot)$, $b(\cdot)$, $c(\cdot)$, and $d(\cdot)$ are ordinary polynomials. Apply Lemma 6.3 with $f_1(x) \equiv |x - \theta|^2$, $g_1(x) \equiv b(x - \theta)$, $f_2(x) \equiv |x|^2$, $g_2(x) \equiv d(x)$, and $\eta(x) \equiv a(x - \theta) + c(x)$, which are all polynomials in x of degrees bounded uniformly in θ. Thus any set of functions $x \mapsto \tau(x - \theta) - \tau(x)$ with such a uniform bound on degrees, for $\theta \in \mathbb{R}^k$, is a VC major class. By a finite Boolean combination with sets $(A_i + \theta) \cap A_j \in \mathcal{C}$, we get that $\mathcal{F}$ is a VC major class, proving Theorem 6.1.

□

6.4 Lemma. Let τ be a piecewise polar polynomial on $\mathbb{R}^k$ where for the Boolean polynomial sets $A_0, \cdots, A_m$ in the definition, all but A_0 are bounded, and on A_0, $\tau(x) \equiv a|x| + b$ for some constants a, b. Then the class $\mathcal{G}$ of all functions

$$x \mapsto H(x, \theta) := [\tau(x - \theta) - \tau(x)]/(|\theta| + 1), \quad \theta \in \mathbb{R}^k,$$

is uniformly bounded.

Proof. For some $r < \infty$, $A_1 \cup \cdots \cup A_m \subset \{x : |x| < r\}$.

Clearly, there is an $M < \infty$ such that $|\tau(x)| \leq M$ for $|x| < r$. Thus $|H(x, \theta)| \leq 2M$ if $|x| < r$ and $|x - \theta| < r$. If $|x| \geq r$ and $|x - \theta| \geq r$ then

$$|H(x, \theta)| = |a(|x - \theta| - |x|)|/(|\theta| + 1) \leq |a\theta|/(|\theta| + 1) < |a|.$$

If $|x| < r$ and $|x - \theta| \geq r$ then

$$|H(x, \theta)| \leq [M + |(a|x - \theta| + b)|]/(|\theta| + 1)$$

$$\leq M + [|a|(r + |\theta|) + |b|]/(|\theta| + 1) \leq M + |a|(r + 1) + |b|.$$

If $|x| \geq r$ and $|x - \theta| < r$ then

$$|H(x, \theta)| \leq |(a|x| + b)|/(|\theta| + 1) + M \leq |a|(|x - \theta| + |\theta|)/(|\theta| + 1) + |b| + M$$

$$\leq |a|(r + 1) + |b| + M,$$

so $\mathcal{G}$ is uniformly bounded.   □

A class $\mathcal{F}$ of functions is a *uniform Donsker class* if, roughly speaking, the central limit theorem for the empirical process $n^{1/2}(P_n - P)$ holds not only uniformly over $\mathcal{F}$ (the Donsker property) but also uniformly over all laws P. Giné and Zinn (1991) gave a precise definition and a characterization (their Theorem 2.3) of uniform Donsker classes under a measurability condition.

6.5 Corollary. If τ satisfies the hypotheses of Theorem 6.1 and Lemma 6.4, then $\mathcal{G}$ of Lemma 6.4 is a uniform Glivenko-Cantelli class and a uniform Donsker class.

Proof. It's straightforward to verify that $\mathcal{G}$ is image admissible Suslin. Since $\mathcal{G}$ is uniformly bounded and VC major, it satisfies Koltchinskii-Pollard entropy bounds (Dudley, 1987, p. 1310) which imply that it is a uniform Glivenko-Cantelli class (Dudley, Giné and Zinn, 1991, Theorem 6) and a uniform Donsker class (Giné and Zinn, 1991).   □

6.6 Theorem. Assume the hypotheses of Corollary 6.5 and that τ is continuous and $a > 0$. Let $\rho(x, \theta) := \tau(x - \theta)$, $a(x) := \tau(x)$, and $h(x, \theta) := \rho(x, \theta) - a(x)$. Then $Ph(\cdot, \theta)$ is finite for all θ and attains its minimum.

If the minimum is attained at a unique point $\theta_0 = \theta_0(P)$, then ρ is adjustable, h is adjusted, and ρ is proper. Moreover, $\rho \in H_S(P, \theta_0)$. Any sequence of weak (resp. strong) approximate M-estimators is weakly (resp. strongly) doubly consistent.

Proof. By Lemma 6.4, for each θ, $h(\cdot, \theta)$ is bounded and measurable, thus integrable, for any law P. Since τ is continuous, $(\theta, x) \mapsto h(x, \theta)$ is jointly continuous. Also, h is bounded uniformly for θ in bounded sets. Thus $Ph(\cdot, \theta)$ is continuous in θ. Since $a > 0$, we have

$$Ph(\cdot, \theta) \to +\infty \quad \text{as} \quad |\theta| \to \infty. \tag{6.7}$$

Thus $Ph(\cdot, \theta)$ attains its minimum.

If the minimum is attained at a unique point θ_0 then ρ is adjustable, h is adjusted, and ρ is proper by continuity, compactness and (6.7).

To show that $\rho \in H_S(P, \theta_0)$, first apply Lemma 6.4. The definition of $H(\mathcal{E}, S)$ holds with $g_\theta \equiv 0$, $c(\theta) := 1 + |\theta|$, and $z_\theta(x) := h(x, \theta)/(1 + |\theta|)$, so $z_\theta \in \mathcal{G}$ of Lemma 6.4. For each neighborhood U of θ_0 there is a $\delta > 0$ with $Pz_\theta > \delta$ for all $\theta \notin U$: this holds for θ bounded by properness, while as $|\theta| \to \infty$, $z_\theta(x)$ converges boundedly for each x to $a > 0$. Thus, any sequence of weak (resp. strong) M-estimators is weakly (resp. strongly) consistent by Theorem 3.3.

To show $\rho \in H(\mathcal{N}, S)$, take a fixed bounded neighborhood V of θ_0 and take $\mathcal{H}$, not depending on $\varepsilon > 0$, to be the set of all functions $h(\cdot, \theta)$ on $\mathbb{R}^k$ for $\theta \in V$. Then $\mathcal{H}$ is a uniformly bounded VC major class and as in Corollary 6.5, a uniform Glivenko-Cantelli class. Let $z_\theta(x) \equiv 0$, $c(\theta) \equiv 1$, and $g_\theta(x) := h(x, \theta)$ for $\theta \in V$, so $\rho \in H(\mathcal{N}, S)$ and $\rho \in H_S(P, \theta_0)$. So, any sequence of weak (resp. strong) approximate M-estimators is weakly (resp. strongly) doubly consistent by Theorem 3.3. □

An example - in fact, the main example in view - of a function ρ satisfying the hypotheses of Theorem 6.6 is ρ_r as defined in Sec. 1. Note that for $r > 0$, the values and gradients agree for $|x| = r$ (although second derivatives differ), so that ρ_r is a C^1 function. Here $m = 1$, P_1 is a polynomial of degree 2 and P_0 is a polar polynomial of degree 1 (in $|x|$) as always.

The function $h(x, \theta) := \rho_r(x - \theta) - \rho_r(x)$ is not proper or even adjustable in all cases. The following facts are known:

(i) For $r \geq 0$ and $k = 1$, $h(x, \theta) := \rho_r(x - \theta) - \rho_r(x)$ is proper for P unless the interval of medians of P has length $> 2r$, which for $r = 0$ just means that the median is not unique.

(ii) For $r = 0$ and $k > 1$, $h(x, \theta) := |x - \theta| - |x|$ is proper for P if and only if either P is not concentrated in any line, or it is concentrated in a line and has a unique median there. When h is proper, the resulting θ_0 is called the *spatial median* of P, e.g. Milasevic and Ducharme (1987).

7. Huber's conditions

First, conditions (A-1) through (A-5) of Huber (1967), and some related conditions, will be listed, then Theorem 3.4 will be proved. We have a sample space $(X, \mathcal{A})$ having a probability measure P for which $X_1, X_2, \cdots$, are i.i.d., and a parameter space Θ which is locally compact with a countable base and so a separable metrizable space. We have a real-valued function ρ on $X \times \Theta$, a measurable function $a(\cdot)$ on X and a positive, continuous function $b(\cdot)$ on Θ. Let $h(x, \theta) \equiv \rho(x, \theta) - a(x)$.

(A-1)   For each fixed θ, $\rho(x, \theta)$ is an $\mathcal{A}$-measurable extended real-valued function of x, and $\rho(x, \theta)$ is a separable process relative to the closed intervals: there is a P-null set N and a countable subset S of Θ such that for every open set $U \subset \Theta$ and every (possibly unbounded) closed interval $F \subset \mathbb{R}$, the sets

$$\{x : \rho(x, \theta) \in F \text{ for all } \theta \in U\} \text{ and } \{x : \rho(x, \theta) \in F \text{ for all } \theta \in U \cap S\}$$

differ at most by a subset of N.

Doob (1953, pp. 51–59) shows that any real-valued stochastic process has a version separable with respect to the closed sets, and so *a fortiori* for the closed intervals, taking values in $[-\infty, +\infty]$.

(A-2)   For P-almost all x, $\rho(x, \theta)$ is a lower semicontinuous function of θ, i.e. for all $\theta \in \Theta$, $\rho(x, \theta) \leq \liminf_{v \to \theta} \rho(x, v)$.

Clearly, (A-2) also holds for h in place of ρ, for any function $a(x)$.

(A-3)   For some measurable function $a(\cdot)$ and $h(x, \theta)$
$\rho(x, \theta) - a(x)$, $Ph(\cdot, \theta)^- < \infty$ for all $\theta \in \Theta$, and $Ph(\cdot, \theta)^+ < \infty$ for some $\theta \in \Theta$.

It follows that $\gamma(\theta) := Ph(\cdot, \theta)$ is well-defined (possibly $+\infty$, but not $-\infty$) for all θ. If (A-3) holds for $a(\cdot) = a_1(\cdot)$ for some $a_1(\cdot)$, then it holds for another $a(\cdot) = a_2(\cdot)$ if and only if $a_2 - a_1$ is integrable for P (Lemma 2.1 above). Thus, as Huber notes, $\{\theta : P|h(\cdot, \theta)| < \infty\}$ doesn't depend on the choice of $a(\cdot)$.

(A-4)   There is a $\theta_0 \in \Theta$ such that $\gamma(\theta) > H_0 := \gamma(\theta_0)$ for all $\theta \neq \theta_0$.

Also, by Lemma 2.1, (A-4) and θ_0 do not depend on the choice of $a(\cdot)$. Conditions (A-3) and (A-4) both hold for $a(x) := h(x, \theta_0)$ (although (A-5) below might not hold in that case), and if so, then $H_0 = 0$. Let $\mathcal{K}$ be the class of all compact subsets of Θ. If Θ is not compact, for a function f on Θ, $\liminf_{\theta \to \infty} f(\theta)$ means $\sup_{K \in \mathcal{K}} \inf_{\theta \notin K} f(\theta)$. It also is the $\liminf$ in the usual sense when θ approaches the point at ∞ in the one-point compactification of Θ. Or, in the above definition of $\liminf$, $\mathcal{K}$ can be replaced by a sequence $\{K_n\}$ of compact sets whose union is Θ such that for all n, K_n is included in the interior of K_{n+1}.

(A-5)   For some continuous function $b(\theta) > 0$, and $a(x)$ for which (A-3) holds,
   (i) $\inf_{\theta \in \Theta} h(x, \theta)/b(\theta) \geq j(x)$ for some P-integrable function j;
   (ii) $\liminf_{\theta \to \infty} b(\theta) > H_0$;
   (iii) $E[\liminf_{\theta \to \infty} h(x, \theta)/b(\theta)] \geq 1$.

If Θ is compact, (ii) and (iii) become vacuous and can be omitted. Huber (1967) treats $a(\cdot)$ as fixed in (A-5). I chose the formulation that there exists such an $a(\cdot)$ since, whenever there exists $a(\cdot) = a_1(\cdot)$ satisfying (A-1) through (A-5), there will also exist $a(\cdot) = a_2(\cdot)$ satisfying (A-1) through (A-4) but not (A-5(ii)) for the same $b(\cdot)$. I don't know whether (A-5) might still hold for a different $b(\cdot)$.

Now we are ready for:

Proof of Theorem 3.4. The proof includes parts of those of Huber (1967, Lemma 1 and Theorem 1).

Condition (A-1), for $x \notin N$, and any non-empty open set $U \subset \Theta$, implies

$$\inf_{\theta \in U} g(x, \theta) = \inf_{\theta \in U \cap S} g(x, \theta),$$

for $g = \rho$, taking F as any closed half-line $[a, \infty)$. Thus, any such infimum is a (completion) measurable function of x. Then, the same holds for $g = h$ since $a(\cdot)$ is measurable and subtracting $a(x)$ can be interchanged with the infima.

Let U be open with compact closure in Θ. Then for some $\kappa > 0$, $b(\theta) \geq \kappa$ for all $\theta \in U$. For any $m = 1, 2, \cdots$, let $U = \bigcup_{i=1}^{k} U_{mi}$ where U_{mi} are open, $k = k(m) < \infty$, and $\sup\{d(\theta, \phi) : \theta, \phi \in U_{mi}\} \leq 1/m$ for each i. Choose $\theta_{mi} \in U_{mi}$ for each m and $i = 1, \cdots, k(m)$. Then $\min_i \inf_{\theta \in U_{mi}} h(x, \theta)/b(\theta_{mi})$ is measurable for each m and converges as $m \to \infty$ to $\inf_{\theta \in U} h(x, \theta)/b(\theta)$, which is thus also measurable. The same holds for an arbitrary open set $U \subset \Theta$ as a countable union of open sets with compact closures.

Take $0 < \varepsilon < 1$ small enough by (A-5(ii)) so that

$$\liminf_{\theta \to \infty} b(\theta) > (H_0 + \varepsilon)/(1 - \varepsilon) \tag{7.1}$$

and if $H_0 < 0$, also $H_0 + \varepsilon < 0$. For any compact set $C \subset \Theta$ let $f_C(x) := \inf_{\theta \notin C} h(x, \theta)/b(\theta)$. Then f_C is measurable. By (A-5(i),(iii)) and monotone convergence, there is a compact C such that $P f_C(x) > 1 - \varepsilon/2$ and for $\theta \notin C$ by (7.1),

$$b(\theta) > (H_0 + \varepsilon)/(1 - \varepsilon). \tag{7.2}$$

For any $\theta \in \Theta$, (A-1), (A-2), (A-5(i)) and monotone convergence imply that as a neighborhood V of θ decreases to $\{\theta\}$,

$$P \inf_{v \in V} h(\cdot, v) \uparrow P h(\cdot, \theta). \tag{7.3}$$

Thus γ is lower semicontinuous.

It will be proved by cases that $\rho \in H(\mathcal{E}, S, P, \theta_0)$.

Case 1. Suppose $\theta \notin C$, $H_0 < 0$ and $b(\theta) \leq 1$. Then also $H_0 + \varepsilon < 0$, and the definition of $H(\mathcal{G}, \cdots)$ with $z_\theta \equiv 0$ will apply as follows. If $P f_C < \infty$ let $\tau := f_C$. If $P f_C = +\infty$ take a function $\tau \leq f_C$ such that $P\tau < \infty$ and

$$P\tau > 1 - \varepsilon/2, \tag{7.4}$$

say $\tau := \min(M, f_C)$ for some large enough M. Then (7.4) also holds when $\tau = f_C$. We have $\tau \in \mathcal{L}^1(P)$ and $\mathcal{G}_1 := \{s\tau : 0 \leq s \leq 1\}$ is clearly a strong Glivenko-Cantelli class. We then have for all x

$$h(x, \theta) \geq b(\theta) f_C(x) \geq g_\theta(x) := b(\theta)\tau(x)$$

where $g_\theta \in \mathcal{G}_1$ and $P g_\theta \geq 0 > H_0 + \varepsilon$. So (3.1) holds with $\delta = \varepsilon$.

Case 2. Suppose $\theta \notin C$ and $H_0 < H_0 + \varepsilon < 0$ but now $b(\theta) > 1$. Let $g_\theta \equiv 0$, $c(\theta) := b(\theta)$, and let $z_\theta = \tau$ for τ as defined in Case 1. Then $\tau \in \mathcal{G}_1$, $h(x, \theta) \geq c(\theta) z_\theta(x)$ for all x, and

$$P g_\theta + P z_\theta = P\tau > 1/2 > 0 > H_0 + \varepsilon.$$

So (3.1) holds with $\delta = \min(\varepsilon, 1/2)$.

Case 3. Suppose $\theta \notin C$ and $H_0 \geq 0$. Then $H_0 + \varepsilon > 0$. Let

$$c(\theta) := b(\theta)(1 - \varepsilon)/(H_0 + \varepsilon) > 1$$

by (7.2), and

$$g_\theta \equiv 0, \ z := z_\theta := (H_0 + \varepsilon)(1 - \varepsilon)^{-1}\tau.$$

Then $\mathcal{G}_3 := \{z\}$ is a strong Glivenko-Cantelli class, $h(x, \theta) \geq b(\theta)\tau(x) = c(\theta)z_\theta(x)$ for all x and $Pz > H_0 + \varepsilon \geq \varepsilon$ by (7.4). So (3.1) holds for $\delta = \varepsilon$.

Case 4. Let U be any neighborhood of θ_0, and $\theta \in C \setminus U$. Now $\inf_{\theta \in C \setminus U} \gamma(\theta)$ is attained at some θ_1 with $\gamma(\theta_1) > \gamma(\theta_0)$ by lower semicontinuity of $\gamma(\cdot)$ and (A-4). Take $0 < \delta_2 < 1$ with $\gamma(\theta_1) \geq H_0 + 3\delta_2$. For each $\theta \in C \setminus U$ take a neighborhood $V = V_\theta$ of θ by (7.3) such that

$$P \inf_{v \in V} h(\cdot, v) > H_0 + 2\delta_2.$$

Then the open cover $\{V_\theta\}$ of the compact set $C \setminus U$ has a finite subcover $V(1)$, $\cdots$, $V(m)$. Let $y(i)(x) := \inf_{v \in V(i)} h(x, v)$ for $i = 1, \cdots, m$. Let $g(i) := y(i)$ if $Py(i) < \infty$, otherwise take $g(i) \leq y(i)$ such that $H_0 + \delta_2 < Pg(i) < \infty$. Then for all $v \in V(i)$, $h(x, v) \geq g_i(x)$, $\mathcal{G}_4 := \{g(i)\}_{i=1}^m$ is a strong Glivenko-Cantelli class and $Pg > H_0 + \delta_2$ for all $g \in \mathcal{G}_4$, so (3.1) holds with $z_\theta \equiv 0$.

Cases 1 through 4 cover all $\theta \notin U$. Letting $\mathcal{G} := \mathcal{G}_1 \cup \mathcal{G}_3 \cup \mathcal{G}_4$ and $\delta := \min(1/2, \varepsilon, \delta_2)$, we have $\rho \in H(\mathcal{E}, S, P, \theta_0)$ as desired. Then, (7.3) for $\theta = \theta_0$ implies that $\rho \in H_S(P, \theta_0)$. So Theorem 3.4 is proved. □

Appendix A: Existence of measurable estimators

A measurable space $(S, \mathcal{U})$ is called *separable* if $\mathcal{U}$ is countably generated and contains all singletons $\{u\}$, $u \in U$. A separable measurable space $(S, \mathcal{U})$ is called *Suslin* if there exists a complete separable metric space (Y, d) and a function T from Y onto S, measurable for the Borel σ-algebra of Y. Or if S is a class $\mathcal{F}$ of measurable functions on a measurable space $(X, \mathcal{A})$, then $\mathcal{F}$ is called *image admissible Suslin* if $(y, x) \mapsto T(y)(x)$ is jointly measurable (e.g. Dudley, 1984, Secs. 10.2, 10.3; the assumption that $(X, \mathcal{A})$ is Suslin turns out to be unnecessary and is now dropped).

If $(X, \mathcal{A})$ is a measurable space and $A \subset X$, A is called *universally measurable (u.m.)* iff for every probability measure μ on $(X, \mathcal{A})$, A is measurable for the completion of μ. A function f from X into W, where $(W, \mathcal{W})$ is a measurable space, is called *universally measurable (u.m.)* iff for each $C \in \mathcal{W}$, $f^{-1}(C)$ is u.m. in X. Recall:

A.1 Selection theorem (Sainte-Beuve). Let $(X, \mathcal{A})$ be any measurable space and $(S, \mathcal{U})$ a Suslin measurable space. Let g be a jointly measurable real-valued function on $S \times X$. Then for any Borel set $A \subset \mathbb{R}$,

$$\Pi_{X,g}(A) := \{x \in X : g(s,x) \in A \text{ for some } s \in S\}$$

is u.m. in X, and there is a u.m. function f from $\Pi_{X,g}(A)$ into S such that $g(f(x), x) \in A$ for all $x \in \Pi_{X,g}(A)$.

The selection theorem follows from results of Sainte-Beuve (1974, Theorems 3,4). Hoffmann-Jørgensen (1994, p. 345) also treats choice of estimators by measurable selection.

A.2 Theorem. If $(X, \mathcal{A})$ is any measurable space and $(\Theta, \mathcal{S})$ is Suslin, and if a sequence of approximate M-estimators exists, then such estimators can be chosen to be u.m.

Proof. Since approximate M-estimators exist,

$$\tau_n(X_1, \cdots, X_n) := \inf_{\theta \in \Theta} P_n h(\cdot, \theta)$$

has finite real values eventually. It follows from Theorem A.1, taking $A := (-\infty, t)$ for any t, that τ_n is a u.m. function of $X_1, \cdots, X_n$. Let

$$A_n := \{\langle x, \theta \rangle \in X^n \times \Theta : -\infty < \tau_n(x_1, \cdots, x_n) \leq P_n h(\cdot, \theta)$$

$$< \tau_n(x_1, \cdots, x_n) + \frac{1}{n} < \infty\}.$$

Then A_n is a u.m. subset of $X^n \times \Theta$, whose projection into X^n is the set of all $(x_1, \cdots, x_n)$ such that $\tau_n(x_1, \cdots, x_n)$ is finite. Defining an estimator $T_n(X_1, \cdots, X_n)$ by measurable selection from A_n, or as a fixed point θ_1 of Θ if it is not defined otherwise, we get a sequence of u.m. approximate M-estimators.

Appendix B: U-statistic Glivenko-Cantelli classes

The contents of this Appendix are entirely due to E. Giné except that I am responsible for any errors. Let $(X, \mathcal{A}, P)$ be a probability space and $(X^k, \mathcal{A}^k, P^k)$ its k-fold Cartesian product. Recalling the definition of U_n^k and related definitions from Sec. 3 above we have the following:

B.1 Theorem. If $\mathcal{F}$ is a weak (resp. strong) Glivenko-Cantelli class of P^k-integrable functions on X^k, for some $F \geq |f|$ for all $f \in \mathcal{F}$, $P^k F < \infty$, and $\mathcal{F}$ is image admissible Suslin, then $\mathcal{F}$ is also a weak (resp. strong) U-Glivenko-Cantelli class.

Proof. Hoeffding (1963, (5.4),(5.5)) observed that for f symmetric,

$$U_n^k(f) \equiv (n!)^{-1} \sum_\sigma W(X_{\sigma(1)}, \cdots, X_{\sigma(n)})(f)$$

where the sum is over all permutations σ of $\{1, \cdots, n\}$,

$$W(x_1, \cdots, x_n)(f) := \frac{1}{m} \sum_{j=0}^{m-1} f(x_{jk+1}, \cdots, x_{(j+1)k})$$

and $m = [n/k]$, $[x]$ being the largest integer $\leq x$. Let $\|A\|_{\mathcal{F}} := \sup_{f \in \mathcal{F}} |A(f)|$. For fixed k, the sequence $\|(P^k)_n - P^k\|_{\mathcal{F}}$ of real random variables is a reversed submartingale (van der Vaart and Wellner, 1996, Lemma 2.4.5) and so is $\|U_n^k - P^k\|_{\mathcal{F}}$ by a similar proof (Nolan and Pollard, 1987, Sec. 3). For real-valued reversed martingales or reversed submartingales, almost sure convergence is equivalent to convergence in $\mathcal{L}^1$ by theorems of Doob, e.g. Dudley (1993, Theorems 10.6.1, 10.6.4). Then for $X_1, \cdots, X_n$ i.i.d. (P),

$$E\|U_n^k(f) - P^k f\|_{\mathcal{F}} \leq E\|(W(X_1, \cdots, X_n) - P^k)(f)\|_{\mathcal{F}}$$

$$= E\|\frac{1}{m} \sum_{j=0}^{m-1} f(X_{jk+1}, \cdots, X_{(j+1)k}) - P^k f\|_{\mathcal{F}} \to 0$$

as $n \to \infty$ since $\mathcal{F}$ is a Glivenko-Cantelli class for P^k, and this implies the conclusion. $\qquad \square$

Acknowledgments. I thank Donald Cohn for a helpful discussion about separability, Evarist Giné for information about U-process laws of large numbers, Adam Jakubowski for information about sequential convergence, Jinghua Qian for many corrections, and Jørgen Hoffmann-Jørgensen for stimulating conversations.

References

Akaike, H. (1974). A new look at the statistical model identification. *IEEE Trans. Auto. Control* **19**, 716–723.

Arcones, M. A., and Giné, E. (1993). Limit theorems for U-processes. *Ann. Probab.* **21**, 1494–1542.

Bickel, P. J., and Doksum, K. A. (1977), *Mathematical Statistics*, Holden-Day, San Francisco.

Doob, J. L. (1953). *Stochastic Processes*. Wiley, New York.

Dudley, R. M. (1984). A course on empirical processes. In *Ecole d'été de probabilités de St.-Flour XII* (1982). *Lecture Notes in Math.* (Springer) **1097**, 1–142.

Dudley, R. M. (1987). Universal Donsker classes and metric entropy. *Ann. Probab.* **15**, 1306–1326.

Dudley, R. M. (1993). *Real Analysis and Probability*. 2d printing, corrected. Chapman and Hall, New York.

Dudley, R. M., Giné, E. and Zinn, J. (1991). Uniform and universal Glivenko-Cantelli classes. *J. Theoretical Probab.* **4**, 485–510.

Efron, B., and Tibshirani, R. J. (1993). *An Introduction to the Bootstrap*. Chapman and Hall, New York.

Giné, E., and Zinn, J. (1984). Some limit theorems for empirical processes (with discussion). *Ann. Probab.* **12**, 929–989.

Giné, E., and Zinn, J. (1990). Bootstrapping general empirical measures. *Ann. Probab.* **18**, 851–869.

Giné, E., and Zinn, J. (1991). Gaussian characterization of uniform Donsker classes of functions. *Ann. Probab.* **19**, 758–782.

Haldane, J. B. S. (1948). Note on the median of a multivariate distribution. *Biometrika* **35**, 414–415.

Haughton, D. M.-A. (1988). On the choice of a model to fit data from an exponential family. *Ann. Statist.* **16**, 342–355.

Hoeffding, W. (1963). Probability inequalities for sums of bounded random variables. *J. Amer. Statist. Assoc.* **58**, 13–30.

Hoffmann-Jørgensen, J. (1994). *Probability with a View Toward Statistics*, vol. **2**, Chapman and Hall, London.

Huber, P. J. (1967). The behavior of maximum likelihood estimates under nonstandard conditions. *Proc. Fifth Berkeley Symp. Math. Statist. Probability* **1**, 221–233. University of California Press, Berkeley and Los Angeles.

Huber, P. J. (1981). *Robust Statistics*. Wiley, New York.

Kantorovich, L. V.; Vulikh, B. Z.; and Pinsker, A. G. (1950). *Functional Analysis in Partially Ordered Spaces* (in Russian). Gostekhizdat, Moscow.

Kantorovich, L. V.; Vulikh, B. Z.; and Pinsker, A. G. (1951). Partially ordered groups and partially ordered linear spaces (in Russian). *Uspekhi Mat. Nauk* **6** no. 3, 331–398; transl. in *Amer. Math. Soc. Transl.* Ser. 2 **27** (1963), 51–124.

Kisyński, J. (1960). Convergence du type L. *Colloq. Math.* **7**, 205–211.

Kullback, S. (1983). Kullback information. In *Encyclopedia of Statistical Sciences*, vol. **4**, pp. 421–425, Eds. S. Kotz, N. L. Johnson. Wiley, New York.

Kullback, S., and Leibler, R. A. (1951). On information and sufficiency. *Ann. Math. Statist.* **22**, 79–86.

Le Cam, L. (1986). *Asymptotic Methods in Statistical Decision Theory*. Springer, New York.

Milasevic, P. and Ducharme, G. R. (1987). Uniqueness of the spatial median. *Ann. Statist.* **15**, 1332–1333.

Nolan, D., and Pollard, D. B. (1987). U-processes: Rates of convergence. *Ann. Statist.* **15**, 780–799.

Nolan, D., and Pollard, D. B. (1988). Functional limit theorems for U-processes. *Ann. Probab.* **16**, 1291–1298.

Ossiander, M. (1987). A central limit theorem under metric entropy with L_2 bracketing. *Ann. Probab.* **15**, 897–919.

Perlman, M. D. (1972). On the strong consistency of approximate maximum likelihood estimators. *Proc 6th Berkeley Symp. Math. Statist. Prob.* **1**, 263–281.

Pitman, E. J. G. (1979). *Some Basic Theory for Statistical Inference*. Chapman and Hall, London.

Poskitt, D. S. (1987). Precision, complexity and Bayesian model determination. *J. Roy. Statist. Soc.* B **49**, 199–208.

Quiroz, A. J., and Dudley, R. M. (1991). Some new tests for multivariate normality. *Probab. Th. Rel. Fields* **87**, 521–546.

Sainte-Beuve, M.-F. (1974). On the extension of von Neumann-Aumann's theorem. *J. Functional Analysis* **17**, 112–129.

Sawa, T. (1978). Information criteria for discriminating among alternative regression models. *Econometrica* **46**, 1273–1291.

Schwarz, G. (1978). Estimating the dimension of a model. *Ann. Statist.* **6**, 461–464.

Shapiro, A. (1989). Asymptotic properties of statistical estimators in stochastic programming. *Ann. Statist.* **17**, 841–858.

Sheehy, A., and Wellner, J. A. (1993). Uniform Donsker classes of functions. *Ann. Probab.* **20**, 1983–2030.

Talagrand, M. (1987). The Glivenko-Cantelli problem. *Ann. Probab.* **15**, 837–870.

van de Geer, S. (1993). Hellinger-consistency of certain nonparametric maximum likelihood estimators. *Ann. Statist.* **21**, 14–44.

van der Vaart, A. W., and Wellner, J. A. (1996). *Weak Convergence and Empirical Processes*. Springer, New York.

Vapnik, V. N., and Červonenkis, A. Ya. (1981). Necessary and sufficient conditions for the uniform convergence of means to their expectations. *Theor. Probability Appl.* **26**, 532–553.

Zacks, S. (1971). *The Theory of Statistical Inference*. Wiley, New York.

R.M. Dudley
Room 2-245, MIT,
Cambridge, MA 02139-4307, USA
rmd@math.mit.edu

Progress in Probability, Vol. 43
© 1998 Birkhäuser Verlag Basel/Switzerland

Small Deviation Probabilities of Sums of Independent Random Variables

T. Dunker[1] and M. A. Lifshits[2] and W. Linde

1. Introduction

Let $\xi_1, \xi_2, \ldots$ be a sequence of independent $\mathcal{N}(0,1)$–distributed random variables (r.v.'s) and let $(\phi(j))_{j=1}^\infty$ be a summable sequence of positive real numbers. The sum $S := \sum_{j=1}^\infty \phi(j)\xi_j^2$ is then well defined and one may ask for the small deviation probability of S, i.e. for the asymptotic behavior of $\mathbb{P}(S < r)$ as $r \to 0$. In 1974 G. N. Sytaya [S] gave a complete description of this behavior in terms of the Laplace transform of S. Recently, this result was considerably extended to sums $S := \sum_{j=1}^\infty \phi(j)Z_j$ for a large class of i.i.d. r.v.'s $Z_j \geq 0$ (cf.[DR], [Lif2]). Yet for concrete sequences $(\phi(j))_{j=1}^\infty$ those descriptions of the asymptotic behavior are very difficult to handle because they use an implicitly defined function of the radius $r > 0$. In 1986 V. M. Zolotarev [Z2] announced an explicit description of the behavior of $\mathbb{P}(\sum_{j=1}^\infty \phi(j)\xi_j^2 < r)$ in the case that ϕ can be extended to a decreasing and logarithmically convex function on $[1, \infty)$. We show that, unfortunately, this result is not valid without further assumptions about the function ϕ (a natural example will be given where an extra oscillating term appears). Our aim is to state and to prove a correct version of Zolotarev's result in the more general setting of [Lif2], and we show how our description applies in the most important specific examples. For other results related to small deviation problems see [A], [I], [KLL], [Li], [LL], [MWZ], [NS] and [Z1].

Our results heavily depend on the assumption $\mathbb{V}_{[0,\infty)}(tf'(t)/f(t)) < \infty$ where f denotes the Laplace transform of Z_j and $\mathbb{V}$ stands for variation. In the last section we apply known results about slowly varying functions to state sufficient conditions for the distribution of Z_j such that its Laplace transform possesses this property.

2. Notations and Basic Results

Let $Z_1, Z_2, \ldots$ be a sequence of non-negative i.i.d. r.v.'s with finite second moment. Given a summable sequence $(\phi(j))_{j=1}^\infty$ of positive real numbers, we define S to be

1)   Research supported by the DFG-Graduiertenkolleg "Analytische und Stochastische Strukturen und Systeme", Universität Jena

2)   Research supported by International Science Foundation (ISF) and Russian Foundation for Basic Research (RFBI) and carried out during the author's sojourn in Strasbourg and Lille-1 universities

$\sum_{j=1}^{\infty} \phi(j) Z_j$. Let F denote the distribution function (d.f.) of the Z_j's, i.e.

$$F(r) := \mathbb{P}(Z_1 < r),$$

and let

$$f(u) := \int_{[0,\infty)} \exp(-ur) \, dF(r)$$

be the Laplace transform of F (or Z_1). Then the following functions derived from f will play an important role in our further investigations:

$$f_1(u) \quad := \quad (\log f)'(u) = \frac{f'(u)}{f(u)} \quad \text{and}$$

$$f_2(u) \quad := \quad (\log f)''(u) = \frac{f''(u)}{f(u)} - \frac{f'^2(u)}{f^2(u)}.$$

Analogously, the Laplace transform Λ of S and the derivatives of $L := \log \Lambda$ are defined by

$$\Lambda(u) \quad := \quad \mathbb{E} \exp(-uS) = \prod_{j=1}^{\infty} f(u\phi(j)),$$

$$L(u) \quad := \quad \log \Lambda(u) = \sum_{j=1}^{\infty} \log f(u\phi(j)),$$

$$L'(u) \quad = \quad \sum_{j=1}^{\infty} \phi(j) f_1(u\phi(j)) \quad \text{and}$$

$$L''(u) \quad = \quad \sum_{j=1}^{\infty} \phi^2(j) f_2(u\phi(j)).$$

Introducing the Esscher transform F_u of F as

$$F_u(r) := \frac{1}{f(u)} \int_{[0,r)} \exp(-us) \, dF(s)$$

we recall (cf. [Lif2]) that $f_2(u)$ coincides with the variance of F_u. Especially, f_2 is strictly positive in the non-degenerated case.

In order to apply the main result of [Lif2] the d.f. F has to possess the following property:

Condition L *The d.f. F satisfies condition L provided that there exist constants $b \in (0, 1)$, $c_1, c_2 > 1$ and $\epsilon > 0$ such that for each $r \leq \epsilon$ the estimates*

$$c_1 F(br) \leq F(r) \leq c_2 F(br)$$

hold.

REMARK 2.1 In the terminology of Karamata theory (cf. [BGT], Ch. 2) condition **L** says nothing else than $F(1/\cdot)$ is of bounded and positive decrease. For example, if $F(1/\cdot) \in \mathbf{R}_\alpha$, the class of regularly varying functions of order α, with $\alpha < 0$, then it satisfies **L**. In different words, every d.f. F with $F(1/x) = x^\alpha l(x)$ for $\alpha < 0$ and l slowly varying (e.g. $l(x) = (\log x)^\beta$ for large x and $\beta \in \mathbb{R}$) possesses property **L**. Furthermore, let us mention that **L** implies $u^2 L''(u) \to \infty$ (cf. [Lif2]).

The following result of [Lif2] is the basic ingredient of our further investigations.

THEOREM 2.2 *Let the d.f. F of Z_1 satisfy **L** and define S with a summable sequence $(\phi(j))_{j=1}^\infty$ as above. Then we have*

$$\mathbb{P}(S < r) \sim \frac{\exp(L(u) + ur)}{\sqrt{2\pi u^2 L''(u)}} \tag{1}$$

where $u = u(r)$ is any function satisfying

$$\lim_{r \to 0} \frac{uL'(u) + ur}{\sqrt{u^2 L''(u)}} = 0.$$

REMARK 2.3 Especially, one may define $u = u(r)$ as unique solution of

$$uL'(u) + ur = 0 \tag{2}$$

for $r < r_0$. Yet, it will turn out to be very useful that asymptotic solutions of (2) work as well.

3. General Result

The functions L, uL' and $u^2 L''$ appearing in Theorem 2.2 are defined by infinite sums. In order to evaluate their asymptotic behavior, we replace the sums by suitable integrals. So we have to assume that ϕ is a positive integrable function defined on the whole interval $[1, \infty)$ with values in $(0, \infty)$, and the corresponding integrals are then defined by

$$
\begin{aligned}
I_0(u) &:= \int_1^\infty \log f(u\phi(t))\, dt, \\
I_1(u) &:= \int_1^\infty u\phi(t) f_1(u\phi(t))\, dt, \\
I_2(u) &:= \int_1^\infty (u\phi(t))^2 f_2(u\phi(t))\, dt.
\end{aligned}
$$

In order to prove the main result about the behavior of $\mathbb{P}(S < r)$ as $r \downarrow 0$ we further restrict the set of d.f.'s F.

Condition I *Let $\mathbb{V}_{[a,b)}\, g$ be the total variation of a function g defined on $[a, b)$. The d.f. F satisfies **I** provided that $\mathbb{V}_{[0,\infty)}\, tf_1(t)$ is finite. Note that this happens iff $\int_{[0,\infty)} |(tf_1(t))'|\, dt < \infty$.*

Obviously, condition **I** implies the existence of the limit

$$\alpha := \lim_{u \to \infty} u f_1(u) \le 0,$$

and by Karamata and Tauberian Theorems for regularly varying functions (see e.g. [BGT], Sect. 1.6 and 1.7) it follows that $F \in \mathbf{I} \cap \mathbf{L}$ yields $F(1/\cdot) \in \mathbf{R}_\alpha$ and $\alpha < 0$.

THEOREM 3.1 *Let* $Z_1, Z_2, \ldots$ *be as above and assume that their d.f.* F *satisfies* **L** *and* **I**. *Let* $\alpha < 0$ *be the index of the regularly varying function* $F(1/\cdot)$ *and suppose that* ϕ *is a positive, logarithmically convex, twice differentiable and integrable function on* $[1, \infty)$. *Under these assumptions we have*

$$\mathbb{P}\left(\sum_{j=1}^{\infty} \phi(j) Z_j < r \right)$$

$$\sim \sqrt{\frac{\Gamma(1 - \alpha) F(1/u\phi(1))}{2\pi I_2(u)}} \, \exp(I_0(u) + \rho(u) + ur)$$

where $u = u(r)$ *is an arbitrary function satisfying*

$$\lim_{r \to 0} \frac{I_1(u) + ur}{\sqrt{I_2(u)}} = 0.$$

The function ρ *is defined by*

$$\rho(u) := \sum_{j=1}^{\infty} \int_0^1 \frac{t - t^2}{2} (\log f(u\phi))''(t + j) \, dt$$

and, moreover, ρ *is bounded.*

Proof. In view of Theorem 2.2 and $I_2(u) \to \infty$ it suffices to show the following:

(a) $L(u) = \frac{1}{2} \log f(u\phi(1)) + I_0(u) + \rho(u)$

(b) $uL'(u) = I_1(u) + O(1)$

(c) $u^2 L''(u) \sim I_2(u)$

(d) $\sup_{u \ge 0} |\rho(u)| < \infty.$

(a): The Euler-MacLaurin's summation formula of second order asserts

$$\sum_{j=1}^{N} h(j) = \int_1^N h(t) \, dt + \frac{1}{2}(h(1) + h(N)) + \sum_{j=1}^{N} \int_0^1 \frac{t - t^2}{2} h''(t + j) \, dt$$

for any twice differentiable function $h : [1, N] \to \mathbb{R}$. Applying this to the function $h := \log f(u\phi)$ (here $u > 0$ is fixed) we immediately obtain the desired equality by taking $N \to \infty$.

(d): An application of $\phi'' = \phi(\log\phi)'' + \phi'^2/\phi$ yields

$$
\begin{aligned}
\frac{d^2}{dt^2}\log f(u\phi) &= f_2(u\phi)(u\phi')^2 + f_1(u\phi)u\phi'' \\
&= [f_2(u\phi)u\phi + f_1(u\phi)]u\phi'\frac{\phi'}{\phi} + u\phi f_1(u\phi)(\log\phi)'',
\end{aligned}
$$

which leads to

$$
\begin{aligned}
\rho(u) &= \sum_{j=1}^{\infty}\int_0^1 \frac{t-t^2}{2}(\log f(u\phi))''(t+j)\,dt \\
&= \sum_{j=1}^{\infty}T_j(u) + \sum_{j=1}^{\infty}\int_0^1 \frac{t-t^2}{2}\{u\phi f_1(u\phi)(\log\phi)''\}(t+j)\,dt \qquad (3)
\end{aligned}
$$

with

$$
T_j(u) = \int_0^1 \frac{t-t^2}{2}\left\{[f_2(u\phi)u\phi + f_1(u\phi)]u\phi'\frac{\phi'}{\phi}\right\}(t+j)\,dt.
$$

Yet

$$
|T_j(u)| \le \frac{1}{8}\sup_{t\in[j,j+1]}\frac{|\phi'(t)|}{\phi(t)}\int_{u\phi(j+1)}^{u\phi(j)}|f_2(s)s + f_1(s)|\,ds, \qquad (4)
$$

hence by logarithmic convexity of ϕ and by **I** it follows

$$
\sum_{j=1}^{\infty}|T_j(u)| \le \frac{1}{8}\frac{|\phi'(1)|}{\phi(1)}\int_0^{\infty}|(tf_1(t))'|\,dt < \infty,
$$

and the first sum in (3) is bounded. Since $tf_1(t)$ is bounded and $(\log\phi)''(t)$ is integrable, Lebesgue's DCT applies to the second term in (3). Consequently, by $tf_1(t)\to\alpha$ this part of (3) converges to αC_ϕ with

$$
\begin{aligned}
C_\phi &:= \sum_{j=1}^{\infty}\int_0^1 \frac{t-t^2}{2}(\log\phi)''(t+j)\,dt \qquad (5) \\
&= \frac{1}{2}\sum_{j=1}^{\infty}\int_0^1 \log\frac{\phi(j)\phi(j+1)}{\phi^2(t+j)}\,dt.
\end{aligned}
$$

This proves the boundedness of ρ and, moreover,

$$
\rho(u) = \sum_{j=1}^{\infty}T_j(u) + \alpha C_\phi + o(1). \qquad (6)
$$

(b): The Euler-MacLaurin's summation formula of first order gives

$$
uL'(u) = I_1(u) + \frac{1}{2}u\phi(1)f_1(u\phi(1)) + \sum_{j=1}^{\infty}\int_0^1 \frac{2t-1}{2}(u\phi f_1(u\phi))'(t+j)\,dt
$$

and it remains to investigate the sum on the right-hand side. Because of

$$\int_0^1 \frac{2t-1}{2} (u\phi f_1(u\phi))'(t+j)\, dt =$$

$$= \int_j^{j+1} \left[\frac{(u\phi f_1(u\phi))(j) + (u\phi f_1(u\phi))(j+1)}{2} - (u\phi f_1(u\phi))(t) \right] dt$$

$$\leq \mathbb{V}_{[u\phi(j+1), u\phi(j))}\, t f_1(t),$$

it follows

$$\left| \sum_{j=1}^{\infty} \int_0^1 \frac{2t-1}{2} (u\phi f_1(u\phi))'(t+j)\, dt \right| \leq \mathbb{V}_{[0,\infty)}\, t f_1(t) < \infty$$

and $uL'(u) = I_1(u) + O(1)$ as claimed above.

(c): Let us fix $\epsilon > 0$. As $\lim_{x \to \infty} x^2 f_2(x) = -\alpha > 0$, we find an $x_0(\epsilon)$ such that $\theta(u,t)$ defined by

$$((u\phi)^2 f_2(u\phi))(t) = -\alpha(1 + \epsilon\theta(u,t))$$

satisfies $|\theta(u,t)| \leq 1$ provided that $u\phi(t) > x_0(\epsilon)$. This implies

$$\left| (u\phi)^{-2} f_2^{-1}(u\phi)(j) \int_j^{j+1} ((u\phi)^2 f_2(u\phi))(t)\, dt - 1 \right| < \epsilon \tag{7}$$

for all $j \in J(u) := \{t : u\phi(t) > x_0(\epsilon)\}$. For the remaining indices j we distinguish two cases.

1. *Case $\phi'/\phi \to 0$:* This implies that $\log \phi(s) - \log \phi(t)$ with $s, t \in [j, j+1]$ can be arbitrarily small for j sufficiently large. Then the uniform continuity of $\log f_2$ on $[0, x_0]$ yields the assertion.

2. *Case $|\phi'|/\phi \geq c > 0$:* Then ϕ decreases exponentially and one can show that the sum over the index set $\mathbb{N} \setminus J(u)$ is finite and, hence, it is negligible with respect to the sum over $J(u)$. The same is true for the corresponding integral. This proves our assertion. $\qquad\square$

COROLLARY 3.2 *Let $Z_1, Z_2, \ldots$ and ϕ be as in Theorem 3.1. If, furthermore,*

$$\lim_{t \to \infty} \frac{\phi'(t)}{\phi(t)} = 0, \tag{8}$$

then we have

$$\mathbb{P}\left(\sum_{j=1}^{\infty} \phi(j) Z_j < r \right)$$

$$\sim \sqrt{\frac{\Gamma(1-\alpha)F(1/u\phi(1))}{2\pi I_2(u)}}\, \exp(I_0(u) + \alpha C_\phi + ur) \tag{9}$$

where $u = u(r)$ is any function satisfying

$$\lim_{r \to 0} \frac{I_1(u) + ur}{\sqrt{I_2(u)}} = 0. \tag{10}$$

The constant C_ϕ is defined by (5).

Proof. By (6) it remains to prove that $\sum_j T_j(u) \to 0$ as $u \to \infty$ or, equivalently, $\lim_{u \to \infty} \rho(u) = \alpha C_\phi$. Let us continue with equation (4). Fixing $\epsilon > 0$ we find a $J \in \mathbb{N}$ such that $\sup_{[J,\infty)} |\phi'|/\phi \le 4\epsilon/(\mathbb{V}_{[0,\infty)} tf_1(t))$ and obtain

$$\sum_{j=1}^{\infty} |T_j(u)| \le \frac{1}{8} \frac{|\phi'(1)|}{\phi(1)} \mathbb{V}_{[u\phi(J),\infty)} tf_1(t) + \frac{\epsilon}{2} \le \epsilon$$

for sufficiently large u, i.e. for $u > 0$ with $\mathbb{V}_{[u\phi(J),\infty)} tf_1(t) \le 4\epsilon\phi(1)/|\phi'(1)|$.   □

REMARK 3.3 Later we shall see that Corollary 3.2 becomes false without assumption (8), i.e. in general the function ρ cannot be replaced by a constant.

4. Examples

4.1. Polynomial Coefficients   We consider now $\phi(t) := t^{-A}$ with $A > 1$, i.e. we are interested in the behavior of

$$\mathbb{P}\left(\sum_{j=1}^{\infty} j^{-A} Z_j < r \right)$$

as $r \downarrow 0$. Since ϕ satisfies (8), we are in the situation of Corollary 3.2. Let us first evaluate the asymptotics of I_0, I_1 and I_2 in this case. We define a constant $K = K_{A,F} > 0$ by

$$K := -\int_0^{\infty} t^{-1/A} f_1(t)\, dt$$

and observe that

$$-\int_u^{\infty} t^{-1/A} f_1(t)\, dt = -(\alpha + o(1)) \int_u^{\infty} t^{-1-1/A}\, dt = -(\alpha + o(1)) A u^{-1/A}$$

since $tf_1(t) \to \alpha$. From this one easily derives

$$I_0(u) \quad = \quad -\log f(u) - K u^{1/A} - \alpha A + o(1)$$

and

$$I_1(u) \quad = \quad -A^{-1} K u^{1/A} - \alpha + o(1).$$

If $a > 1$ is defined by $a = A/(A-1)$, in view of $u^{1/a} f_1(u) \to 0$ we obtain

$$\int_0^u t^{1/a} f_2(t)\, dt = u^{1/a} f_1(u) - \frac{1}{a} \int_0^u t^{-1/A} f_1(t)\, dt \longrightarrow \frac{K}{a}$$

and, consequently,

$$I_2(u) \quad \sim \quad \frac{K}{Aa} u^{1/A}.$$

The constant C_ϕ appearing in (9) can be determined by

$$C_{t-A} \quad = \quad -A\left(\frac{1}{2}\log 2\pi - 1\right)$$

using Stirling's formula. If we define the function $u := u(r)$ by

$$u(r) := \left(\frac{K}{A}\right)^a r^{-a},$$

then it follows

$$ur = \frac{K}{A}u^{1/A} = -I_1(u) + O(1)$$

and, consequently, u satisfies (10). Thus it can be used for the description of the asymptotic (9), and this leads to the following result:

PROPOSITION 4.1 *Let $A > 1$ be given and let $Z_1, Z_2, \ldots$ be an i.i.d sequence with d.f. F satisfying $\mathbf{L}$ and $\mathbf{I}$. Especially, there exist $\alpha < 0$ and a slowly varying function l such that $F(1/x) = x^\alpha l(x)$ for $x \to \infty$. Then we have*

$$\mathbb{P}\left(\sum_{j=1}^{\infty} j^{-A} Z_j < r\right)$$

$$\sim \quad \frac{a^{\frac{1}{2}} A^{a\frac{1+\alpha}{2}}}{(2\pi)^{\frac{1+\alpha A}{2}} \Gamma^{\frac{1}{2}}(-\alpha+1) K^{a\frac{1+\alpha}{2}}} \frac{r^{\frac{(1+\alpha A)}{2(A-1)}}}{l^{\frac{1}{2}}(r^{-a})} \exp\left(-\frac{(A-1)K^a}{A^a} r^{-\frac{1}{A-1}}\right)$$

with constants K, a defined above.

REMARK 4.2 Observe that α and l are completely determined by the lower tail behavior of Z_1, while the constant K depends on the whole distribution of Z_1.

Example: Let $Z_1, Z_2, \ldots$ be defined by $Z_j = |\xi_j|^p$ with $\xi_j \sim \mathcal{N}(0, 1)$ and $p > 0$. Then it is easy to see that $F(1/\cdot) \in \mathbf{R}_\alpha$ with index $\alpha = -1/p$ and $l(r) \sim \sqrt{2/\pi}$. Moreover, F satisfies condition $\mathbf{I}$. For $p = 2$ this follows directly by $f(t) = (1 + 2t)^{-\frac{1}{2}}$ and for $p \neq 2$ one may apply Corollary 5.5 from below. Thus Proposition 4.1 gives the exact behavior of $\mathbb{P}(\sum_{j=1}^{\infty} j^{-A}|\xi_j|^p < r)$ as $r \to 0$, where all constants are known besides K (see [Z1]). Only for $p = 2$ we are able to calculate the exact value of K, namely,

$$K = 2^{-\frac{1}{a}} \int_0^\infty \frac{s^{\frac{1}{a}-1}}{1+s} ds = 2^{-\frac{1}{a}} \frac{\pi}{\sin\left(\frac{\pi}{a}\right)} = 2^{-\frac{1}{a}} \frac{\pi}{\sin\left(\frac{\pi}{A}\right)}.$$

So we obtain the following result for $p = 2$:

COROLLARY 4.3 [Z2] *If $\xi_1, \xi_2, \ldots$ are i.i.d. $\mathcal{N}(0,1)$–distributed and $A > 1$, then*

$$\mathbb{P}\left(\sum_{j=1}^{\infty} j^{-A}\xi_j^2 < r\right)$$

$$\sim \frac{2^{\frac{A}{4}}a^{\frac{1}{2}}A^{\frac{a}{4}}\left(\sin\frac{\pi}{A}\right)^{\frac{a}{4}}}{\pi^{\frac{a^2+a-A}{4a}}} r^{\frac{a-A}{4A}} \exp\left(-\frac{A-1}{2}\left(\frac{\pi}{A\sin\frac{\pi}{A}}\right)^a r^{-\frac{1}{A-1}}\right).$$

For $A = 2$ this implies the small deviation probability of the $L^2([0,1])$–norm of the Brownian bridge $(B_t)_{t\in[0,1]}$. Recall that

$$\|B\|_2^2 = \int_0^1 B_t^2 \, dt = \sum_{j=1}^{\infty}(\pi j)^{-2}\xi_j^2,$$

hence Corollary 4.3 leads to the well known asymptotic behavior

$$\mathbb{P}(\|B\|_2^2 < r) = \mathbb{P}\left(\sum_{j=1}^{\infty} j^{-2}\xi_j^2 < \pi^2 r\right) \sim 2\sqrt{\frac{2}{\pi}}\exp\left(-\frac{1}{8r}\right)$$

(cf. [AD] or [Lif1], Ch. 18).

4.2. The Exponential Case The aim of this section is to investigate the small deviation probability for $S := \sum_{j=0}^{\infty}\exp(-j)Z_j$, i.e. the function ϕ is defined by $\phi(t) = \exp(1-t)$. This function is of special interest because it does not satisfy (8), hence we have to use here Theorem 3.1 directly (and not Corollary 3.2).

PROPOSITION 4.4 *Let $Z_0, Z_1, \ldots$ be as above and let f be as before the Laplace transform of Z_0. Then there exists a bounded, continuous and 1-periodic function ψ on $\mathbb{R}$ such that*

$$\mathbb{P}\left(\sum_{j=0}^{\infty}\exp(-j)Z_j < r\right)$$

$$\sim \sqrt{\frac{f(u)}{-2\pi\log f(u)}}\exp\left(\int_0^u \frac{\log f(s)}{s}\,ds + ur + \psi(\log u)\right)$$

where $u = u(r)$ is an arbitrary function satisfying

$$\lim_{r\to 0}\frac{ur + \log f(u)}{\sqrt{-\log f(u)}} = 0.$$

Proof. Substituting and integration by parts easily give

$$I_2(u) \quad\sim\quad -\log f(u), \tag{11}$$

$$I_1(u) \quad=\quad \log f(u), \tag{12}$$

$$I_0(u) \quad=\quad \int_0^u \frac{\log f(s)}{s}\,ds \tag{13}$$

and $C_\phi = 0$. Hence, in view of Theorem 3.1 it remains to construct a bounded, continuous and 1-periodic function ψ such that $\rho(u) = \psi(\log u) + o(1)$. To do so, we define an auxiliary function g by

$$g(t) := t(tf_2(t) + f_1(t))$$

and easily get

$$\rho(u) = \sum_{j=1}^{\infty} \int_0^1 \frac{t - t^2}{2} g(u \exp(1 - j - t)) \, dt$$

in this case. Given $s \in \mathbb{R}$ and $k \in \mathbb{N}$ this implies

$$
\begin{aligned}
\rho(\exp(s - 1 + k)) &= \sum_{j=1}^{\infty} \int_0^1 \frac{t - t^2}{2} g(\exp(s - t - j + k)) \, dt \\
&= \sum_{j=-\infty}^{k-1} \int_0^1 \frac{t - t^2}{2} g(\exp(s - t + j)) \, dt,
\end{aligned}
$$

and

$$
\begin{aligned}
\sum_{j=k}^{\infty} \int_0^1 \frac{t - t^2}{2} |g(\exp(s - t + j))| \, dt &\leq \frac{1}{8} \int_k^{\infty} |g(\exp(s - t))| \, dt \\
&= \frac{1}{8} \int_0^{\exp(s-k)} |(xf_1(x))'| \, dx.
\end{aligned}
$$

Thus, under assumption **I** the limit

$$\psi(s) := \lim_{k \to \infty} \rho(\exp(s - 1 + k))$$

always exists. Moreover, the convergence is uniform for $s \in [0, 1]$. This uniform convergence clearly implies $\rho(u) = \psi(\log u) + o(1)$. Of course, ψ is 1-periodic and, since ρ is bounded and continuous, ψ possesses the same properties. $\square$

It remains to show that there exist cases where ψ is non-constant. To investigate this question, let us introduce the function

$$g_0(t) = g(\exp(t)) = \exp(t)(\exp(t)f_2(\exp(t)) + f_1(\exp(t))).$$

Then we have:

PROPOSITION 4.5 *It holds*

$$\psi(s) = \frac{\alpha}{12} - \frac{1}{4\pi^2} \sum_{k \in \mathbb{Z} \setminus \{0\}} \frac{\mathcal{F}(g_0)(-2\pi k)}{k^2} \exp(2\pi iks)$$

whereas $\mathcal{F}(g_0)(t) = \int_{-\infty}^{\infty} g_0(x) \exp(itx) \, dx$ *denotes the Fourier transform of* $g_0 \in L^1(\mathbb{R})$. *In particular,* ψ *is non-constant iff* $\mathcal{F}(g_0)(2\pi k) \neq 0$ *for some* $k \in \mathbb{Z} \setminus \{0\}$.

Proof. Expand ψ as Fourier series, i.e.

$$\psi(s) = \sum_{k=-\infty}^{\infty} \widehat{\psi}(k) \exp(2\pi i k s) \quad \text{with} \quad \widehat{\psi}(k) = \int_0^1 \psi(x) \exp(-2\pi i k x) \, dx.$$

Since $\psi(s)$ is the uniform limit of $\rho(\exp(s + k))$ as $k \to \infty$, it follows

$$
\begin{aligned}
\widehat{\psi}(k) &= \int_0^1 \frac{t - t^2}{2} \sum_{j=-\infty}^{\infty} \int_0^1 g_0(x - t + j) \exp(-2\pi i k x) \, dx \, dt \\
&= \int_0^1 \frac{t - t^2}{2} \exp(-2\pi i k t) \, dt \int_{-\infty}^{\infty} g_0(x) \exp(-2\pi i k x) \, dx \\
&= \begin{cases} \frac{1}{12} \mathcal{F}(g_0)(0) & k = 0 \\ \frac{1}{(2\pi k)^2} \mathcal{F}(g_0)(-2\pi k) & k \neq 0 \end{cases}.
\end{aligned}
$$

Yet $\mathcal{F}(g_0)(0) = \int_{-\infty}^{\infty} g_0(x) \, dx = \int_0^{\infty} (t f_1(t))' \, dt = \alpha$, and this completes the proof. $\qquad\square$

Let us apply these results now to the Gaussian case. It will turn out that in this case ψ is indeed non-constant. Especially, this tells us that in general the function ρ in Theorem 3.1 cannot be replaced by a constant thus disproving a result of [Z2].

THEOREM 4.6 *Let $\xi_0, \xi_1, \ldots$ be independent $\mathcal{N}(0, 1)$–distributed r.v.'s. Then we have*

$$\mathbb{P}\left(\sum_{j=0}^{\infty} \exp(-j) \xi_j^2 < r \right)$$

$$\sim \frac{\exp\left(-\frac{\pi^2}{12} - \frac{1}{4} \left(\log \left(\frac{1}{r} \log \frac{1}{r} \right) \right)^2 + \psi_0 \left(\log \left(\frac{1}{r} \log \frac{1}{r} \right) \right) \right)}{\pi^{1/2} r^{1/4} \left(\log \frac{1}{r} \right)^{3/4}}$$

where ψ_0 is a non-constant, 1-periodic and bounded function.

Proof. Applying (11) - (13) to $f(t) = (1 + 2t)^{-1/2}$ easily gives

$$
\begin{aligned}
I_0(u) &= -\frac{(\log(1 + 2u))^2}{4} - \frac{\pi^2}{12} + o(1) = -\frac{(\log 2u)^2}{4} - \frac{\pi^2}{12} + o(1), \\
I_1(u) &= -\frac{\log(1 + 2u)}{2} = -\frac{\log u}{2} + O(1) \quad \text{and} \\
I_2(u) &\sim \frac{\log(1 + 2u)}{2} \sim \frac{\log u}{2},
\end{aligned}
$$

and by Proposition 4.4 we obtain

$$\mathbb{P}\left(\sum_{j=0}^{\infty} \exp(-j) \xi_j^2 < r \right)$$

$$\sim \sqrt{\frac{(1 + 2u)^{-1/2}}{\pi \log u}} \exp\left(-\frac{\pi^2}{12} - \frac{1}{4} (\log(2u))^2 + ur + \psi(\log u) \right) \qquad (14)$$

provided that

$$\lim_{r \to 0} \frac{ur - \frac{1}{2}\log u}{\sqrt{\log u}} = 0. \tag{15}$$

Define $u = u(r)$ by

$$u(r) := \frac{1}{2r}\log\frac{1}{r}.$$

Then u satisfies (15), and using (14), we finally get

$$\mathbb{P}\left(\sum_{j=0}^{\infty}\exp(-j)\xi_j^2 < r\right)$$

$$\sim \frac{\exp\left(-\frac{\pi^2}{12} - \frac{1}{4}\left(\log\left(\frac{1}{r}\log\frac{1}{r}\right)\right)^2 + \psi\left(\log\left(\frac{1}{r}\log\frac{1}{r}\right) - \log 2\right)\right)}{\pi^{1/2}r^{1/4}\left(\log\frac{1}{r}\right)^{3/4}}$$

proving the statement of the Theorem with periodic component

$$\psi_0(x) := \psi(x - \log 2).$$

So it remains to show that ψ_0 is non-constant. We easily see that the function g_0 defined above admits the representation

$$g_0(x) = \frac{-\exp(x)}{(1 + 2\exp(x))^2} = -\frac{1}{8}\cosh^{-2}\left(\frac{x + \log 2}{2}\right).$$

Therefore, by Proposition 4.5

$$\begin{aligned}
\psi_0(x) &= -\frac{1}{24} + \frac{1}{8}\sum_{k \in \mathbb{Z}\setminus\{0\}}\frac{\mathcal{F}\left(\cosh^{-2}\left(\frac{\cdot}{2}\right)\right)(-2\pi k)}{(2\pi k)^2}\exp(2\pi i k x) \\
&= -\frac{1}{24} + \frac{1}{8}\sum_{k=1}^{\infty}\frac{\mathcal{F}\left(\cosh^{-2}\right)(4\pi k)}{(\pi k)^2}\cos(2\pi k x)
\end{aligned}$$

is the Fourier expansion of ψ_0. Using properties of the Fourier transform it follows that $\mathcal{F}(\cosh^{-2})(t) > 0$ for all $t \in \mathbb{R}$. Hence all Fourier coefficients of ψ_0 are positive and, consequently, ψ_0 is non-constant. $\square$

REMARK 4.7 It follows from the Fourier expansion that $\psi_0(x) = \psi_0(-x)$, hence ψ_0 on $[0, 1]$ is symmetric around $1/2$, i.e. $\psi_0(1 - x) = \psi_0(x)$. Moreover, $\psi_0'(0) = \psi_0'(1/2) = 0$ and by $\mathcal{F}(\cosh^{-2})(t) > 0$ we have $\psi_0''(0) < 0$, so ψ_0 has a strict local maximum at zero.

Discussion: Given a sequence $(r_n)_{n \in \mathbb{N}}$ tending to zero, we define $s_n \in [0, 1)$ by

$$s_n := \frac{1}{r_n}\log\frac{1}{r_n} - \left[\frac{1}{r_n}\log\frac{1}{r_n}\right].$$

Setting

$$\Psi(r) := \frac{\exp\left(-\frac{\pi^2}{12}\right)}{\pi^{1/2}} \frac{\exp\left(-\frac{1}{4}\left(\log\left(\frac{1}{r}\log\frac{1}{r}\right)\right)^2\right)}{r^{1/4}\left(\log\frac{1}{r}\right)^{3/4}},$$

Theorem 4.6 asserts

$$\mathbb{P}\left(\sum_{j=0}^{\infty}\exp(-j)\xi_j^2 < r_n\right) \sim \Psi(r_n)\exp(\psi_0(s_n)),$$

i.e. the asymptotic behavior of $\mathbb{P}\left(\sum_{j=0}^{\infty}\exp(-j)\xi_j^2 < r_n\right)$ depends on the way how the r_n's tend to zero. For example, if $s_n = s$ for all $n \in \mathbb{N}$, then

$$\lim_{n\to\infty} \frac{\mathbb{P}\left(\sum_{j=0}^{\infty}\exp(-j)\xi_j^2 < r_n\right)}{\Psi(r_n)} = \exp(\psi_0(s)) \tag{16}$$

exists, while the limit does not exist if $s_{2n+1} = 0$ and $s_{2n} = \epsilon$ with ϵ chosen in a sufficiently small neighborhood of zero such that $\psi_0(0) \neq \psi_0(\epsilon)$. Numerical experiments show that ψ_0 has a global maximum at zero and a global minimum at $\frac{1}{2}$. Then (16) becomes minimal for $s = \frac{1}{2}$, i.e. if $(1/r_n)\log(1/r_n)+1/2 \in \mathbb{N}$, and maximal for $(1/r_n)\log(1/r_n) \in \mathbb{N}$.

5. Slow Variation and Bounded Variation

Our investigations heavily depended on the assumptions that the d.f. F of the r.v.'s $Z_1, Z_2, \ldots$ possesses properties $\mathbf{I}$ and $\mathbf{L}$. While $\mathbf{L}$ is easy to check (it is a condition on F), property $\mathbf{I}$ is much harder to handle. Recall that $\mathbf{I}$ is a condition on the derivative of the Laplace transform of F. So our aim is to find conditions on F (and not on its Laplace transform) that insure the validity of $\mathbf{I}$. Since $\mathbf{I}$ and $\mathbf{L}$ together imply $F(1/\cdot) \in \mathbf{R}_\alpha$ for some $\alpha < 0$, we may restrict our attention to d.f.'s F possessing this additional property. More precisely, given $\alpha < 0$ we are looking for those slowly varying functions l such that $F(1/x) = x^\alpha l(x)$ is a d.f. satisfying $\mathbf{I}$. We first reformulate $\mathbf{I}$ into a condition on l. The Laplace transform of F with $F(1/x) = x^\alpha l(x)$ can be written as

$$f(u) = u^\alpha l(u) I_\alpha(u)$$

with

$$I_\alpha(u) := \int_0^\infty \exp\left(-\frac{1}{t}\right) t^{-2+\alpha} \frac{l(tu)}{l(u)}\, dt.$$

Analogously, we obtain

$$f'(u) = u^{\alpha-1} l(u)(I_\alpha(u) - I_{\alpha-1}(u)) \quad \text{and}$$
$$f''(u) = u^{\alpha-2} l(u)(I_{\alpha-2}(u) - 2I_{\alpha-1}(u)),$$

and the function $(uf_1(u))'$ in the definition of property **I** equals

$$
\begin{aligned}
(uf_1(u))' &= uf_2(u) + f_1(u) \\
&= u^{-1}\frac{I_{\alpha-2}(u)I_\alpha(u) - I_{\alpha-1}(u)I_\alpha(u) - I_{\alpha-1}^2(u)}{I_\alpha^2(u)}.
\end{aligned} \tag{17}
$$

Since $(uf_1(u))'$ is continuous on $[0,\infty)$, its integrability is determined by its behavior for $u \to \infty$, thus in view of (17) by the behavior of $I_\alpha(u)$, $I_{\alpha-1}(u)$ and $I_{\alpha-2}(u)$ as $u \to \infty$. The Karamata Tauberian Theorem (cf. [BGT], Th. 1.7.1) asserts that $I_\alpha(u) \to \Gamma(1-\alpha)$ for $u \to \infty$. Yet this does not suffice for our purposes because we need more exact information about the rate of this convergence. There exist several approaches how to investigate the convergence of $l(tu)/l(u) - 1$ and $\log l(tu) - \log l(u)$, respectively (see, i.e. [BGT], [GS] or [OW]). We chose two of them in order to derive sufficient conditions for **I**.

DEFINITION 5.1 *The function l belongs to $SR2(g)$ with $g \in \mathbf{R}_\rho$ ($\rho \leq 0$) provided there exists a function k such that for all $t \geq 1$*

$$
\frac{l(tu)}{l(u)} - 1 \sim k(t)g(u) \tag{18}
$$

as $u \to \infty$. Note that then necessarily

$$
k(t) = \begin{cases} c\log t & \rho = 0 \\ c\frac{t^\rho - 1}{\rho} & \rho < 0 \end{cases}
$$

with some $c \in \mathbb{R}$ and, moreover, (18) also holds for $t \in (0,1)$. We shall say that l is of g–index c.

In the case $\rho = 0$ this class is not always sufficient for our purpose and we will need the following extension from [OW].

DEFINITION 5.2 *The function $\log l$ belongs to the class $\Pi R1(g,h)$ with $g \in \mathbf{R}_0$ and $h(u) = o(g(u))$ provided that $l \in SR2(g)$ with g–index $c \neq 0$ and*

$$
\log l(tu) - \log l(u) - cg(u)\log t = O(h(u))
$$

as u tends to infinity.

REMARK 5.3 We may change $g(u)$ on any compact interval of $[0,\infty)$ without effecting the above definition. In addition, we know that g and $1/g$ are bounded on any compact interval far enough to the right of the real axes. Consequently, without losing generality we will always assume g and $1/g$ to be bounded on compact sets of $[0,\infty)$.

Defining

$$
J_n(1-\alpha) := \int_0^\infty \exp(-s)\, s^{-\alpha}\, (-\log s)^n \, ds
$$

we can describe the asymptotic behavior of $I_\alpha(u)$ as follows:

PROPOSITION 5.4 *For $l \in SR2$ with g-index c and $\alpha < 0$ we have*

$$I_\alpha(u) = \begin{cases} \Gamma(1-\alpha) + cJ_1(1-\alpha)g(u) + o(g(u)) & \rho = 0 \\ \Gamma(1-\alpha) + c\dfrac{\Gamma(1-\alpha-\rho) - \Gamma(1-\alpha)}{\rho}g(u) + o(g(u)) & \rho < 0 \end{cases}$$

as $u \to \infty$. If furthermore $\log l \in IIR1(g, g^n)$ for some $n > 1$, then we obtain

$$I_\alpha(u) = \sum_{j=0}^{n-1} \frac{c^j J_j(1-\alpha)}{j!} g^j(u) + O(g^n(u)).$$

The proof is based on the ideas from [GS], section 2.5, which can be extended to the above situation and the setting of [OW].
Substitution in equation (17) yields the estimates $(uf_1(u))' = u^{-1}O(g(u))$ and $(uf_1(u))' = u^{-1}O(g^n(u))$, respectively.

COROLLARY 5.5 *Each of the following conditions on l ensures that $\mathbf{I}$ holds for the d.f. F defined by $F(1/x) := x^\alpha l(x)$ with $\alpha < 0$.*
  a) $l \in SR2(g)$ with $g \in \mathbf{R}_\rho$ and $\rho < 0$.
  b) $l \in SR2(g)$ with $g \in \mathbf{R}_0$ and $u^{-1}g(u)$ is integrable on $[1, \infty)$.
  c) $\log l \in IIR1(g, g^n)$ with $g \in \mathbf{R}_0$ and $u^{-1}g^n(u)$ is integrable on $[1, \infty)$.

Examples: Finally, let us list some l that fulfill the above conditions. The definitions below have to be understood to be valid for sufficiently large u.
  1. For $l(u) := a + u^\rho$ ($a \in (0, \infty)$, $\rho < 0$) one can verify (a) with $g(u) = u^\rho$.
  2. The function $l(u) := a + (\log u)^\beta$ ($\beta < 0$) is an example for (b) with $g(u) = (\log u)^{\beta-1}$.
  3. The example $l(u) := (\log u)^\gamma$ ($\gamma \in \mathbb{R}$) requires the application of (c) since $u^{-1}g(u)$ with $g(u) = \log u$ is not integrable. It can be checked easily that $\log(\log u)^\gamma \in IIR1(g, g^2)$.
  4. Higher order terms can be observed if we consider $l(u) := \exp(\pm(\log u)^{1-\delta})$ ($\delta \in (0, 1)$). Then $g(u) = (\log u)^{-\delta}$ and one can show that $\log l \in IIR1(g, g^n)$ where n satisfies $(n-1)\delta \leq 1 < n\delta$.

References

[A]   J. M. P. Albin. Minima of H–valued Gaussian Processes. *Ann. Probab.*, 24:788–824, 1996.
[AD] T. W. Anderson and D. A. Darling. Asymptotic theory of certain 'goodness of fit' criteria based on stochastic processes. *Ann. Math. Statist.*, 23:193–212, 1952.
[BGT] N. H. Bingham, C. M. Goldie and J. L. Teugels. *Regular Variation*. Cambridge University Press, 1987.
[DR] R. Davis and S. Resnick. Extremes of moving averages of random variables with finite endpoint. *Ann. Probab.*, 19:312–328, 1991.
[GS] C. M. Goldie and R. L. Smith. Slow variation with remainder: Theory and applications. *Quart. J. Math. Oxford*, 38:45–71, 1987.
[I]   I. A. Ibragimov. On the probability that a Gaussian vector with values in a Hilbert space hits a sphere of small radius. *J. Sov. Math.*, 20:2164–2174, 1982.

[KLL]   J. Kuelbs, W. V. Li and W. Linde. The Gaussian measure of shifted balls. *Probab. Theory Rel. Fields*, 98:143–162, 1994.

[Li]   W. V. Li. On the lower tail of Gaussian measures. In *Probability in Banach Spaces VIII, ser. Progress in Probability*, 30:106–117, 1992.

[Lif1] M. A. Lifshits. *Gaussian Random Functions*. Kluwer, 1995.

[Lif2] M. A. Lifshits. On the lower tail probability of some random series. *Ann. Probab.*, 25, 1997.

[LL]   W. V. Li and W. Linde. Small ball problems for non-centered Gaussian measures. *Prob. Math. Stat.*, 14:231–251, 1993.

[MWZ]   E. Mayer-Wolf and O. Zeitouni. The probability of small Gaussian ellipsoids and associated conditional moments. *Ann. Probab.*, 21:14–24, 1993.

[NS]   A. V. Nagaev and A. N. Startsev. Asymptotic properties of the distribution function of an infinite quadratic form of Gaussian random variables. In *Limit theorems for stochastic processes and statistical conclusions, Work Collect., Tashkent*, p.144–160, 1981.

[OW]   E. Omey and E. Willekens. Π–variation with remainder. *J. London Math. Soc.* (2), 37:105–118, 1988.

[S]   G. N. Sytaya. On the asymptotic representation of the Gaussian measure in a Hilbert space. In *Theory of Stochastic Processes*, 2: 94–102, 1974.

[Z1]   V. M. Zolotarev. Gaussian measure asymptotic in l_p on a set of centered spheres with radii tending to zero. In *12th Europ. Meeting of Statisticians, Varna*, p.254, 1979.

[Z2]   V. M. Zolotarev. Asymptotic behavior of Gaussian measures in l_2. *J. Sov. Math.*, 24:2330–2334, 1986.

M. A. Lifshits   T. Dunker and W. Linde
Mancomtech Center   Friedrich Schiller Universität
St.-Petersburg, Russia   Jena, Germany

Progress in Probability, Vol. 43
© 1998 Birkhäuser Verlag Basel/Switzerland

Strong Approximations to the Local Empirical Process

UWE EINMAHL* AND DAVID M. MASON*

ABSTRACT. In a previous paper we introduced a notion of a local empirical process indexed by functions which has useful applications to density and regression function estimation among other areas. We have shown that if the function class is uniformly bounded, one can obtain a strong approximation to this process by a suitable Gaussian process. We now prove such a result when the underlying function class is unbounded, but has an envelope function with a finite p-th moment for some $p > 2$. Among other applications, our new strong invariance principle for the unbounded case can be used to prove laws of the iterated logarithm for the kernel regression function estimator under mild conditions.

1. Introduction and Statements of Main Results

Einmahl and Mason (1997) introduced a notion of a local empirical process indexed by functions extending an earlier notion due to Deheuvels and Mason (1994). They established via a strong approximation a compact law of the iterated logarithm [LIL] for this process when the indexing class of functions is uniformly bounded. As applications of their result they obtained LIL's for kernel density estimators and the conditional empirical process. Our aim is to establish a version of their strong approximation when the index set is not necessarily a uniformly bounded class of functions. It will be seen that such a result is well suited for studying the rate of strong consistency of kernel regression function estimators. We shall use the compact LIL's that result from our strong approximations to prove LIL's for the Nadaraya-Watson kernel regression function estimator. As far as we know, these are the only LIL's for this estimator in the literature with valid proofs. Consult Remark 5 below.

We shall first define the local empirical process and then state the Einmahl and Mason (1997) strong Gaussian approximation for this process when the index set is a class of bounded functions. Let us begin by recalling the basic notation and assumptions of Einmahl and Mason (1997).

Let $X_1, X_2, \ldots,$ be a sequence of i.i.d. $R^d, d \geq 1$, valued random vectors with distribution $\mathbf{P}$ on the Borel sets $\mathcal{B}$ of R^d. Given any $x \in R^d$, we set for any invertible bimeasurable transformation $h : R^d \to R^d$ and any measurable subset $J \subseteq R^d$,

$$A(h) = x + hJ. \tag{1.1}$$

*) Research partially supported by the Volkswagenstiftung (RiP program at Oberwolfach) and an NSF Grant.

Let $(h_n)_{n\geq 1}$ be a sequence of invertible bimeasurable transformations from $R^d \to R^d$ and assume that with $A_n := A(h_n)$ and $a_n := \mathbf{P}(A_n), n \geq 1$,

$$a_n > 0, \text{ for } n \geq 1, \tag{A.i}$$

$$na_n \to \infty, \text{ as } n \to \infty, \tag{A.ii}$$

$$a_n \to 0, \text{ as } n \to \infty. \tag{A.iii}$$

Define the probability measure P_n on $(R^d, \mathcal{B})$,

$$P_n(B) = \mathbf{P}(x + h_n(J \cap B))/a_n, B \in \mathcal{B}. \tag{1.2}$$

Let $\mathcal{F}$ denote a class of square $\mathbf{P}$-integrable functions on R^d with supports contained in J. To avoid measurability problems we shall assume that there exists a countable subclass of $\mathcal{F}_c$ of $\mathcal{F}$ and a measurable set D with $P_n(D) = 0$ for all $n \geq 0$ such that for any $x_1, \ldots, x_m \in R^d - D$ and $f \in \mathcal{F}$ there exists a sequence $\{f_j\}_{j\geq 1} \subset \mathcal{F}_c$ satisfying

$$\lim_{j\to\infty} f_j(x_k) = f(x_k), \text{ for } k = 1, \ldots, m; \tag{S.i}$$

$$\lim_{j\to\infty} P_n(f_j) = P_n(f), \text{ for each } n \geq 1; \tag{S.ii}$$

$$\lim_{j\to\infty} P_n(f_j^2) = P_n(f^2), \text{ for each } n \geq 1. \tag{S.iii}$$

Given any integer $n \geq 1$ and invertible bimeasurable transformation $h : R^d \to R^d$, we introduce the *local empirical process at x indexed by $\mathcal{F}$*

$$L_n(f, h) = \sum_{i=1}^{n} \frac{f(h^{-1}(X_i - x)) - Ef(h^{-1}(X_i - x))}{\sqrt{n\mathbf{P}(A(h))}}. \tag{1.2}$$

For integers $m \geq 1$ and $n \geq 1$ define the empirical process indexed by $\mathcal{F}$

$$\alpha_m^{(n)}(f) := \sum_{i=1}^{m} \frac{f(Y_i^{(n)}) - P_n(f)}{\sqrt{m}}, \quad f \in \mathcal{F}, \tag{1.3}$$

where $Y_1^{(n)}, \ldots, Y_m^{(n)}$ are assumed to be i.i.d. P_n.

Let

$$\mathcal{F}' = \{f - g : f, g \in \mathcal{F}\}, \mathcal{F}^2 = \{f^2 : f \in \mathcal{F}\},$$
$$(\mathcal{F}')^2 = \{(f - g)^2 : f, g \in \mathcal{F}\} \text{ and } \mathcal{G} = \mathcal{F} \cup \mathcal{F}^2 \cup \mathcal{F}' \cup (\mathcal{F}')^2.$$

For any functional T defined on a subset $\mathcal{H}$ of the real valued functions on J we denote

$$\| T \|_{\mathcal{H}} := \sup\{| T(f) |: f \in \mathcal{H}\} \tag{1.4}$$

We shall require the following additional assumption on the class of functions $\mathcal{F}$ and the sequence of probability measures $(P_n)_{n\geq 1}$. There exists a probability measure P_0 such that P_n converges weakly to P_0 and

$$\| P_n - P_0 \|_{\mathcal{G}} \to 0 \text{ as } n \to \infty. \tag{F.i}$$

We shall denote by

$$(B_0(f))_{f \in \mathcal{F}} \tag{1.5}$$

a P_0-Brownian bridge indexed by $\mathcal{F}$, that is, B_0 is a Gaussian process indexed by $\mathcal{F}$ with mean zero and covariance function

$$Cov(B_0(f), B_0(g)) = P_0(fg) - P_0(f)P_0(g), \quad f, g \in \mathcal{F}. \tag{1.6}$$

We will assume that B_0 has uniformly ρ_0 continuous sample paths, where

$$\rho_0^2(f, g) = Var(B_0(f) - B_0(g)), \quad f, g \in \mathcal{F}. \tag{1.7}$$

Such a process B_0 exists whenever the class $\mathcal{F}$ is P_0-pregaussian. (For the definition of "P_0-pregaussian" refer to Ledoux and Talagrand (1991).) Let Z be a standard normal random variable independent of B_0 and introduce the P_0-Brownian motion W_0 indexed by $\mathcal{F}$

$$W_0(f) := B_0(f) + ZP_0(f), \quad f \in \mathcal{F}. \tag{1.8}$$

Note that W_0 has covariance function $P_0(fg)$, $f, g \in \mathcal{F}$.

Einmahl and Mason (1997) established the following strong approximation and compact LIL for the local empirical process. (For the definition of the term "reproducing kernel Hilbert space" used in the statement of Theorem 3 refer to Ledoux and Talagrand (1991).)

We shall use the notation $Lx = \log(e \vee x)$ and $[x] = $ integer part of x.

Theorem 1. *Assume the measurability conditions* (S.i), (S.ii) *and* (S.iii) *along with* (A.i), (A.ii) *and* (A.iii). *Also assume that as* $n \to \infty$

$$a_n \sim d_n \text{ with } d_n \searrow 0, nd_n \nearrow \infty \text{ and } nd_n/LLn \to \infty. \tag{A.iv}$$

Further assume (F.i) *and*

$$| f | \le M \text{ for all } f \in \mathcal{F} \text{ and some } M \ge 1. \tag{F.ii}$$

In addition, assume for each $n \ge 1$ *and* $m \ge n$, *both*

$$f \circ h_n^{-1} \text{ and } f \circ h_m^{-1} \circ h_n \in \mathcal{F}; \tag{F.iii}$$

there exists a sequence of positive constants $(b_n)_{n \ge 1}$ *such that for all* $f, g \in \mathcal{F}$ *and* $n \ge 1$

$$\int_{R^d} f(h_n z)g(h_n z)dP_0(z) = b_n^{-1} \int_{R^d} f(z)g(z)dP_0(z), \tag{F.iv}$$

where $a_n/b_n \to 1$, *as* $n \to \infty$. *Further assume that with* $k(n) = [na_n]$

$$\| \alpha_{k(n)}^{(n)} \|_{\mathcal{F}} / \sqrt{LLn} \to_P 0, \text{ as } n \to \infty; \tag{F.v}$$

$$\mathcal{F} \text{ is } P_0\text{--pregaussian.} \tag{F.vi}$$

Then one can construct $X_1, X_2, \ldots,$ i.i.d. $\mathbf{P}$ and a sequence $W_1, W_2, \ldots,$ of independent P_0-Brownian motions indexed by $\mathcal{F}$ on the same probability space, such that with probability one, as $n \to \infty,$

$$\sup_{f \in \mathcal{F}} |L_n(f, h_n) - \frac{1}{\sqrt{nb_n}} \sum_{i=1}^{n} W_i(f \circ h_n^{-1})|/\sqrt{LLn} \to 0. \qquad (1.9)$$

Einmahl and Mason (1997) derive as a corollary to Theorem 1 the following compact LIL for the local empirical process.

Corollary 1. *In addition to the assumptions of Theorem 1 assume that for all f, g such that $P_0(f - g)^2 < \infty$, $n \geq 1$ and $2n \geq m \geq n$, we have for some $L > 0$,*

$$\int_{R^d} \{f(h_m^{-1}(h_n(z))) - g(h_m^{-1}(h_n(z)))\}^2 dP_0(z)$$

$$\leq L \int_{R^d} \{f(z) - g(z)\}^2 dP_0(z); \qquad (\text{F.vii})$$

and for every compact subset $A \subset R^d$ and $\delta > 0$ there exists a $q_0 > 1$ such that for all $1 < q < q_0$ with $n_k = [q^k]$

$$\max_{n_k \leq m \leq n_{k+1}} \sup_{z \in A} | z - h_m^{-1}(h_{n_k}(z)) | \leq \delta \qquad (\text{F.viii})$$

for all large enough k depending on A and $\delta > 0$. Then with probability one the sequence of processes

$$\left(L_n(f, h_n)/\sqrt{2LLn} \right)_{f \in \mathcal{F}} \qquad (1.10)$$

is relatively compact in $\mathcal{B}(\mathcal{F})$ (the set of bounded real valued functions defined on $\mathcal{F}$) with set of limit points equal to $\mathcal{K}$, where $\mathcal{K}$ is the reproducing kernel Hilbert space pertaining to the covariance function $P_0(fg)$ of W_0.

Our goal is to obtain the following strong approximation to the local empirical process when the indexing class can have an unbounded envelope function.

Theorem 2. *Assume all the conditions of Theorem 1, except for (F.ii). We replace (F.ii) by $| f | \leq F$ for all $f \in \mathcal{F}$, where F is a measurable function satisfying for some $p > 2$*

$$\lim_{\lambda \to \infty} \limsup_{n \to \infty} P_n F^p 1(F > \lambda) = 0. \qquad (\text{F.ix})$$

Further, assume that for the same $p > 2$ as in (F.ix) we have

$$(na_n)^{-1} = O\left((c_n \log n)^{1/(1-p/2)} LLn \right), \qquad (\text{p})$$

for some sequence $(c_n)_{n \geq 1}$ satisfying $\sum_{r=1}^{\infty} (rc_{2^r} \log r)^{-1} < \infty$. Then one can construct $X_1, X_2, \ldots,$ i.i.d. $\mathbf{P}$ and a sequence $W_1, W_2, \ldots,$ of independent P_0-Brownian motions indexed by $\mathcal{F}$ on the same probability space, such that with probability one, as $n \to \infty$, (1.9) holds.

Theorem 2 leads immediately to the following compact LIL.

Corollary 2. *In addition to the assumptions of Theorem 2, assume* (F.vii) *and* (F.viii). *Then the conclusion of Corollary 1 holds.*

Corollaries 1 and 2 are immediately applicable to proving LIL's for kernel regression function estimators.

LIL's for Kernel Regression Function Estimators

Let $(X, Y), (X_1, Y_1), (X_2, Y_2), \ldots$, be i.i.d. $R^d \times R$, $d \geq 1$, valued random variables with common joint Lebesgue density g_{XY} and marginals g_X and g_Y. Assume that for a chosen $x \in R^d$ the regression function

$$m(x) := E(Y \mid X = x) \text{ exists.} \tag{1.11}$$

Nadaraya (1964) and Watson (1964) introduced independently the nonparametric kernel estimator of $m(x)$

$$\widehat{m}_n(x) = \frac{\sum_{i=1}^{n} Y_i K((X_i - x)/\gamma_n^{1/d})}{\sum_{i=1}^{n} K((X_i - x)/\gamma_n^{1/d})}, \tag{1.12}$$

based upon the sample $(X_1, Y_1), \ldots, (X_n, Y_n)$, where $(\gamma_n)_{n \geq 1}$ is a sequence of positive constants converging to zero as $n \to \infty$, and K is a kernel function, which we shall assume satisfies the following conditions:

$$K \text{ is continuous on } [-1/2, 1/2]^d; \tag{K.i}$$

$$K(s) = 0 \text{ for } s \notin [-1/2, 1/2]^d; \tag{K.ii}$$

$$\int_{R^d} K(s) ds = 1; \tag{K.iii}$$

$$\{K(t\cdot) : t \geq 1\} \text{ is a VC subgraph class.} \tag{K.iv}$$

Remark 1. For definitions of a VC subgraph class refer to Giné and Zinn (1984), Pollard (1984) and especially the recent monograph of van der Vaart and Wellner [V-W](1996). Here are two important situations when (K.iv) is satisfied.
(I) Condition (K.iv) holds in the case $d = 1$ whenever K is of bounded variation on R. This can be readily inferred from properties of VC subgraph classes of functions as detailed in Lemma 2.6.18 of [V-W] (1996).
(II) Condition (K.iv) holds whenever $K(s) = \psi(s^T B s)$ for some monotone function ψ and $d \times d$ matrix B. This result is attributed in the literature to Pollard (1982).

Towards stating our LIL's for $\widehat{m}_n(x)$ we set

$$\widehat{r}_n(x) = (n\gamma_n)^{-1} \sum_{i=1}^{n} Y_i K((X_i - x)/\gamma_n^{1/d}) \text{ and } r_n(x) = E\widehat{r}_n(x); \tag{1.13}$$

$$\widehat{g}_n(x) = (n\gamma_n)^{-1} \sum_{i=1}^{n} K((X_i - x)/\gamma_n^{1/d}) \text{ and } g_n(x) = E\widehat{g}_n(x). \tag{1.14}$$

Further, we let

$$\sigma^2(x) = Var(Y \mid X = x); \tag{1.15}$$

$$\| K \|_2^2 = \int_{R^d} K^2(s)ds. \tag{1.16}$$

We shall prove the following two LIL's for $\widehat{m}_n(x)$.

Theorem 3. *Assume that for some $\epsilon > 0$*

$$g_{XY} \text{ is continuous on } \{ x + (-\epsilon, \epsilon)^d \} \times (-\infty, \infty), \tag{M.i}$$

$$g_X \text{ is continuous and positive on } x + (-\epsilon, \epsilon)^d \tag{M.ii}$$

and with $H(s,y) := \mid y \mid 1\{s \in x + (-\epsilon, \epsilon)^d\}$ for some $M > 0$ on the support of g_{XY}

$$\mid H \mid \le M. \tag{M.iii}$$

Also impose on the sequence $(\gamma_n)_{n \ge 1}$ the growth conditions

$$\gamma_n \sim d_n \text{ with } d_n \searrow 0, \ n d_n \nearrow \infty; \tag{g.i}$$

$$(n\gamma_n)^{-1} = o((LLn)^{-1}). \tag{g.ii}$$

Then for any kernel K satisfying (K.i), (K.ii),(K.iii) and (K.iv), with probability one,

$$\limsup_{n \to \infty} \pm \frac{\sqrt{n\gamma_n}}{\sqrt{2LLn}} (\widehat{m}_n(x) - r_n(x)/g_n(x)) = \frac{\sigma(x) \| K \|_2}{\sqrt{g_X(x)}}. \tag{1.17}$$

Our next LIL result allows H to be unbounded on the support of g_{XY}.

Theorem 4. *Assume that for some $\epsilon > 0$ conditions (M.i) and (M.ii) hold and for some measurable function h and $p > 2$ we have for all $\gamma > 0$ small enough, uniformly in y*

$$\int_{[-1/2,1/2]^d} g_{XY}(\gamma^{1/d}s + x, y)ds \le h(y), \tag{M.iv}$$

where $\int_R \{1 + \mid y \mid^p\}h(y)dy < \infty$. Also impose on the sequence $(\gamma_n)_{n \ge 1}$ the growth conditions (g.i) and

$$(n\gamma_n)^{-1} = O\left((c_n \log n)^{1/(1-p/2)} LLn\right), \tag{g.iii}$$

for some sequence $(c_n)_{n \ge 1}$ satisfying $\sum_{r=1}^{\infty}(rc_{2^r} \log r)^{-1} < \infty$. Then for any kernel K satisfying (K.i), (K.ii), (K.iii) and (K.iv), we have (1.17) with probability one.

Remark 2. An easy sufficient condition for (M.iv) is the following: assume that $E(|Y|^p|X = x) < \infty$ and there exists a constant $C > 0$ and an $\epsilon > 0$ such that for all $(u,y) \in \{x + (-\epsilon, \epsilon)^d\} \times (-\infty, \infty)$ we have $Cg_{XY}(x,y) \ge g_{XY}(u,y)$. In this case we can choose $h(y) = Cg_{XY}(x,y)$.

Remark 3. Note that a sufficient condition for (g.iii) is

$$n\gamma_n/(\log n)^q \to \infty, \text{ as } n \to \infty, \tag{g.iv}$$

for some $q > 2/(p-2)$. Most likely a truncation argument would lead to a slight improvement of condition (g.iii), however the present condition should be more than sufficient for statistical purposes.

Remark 4. Deheuvels and Mason (1994) established a functional LIL for the local empirical process indexed by sets. As an application of their result they derived an LIL for the kernel density estimator $\widehat{g}_n(x)$. In the case $d = 1$ they showed that if K satisfies (I) and conditions (M.ii), (g.i) and (g.ii) hold, then with probability 1

$$\limsup_{n \to \infty} \pm \frac{\sqrt{n\gamma_n}}{\sqrt{LLn}} (\widehat{g}_n(x) - g_n(x)) = \sqrt{g_X(x)} \parallel K \parallel_2 . \qquad (1.18)$$

Condition (g.ii) is needed for $\widehat{g}_n(x)$ to be an almost surely consistent estimator of $g_X(x)$. Refer to Deheuvels and Mason (1995). It is natural then to conjecture that condition (g.ii) is also necessary for $\widehat{m}_n(x)$ to be a strongly consistent estimator of $m(x)$.

Remark 5. Earlier, Hall (1981) had proved that (1.18) holds in the case $d = 1$ under the somewhat more restrictive assumption that

$$\frac{n\gamma_n LLn}{\log^4 n} \to \infty \text{ as } n \to \infty.$$

His proof was based on the Komlós, Major and Tusnády (1975) Kiefer process strong approximation to the empirical process indexed by intervals. A number of authors have attempted to adapt Hall's method of proof to study the rate of strong consistency of nonparametric kernel-type estimators of $m(x)$ in the case $d = 1$. We mention Mack and Silverman (1982), Haerdle (1984) and Hong (1993). In particular, Haerdle (1984) is concerned with LIL's for $\widehat{m}_n(x)$. Unfortunately all of their proofs are based on a misuse of a strong approximation of the bivariate uniform empirical process by a bivariate Brownian bridge announced with a short outline of proof by Tusnády (1977). These authors mistakenly assumed that the Tusnády Brownian bridge approximation holds along a sequence $U_1, U_2, \ldots$ of i.i.d. uniform random variables on the unit square. However, such an approximation is impossible. Tusnády's approximation can only be valid for each fixed sample $U_1, \ldots, U_n$, $n \geq 1$, and thus cannot be used directly to infer almost sure limiting results for functionals of the bivariate uniform empirical process.

Proofs of Theorems 3 and 4
We shall see that Theorem 3 is derived from Corollary 1 and Theorem 4 from Corollary 2 with the notation $(X_1, Y_1), (X_2, Y_2), \ldots$ replacing $X_1, X_2, ..$ and $d+1$ replacing d.

Set $J = [-1/2, 1/2]^d \times R$, and for each $n \geq 1$, let $h_n(u, v) = (\gamma_n^{1/d} u, v)$ and $A(h_n) = (x, 0) + h_n J$. Further, let P_n be the probability measure defined on the Borel subsets B of R^d

$$P_n(B) = \mathbf{P}((x, 0) + h_n(J \cap B))/\mathbf{P}(A(h_n)). \qquad (1.19)$$

Notice that P_n has density

$$p_n(u, y) = \frac{g_{XY}(\gamma_n^{1/d} u + x, y) 1_J(u, y)}{\int_{[-1/2, 1/2]^d} g_X(\gamma_n^{1/d} s + x) ds}.$$

Clearly by (M.i) and (M.ii) we have $p_n \to p_0$ a.s., as $n \to \infty$, where

$$p_0(u, y) = \frac{g_{XY}(x, y) 1_J(u, y)}{g_X(x)}. \tag{1.20}$$

Let P_0 denote the probability measure on R^{d+1} with density p_0. By Scheffé's theorem, we have

$$\| P_n - P_0 \|_{\mathcal{G}} \to 0, \text{ as } n \to \infty, \tag{1.21}$$

where $\mathcal{G}$ denotes any uniformly bounded class of measurable functions on J. Introduce the function defined on R with $p > 2$

$$\Psi_n(y) = \frac{|y|^p \int_{[-1/2, 1/2]^d} g_{XY}(\gamma_n^{1/d} s + x, y) ds}{\int_{[-1/2, 1/2]^d} g_X(\gamma_n^{1/d} s + x) ds}. \tag{1.22}$$

Notice whenever (M.iv) holds that due to (M.i) for all large enough n uniformly in y

$$\Psi_n(y) \le 2|y|^p \int_{[-1/2, 1/2]^d} g_{XY}(\gamma_n^{1/d} u + x, y) du / g_X(x) \le \frac{2|y|^p h(y)}{g_X(x)}, \tag{1.23}$$

where

$$\int_R \{1 + |y|^p\} h(y) dy < \infty. \tag{1.24}$$

Let $\mathcal{F}$ be any class of measurable functions defined on R^{d+1} with support in J and envelope function F satisfying

$$F(u, y) \le k_1 \mid y \mid + k_2 \tag{1.25}$$

for constants $0 < k_1, k_2 < \infty$. From (1.21) we see that whenever (M.iii) is satisfied then (F.i) holds for $\mathcal{F}$. Further, an elementary argument based on (1.21), (1.23) and (1.24) in conjunction with the Lebesgue dominated convergence theorem shows that whenever (M.iv) holds then (F.i) and (F.ix) are also fulfilled by $\mathcal{F}$.

Consider the class of functions

$$\mathcal{F} = \{(u, y) \to (y - m(x)) \frac{K(tu)}{g_X(x)}, t \ge 1\}. \tag{1.26}$$

Since $\{K(t\cdot), t \ge 1\}$ is a VC subgraph classes, it follows that the class $\mathcal{F}$ has this property too. (Use Lemma 2.6.18 of [V-W] (1996).) It is readily shown that the class $\mathcal{F}$ also satisfies the measurability condition (S). Therefore using a standard fact about VC subgraph classes, we see that whenever (F.ii) or (F.ix) hold, our class $\mathcal{F}$ is a uniform Donsker class, which trivially implies (F.v) and (F.vi).

Set

$$f(u, y) = (y - m(x)) \frac{K(u)}{g_X(x)}. \tag{1.27}$$

Observe that

$$P_0 f^2 = \sigma^2(x) \frac{\| K \|_2^2}{g_X^2(x)}. \tag{1.28}$$

Trivially the assumptions on $(\gamma_n)_{n\geq 1}$ imply that (A.i), (A.ii), (A.iii) and (A.iv) hold, as well as (p) when appropriate. It is also routine to verify that (F.iii), (F.iv), (F.vi), (F.vii) and (F.viii) are satisfied. Thus applying the compact LIL given in Corollary 1 or 2 in combination with the continuous mapping theorem (see Lemma 2.2 of Wichura (1973)) applied to the evaluation functional corresponding to the choice $t = 1$, we get after a little calculation that

$$\limsup_{n\to\infty} \pm \frac{L_n(f, h_n)}{\sqrt{2LLn}} = \sigma(x)\frac{\| K \|_2}{g_X(x)} \ , \ \text{a.s.} \tag{1.29}$$

Recall the definitions of $\widehat{r}_n(x)$, $r_n(x)$, $\widehat{g}_n(x)$ and $g_n(x)$ in (1.13) and (1.14). Notice that we can write

$$L_n(f, h_n) = \gamma_n\sqrt{\frac{n}{a_n}}\frac{(\widehat{r}_n(x) - r_n(x) - m(x)\{\widehat{g}_n(x) - g_n(x)\})}{g_X(x)}. \tag{1.30}$$

Hence, since $\sqrt{a_n/\gamma_n} \to \sqrt{g_X(x)}$ as $n \to \infty$, we get from (1.29) that with probability 1

$$\limsup_{n\to\infty} \pm \frac{\sqrt{n\gamma_n}\,(\widehat{r}_n(x) - r_n(x) - m(x)\{\widehat{g}_n(x) - g_n(x)\})}{\sqrt{2LLn}g_X(x)} = \frac{\sigma(x)\| K \|_2}{\sqrt{g_X(x)}}. \tag{1.31}$$

Since we showed that under the assumptions of Theorem 3 or 4 any class of functions satisfying (1.25) also fulfills (F.i), we have, in particular, as $n \to \infty$

$$g_n(x) \to g_X(x) \text{ and } r_n(x) \to m(x)g_X(x). \tag{1.32}$$

Moreover, by arguing just as above, the LIL (1.18) holds for $\widehat{g}_n(x)$, as it does analogously for $\widehat{r}_n(x)$. Combining this with (1.32) we obtain that with probability 1 as $n \to \infty$

$$\widehat{g}_n(x) \to g(x) \text{ and } \widehat{r}_n(x) \to m(x)g_X(x). \tag{1.33}$$

Using (1.31), (1.32) and (1.33) we now get after some routine manipulation that

$$\widehat{m}_n(x) - r_n(x)/g_n(x) =$$

$$(\widehat{r}_n(x) - r_n(x) - m(x)\{\widehat{g}_n(x) - g_n(x)\})/g_X(x) \ + o\left(\sqrt{LLn/(n\gamma_n)}\right) \text{ a.s.},$$

from which we immediately conclude (1.17). □

2. Some Useful Coupling Inequalities

One of the essential tools in the proof of the Einmahl and Mason (1997) strong approximation was a result of Zaitsev (1987) on the rate of convergence in the central limit theorem for sums of multidimensional random vectors. Zaitsev's result led to an important coupling inequality for the empirical process especially useful for obtaining a strong approximation to the local empirical process under optimal conditions when the indexing class is uniformly bounded. We shall establish a coupling inequality that will allow us to obtain a useful strong approximation when the indexing class is not uniformly bounded. (Note that throughout $|\cdot|$ always denotes the Euclidean norm.)

Proposition 1. *Let* $X_1, \ldots, X_n, n \geq 1$, *be independent random vectors in* R^d *with common covariance matrix* $C = B^2$. *Let* σ *be the largest eigenvalue of* B. *Given* $p > 2, 0 < \gamma < \sigma$ *and* $x > 0$, *one can construct* d-*dimensional standard normal random vectors* $Z_1, \ldots, Z_n$ *such that*

$$P\left\{\max_{1 \leq j \leq n} |\sum_{i=1}^{j}(X_i - B \cdot Z_i)| \geq c_1\sqrt{n}\sigma + x\right\}$$

$$\leq c_2\{\exp(-x^2/576n\gamma^2) + (\sigma/\gamma)^p x^{-p} \sum_{i=1}^{n} E|X_i|^p\},$$

where c_1, c_2 *are positive constants depending only on* p *and* d.

In order to prove our proposition, we shall need the following two inequalities.

Inequality 1. *Let* $X_1, \ldots, X_n, n \geq 1$, *be independent* B-*valued random variables such that for some* $p > 2$, $E\|X_i\|^p < \infty, 1 \leq i \leq n$. *Suppose that*

$$E\|\sum_{j=k}^{m} X_j\| \leq \mu_n, 1 \leq k \leq m \leq n, \tag{2.1}$$

Then we have for all $t > 0$,

$$P\left\{\max_{1 \leq k \leq n} \|\sum_{i=1}^{k} X_i\| \geq t + 38p^2\mu_n\right\}$$

$$\leq 22\exp(-t^2/144\Lambda_n) + c\sum_{j=1}^{n} E\|X_j\|^p/t^p, \tag{2.2}$$

where $\Lambda_n = \sup\{\sum_{i=1}^{n} Ef^2(X_i) : f \in B^*, \|f\| \leq 1\}$ *and* c *is a positive constant depending only on* p.

Inequality 2. *Let* $Y_1, \ldots, Y_n$ *be centered Gaussian random variables taking values in a separable Banach space. Then we have for all* $t > 0$,

$$P\left\{\|\sum_{i=1}^{n} Y_i\| \geq 2E\|\sum_{i=1}^{n} Y_i\| + t\right\} \leq \exp(-t^2/2\Lambda_n), \tag{2.3}$$

where $\Lambda_n = \sup\{\sum_{i=1}^{n} Ef^2(Y_i) : f \in B^*, \|f\| \leq 1\}$.

Inequality 1 is a Banach space version of an inequality of Fuk-Nagaev due to Einmahl (1993) (see his Lemma 3) and Inequality 2 follows readily from Lemma 3.1 in Ledoux and Talagrand (1991).

We shall also require the following consequence of Proposition 1 of Einmahl (1987).

Fact. *Let $X_1, \ldots, X_n, n \geq 1$, be independent d-dimensional random vectors with zero means and each with identity covariance matrix I. If the underlying probability space is rich enough one can construct independent d-dimensional standard normal random vectors $Z_1, \ldots, Z_n$ such that*

$$P\left\{ \max_{1 \leq j \leq n} \left| \sum_{i=1}^{j} (X_i - Z_i) \right| \geq \bar{c}_1 \sqrt{n} + x \right\} \leq \bar{c}_2 x^{-p} \sum_{i=1}^{n} E|X_i|^p, \qquad (2.4)$$

where c_1 and c_2 are constants depending on p and d only.

Statement (2.4) follows after a little analysis by setting $s = p$, $\delta = c_{34}^{-1}(\bar{c}_1 \sqrt{n} + x)$ and choosing $\bar{c}_1$ large enough so that $\bar{c}_1/c_{34} \geq d^{1/(p-2)}$ in Proposition 1 of Einmahl (1987).

Proof of the Proposition 1
Step (i): Let $\sigma = \sigma_1 \geq \sigma_2 \geq \ldots \geq \sigma_d$ be the eigenvalues of the matrix B arranged in a nonincreasing order and taking into account multiplicities. Let σ_{d_1} be the smallest eigenvalue which is greater than or equal to γ. We then have

$$\sigma_{d_1} \geq \gamma \qquad (2.5)$$

and if $d > d_1$,

$$\sigma_i < \gamma, d_1 < i \leq d. \qquad (2.6)$$

Let A be the orthogonal matrix for which

$$\text{cov}(A \cdot X_i) = ACA' = D^2, 1 \leq i \leq n, \qquad (2.7)$$

where $D = \text{diag}(\sigma_1, \ldots, \sigma_d)$.

Further let A_1 be the $d_1 \times d$ matrix which consists of the first d_1 rows of A, and set

$$Y_i := D_1^{-1} \cdot A_1 \cdot X_i, 1 \leq i \leq n, \qquad (2.8)$$

where $D_1 = \text{diag}(\sigma_1, \ldots, \sigma_{d_1})$.

Clearly, $Y_1, \ldots, Y_n$ are independent d_1-dimensional random vectors with mean zero and $\text{cov}(Y_i) = I_1 =(d_1$-dimensional identity matrix$)$, $1 \leq i \leq n$. Now applying our Fact, we can find independent d_1-dimensional standard normal random vectors $Z_1', \ldots, Z_n'$ so that

$$P\left\{ \max_{1 \leq j \leq n} \left| \sum_{i=1}^{j} (Y_i - Z_i') \right| \geq \bar{c}_1 \sqrt{n} + \frac{x}{\sigma \sqrt{2}} \right\} \leq \bar{c}_2 \sqrt{2}^p \sigma^p \sum_{i=1}^{n} E|Y_i|^p/x^p,$$

which by using (2.5) and (2.8) is

$$\leq \bar{c}_2 \sqrt{2}^p (\sigma/\gamma)^p \sum_{i=1}^{n} E|X_i|^p/x^p. \qquad (2.9)$$

Using the trivial inequality

$$\max_{1 \leq j \leq n} \left| \sum_{i=1}^{j} D_1 (Y_i - Z_i') \right| \leq \sigma \max_{1 \leq j \leq n} \left| \sum_{i=1}^{j} (Y_i - Z_i') \right| \qquad (2.10)$$

we readily obtain from (2.8) and (2.9)

$$P\left\{\max_{1\leq j\leq n}|\sum_{i=1}^{j}(A_1\cdot X_i - D_1\cdot Z_i')| \geq \bar{c}_1\sigma\sqrt{n} + \frac{x}{\sqrt{2}}\right\}$$

$$\leq \bar{c}_2\sqrt{2}^p(\sigma/\gamma)^p\sum_{i=1}^{n} E|X_i|^p/x^p. \tag{2.11}$$

Step (ii): This step is necessary if $d_1 < d$. If $d_1 = d$, we can go immediately to step (iii) below.

Assuming $d_1 < d$, let A_2 be the $(d-d_1)\times d$ matrix which consists of the last $d - d_1$ rows of A. Note that, in view of (2.6), all the eigenvalues of $\text{cov}(A_2\cdot X)$ are bounded by γ^2.

Further, note that

$$E|\sum_{i=1}^{n} A_2 X_i| \leq \left(E(\sum_{i=1}^{n} X_i)^2\right)^{1/2} \leq \sigma\sqrt{dn}.$$

Let $Z_1'', \ldots, Z_n''$ be independent $d-d_1$-dimensional standard normal random vectors and let $D_2 = \text{diag}(\sigma_{d_1+1}, \ldots, \sigma_d)$. Then using Inequalities 1 and 2, along with the last inequality, it is easy to see that

$$P\left\{\max_{1\leq j\leq n}|\sum_{i=1}^{j}(A_2\cdot X_i - D_2\cdot Z_i'')| \geq (38p^2 + 2)\sigma\sqrt{nd} + \frac{x}{\sqrt{2}}\right\} \tag{2.12}$$

$$\leq P\left\{\max_{1\leq j\leq n}|\sum_{i=1}^{j} A_2\cdot X_i| \geq 38p^2\sigma\sqrt{nd} + \frac{x}{2}\right\}$$

$$+P\left\{\max_{1\leq j\leq n}|\sum_{i=1}^{j} D_2\cdot Z_i''| \geq 2\sigma\sqrt{nd} + \frac{x}{5}\right\}$$

$$\leq 22\exp(-x^2/576\gamma^2) + c2^p\sum_{i=1}^{n} E||X_i||^p/x^p + 2\exp(-x^2/50\gamma^2).$$

Step (iii): We are now ready to complete the proof. Let $\overline{Z}_i$ be the d-dimensional random vector which is obtained by combining the components of Z_i' and Z_i'', $1 \leq i \leq n$. We can assume that Z_i' and Z_i'' are independent of each other, and we thus have obtained independent d-dimensional standard normal random vectors. Set

$$Z_i = A'\cdot\overline{Z}_i, 1 \leq i \leq n.$$

Since A is orthogonal, we still have standard normal random vectors. Moreover, it is easy to see that with $B = A'DA$,

$$P\left\{\max_{1\leq j\leq n}|\sum_{i=1}^{j}(X_i - B\cdot Z_i)| \geq c_1\sigma\sqrt{n} + x\right\} \tag{2.13}$$

$$= P\left\{\max_{1\leq j\leq n}|\sum_{i=1}^{j}(A\cdot X_i - D\cdot \overline{Z}_i)|\geq c_1\sigma\sqrt{n}+x\right\}$$

$$\leq P\left\{\max_{1\leq j\leq n}|\sum_{i=1}^{j}(A_1\cdot X_i - D_1\cdot Z_i')|\geq c_1\sigma\sqrt{n/2}+x/\sqrt{2}\right\}$$

$$+P\left\{\max_{1\leq j\leq n}|\sum_{i=1}^{j}(A_2\cdot X_i - D_2\cdot Z_i'')|\geq c_1\sigma\sqrt{n/2}+x/\sqrt{2}\right\}.$$

Recalling (2.11) and (2.12) it is now obvious that the assertion of Proposition 1 holds for a suitable choice of c_1. □

For $f,g\in\mathcal{F}$ define for any $n\geq 0$, $\rho_n^2(f,g) = Var(f(Y_1^{(n)}) - g(Y_1^{(n)}))$ and $e_n^2(f,g) = E(f(Y_1^{(n)}) - g(Y_1^{(n)}))^2$. Further for any $n\geq 0$ and $\delta > 0$ set

$$\mathcal{F}_n'(\delta) = \{(f,g)\in\mathcal{F}\times\mathcal{F} : \rho_n(f,g)\leq\delta\}$$

and for any real valued functional on $\mathcal{F}$ set

$$||T||_{\mathcal{F}_n'(\delta)} = \sup\{|T(f)-T(g)| : \rho_n(f,g)\leq\delta\}.$$

Proposition 2. *Let $\mathcal{F}$ be any class of functions satisfying (S.i), (S.ii) and (S.iii) and $\{P_n\}_{n\geq 1}$ be a sequence of probability measures on $(R^d,\mathcal{B})$ for which (F.ix) holds for a probability measure P_0. Assume further that $\mathcal{F}$ is P_0−pregaussian. Then given any $0 < \delta < 1$ and sequence of positive integers $\{k_n\}_{n\geq 1}$ there exists an $n(\delta) > 0$ such that for every $n\geq n(\delta)$ and $u > 0$ one can construct empirical processes $(\alpha_m^{(n)}(f))_{f\in\mathcal{F}}, 1\leq m\leq k_n$ and independent P_0−Brownian Bridges $(\overline{B}_m(f))_{f\in\mathcal{F}}, 1\leq m\leq k_n$ on the same probability space such that*

$$P\left\{\max_{1\leq m\leq k_n}||\sqrt{m}\alpha_m^{(n)} - \sum_{i=1}^{m}\overline{B}_i||_\mathcal{F}\geq u+\beta_{n,k_n}(\delta)\right\} \qquad (2.14)$$

$$\leq K_1\exp(-u^2/A_0^2 k_n\delta^2) + K_2 k_n P_0 F^p u^{-p},$$

where $\beta_{n,k_n}(\delta) := K_3\sqrt{k_n} + A_1(\sqrt{k_n}(E||\overline{B}||_\mathcal{F} + E||\alpha_{k_n}^{(n)}||_{\mathcal{F}_n'(\delta)}))$, $K_1 = K_1(\delta)$, $K_2 = K_2(\delta)$ and $K_3 = K_3(\delta)$ are constants depending on δ, and A_0 and A_1 are absolute constants.

Proof. Our proof closely parallels that of Proposition 2.2 of Einmahl and Mason (1997). First note that since $\mathcal{F}$ is P_0−pregaussian, it is totally bounded with repect to ρ_0. Condition (F.i) permits us to find for any $0 < \delta < 1$ a subclass $\{f_1,\ldots,f_r\}$ of $\mathcal{F}$ where r depends on δ such that for all $n\geq n_1(\delta)$, for some $n_1(\delta) > 0$,

$$\min_{1\leq i\leq r}\sup_{f\in\mathcal{F}}\rho_n(f,f_i)\leq\delta. \qquad (2.15)$$

Thus if the sequence $(\alpha_m^{(n)}(f))_{f\in\mathcal{F}}, 1\leq m\leq k_n, n\geq n_1(\delta)$, is given, we set for $1\leq j\leq k_n$,

$$X_j^{(i)} = f_i(Y_j^{(n)}) - P_n(f_i), \quad i=1,\ldots r,$$

and we clearly have for $1 \leq m \leq k_n$

$$(\sqrt{m}\alpha_m^{(n)}(f_1), \ldots, \sqrt{m}\alpha_m^{(n)}(f_r)) = (\sum_{j=1}^{m} X_j^{(1)}, \ldots, \sum_{j=1}^{m} X_j^{(r)}). \qquad (2.16)$$

Define for $1 \leq j \leq k_n$ the random r−vectors

$$X_j = (X_j^{(1)}, \ldots, X_j^{(r)})'.$$

Using Proposition 1 we couple standard normal random r−vectors $Z_1, \ldots, Z_{k_n}$ on the same probability space as the random r−vectors $X_1, \ldots, X_{k_n}$ such that

$$P\left\{ \max_{1 \leq m \leq k_n} |\sum_{i=1}^{m}(X_i - B_{(n)} \cdot Z_i)| \geq c_1\sqrt{k_n}\sigma(n) + u/4 \right\} \qquad (2.17)$$

$$\leq c_2\left\{ \exp(-(u/4)^2/576k_n\gamma(n)^2) + (\sigma(n)/\gamma(n))^p(u/4)^{-p}\sum_{i=1}^{k_n} E|X_i|^p \right\},$$

where $C_{(n)} =: B_{(n)}^2$ is the covariance matrix of X_1, $\sigma(n)$ is the largest eigenvalue of $B_{(n)}$, $0 < \gamma(n) < \sigma(n)$ and c_1, c_2 depend only on p and r.

By (F.ix) the right hand side of the above inequality is less than or equal to

$$c_2\left\{ \exp(-(u/4)^2/576k_n\gamma(n)^2) + (\sigma(n)/\gamma(n))^p(u/4)^{-p}c_3k_nP_0F^p \right\} \qquad (2.18)$$

for some constant $c_3 = c_3(p, r)$. Now let

$$C := \operatorname{cov}(\overline{B}(f_1), \ldots, \overline{B}(f_r))'.$$

Since

$$(B \cdot Z_i)_{1 \leq i \leq k_n} =_d ((\overline{B}_i(f_1), \ldots, \overline{B}_i(f_r))')_{1 \leq i \leq k_n},$$

where $C = B^2$ and $\overline{B}_1, \ldots \overline{B}_{k_n}$ are i.i.d. P_0−Brownian bridges, we can assume without loss of generality that

$$B \cdot Z_i = (\overline{B}_i(f_1), \ldots, \overline{B}_i(f_r))', \quad i = 1, \ldots, k_n.$$

Now since $C_{(n)} \to C$ as $n \to \infty$, we have $\sigma(n) \to \sigma$ as $n \to \infty$, where σ is the largest eigenvalue of B, and hence for all $n \geq n_2(\delta)$, for some $n_2(\delta) > 0$, $\sigma(n) \leq \sigma(1 + \delta)$. Thus for all $n \geq n_1(\delta) \vee n_2(\delta)$, we see from (2.18) by choosing $\gamma(n) = \min\{\delta, \sigma(n)/2\}$ and $K_3 > c_1\sigma(1 + \delta)$ that

$$\Delta_1 := P\left\{ \max_{1 \leq m \leq k_n} |\sum_{i=1}^{m}(X_i - B_{(n)} \cdot Z_i)| \geq K_3\sqrt{k_n} + u/4 \right\} \qquad (2.19)$$

$$\leq c_2\{\exp(-(u/4)^2/576k_n\delta^2) + c_3(8 + 4\sigma(1 + \delta)/\delta)^pk_nP_0F^pu^{-p}\}.$$

Also since $B_{(n)} \to B$,as $n \to \infty$, we have using symmetry of the Z_i's and a standard bound on the tail of the normal distribution that for all $u > 0$ and $n \geq n_3(\delta)$, for some $n_3(\delta) > 0$

$$\Delta_2 := P\left\{\max_{1 \leq m \leq k_n} |\sum_{i=1}^{m}(B \cdot Z_i - B_{(n)} \cdot Z_i)|_+ \geq u/4\right\} \leq r\exp(-u^2/\delta^2 k_n), \quad (2.20)$$

where $|\cdot|_+$ denotes the maximal norm on R^r.

Finally, by (F.i) we have for all $n \geq n(\delta) = n_1(\delta) \vee n_2(\delta) \vee n_3(\delta) \vee n_4(\delta)$ for some $n_4(\delta) > 0$, $\mathcal{F}'_n(\delta) \subset \mathcal{F}'_0(2\delta)$ and thus

$$P\left\{\max_{1 \leq m \leq k_n} ||\sqrt{m}\alpha_m^{(n)} - \sum_{i=1}^{m}\overline{B}_i||_{\mathcal{F}} \geq u + \beta_{n,k_n}(\delta)\right\} \leq \Delta_1 + \Delta_2 \quad (2.21)$$

$$+P\left\{\max_{1 \leq m \leq k_n} ||\sqrt{m}\alpha_m^{(n)}||_{\mathcal{F}'_n(\delta)} \geq u/4 + A_1\sqrt{k_n}E||\alpha_{k_n}^{(n)}||_{\mathcal{F}'_n(\delta)}\right\}$$

$$+P\left\{\max_{1 \leq m \leq k_n} ||\sum_{i=1}^{m}\overline{B}_i||_{\mathcal{F}'_0(2\delta)} \geq u/4 + A_1\sqrt{k_n}E||\overline{B}||_{\mathcal{F}}\right\}$$

$$=: \Delta_1 + \Delta_2 + \Delta_3 + \Delta_4.$$

Using (F.ix) and Inequality 1 we see that with c_3 as in (2.18)

$$P\left\{\max_{1 \leq m \leq k_n} ||\sqrt{m}\alpha_m^{(n)}||_{\mathcal{F}'_n(\delta)} \geq u/4 + 38p^2E||\sqrt{k_n}\alpha_{k_n}^{(n)}||_{\mathcal{F}'_n(\delta)}\right\}$$

$$\leq 22\exp(-(u/4)^2/144k_n\delta^2) + 4^p c_3 k_n P_0 F^p/u^p.$$

Therefore as long as $A_1 \geq 38p^2$

$$\Delta_3 \leq 22\exp(-(u/4)^2/144k_n\delta^2) + 4^p c_3 P_0 F^p/u^p. \quad (2.22)$$

Finally, using Inequality 2, along with Lévy's inequality, Lemma 3.1.11 of Dudley (1984), we find that

$$\Delta_4 \leq 2\exp(-u^2/128k_n\delta^2), \quad (2.23)$$

provided we have chosen $A_1 \geq 2$. Combining (2.19), (2.20), (2.22) and (2.23) we get (2.14). $\qquad\square$

3. Proofs of Main Results

Given any integer $n \geq 1$ and invertible bimeasurable tranformation $h_n : R^d \to R^d$, let $Y_1^{(n)}, Y_2^{(n)}, \ldots,$ be i.i.d. P_n. Set for $f \in \mathcal{F}, n \geq 1$ and $j \geq 1$

$$S_j^{(n)}(f) = \sum_{i=1}^{j} f(h_n^{-1}(X_i - x)) - ja_n P_n(f). \quad (3.1)$$

Proof of Theorem 2

Following with obvious modifications the same steps and lemmas that yielded Proposition 3.3 of Einmahl and Mason (1997), replacing the use of their Proposition 2.2 in Lemma 3.5 by our Proposition 2, it is straightforward to obtain the following proposition.

Proposition 3. *Under the assumptions of Theorem 2, given $0 < \delta < 1$, there exists an $m(\delta) > 0$ such that for each $n \geq m(\delta)$, one can construct independent P_0-Brownian motions $\overline{W}_i$, $1 \leq i \leq n$, indexed by $\mathcal{F}$ so that*

$$P\left\{\max_{1 \leq j \leq n} \|S_j^{(n)}(f) - \sqrt{b_n} \sum_{i=1}^{j} \overline{W}_i\|_{\mathcal{F}} \geq D(\delta\sqrt{na_n LLn} + \widetilde{\gamma}_n + D_1\widetilde{\beta}_{n,[na_n]})\right\}$$

$$\leq D_2(na_n \exp(-D_3\sqrt{na_n LLn}) + (Ln)^{-1.5}$$

$$+ (Ln)^{-(\delta/a_n)^2} + (na_n)^{1-p/2}(LLn)^{-p/2}), \qquad (3.2)$$

where b_n is as in (F.iv), $D \geq 1$ is an absolute constant, $D_1 = D_1(\delta)$, $D_2 = D_2(\delta)$ and $D_3 = D_3(\delta)$,

$$\widetilde{\gamma}_n = 12E\|\sqrt{l_n}\alpha_{l_n}^{(n)}\|_{\mathcal{F}}, \qquad (3.3)$$

with $l_n = [4\sqrt{na_n LLn}]$, and

$$\widetilde{\beta}_{n,[na_n]} = \sqrt{na_n}\{E\|\alpha_{[na_n]}^{(n)}\|_{\mathcal{F}} + E\|B_0\|_{\mathcal{F}} + E\|W_0\|_{\mathcal{F}}\}. \qquad (3.4)$$

The proof of Theorem 2 proceeds from Proposition 3 using a now standard blocking argument due to Major (1976). For details refer to Einmahl and Mason (1997) to see how their Theorem 1.2 follows from their Proposition 3.3. In our present situation assumption (p) comes into play.

Proof of Corollary 2

Corollary 2 is derived from Theorem 2 in the same way that Corollary 1.1 of Einmahl and Mason (1997) follows from their Theorem 1.2. However their Lemma 3.10 must be replaced by the following lemma whose proof is a minor modification of that lemma.

Lemma 1. *Let $\mathcal{F}$ be a class of measurable functions with support in J, totally bounded for ρ_0 and satisfying (F.iii). In addition assume that (F.vii), (F.viii) and (F.ix) hold for a given sequence of bimeasurable invertible transformations $h_n : R^d \to R^d$, then*

$$\limsup_{n \to \infty} \sup_{f \in \mathcal{F}} \max_{n_k \leq n \leq n_{k+1}} e_0(f_{n,k}, f) \leq \delta(q), \qquad (3.5)$$

with $f_{n,k} = f \circ h_n^{-1} \circ h_{n_k}$, where $\delta(q) \to 0$ as $q \downarrow 1$.

References

Deheuvels, P. and Mason, D.M. (1994). Functional laws of the iterated logarithm for local empirical processes indexed by sets. *Ann. Probab.* **22** 1619–1661.

Deheuvels, P. and Mason, D.M. (1995). Nonstandard local empirical processes indexed by sets. *J. Stat. Plan. Inf.* **45** 91–112.

Dudley, R.M. (1984). A course on empirical processes (École d'Été de Probabilités de Saint-Flour XII-1982) *Lecture Notes in Mathematics* **1097** 2–141, (ed. P.L. Hennequin) Springer-Verlag, New York

Einmahl, U. (1987). A useful estimate in the multidimensional invariance principle. *Probab. Th. Re. Fields* **76** 81–101.

Einmahl, U. (1993). Toward a general law of the iterated logarithm in Banach space. *Ann. Probab.* **21** 2012–2045.

Einmahl, U. and MASON, D.M. (1997). Gaussian approximation of local empirical processes indexed by functions. *Probab. Th. Re. Fields* **107** 283–301.

Giné, E. and ZINN, J. (1984). Some limit theorems for empirical processes. *Ann. Probab.* **12** 929–989.

Haerdle, W. (1984). A law of the iterated logarithm for nonparametric regression function estimators. *Ann. Statist.* **12** 624–635.

Hall, P. (1981). Laws of the iterated logarithm for nonparametric density estimators. *Z. Wahrsch. verw. Gebiete.* **56** 47–61.

Hong, S. Y. (1992). A law of the iterated logarithm for neighour-type regression function estimator. *Chin. Ann. Math.* **13** 180–195.

Komlós, J., Major, P. and Tusnády, G. (1975). An approximation of partial sums of independent rv's and the sample df I. *Z. Wahrsch. verw. Gebiete.* **32** 111–131.

Ledoux, M. and Talagrand, M. (1991). *Probability on Banach Spaces*, Springer-Verlag, Berlin.

Mack, Y.P. and Silverman, B.W. (1982). Weak and strong uniform consistency of kernel regression estimates. *Z. Wahrsch. verw. Gebiete.* **61** 405–415.

Major, P. (1976). Approximations of partial sums of i.i.d. r.v.s when the summands have only two moments. *Z. Wahrscheinlichkeitstheorie verw. Gebiete* **35** 221–229.

Nadaraya, E.A. (1964). On estimating regression. *Theor. Probab. Appl.* **9** 141–142.

Pollard, D. (1982). *Rates of strong uniform convergence*, Preprint

Pollard, D. (1984). *Convergence of Stochastic Processes.* Springer-Verlag, New York.

Tusnády, G. (1977). A remark on the approximation of the sample df in the multidimensional case. *Period. Math. Hung.* **8** 53–55.

Van der Vaart, A.W. and Wellner, J.A. (1996). *Weak Convergence and Empirical Processes.* Springer-Verlag, New York.

Watson, G.S. (1964). Smooth regression analysis. *Sankhyā* A **26** 359–372.

Wichura, M.J. (1974). Some Strassen-type laws of the iterated logarithm for multiparameter stochastic processes with independent increments. *Ann. Probab.* **1** 272–296.

Zaitsev, A.Yu. (1987). On the Gaussian approximation of convolutions under multidimensional analogues of S.N. Bernstein's inequality conditions. *Probab. Th. Re. Fields* 74, 534–566.

Uwe Einmahl
Departement Wiskunde
Vrije Universiteit Brussel
Pleinlaan 2
1050 Brussels, Belgium

David M. Mason
Department of Mathematical Sciences
501 Ewing Hall
University of Delaware
Newark, DE 19716, USA

Progress in Probability, Vol. 43
© 1998 Birkhäuser Verlag Basel/Switzerland

On Random Measure Processes with Application to Smoothed Empirical Processes

PETER GAENSSLER, DANIEL ROST AND KLAUS ZIEGLER

1. Introduction

We consider function-indexed so-called Random Measure Processes (RMP's) and focus especially on a uniform law of large numbers (ULLN) for RMP's. Demonstrating both its power and its generality we apply it to derive a ULLN for smoothed empirical processes covering a former result by Yukich (1989). Finally we also present a functional central limit theorem (FCLT) for smoothed empirical processes under conditions different from those found in the literature.

The context is as follows: Let $X = (X, \mathcal{X})$ be an arbitrary measurable space (in Section 3 we shall then specialize to a linear metric space X with its Borel σ-algebra $\mathcal{X}$) and let $\mathcal{F}$ be a class of measurable functions $f : X \longrightarrow \mathbb{R}$ with measurable envelope $F : X \longrightarrow \mathbb{R}$, i.e. $\sup_{f \in \mathcal{F}} |f(x)| \leq F(x)$ for all $x \in X$. Let $(w_{nj})_{1 \leq j \leq j(n)}, n \in \mathbb{N}$, be a triangular array of random probability measures on $(X, \mathcal{X})$ and $(\xi_{nj})_{1 \leq j \leq j(n)}, n \in \mathbb{N}$, be a triangular array of real-valued random variables. We define Random Measure Processes (indexed by $\mathcal{F}$) $S_n = (S_n(f))_{f \in \mathcal{F}}$ by

$$S_n(f) := \sum_{j \leq j(n)} w_{nj}(f) \cdot \xi_{nj} \text{ for } f \in \mathcal{F}, \tag{1.1}$$

where $w_{nj}(f) := \int_X f \, d \, w_{nj}$. We tacitly assume regularity conditions such as measurability and finiteness of $w_{nj}(F), 1 \leq j \leq j(n), n \in \mathbb{N}$.
We assume throughout that

$$\text{for all } n \in \mathbb{N} \text{ the sequence of pairs}$$
$$(w_{n1}, \xi_{n1}), \ldots, (w_{nj(n)}, \xi_{nj(n)}) \text{ is independent.} \tag{1.2}$$

However, we do not assume that the pairs are identically distributed; also dependence within each pair is allowed.

In this general form such processes were first studied by Ziegler (1994). They typically occur in the area of

- partial-sum processes, where e.g. $X = [0,1]^d, d \geq 1, w_{nj} := \delta_{\underline{j}/n}$ and $\xi_{nj} = \frac{1}{n^d} \cdot \xi_{\underline{j}}$ with $\underline{j}$ ranging over the d-dimensional lattice $\mathbb{N}^d$

or

- empirical processes, where $w_{nj} := \delta_{\eta_{nj}}, \eta_{nj}$ being random elements in $(X, \mathcal{X})$ and $\xi_{nj} = \frac{1}{j(n)}$

and, as we shall see,

- smoothed empirical processes which will be studied in Section 3.

In this paper we shall focus especially on uniform laws of large numbers, i.e. we seek for conditions under which

$$\sup_{f \in \mathcal{F}} |S_n(f) - \mathbb{E}(S_n(f))| \xrightarrow{\text{L}_p} 0 \qquad (1.3)$$

holds with $p \geq 1$, where $\xrightarrow{\text{L}_p}$ denotes convergence with respect to the L_p-metric. The ULLN for RMP's is given by Theorem 2.1 in Section 2, its proof being deferred to Section 4. In Section 3 we consider smoothed empirical processes and show how they fit into the framework of RMP's. Thus, by a reformulation of Theorem 2.1 we obtain a ULLN for smoothed empirical processes. Its conditions are discussed and compared with those appearing in Yukich (1989). In addition we state a new FCLT for smoothed empirical processes. As to applications concerning ULLN's and FCLT's for empirical and partial-sum processes we refer to Ziegler (1994,1996) and Gaenssler and Ziegler (1994).

2. A ULLN for RMP's

In order to formulate our ULLN for RMP's we have to introduce some more notation.
Given S_n as in (1.1), for any $\delta > 0$ let

$$\mu_{n\delta} := \sum_{j \leq j(n)} w_{nj} \cdot |\xi_{nj}| \cdot I(w_{nj}(F)|\xi_{nj}| \leq \delta)$$

and let $d_{\mu_{n\delta}}^{(1)}$ be the random L_1-pseudometric on $\mathcal{F}$ defined by

$$d_{\mu_{n\delta}}^{(1)}(f,g) := \mu_{n\delta}(|f-g|) \text{ for } f,g \in \mathcal{F}.$$

Finally, for any $\varepsilon > 0$ let

$$N(\varepsilon, \mathcal{F}, d_{\mu_{n\delta}}^{(1)}) := \min\{m \in \mathbb{N} : \exists f_1, \ldots, f_m \in \mathcal{F} \text{ such that for all } f \in \mathcal{F}$$
$$d_{\mu_{n\delta}}^{(1)}(f, f_i) < \varepsilon \text{ for some } f_i\}$$

and let

$$H(\cdot, \mathcal{F}, d_{\mu_{n\delta}}^{(1)}) := \log N(\cdot, \mathcal{F}, d_{\mu_{n\delta}}^{(1)})$$

be the metric entropy of $(\mathcal{F}, d_{\mu_{n\delta}}^{(1)})$.

In order to avoid measurability questions (which can be dealt with) we assume here for simplicity that the index set $\mathcal{F}$ is countable.

Theorem 2.1 *Assume that* (2.1)–(2.3) *hold, where*

$$\lim_{n \to \infty} \sum_{j \leq j(n)} \mathbb{E}^{\frac{1}{p}}\left(w_{nj}(F)^p \cdot |\xi_{nj}|^p \cdot I(w_{nj}(F)|\xi_{nj}| > \delta)\right) = 0 \text{ for all } \delta > 0 \qquad (2.1)$$

$$\sup_{n \in \mathbb{N}} \sum_{j \leq j(n)} \mathbb{E}\Big(w_{nj}(F) \cdot |\xi_{nj}| \cdot I(w_{nj}(F)|\xi_{nj}| \leq \delta_1) \Big) < \infty \text{ for some } \delta_1 > 0 \quad (2.2)$$

$$\text{For all } \tau > 0 \text{ there exists } \delta \equiv \delta(\tau) > 0 \text{ such that}$$
$$\Big(H(\tau \cdot \mu_{n\delta}(F), \mathcal{F}, d^{(1)}_{\mu_{n\delta}}) \Big)_{n \in \mathbb{N}} \text{ is stochastically bounded.} \quad (2.3)$$

Then

$$\sup_{f \in \mathcal{F}} |S_n(f) - \mathbb{E}(S_n(f))| \xrightarrow{L_p} 0 .$$

$\Big(H(\tau \cdot \mu_{n\delta}(F), \mathcal{F}, d^{(1)}_{\mu_{n\delta}}) \Big)_{n \in \mathbb{N}}$ *stochastically bounded means that for all $\rho > 0$ there exists an $M \equiv M(\rho) < \infty$ such that*

$$\limsup_{n \to \infty} \mathbb{P}^* \Big(H(\tau \cdot \mu_{n\delta}(F), \mathcal{F}, d^{(1)}_{\mu_{n\delta}}) > M \Big) < \rho$$

(where $\mathbb{P}^*$ denotes outer probability).

In proving Theorem 2.1 we apply Hoffmann-Jørgensen's inequality, symmetrization techniques and a maximal inequality for Rademacher averages. These tools as well as the proof of Theorem 2.1 are presented in Section 4.

3. A ULLN and a FCLT for Smoothed Empirical Processes

Now let X be a linear metric space with $\mathcal{X}$ denoting its Borel σ-algebra and let $\eta_j, j \in \mathbb{N}$, be i.i.d. random elements taking its values in $(X, \mathcal{X})$. Let ν denote the law of η_1 and ν_n be the empirical measure based on $\eta_1, \dots, \eta_n$, i.e. $\nu_n := \frac{1}{n} \sum_{j \leq n} \delta_{\eta_j}$. ν_n is the non-parametric maximum likelihood estimator for ν. If the underlying measure ν is "smooth" it is natural to use a "smoothed" version $\tilde{\nu}_n$ of ν_n as an estimator for ν, rather than the empirical measure itself. Following Yukich (1989) we consider "smoothing through convolution" as follows:
Let $\mu_n, n \in \mathbb{N}$, be a sequence of probability measures on $(X, \mathcal{X})$ with $\mu_n \longrightarrow \delta_0$ weakly and let

$$\tilde{\nu}_n := \nu_n \star \mu_n$$

be the so-called "smoothed" version of ν_n. The corresponding "smoothed empirical process" (indexed by $\mathcal{F}$) $\tilde{\nu}_n \equiv (\tilde{\nu}_n(f))_{f \in \mathcal{F}}$ is then defined by

$$\tilde{\nu}_n(f) := \nu_n \star \mu_n(f) = \int_X \int_X f(y + z) \mu_n(dy) \nu_n(dz) = \frac{1}{n} \sum_{j \leq n} \int_X f(y + \eta_j) \mu_n(dy).$$
$$(3.1)$$

(3.1) also includes kernel smoothing by choosing (let us take $X = \mathbb{R}$ for simplicity) $\mu_n((-\infty, t]) := W(\frac{t}{b_n})$ with $b_n > 0, b_n \to 0$ and $W(t) := \int_{-\infty}^t w(u) du$.
Considering $X = \mathbb{R}^k, k \in \mathbb{N}$, Winter (1973) and Yamato (1973) were the first who obtained ULLN-results for smoothed empirical processes (see also Prakasa Rao (1981)). From Yukich (1989) we know the following result in the case $X = \mathbb{R}^k$:

Theorem 3.1 (Yukich) *Let $X = \mathbb{R}^k, k \in \mathbb{N}$, and assume that*

$$\mathcal{F} \text{ is uniformly bounded (i.e. } \sup_{f \in \mathcal{F}} \sup_{x \in X} |f(x)| \leq K < \infty) \quad (3.2)$$

and

$$N^{[]}(\tau, \mathcal{F}, \nu) < \infty \quad \text{for all } \tau > 0 \tag{3.3}$$

where $N^{[]}(\tau, \mathcal{F}, \nu) := \min\{m \in \mathbb{N} : \exists\, f_1, \ldots, f_m \in L_1(\nu), f_i \text{ continuous, such that}$
for all $f \in \mathcal{F}$ *there exist* f_i, f_j *with* $f_i \leq f \leq f_j$ *and* $\nu(f_j - f_i) < \tau\}$.
Then

$$\sup_{f \in \mathcal{F}} |\tilde{\nu}_n(f) - \nu(f)| \xrightarrow{L_p} 0. \tag{3.4}$$

A result like (3.4) *will be called a ULLN for smoothed empirical processes.*

Returning to our general setting it is basic to realize that the definition of $\tilde{\nu}_n$ (according to (3.1)) fits into our framework developed in Section 1: Not only ν_n, but $\tilde{\nu}_n$, too, is a RMP, i.e. $\tilde{\nu}_n = \sum_{j \leq n} w_{nj} \cdot \xi_{nj}$ with $\xi_{nj} := \frac{1}{n}, 1 \leq j \leq n$, and $w_{nj}(B) := \mu_n(B - \eta_j), B \in \mathcal{X}, 1 \leq j \leq n$ (note that $B - z \in \mathcal{X}$ for all $z \in X$ with X being a topological vector-space, and that (1.2) is satisfied, too). Therefore – in view of the decomposition

$$\tilde{\nu}_n - \nu = \tilde{\nu}_n - \nu \star \mu_n + \nu \star \mu_n - \nu$$

and the fact that $\mathbb{E}(\tilde{\nu}_n(f)) = \nu \star \mu_n(f)$ for all $f \in \mathcal{F}$ and $n \in \mathbb{N}$ – Theorem 2.1 immediately yields the following ULLN for smoothed empirical processes (using the notation of Section 2).

Theorem 3.2 *Let* X *be an arbitrary linear metric space and* F *be a measurable envelope of* $\mathcal{F}$ *as in Section 1. Assume that* (3.5)–(3.8) *hold, where*

$$\lim_{n \to \infty} \mathbb{E}\left(w_{n1}(F)^p \cdot I(\frac{1}{n} w_{n1}(F) > \delta)\right) = 0 \text{ for all } \delta > 0 \tag{3.5}$$

$$\sup_{n \in \mathbb{N}} \mathbb{E}\left(w_{n1}(F) \cdot I(\frac{1}{n} w_{n1}(F) \leq \delta_1)\right) < \infty \text{ for some } \delta_1 > 0 \tag{3.6}$$

For all $\tau > 0$ *there exists* $\delta \equiv \delta(\tau) > 0$ *such that*

$$\left(H(\tau \cdot \mu_{n\delta}(F), \mathcal{F}, d^{(1)}_{\mu_{n\delta}})\right)_{n \in \mathbb{N}} \text{ is stochastically bounded.} \tag{3.7}$$

$$\sup_{f \in \mathcal{F}} |\nu \star \mu_n(f) - \nu(f)| \longrightarrow 0. \tag{3.8}$$

Then

$$\sup_{f \in \mathcal{F}} |\tilde{\nu}_n(f) - \nu(f)| \xrightarrow{L_p} 0.$$

In the case of uniformly bounded $\mathcal{F}$ we have $\{\frac{1}{n} w_{n1}(F) > \delta\} = \emptyset$ and $\{\frac{1}{n} w_{n1}(F) \leq \delta\} = \Omega$ for all $\delta > 0$ and large enough n, hence (3.5) and (3.6) are fulfilled, and since $\mu_{n\delta} = \tilde{\nu}_n$ for each $\delta > 0$ and large enough n, we have

$$\sup_{f \in \mathcal{F}} |\tilde{\nu}_n(f) - \nu(f)| \xrightarrow{L_p} 0$$

under the only conditions (3.8) and

$$\left(H(\tau, \mathcal{F}, d^{(1)}_{\tilde{\nu}_n})\right)_{n \in \mathbb{N}} \text{ stochastically bounded for all } \tau > 0. \tag{3.9}$$

Now it is known that (3.9) is fulfilled when $\mathcal{F}$ is a uniformly bounded VC subgraph class (for the definition of VC subgraph classes see e.g. v.d.Vaart and Wellner (1996)), and thus we get the following result:

Theorem 3.3 *Let X be an arbitrary linear metric space and $\mathcal{F}$ be a uniformly bounded VC subgraph class. Suppose (3.8) holds. Then*

$$\sup_{f\in\mathcal{F}}|\tilde{\nu}_n(f)-\nu(f)| \xrightarrow{\ \mathrm{L}_p\ } 0\ .$$

We remark here that under further regularity assumptions on the envelope function F of $\mathcal{F}$ a similar result can be obtained for not necessarily uniformly bounded $\mathcal{F}$.

Our ULLN-results for smoothed empirical processes are fairly general. For $X = \mathbb{R}^k$ and uniformly bounded $\mathcal{F}$, Theorem 3.1 (Yukich's result) follows according to the following Lemma:

Lemma 3.4 *Let $X = \mathbb{R}^k, k \in \mathbb{N}$. Suppose (3.2) and (3.3) are fulfilled. Then (3.8) and (3.9) hold true.*

Proof. Given any $\tau > 0$ let $f_1,\ldots,f_m$ be continuous, ν-integrable and bounded (note that $\mathcal{F}$ is assumed to be uniformly bounded) such that for all $f \in \mathcal{F}$ there exist f_i, f_j with $f_i \le f \le f_j$ and $\nu(f_j - f_i) < \tau$ (note that $N^{[]}(\tau,\mathcal{F},\nu) < \infty$). Now for all f_i, f_j with

$$[f_i, f_j] := \{f \in \mathcal{F} : f_i \le f \le f_j\} \neq \emptyset$$

choose a $g_{ij} \in [f_i, f_j]$. Then, given $f \in \mathcal{F}$ and f_i, f_j with $f \in [f_i, f_j]$ and $\nu(f_j-f_i) < \tau$, we have

$$\tilde{\nu}_n(|f - g_{ij}|) \le \tilde{\nu}_n(f_j - f_i)$$
$$= \nu_n \star \mu_n(f_j - f_i) \longrightarrow \nu \star \delta_0(f_j - f_i) \text{ a.s. } (n \to \infty)$$

since $\mu_n \longrightarrow \delta_0$ weakly and $\nu_n \longrightarrow \nu$ weakly a.s. Note that $f_j - f_i$ is bounded and continuous.

Since $\nu \star \delta_0(f_j - f_i) = \nu(f_j - f_i) < \tau$, it follows that

$$\limsup_{n\to\infty} H(\tau, \mathcal{F}, \delta_{\tilde{\nu}_n}^{(1)}) \le \log(m^2) \quad \text{a.s.}$$

So we conclude that (3.9) holds.

Now, from $f \in [f_i, f_j]$ and $\nu(f_j - f_i) < \tau$ we can also conclude that

$$|\nu \star \mu_n(f) - \nu(f)| \le \max\{|\nu \star \mu_n(f_j) - \nu(f_j)| + |\nu(f_j) - \nu(f)|,$$
$$|\nu \star \mu_n(f_i) - \nu(f_i)| + |\nu(f_i) - \nu(f)|\},$$

thus

$$\sup_{f\in\mathcal{F}}|\nu \star \mu_n(f) - \nu(f)| \le \max\{|\nu \star \mu_n(f_j) - \nu(f_j)| : 1 \le j \le m\} + \tau.$$

But $\nu \star \mu_n(f_j) - \nu(f_j) \longrightarrow 0$ $(n \to \infty)$ for all $j = 1, \ldots, m$, since $\nu \star \mu_n \longrightarrow \nu$ weakly and the f_j's are bounded and continuous. So we get

$$\limsup_{n \to \infty} \sup_{f \in \mathcal{F}} |\nu \star \mu_n(f) - \nu(f)| \leq \tau.$$

Since $\tau > 0$ was arbitrary, this gives (3.8). □

From the following example it is easily seen that condition (3.8) cannot be dispensed with in general. In fact, the example shows that (3.5)–(3.7) may hold, but not (3.8).

Example 3.5 Take $X = \mathbb{R}, \mathcal{F} = \{I(x \leq t) : t \in \mathbb{R}\}, \nu = \delta_0$ and $\mu_n := \delta_{\frac{1}{n}}, n \in \mathbb{N}$. Here (3.5)–(3.7) hold, but (3.8) does not!

Based on a FCLT for RMP's (cf. Ziegler (1996), 6.2), it is also possible to obtain the following FCLT for smoothed empirical processes. The conditions differ substantially from those appearing in Yukich (1992) and v.d.Vaart (1994).

Theorem 3.6 *Let X be an arbitrary linear metric space and let $\mathcal{F}$ have uniformly integrable entropy (see e.g. v.d.Vaart and Wellner (1996) (2.1.7) for a definition). Assume that the following conditions (3.10)–(3.12) are fulfilled:*

For each $\rho > 0$ there exists $\delta_n \equiv \delta_n(\rho), n \in \mathbb{N},\ \delta_n \to 0$ such that

$$\limsup_{n \to \infty} \sqrt{n} \cdot \mathbb{E}\left(w_{n1}(F)^p\, I(w_{n1}(F) > \delta_n \sqrt{n})\right) \leq \rho \qquad (3.10)$$

$$\limsup_{n \to \infty} \sup_{d_\nu^{(2)}(f,g) \leq \alpha} \mathbb{E}\left(w_{n1}((f-g)^2) \cdot I(w_{n1}(F) \leq \delta_n \sqrt{n})\right) \to 0 \ for\ \alpha \to 0 \quad (3.11)$$

where

$$d_\nu^{(2)}(f,g) := \nu((f-g)^2)^{\frac{1}{2}} \ for\ f, g \in \mathcal{F}$$

$$\sup_{n \in \mathbb{N}} \mathbb{E}\left(w_{n1}(F^2) \cdot I(w_{nj}(F) \leq \delta_n \sqrt{n})\right) < \infty \qquad (3.12)$$

Suppose further that

$$\sup_{f \in \mathcal{F}} \sqrt{n}|\nu \star \mu_n(f) - \nu(f)| \longrightarrow 0 . \qquad (3.13)$$

Assume in addition that there exists a mean zero Gaussian process $\bar{G} = (\bar{G}(f))_{f \in \mathcal{F}}$ such that the finite dimensional distributions of $\sqrt{n}(\tilde{\nu}_n - \nu \star \mu_n)$ converge to those of $\bar{G}$. Then there exists a mean zero Gaussian process $G = (G(f))_{f \in \mathcal{F}}$ with bounded and uniformly $d_\nu^{(2)}$- continuous sample paths such that

$$\sqrt{n}(\tilde{\nu}_n - \nu) \xrightarrow{\mathcal{L}_p} G \quad in\ l^\infty(\mathcal{F})$$

where $l^\infty(\mathcal{F})$ denotes the space of all bounded functions on $\mathcal{F}$ endowed with the sup-norm and where $\xrightarrow{\mathcal{L}_p}$ denotes convergence in "law" in the sense of Hoffmann-Jørgensen (1984, 1991).

It is important to note that we do not assume $\mathcal{F}$ to be invariant under translation as done by Yukich (1992) and v.d.Vaart (1994), from which it follows that $\mathcal{F}$ is uniformly bounded. Various other results, especially on FCLT's for smoothed empirical processes will be contained in a paper under preparation by the second author.

4. Proof of Theorem 2.1

For the proof of Theorem 2.1 some techniques and inequalities commonly used in modern probability theory are needed:

Symmetrization lemma 4.1.   Let $X_1, \ldots, X_n$ be independent stochastic processes indexed by an arbitrary set T and let $\varepsilon_1, \ldots, \varepsilon_n$ be a Rademacher sequence independent of $X_1, \ldots, X_n$. Here independence is to be understood that all processes and variables are defined on a common product space via coordinate projections. Let $\Psi : \mathbb{R}_+ \longrightarrow \mathbb{R}_+$ be a convex and nondecreasing function. Then

$$\mathbb{E}^* \left(\Psi(\sup_{t \in T} | \sum_{i=1}^{n} (X_i(t) - \mathbb{E}(X_i(t))| \right) \leq \mathbb{E}^* \left(\Psi(2 \cdot \sup_{t \in T} | \sum_{i=1}^{n} \varepsilon_i \cdot X_i(t)| \right)$$

with $\mathbb{E}^*$ denoting outer expectation.

Maximal inequality for Rademacher averages 4.2.   Let T be an arbitrary index set and $x_i \in \mathbb{R}^T, i = 1, \ldots, n$. Put $d_1(s, t) := \sum_{i=1}^{n} |x_i(s) - x_i(t)|, s, t \in T$. Then there exists a universal constant $0 < K_1 < \infty$ such that for any Rademacher sequence $\varepsilon_1, \ldots, \varepsilon_n$ and for all $\gamma > 0$

$$\mathbb{E} \left(\sup_{t \in T} | \sum_{i=1}^{n} \varepsilon_i \cdot x_i(t)| \right) \leq \gamma + K_1 (1 + H(\gamma, T, d_1))^{\frac{1}{2}} \cdot \sup_{t \in T} \left(\sum_{i=1}^{n} x_i^2(t) \right)^{\frac{1}{2}}$$

where the metric entropy $H(\gamma, T, d_1)$ is defined as in Section 2 with T instead of $\mathcal{F}$ and d_1 replacing $d_{\mu_{n\delta}}^{(1)}$.

For the proof of 4.1 we refer to v.d.Vaart and Wellner (1996), Lemma 2.3.6. We only sketch the proof of 4.2:

First, consider the case when T is finite, $|T| = m, m \in \mathbb{N}$.
Put $\psi(x) := \exp(x^2) - 1, x \geq 0$, and let $|| \cdot ||_\psi$ be the corresponding Orlicz-norm, i.e.

$$||X||_\psi := \inf\{C > 0 : \mathbb{E}(\psi(\frac{|X|}{C})) \leq 1\}$$

for a random variable X. Since $x^2 \leq \psi(x)$ for $x \geq 0$ we have

$$\mathbb{E} \left(\max_{t \in T} | \sum_{i=1}^{n} \varepsilon_i \cdot x_i(t)| \right) \leq || \max_{t \in T} | \sum_{i=1}^{n} \varepsilon_i \cdot x_i(t)| ||_\psi.$$

From a maximal inequality for Orlicz-norms we get

$$|| \max_{t \in T} | \sum_{i=1}^{n} \varepsilon_i \cdot x_i(t)| ||_\psi \leq K \cdot \psi^{-1}(m) \cdot \max_{t \in T} || \sum_{i=1}^{n} \varepsilon_i \cdot x_i(t)||_\psi$$

for some constant $K > 0$ (cf. v.d.Vaart and Wellner (1996), Lemma 2.2.2).

Now the exponential tail bound for Rademacher sums (cf. v.d.Vaart and Wellner (1996), Lemma 2.2.7) leads to $\sqrt{6} \cdot \left(\sum_{i=1}^n x_i^2(t) \right)^{\frac{1}{2}}$ as an upper bound for $\| \sum_{i=1}^n \varepsilon_i \cdot x_i(t) \|_\psi$. Together with $\psi^{-1}(m) \leq (1 + \log m)^{\frac{1}{2}}$ we get

$$\mathbb{E}\left(\max_{t \in T} | \sum_{i=1}^n \varepsilon_i \cdot x_i(t)| \right) \leq K_1 \cdot (1 + \log m)^{\frac{1}{2}} \cdot \max_{t \in T} \left(\sum_{i=1}^n x_i^2(t) \right)^{\frac{1}{2}}.$$

The assertion in the case of an arbitrary index set T now easily follows (cf. Gaenssler and Ziegler (1994), Lemma 2.3).

Another important tool used in our proof is Hoffmann-Jørgensen's inequality. For this we refer to v.d.Vaart and Wellner (1996), A.1.5.

Now we prove Theorem 2.1:

First, by the Symmetrization lemma (take $\Psi(x) := x^p, x \in \mathbb{R}_+$), it suffices to show that

$$\mathbb{E}\left(\sup_{f \in \mathcal{F}} | \sum_{j \leq j(n)} \varepsilon_j w_{nj}(f) \xi_{nj}|^p \right) \longrightarrow 0 \quad \text{as } n \to \infty$$

where (ε_j) is a Rademacher sequence being independent of both arrays (w_{nj}) and (ξ_{nj}).

By (2.1) there exists a sequence (δ_n) of positive real numbers with $\delta_n \to 0$ and

$$\sum_{j \leq j(n)} \mathbb{E}^{\frac{1}{p}} \left(\mu_{nj}(F)^p \cdot I(\mu_{nj}(F) > \delta_n) \right) \longrightarrow 0 \quad (n \to \infty)$$

where we put $\mu_{nj} := w_{nj} \cdot |\xi_{nj}|$ for short. Hence it suffices to show that

$$\mathbb{E}\left(\sup_{f \in \mathcal{F}} |S_{n\delta_n}(f)|^p \right) \longrightarrow 0 \quad (n \to \infty)$$

where $S_{n\delta_n}(f) := \sum_{j \leq j(n)} \varepsilon_j w_{nj}(f) \xi_{nj} \cdot I(\mu_{nj}(F) \leq \delta_n)$.

Since the summands of $S_{n\delta_n}(f)$ are bounded by δ_n, it follows by application of Hoffmann-Jørgensen's inequality that it suffices to show that

$$\sup_{f \in \mathcal{F}} |S_{n\delta_n}(f)| \longrightarrow 0 \quad \text{in probability } (n \to \infty). \tag{4.1}$$

To prove (4.1), let $\beta > 0$ and $\varepsilon > 0$ be fixed. Denote (cf. (2.2))

$$C := \sup_{n \in \mathbb{N}} \sum_{j \leq j(n)} \mathbb{E}\left(\mu_{nj}(F) \cdot I(\mu_{nj}(F) \leq \delta_1) \right)$$

and choose $\rho := \varepsilon/2, \tau := \varepsilon\beta/2C$ and $\delta \equiv \delta(\tau)$ according to (2.3).

Let $A_n := \{H(\tau\mu_{n\delta}(F), \mathcal{F}, d^{(1)}_{\mu_{n\delta}})^* > M(\rho)\}$ where the star $(\star)$ denotes the measurable cover function and $M(\rho)$ is such that $\limsup_{n\to\infty} \mathbb{P}(A_n) < \rho$. Then by the Markov inequality and Fubini's theorem it follows that

$$\mathbb{P}(\sup_{f\in\mathcal{F}} |S_{n\delta_n}(f)| > \beta) \le \mathbb{P}(A_n) + \beta^{-1}\mathbb{E}(1_{CA_n}\mathbb{E}_\varepsilon(\sup_{f\in\mathcal{F}} |S_{n\delta_n}(f)|)), \qquad (4.2)$$

where $\mathbb{E}_\varepsilon$ denotes integration with respect to the Rademacher sequence.

For n large enough such that $\delta_n \le \delta$ we obtain by the maximal inequality 4.2 with a universal constant $0 < K_1 < \infty$

$$\mathbb{E}_\varepsilon(\sup_{f\in\mathcal{F}} |S_{n\delta_n}(f)|)$$

$$\le \tau\mu_{n\delta}(F)$$
$$+ K_1(1 + H(\tau\mu_{n\delta}(F), \mathcal{F}, d^{(1)}_{\mu_{n\delta}}))^{\frac{1}{2}} \sup_{f\in\mathcal{F}} | \sum_{j\le j(n)} w^2_{nj}(f)\varsigma^2_{nj} \cdot I(\mu_{nj}(F) \le \delta_n)|^{\frac{1}{2}}$$

$$\le \tau\mu_{n\delta}(F)$$
$$+ \delta_n^{\frac{1}{2}} K_1(1 + H(\tau\mu_{n\delta}(F), \mathcal{F}, d^{(1)}_{\mu_{n\delta}}))^{\frac{1}{2}} (\sum_{j\le j(n)} \mu_{nj}(F) \cdot I(\mu_{nj}(F) \le \delta_n))^{\frac{1}{2}},$$

whence by definition of A_n it follows that

$$\mathbb{E}(1_{CA_n}\mathbb{E}_\varepsilon(\sup_{f\in\mathcal{F}} |S_{n\delta_n}(f)|)) \le \tau\mathbb{E}(\mu_{n\delta}(F))$$
$$+ \delta_n^{\frac{1}{2}} K_1(1 + M(\rho))^{\frac{1}{2}}\mathbb{E}^{\frac{1}{2}}(\sum_{j\le j(n)} \mu_{nj}(F) \cdot I(\mu_{nj}(F) \le \delta_n)). \qquad (4.3)$$

Observe that by (2.1) we have

$$\limsup_{n\to\infty} \mathbb{E}(\mu_{n\delta}(F)) \le C + \limsup_{n\to\infty} \sum_{j\le j(n)} \mathbb{E}(\mu_{nj}(F) \cdot I(\mu_{nj}(F) > \delta_1)) = C. \quad (4.4)$$

Hence we obtain by (2.2),(2.3),(4.2),(4.3) and (4.4) that

$$\limsup_{n\to\infty} \mathbb{P}(\sup_{f\in\mathcal{F}} |S_{n\delta_n}(f)| > \beta) \le \rho + \beta^{-1}\tau C = \varepsilon$$

by the choice of ρ and τ. Since ε and β were arbitrary, (4.1) is proved. $\qquad\square$

References

Gaenssler, P. and Ziegler, K. (1994). A uniform Law of Large Numbers for set-indexed Processes with applications to Empirical and Partial-Sum Processes. In: Probability in Banach Spaces, **9**, J. Hoffmann-Jørgensen, J. Kuelbs, M.B. Marcus (Eds.), pp. 385–400, Birkhäuser, Boston.

Hoffmann-Jørgensen, J. (1984). Stochastic Processes on Polish Spaces. Unpublished manuscript. Published in 1991 as Vol. 39 of the Various Publication Series, Matematisk Institute, Aarhus Universitet.

Prakasa Rao, B.L.S. (1981). Convergence of smoothed empirical processes for independent random variables. Aligarh J. of Statist. **1**, 13–18.

van der Vaart, A.W. (1994). Weak convergence of smoothed empirical processes. Scand. J. Statist. **21**, 501–504.

van der Vaart, A.W. and Wellner, J.A. (1996). Weak convergence and empirical processes. Springer Series in Statistics, Springer-Verlag New York, Inc.

Winter, B.B. (1973). Strong uniform consistency of integrals of density estimators. Canadian J. of Statist. **1**, 247–253.

Yamato, H. (1973). Uniform convergence of an estimator of a distribution function. Bull. Math. Statist. **15**, 69–78.

Yukich, J.E. (1989). A note on limit theorems for perturbed empirical processes. Stoch. Processes and Their Applications. **33**, 163–173.

Yukich, J.E. (1992). Weak convergence of smoothed empirical processes. Scand. J. Statist. **19**, 271–279.

Ziegler, K. (1994). On functional central limit theorems and uniform laws of large numbers for sums of independent processes. Diss., LMU Muenchen.

Ziegler, K. (1996). Functional Central Limit Theorems for Triangular Arrays of Function-indexed Processes under uniformly integrable entropy conditions. Submitted for publication.

Peter Gaenssler, Daniel Rost and
Klaus Ziegler
Math. Institute
University of Munich
D-80333 Munich, Germany

Progress in Probability, Vol. 43
© 1998 Birkhäuser Verlag Basel/Switzerland

A Consequence For Random Polynomials
of a Result of De La Peña and Montgomery-Smith

EVARIST GINÉ*

1. Introduction

The object of this note is to bring attention to the fact that the remarkable theorem of de la Peña and Montgomery-Smith (1995) on decoupling of tail probabilities of U-statistics implies results, which are best possible up to constants, on decoupling inequalities for tail probabilities of 1) general polynomials in independent normal variables and 2) tetrahedral polynomials in any set of not necessarily symmetric independent random variables. Although these results follow extremely easily from their theorem, I believe they should be recorded because they constitute essentially final answers to questions that have been extensively treated in the literature.

Here we state their decoupling theorem for ease of reference. In what follows I_n^m denotes the set of integers

$$I_n^m := \{(i_1, \ldots, i_m) : 1 \le i_j \le n, \ i_j \ne i_k \text{ if } j \ne k\}.$$

1.1. THEOREM. (de la Peña and Montgomery-Smith, 1995). *For natural numbers $n \ge m$, let $\{X_i\}_{i=1}^n$ be n independent random variables with values in a measurable space $(S, \mathcal{S})$, and let $\{X_i^k\}_{i=1}^n$, $k = 1, \ldots, m$, be m independent copies of this sequence. There are constants $C_m \in (0, \infty)$, depending only on m, such that if B is a separable Banach space and $h_{i_1 \ldots i_m} : S^m \to B$, $(i_1, \ldots, i_m) \in I_n^m$, are measurable functions, then, for all $t > 0$,*

$$\Pr\left\{\left\|\sum_{I_n^m} h_{i_1 \ldots i_m}(X_{i_1}, \ldots, X_{i_m})\right\| > t\right\}$$
$$\le C_m \Pr\left\{C_m \left\|\sum_{I_n^m} h_{i_1 \ldots i_m}(X_{i_1}^1, \ldots, X_{i_m}^m)\right\| > t\right\}. \tag{1.1}$$

If, moreover, the functions $h_{i_1 \ldots i_m}$ are symmetric in the sense that, for all $x_1, \ldots, x_m \in S$ and all permutations s of $1, \ldots, m$,

$$h_{i_1 \ldots i_m}(x_1, \ldots, x_m) = h_{i_{s_1} \ldots i_{s_m}}(x_{s_1}, \ldots, x_{s_m}), \tag{1.2}$$

then the reverse inequality holds, that is, there are constants $D_m \in (0, \infty)$ depending only on m, such that, for all $t > 0$,

$$\Pr\left\{\left\|\sum_{I_n^m} h_{i_1 \ldots i_m}(X_{i_1}^1, \ldots, X_{i_m}^m)\right\| > t\right\}$$
$$\le D_m \Pr\left\{D_m \left\|\sum_{I_n^m} h_{i_1 \ldots i_m}(X_{i_1}, \ldots, X_{i_m})\right\| > t\right\}. \tag{1.3}$$

*)   Partially supported by NSF Grant No. DMS-9625457.

2. Gaussian Chaos

As far as we know, the first instance of comparison between the tails of a multilinear form and the tails of its decoupled version appear in Kwapień (1987), for B-valued tetrahedral Gaussian multilinear forms, where B is a Banach space, that is, for objects of the form

$$\sum_{(i_1,\ldots,i_m)\in I_n^m} x_{i_1,\ldots,i_m} g_{i_1} \cdots g_{i_m}$$

with $x_{i_1,\ldots,i_m} \in B$ and g_i i.i.d. $N(0,1)$. Very closely following Kwapień, Arcones and Giné (1993) observed that the comparison theorem extends to all B-valued homogeneous chaos variables (our main point consisting in finding the adequate definition of decoupling when repetition of indices is allowed). We asked in this article whether the result extends to non-homogeneous chaos, and the object of this section is to observe that, indeed, it does, as a consequence of the decoupling inequality of de la Peña and Montgomery-Smith (1995) in conjunction with the central limit theorem for U-statistics.

Let $G := \{g_i : i \in \mathbb{N}\}$ be an orthogaussian sequence. Any polynomial $Q_{(m)}$ of degree m in the elements of G with coefficients in B admits the decomposition

$$Q_{(m)} = \sum_{k=0}^{m} Q_k \tag{2.1}$$

with $Q_0 = h_0 \in B$ and

$$Q_k = \frac{1}{(k!)^{1/2}} \sum_{1 \leq i_1,\ldots,i_k \leq N} x_{\mathbf{i}} \prod_{j \leq N} H_{j(\mathbf{i})}(g_j), \quad k \geq 1, \tag{2.2}$$

for some $N < \infty$, where: $\mathbf{i}$ is shorthand for the generic multiindex $(i_1,\ldots,i_k)$, $j(\mathbf{i}) = \sum_{r=1}^{k} I(i_r = j)$ is the number of times j appears as a coordinate in the multiindex $\mathbf{i}$, H_k is the Hermite polynomial of degreee k and leading coefficient equal to 1 (that is, H_k is defined by the equation $\exp(ux - u^2/2) = \sum_{n=0}^{\infty} u^n H_n(x)/n!$, $u, x \in \mathbb{R}$) and $x_{\mathbf{i}}$ is a constant (depending on k) times the L_2 inner product of Q_k and $\prod_{j \leq n} H_{j(\mathbf{i})}(g_j)$. (This is classical; see e.g. Neveu, 1968.) In particular, the coefficients $x_{\mathbf{i}}$ are invariant with respect to permutations of the coordinates of the multiindex, that is, they are *symmetric*.

Let ϕ_i, $i = 0, 1, \ldots$ be an orthonormal system of $L_2([0,1], \mathcal{B}, \lambda)$ with $\phi_0 \equiv 1$ (e.g., the Haar system). Given $Q_{(m)}$ as in (2.1), (2.2), consider the symmetric kernels

$$h_k(t_1,\ldots,t_k) = \sum_{1 \leq i_1,\ldots,i_k \leq N} x_{i_1,\ldots,i_k} \phi_{i_1}(t_1) \cdots \phi_{i_k}(t_k), \quad k \geq 1.$$

Note that, the functions ϕ_i being centered for $i > 1$, h_k is a canonical or completely degenerate kernel for $k \geq 1$. Keeping with standard notation, given a symmetric kernel h of k variables and an i.i.d. sequence $\mathbf{X} = (X_i : i \in \mathbb{N})$, we denote

$$U_n^k(h) := U_n^k(h, \mathbf{X}) := \binom{n}{k}^{-1} \sum_{1 \leq i_1 < \ldots < i_k \leq n} h(X_{i_1}, \ldots, X_{i_k})$$

the U-statistic of order k and kernel h based on $\mathbf{X}$, with the convention $U_n^0(c) = c$. In what follows $\mathbf{X}$ is a sequence of i.i.d. random variables uniform on $[0,1]$. Then, the central limit theorem for degenerate U-statistics taking values in a finite dimensional space (in our case, in the span of the vectors $x_{\mathbf{i}}$) gives

$$\sum_{k=0}^m \binom{n}{k}^{1/2} U_n^k(h_k) \to_d Q_{(m)} \tag{2.3}$$

(Bretagnolle, 1983; Dynkin and Mandelbaum, 1983; Rubin and Vitale, 1980; see also Arcones and Giné, 1992). As indicated in Arcones and Giné (1993) the left side of (2.3) can be written as a U-statistic of order m: clearly, if h_k is a symmetric function of k variables, $k \leq m$, then $U_n^k(h) = U_n^m(H_{k,m})$ for the kernel

$$H_{k,m}(t_1,\dots,t_m) := \frac{(m-k)!}{m!} \sum_{(j_1,\dots,j_k)\in I_m^k} h_k(t_{j_1},\dots,t_{j_k}), \quad 1 \leq k \leq m,$$

and, also letting $H_{0,m} := h_0 \in B$, (2.3) becomes

$$U_n^m\left(\sum_{k=0}^m \binom{n}{k}^{1/2} H_{k,m}\right) \to_d Q_{(m)}. \tag{2.3'}$$

The kernel

$$H_{(n)} := \sum_{k=0}^m \binom{n}{k}^{1/2} H_{k,m}$$

is symmetric i.e., it satisfies condition (1.2) of Theorem 1.1. Let now X_i^j, $i \in \mathbb{N}$, $1 \leq j \leq m$, be i.i.d. uniform on $[0,1]$ random variables. To conform with Theorem 1.1, we define

$$U_n^{(m,dec)}(H_{(n)}) := \frac{(n-m)!}{n!} \sum_{\mathbf{i}\in I_n^m} H_{(n)}(X_{i_1}^1,\dots,X_{i_m}^m). \tag{2.4}$$

Theorem 1.1 then asserts that *the tail probabilities of* $\|U_n^{(m)}(H_{(n)})\|$ *and* $\|U_n^{(m,dec)}(H_{(n)})\|$ *are comparable.* In order to translate this into a comparison result for Gaussian polynomials, we compute the limiting distribution of $U_n^{(m,dec)}(H_{(n)})$, slightly expanding on Arcones and Giné (1993). According to definition (2.4) and the definitions of $H_{(n)}$, $H_{k,m}$ and h_k, and letting $|\mathbf{i}| = \max_{\ell \leq k} i_\ell$, we have

$$U_n^{(m,dec)}(H_{(n)}) = h_0 + \frac{(n-m)!}{n!} \sum_{k=1}^m \binom{n}{k}^{1/2} \frac{(m-k)!}{m!} \sum_{|\mathbf{i}|\leq N} x_{\mathbf{i}} \sum_{\mathbf{j}\in I_m^k}$$
$$\sum_{\mathbf{r}\in I_n^m} \phi_{i_1}(X_{r_{j_1}}^{j_1}) \cdots \phi_{i_k}(X_{r_{j_k}}^{j_k})$$

$$= h_0 + \frac{(n-m)!}{n!} \sum_{k=1}^{m} \binom{n}{k}^{1/2} \frac{(m-k)!}{m!} \frac{(n-k)!}{(n-m)!} \sum_{|i|\leq N} x_i \sum_{j\in I_m^k}$$

$$\sum_{r\in I_n^k} \phi_{i_1}(X_{r_1}^{j_1}) \cdots \phi_{i_k}(X_{r_k}^{j_k})$$

$$\asymp h_0 + \sum_{k=1}^{m} \frac{1}{n^{k/2}} \frac{(m-k)!}{(k!)^{1/2}m!} \sum_{|i|\leq N} x_i \sum_{j\in I_m^k}$$

$$\sum_{r\in I_n^k} \phi_{i_1}(X_{r_1}^{j_1}) \cdots \phi_{i_k}(X_{r_k}^{j_k}). \tag{2.5}$$

The first identity in (2.5) is obvious from the definitions; the second identity follows by consolidating repetitions in the sum over $\mathbf{r} \in I_n^m$, according to the equation

$$\sum_{r\in I_n^m} g(X_{r_{j_1}}^{j_1}, \ldots, X_{r_{j_k}}^{j_k}) = \frac{(n-k)!}{(n-m)!} \sum_{r\in I_n^k} g(X_{r_1}^{j_1}, \ldots, X_{r_k}^{j_k}),$$

which is easily verified; and in the last line of (2.5) we simply consolidate numerical coefficients. Now, since for every subset $I \subset I_n^k$,

$$\mathbb{E}\Big[\sum_{r\in I} \phi_{i_1}(X_{r_1}^1) \cdots \phi_{i_k}(X_{r_k}^k)\Big]^2 = \#(I)$$

and $n^k - \#(I_n^k) = O(n^{k-1})$, we have

$$\mathbb{E}\Big[\frac{1}{n^{k/2}}\Big(\sum_{|r|\leq n} - \sum_{r\in I_n^k}\Big)\phi_{i_1}(X_{r_1}^{j_1}) \cdots \phi_{i_k}(X_{r_k}^{j_k})\Big]^2 = \frac{n^k - \#(I_n^k)}{n^k} \to 0.$$

This and (2.5) then give:

$$U_n^{(m,dec)}(H_{(n)})$$

$$\asymp h_0 + \sum_{k=1}^{m} \frac{1}{n^{k/2}} \frac{(m-k)!}{(k!)^{1/2}m!} \sum_{|i|\leq N} x_i \sum_{j\in I_m^k} \sum_{|r|\leq n} \phi_{i_1}(X_{r_1}^{j_1}) \cdots \phi_{i_k}(X_{r_k}^{j_k})$$

$$= h_0 + \sum_{k=1}^{m} \frac{(m-k)!}{(k!)^{1/2}m!} \sum_{|i|\leq N} x_i \sum_{j\in I_m^k} \Big(\frac{\sum_{r=1}^{n}\phi_{i_1}(X_r^{j_1})}{\sqrt{n}}\Big) \cdots \Big(\frac{\sum_{r=1}^{n}\phi_{i_k}(X_r^{j_k})}{\sqrt{n}}\Big).$$

Let now $g_i^{(j)}$, $1 \leq i \leq N$, $1 \leq j \leq m$, denote i.i.d. $N(0,1)$ random variables. Clearly, since the variables X_r^s are independent and the functions ϕ_i are orthonormal for the law of the X's, the vectors $(g_i^{(j)} : 1 \leq i \leq N, 1 \leq j \leq m)$ and $(\phi_i(X_1^j) : 1 \leq i \leq N, 1 \leq j \leq m)$ have the same covariance. Hence, the central limit theorem in $\mathbb{R}^{Nm}$ together with the continuous mapping theorem give that

$$U_n^{(m,dec)}(H_{(n)}) \to_d h_0 + \sum_{k=1}^{m} \frac{(m-k)!}{(k!)^{1/2}m!} \sum_{|i|\leq N} x_i \sum_{j\in I_m^k} g_{i_1}^{(j_1)} \cdots g_{i_k}^{(j_k)}. \tag{2.6}$$

The limit in (2.6) is a 'decoupled' Gaussian chaos polynomial. If we define

$$Q_{(m)}^{dec} := h_0 + \sum_{k=1}^{m} \frac{(m-k)!}{(k!)^{1/2}} \sum_{|\mathbf{i}| \leq N} x_{\mathbf{i}} \sum_{\mathbf{j} \in I_m^k} g_{i_1}^{(j_1)} \cdots g_{i_k}^{(j_k)}, \qquad (2.7)$$

then Theorem 1.1 and the limits (2.3') and (2.6) readily give the following:

2.1. THEOREM. *For every* $m \in \mathbb{N}$ *there exists* $C_m \in (0, \infty)$ *such that if* $Q_{(m)}$ *is a (not necessarily homogeneous) Gaussian polynomial of degree* m *with coefficients in a Banach space* B, *as in* (2.1), (2.2), *and if* $Q_{(m)}^{dec}$ *is its decoupled version, as in* (2.7), *then, for all* $t > 0$,

$$\frac{1}{C_m} \Pr\left\{ \|Q_{(m)}^{dec}\| > C_m t \right\} \leq \Pr\left\{ \|Q_{(m)}\| > t \right\} \leq C_m \Pr\left\{ \|Q_{(m)}^{dec}\| > \frac{t}{C_m} \right\}. \qquad (2.8)$$

The factors $1/k!^{1/2}$ in the defintions (2.1), (2.2) of Q and (2.7) of Q^{dec} are obviously superfluous.

2.2. REMARKS. 1) The proof of Theorem 2.1 is taken almost verbatim from Arcones and Giné (1993, arguments on page 109): Theorem 1.1 is the only new ingredient.

2) Since the constants in Theorem 2.1 do not depend on N, the decoupling result remains valid for $N = \infty$ as long as the corresponding series are convergent, that is, *Theorem 2.2 holds for any element of the (non-homogeneous)* B-*valued Gaussian chaos of order* m. [This should be compared with the previously mentioned results of Kwapień (1987) and Arcones and Giné (1993).]

3. Tetrahedral Polynomials in Independent Random Variables

A B-valued tetrahedral k-linear form in an independent sequence of random variables $\mathbf{X} = (X_i : 1 \leq i \leq N)$ is an expression of the form

$$Q_k(\mathbf{X}, \ldots, \mathbf{X}) := \sum_{I_N^k} a_{i_1,\ldots,i_k} X_{i_1} \cdots X_{i_k}, \qquad (3.1)$$

where N is arbitrary and the coefficients $a_{i_1,\ldots,i_k}$ take values in a Banach space B. The coefficients $a_{i_1,\ldots,i_k}$ can be replaced by $\sum_s a_{i_{s(1)},\ldots,i_{s(k)}}/k!$, the sum extended to the permutations s of k elements. We assume in all that follows that this replacement is in place so that the coefficients $a_{i_1,\ldots,i_k}$ are invariant under permutations of the coordinates of the subindex, that is, they are symmetric. The decoupled version of Q_k, assuming as we do that the coefficients are symmetric, is defined as

$$Q_k(\mathbf{X}^1, \ldots, \mathbf{X}^k) := \sum_{\mathbf{i} \in I_n^k} a_{i_1,\ldots,i_k} X_{i_1}^1 \cdots X_{i_k}^k, \qquad (3.2)$$

where $\mathbf{X}^j = (X_i^j : 1 \leq i \leq N)$, $1 \leq j \leq k$, are i.i.d. random vectors with the same law as $\mathbf{X}$. McConnell and Taqqu (1986) introduced the problem of comparison of the distributions of the random variables (3.1) and (3.2), and proved that the moments of any symmetric tetrahedral m-linear form in a sequence of independent symmetric random variables are dominated, up to universal constants, by the corresponding moments of its decoupled version. See Kwapień and Woyczynski (1992) for some history on this type of results; their book contains a result of Kwapień (1987) to the effect that the moments of a not necessarily homogeneous polynomial and those of its decoupled version are comparable if the independent variables X_i are symmetric. It also contains (Theorem 6.9.2) a theorem on comparison of tail probabilities for tetrahedral bilinear forms in symmetric independent random variables. This result was partially extended by de la Peña, Montgomery-Smith and Szulga (1994, Theorem 3.8 A") to multilinear forms in symmetric independent random variables. We will show here that, as a consequence of Theorem 1.1, the tail probabilities of (not necessarily homogeneous) symmetric tetrahedral polynomials in independent but not necessarily symmetric random variables X_i are comparable to those of their decoupled versions.

Let $Q_{(m)}$ be a B-valued tetrahedral polynomial of degree m on $\mathbb{R}^N$, that is, if $\mathbf{x} = (x_1, \ldots, x_N) \in \mathbb{R}^N$ denotes a generic point in $\mathbb{R}^N$, then

$$Q_{(m)}(\mathbf{x}) = a_0 + \sum_{k=1}^{m} \sum_{\mathbf{i} \in I_N^k} a_{i_1,\ldots,i_k} x_{i_1} \cdots x_{i_k}, \tag{3.3}$$

where $a_{\mathbf{i}} \in B$ and B is a Banach space. Moreover, we assume, as explained above, that the coefficients $a_{\mathbf{i}}$ are *symmetric* in their subindices. Then, $Q_{(m)}(\mathbf{x})$ can be written in the form of a generalized U-statistic of order m,

$$Q_{(m)}(\mathbf{x}) = \sum_{\mathbf{i} \in I_N^m} H_{i_1,\ldots,i_m}^{N}(x_{i_1}, \ldots, x_{i_m}) \tag{3.3'}$$

with

$$
\begin{aligned}
H_{i_1,\ldots,i_m}^{N}(x_1, \ldots, x_m) &:= \frac{(N-m)!}{N!} a_0 \\
&+ \sum_{k=1}^{m} \frac{(N-m)!\,(m-k)!}{(N-k)!\,m!} \sum_{\mathbf{r} \in I_m^k} a_{i_{r_1},\ldots,i_{r_k}} x_{r_1} \cdots x_{r_k}.
\end{aligned}
\tag{3.4}
$$

(See the last section for an analogous observation.) Moreover, the symmetry of the coefficients $a_{\mathbf{i}}$ implies that

$$H_{i_1,\ldots,i_m}^{N}(x_1, \ldots, x_m) = H_{i_{s(1)},\ldots,i_{s(m)}}^{N}(x_{s(1)}, \ldots, x_{s(m)})$$

for all permutations s of $1, \ldots, m$, i.e., these kernels satisfy conditions (1.2). Hence, by Theorem 1.1, the tail probabilities of the random polynomial

$$Q_{(m)}(\mathbf{X}) = \sum_{\mathbf{i} \in I_N^m} H_{i_1,\ldots,i_m}^{N}(X_{i_1}, \ldots, X_{i_m})$$

are comparable to the tail probabilities of its α

$$\tilde{Q}_{(m)}(\mathbf{X}^1,\dots,\mathbf{X}^m) := \sum_{\mathbf{i}\in I_N^m} H_{i_1,\dots,i_m}^N$$

which develops as

$$\tilde{Q}_{(m)}(\mathbf{X}^1,\dots,\mathbf{X}^m) = a_0 + \sum_{k=1}^m \frac{(m-k)!}{m!}\frac{(N-m)!}{(N-k)!} \sum_{\mathbf{i}\in I_N^m}\sum_{\mathbf{r}\in I_m^k}$$

$$= a_0 + \sum_{k=1}^m \frac{(m-k)!}{m!} \sum_{\mathbf{i}\in I_N^k}\sum_{\mathbf{r}\in I_m^k} a_{i_1,\dots,i_k} X_{i_1}^{r_1}\cdots \qquad (3.6)$$

Note that, if Q is homogeneous of degree m, then (3.6) coincides with (3.2), the usual decoupling. So, we have:

3.1. THEOREM. *For every $m \in \mathbb{N}$, there exists $C_m \in (0,\infty)$ such that if $\mathbf{X} = (X_1,\dots X_N)$ is a vector of N independent random variables with $N > m$, $\mathbf{X}^j$, $j = 1,\dots,m$, are m independent copies of $\mathbf{X}$, $Q_{(m)}(\mathbf{X})$, $N > m$, is a (not necessarily homogeneous) tetrahedral polynomial in $\mathbf{X}$ of degree m with coefficients in a Banach space B, as in (3.3) (with $\mathbf{x} = \mathbf{X}$), and if $\tilde{Q}_{(m)}(\mathbf{X}^1,\dots,\mathbf{X}^m)$ is its decoupled version, as in (3.6), then, for all $t > 0$,*

$$\frac{1}{C_m}\Pr\left\{\|\tilde{Q}_{(m)}(\mathbf{X}^1,\dots,\mathbf{X}^m)\| > C_m t\right\}$$

$$\leq \Pr\left\{\|Q_{(m)}(\mathbf{X})\| > t\right\} \qquad (3.7)$$

$$\leq C_m \Pr\left\{\|\tilde{Q}_{(m)}(\mathbf{X}^1,\dots,\mathbf{X}^m)\| > \frac{t}{C_m}\right\}.$$

This theorem removes the assumptions of symmetry of the variables and homogeneity of the polynomials in the above mentioned results of Woyczynski and Kwapień (1992) and de la Peña, Montgomery-Smith and Szulga (1994). Also, it extends to comparison of tail probabilities the general theorem of Kwapień (1987, Theorem 2) on comparison of moments, without assuming symmetry of the variables. We have not compared the constants C_m in (3.7) with those in the results of these authors, but theirs are probably better.

References

[1] Arcones, M. and Giné, E. *On the bootstrap of U and V statistics.* Ann. Statist. **20** (1992), 655–674.

[2] Arcones, M. and Giné, E. *On decoupling, series expansions and tail behavior of chaos processes.* J. Theoret. Probab. **6** (1993), 101–122.

[3] Bretagnolle, J. *Lois limites du bootstrap de certaines fonctionelles.* Ann. Inst. H. Poincaré Sect. B **3** (1983), 281–296.

[4] de la Peña, Víctor H., and Montgomery–Smith, S. *Decoupling inequalities for the tail probabilities of multivariate U–statistics.* Ann. Probab. **23** (1995), 806–816.

...tor H., Montgomery–Smith, S. and Szulga, J. *Contraction and decou-*
...ualities for multilinear forms and U–statistics. Ann. Probab. **22** (1994),
...1765.

Dynkin, E. B. and Mandelbaum, A. *Symmetric statistics, Poisson point processes*
and multiple Wiener integrals. Ann. Statist. **11** (1983), 739–745.

[7] Kwapień, S. *Decoupling inequalities and polynomial chaos.* Ann. Probab. **15** (1987),
1062–1071.

[8] Kwapień, S. and Woyczynski, W. *Random Series and Stochastic Integrals: Single*
and Multiple. Birkhäuser (1992), Boston.

[9] McConnell, T. and Taqqu, M. *Decoupling inequalities for multilinear forms in in-*
dependent symmetric random variables. Ann. Probab. **14** (1986), 943–954.

[10] Neveu, J. *Processus Aléatoires Gaussiens.* Les Presses de l'Univ. de Montréal,
Montréal (1968), Canada.

[11] Rubin, M. and Vitale, R. A. *Asymptotic distribution of symmetric statistics.* Ann.
Statist. **8** (1980), 165–170.

Department of Mathematics
Department of Statistics
University of Connecticut
Storrs, CT 06269
gine@uconnvm.uconn.edu

Progress in Probability, Vol. 43
© 1998 Birkhäuser Verlag Basel/Switzerland

Distinctions Between the Regular and Empirical Central Limit Theorems for Exchangeable Random Variables

MARJORIE G. HAHN[1] AND GANG ZHANG[2*]

ABSTRACT. The central limit behavior of sums of exchangeable random variables with symmetric mixands, via either constant norming or self-norming, is examined. It will be shown that, unlike the i.i.d. situation, for sums of exchangeable random variables self-norming is more robust than constant norming. Necessary and sufficient conditions are provided for self-normalizability to a normal limit and constant normalizability to limits which are either normal or a mixture of normals. An example shows that self-norming can lead to asymptotic normality even when constant norming fails to yield a non-degenerate limit.

1. Introduction

For any sequence of random variables $X_1, X_2, \ldots$, let

$$S_n = \sum_{i=1}^n X_i, \quad V_n^2 = \sum_{i=1}^n X_i^2, \quad X_n^{(1)} = \max\{|X_1|, \ldots, |X_n|\}, \quad n \in \mathbf{N}$$

and let Z be a standard normal random variable.

DEFINITION 1.1 *The Regular Central Limit Theorem (Regular CLT) is said to hold for the sequence $X_1, X_2, \ldots$ if there exists $b_n > 0$ and $a_n \in \mathbf{R}$ such that*

$$\mathcal{L}\left(\frac{S_n - a_n}{b_n}\right) \to \mathcal{L}(Z).$$

In statistical applications, when the underlying distribution is unknown, one must attempt to construct the normalizers from the data and a natural candidate is V_n.

DEFINITION 1.2 *The Empirical or Self-normalized Central Limit Theorem (ECLT) is said to hold for the sequence $X_1, X_2, \ldots$ if*

$$\mathcal{L}\left(\frac{S_n}{V_n}\right) \to \mathcal{L}(Z).$$

1) Partially supported by NSF grant DMS-92-04333

2) Partially supported by summer student support under NSF grant DMS-92-04333

*) This paper is based on the Ph.D. thesis of Gang Zhang written under the direction of Marjorie G. Hahn

The recent paper of Giné, Götze, and Mason (1996) verifies the long-standing conjecture that the regular and empirical central limit theorems are equivalent for sequences of independent and identically distributed (i.i.d.) mean zero random variables.

THEOREM 1.1 (GINÉ, GÖTZE, AND MASON (1996)) *Let* $X, X_1, X_2, \ldots$ *be i.i.d. random variables. Then*

$$\mathcal{L}\left(\frac{S_n}{V_n}\right) \to \mathcal{L}(Z)$$

if and only if
   (a) *there exist* $b_n > 0$ *and* $a_n \in \mathbf{R}$ *such that*

$$\mathcal{L}\left(\frac{S_n - a_n}{b_n}\right) \to \mathcal{L}(Z)$$

   and
   (b) $EX = 0$.

An important implication is that, for i.i.d. mean zero random variables, the numerous characterizations of the regular central limit theorem are also characterizations of the empirical central limit theorem. The purpose of this paper is to show that the empirical and regular central limit theorems are not equivalent when identical distribution of the random variables is retained but the dependence is weakened mildly to exchangeability, even if the exchangeable sequences are symmetric in a very strong sense. Moreover, some of the characterizations usually associated with the regular central limit theorem are more naturally characterizations of the empirical central limit theorem.

DEFINITION 1.3 *An infinite sequence* $\tilde{X} \equiv (X_1, X_2, \ldots)$ *is called exchangeable if for each* n

$$(X_1, X_2, \ldots, X_n) \stackrel{D}{=} (X_{\pi(1)}, X_{\pi(2)}, \ldots, X_{\pi(n)})$$

for any permutation π *of* $1, 2, \ldots n$.

Infinite exchangeable sequences are conditionally i.i.d. by de Finetti's theorem.

THEOREM 1.2 (DE FINETTI'S THEOREM (1937)) *An infinite sequence of exchangeable random variables* $\tilde{X} = (X_1, X_2, \ldots)$ *is a mixture of i.i.d. sequences. That is, there exists a probability space* $(\mathbf{U}, \Theta)$ *such that*

$$P(\tilde{X} \in B) = \int_{\mathbf{U}} P(\tilde{X}(u) \in B)\, \Theta(du), \qquad (1.1)$$

where $\tilde{X}(u) = (X_1(u), X_2(u), \ldots)$ *is a sequence of i.i.d. random variables (called the mixands) and* $\Theta(\cdot)$ *is a probability measure, called the mixing measure.*

One might expect conditional i.i.d. sequences to mimic many properties of i.i.d. sequences. However, the next theorem of Klass and Teicher says that asymptotic normality for exchangeable sequences arises in two ways, one of which is substantially different from that for i.i.d. sequences.

THEOREM 1.3 (THEOREM 2 IN KLASS AND TEICHER (1987)) *Let* $X_1, X_2, \ldots$ *be an infinite exchangeable sequence of random variables. Then there exist* $a_n \in \mathbf{R}$ *and* $0 < b_n \to \infty$ *such that*

$$\frac{S_n}{b_n} \xrightarrow{\mathcal{D}} N(0,1)$$

if and only if there exist $\varepsilon_n \downarrow 0$ *such that*

$$nP^u(|X_1(u)| > \varepsilon_n b_n) \xrightarrow{\Theta} 0 \tag{1.2}$$

and either
  (i) $b_n/\sqrt{n}$ *is slowly varying with*

$$\frac{b_n(u)}{b_n} \xrightarrow{\Theta} 1 \quad \text{and} \quad \frac{a_n(u) - a_n}{b_n} \xrightarrow{\Theta} 0,$$

  where

$$b_n^2(u) = nE^u X_1^2(u) I(|X_1(u)| \leq \varepsilon_n b_n) \tag{1.3}$$

  and

$$a_n(u) = nE^u X_1(u)(|X_1(u)| \leq \varepsilon_n b_n); \tag{1.4}$$

  or
 (ii) b_n/n *is slowly varying with*

$$\frac{b_n(u)}{b_n} \xrightarrow{\Theta} 0 \quad \text{and} \quad \mathcal{L}\left(\frac{a_n(u) - a_n}{b_n}\right) \to N(0,1).$$

For exchangeable random variables with symmetric mixands, asymptotic normality cannot arise from the weak limiting behavior of the centering constants. The goal of this paper is to show that there are still major points of difference between the regular and empirical central limit theorems when restricting to exchangeable sequences with symmetric mixands. In particular, the combined convergence in probability of the joint tails $nP^u(|X_1(u)| > \varepsilon_n b_n) \xrightarrow{\Theta} 0$ and the condition $\frac{b_n(u)}{b_n} \xrightarrow{\Theta} 0$ have significant implications which prevent many of the characterizations of the Regular CLT for i.i.d. random variables from extending to exchangeable random variables. Hence, throughout the remainder of the paper assume that $\mathcal{L}(X_1(u))$ is symmetric.

The following theorem, presented in Griffin and Mason (1991) without (iii), summarizes the major equivalences of the Regular CLT for i.i.d. sequences:

THEOREM 1.4 *Let* $X_1, X_2, \ldots$ *be a sequence of i.i.d. symmetric random variables. Then the following are equivalent:*

(i) $\mathcal{L}\left(\dfrac{S_n}{b_n}\right) \to \mathcal{L}(Z);$

(ii) $\lim\limits_{x\to\infty} \dfrac{x^2 P(|X| > x)}{EX^2 I(|X| \le x)} = 0;$

(iii) *Central Convergence Criterion (CCC) holds: for any $\varepsilon > 0$*

$$nP(|X_1| > \varepsilon b_n) \to 0 \quad \text{and} \quad \frac{n}{b_n^2} EX_1^2 I(|X_1| \le \varepsilon b_n) \to 1;$$

(iv) $\mathcal{L}\left(\dfrac{V_n}{b_n}\right) \to \mathcal{L}(1);$

(v) $\mathcal{L}\left(\dfrac{S_n}{V_n}\right) \to \mathcal{L}(Z);$

(vi) $\mathcal{L}\left(\dfrac{X_n^{(1)}}{S_n}\right) \to \mathcal{L}(0);$

(vii) $\mathcal{L}\left(\dfrac{X_n^{(1)}}{V_n}\right) \to \mathcal{L}(0).$

Moreover, b_n can be chosen as

$$b_n = \sup\{b : nEX_1^2 I(|X_1| \le b) \ge b^2\}.$$

Condition (ii) is the analytic condition for the domain of attraction (DOA) of the normal; (iii) is the central convergence criterion for symmetric random variables; (iv) says that the self-normalizers are comparable to a constant sequence; (vi) is Paul Levy's famous asymptotic negligibility of the maximal term to the partial sum; and (vi) gives the asymptotic negligibility of the maximal term to the self-normalizers.

Now, suppose $\tilde{X} = (X_1, X_2, \ldots)$ is an exchangeable sequence. Via de Finetti's Theorem and dominated convergence, the regular CLT holds if the conditionally i.i.d. mixand sequences a.s. satisfy the regular CLT. One might initially suspect this is also a necessary condition. However, Klass and Teicher (1987) give an example of an exchangeable sequence with symmetric mixands and b_n such that $\dfrac{S_n}{b_n} \xrightarrow{\mathcal{D}} N(0, 1)$ but $X_1(u)$ is not in the DOA of any law for all u in a set of Θ-measure 1. Thus, the requirement that the conditional i.i.d. random variables be in the DOA of the normal almost surely is too strong an imposition for the CLT to hold.

Furthermore, even if for each fixed u, $X_1(u)$ is in the DOA of the normal law, the norming constants might vary from u to u. This can prevent the existence of a single norming sequence that is suitable for almost all u. (See Example 5.1.)

Therefore, determining the behavior of the conditional i.i.d. random variables when assuming weak convergence of the constant-normalized partial sums is not as trivial as one might first surmise. The correct notion is to require that $\mathcal{L}\left(\dfrac{S_n(u)}{b_n}\right)$ converges to the normal law in Θ-probability. This is made precise by the following definition of convergence in probability of random probability laws.

DEFINITION 1.4 *Given random variables* $Z_n(u,\omega)$ *and* $Z(u,\omega) : (\mathbf{U} \times \Omega, \Theta \times P) \to \mathbf{R}$, *let* $\mathcal{L}(Z_n(u)) \equiv P(\omega : Z_n(u,\omega) \in \cdot)$ *and* $\mathcal{L}(Z(u)) \equiv P(\omega : Z(u,\omega) \in \cdot)$. *Then* $\mathcal{L}(Z_n(u))$ *is said to converge to* $\mathcal{L}(Z(u))$ *in* Θ-*probability, denoted by*

$$\mathcal{L}(Z_n(u)) \overset{\Theta}{\to} \mathcal{L}(Z(u)),$$

if $\forall \varepsilon > 0$, *as* $n \to \infty$,

$$\Theta \{u : d\left(\mathcal{L}(Z_n(u)), \mathcal{L}(Z(u))\right) > \varepsilon\} \to 0,$$

where $d(\cdot, \cdot)$ *is any metric that metrizes the weak convergence, e.g., the Lévy distance or* d_3 *distance as in Araujo and Giné (1980).*

The notion of random probability laws converging in probability has also appeared elsewhere. For instance, Brown and Eagleson (1971), Eagleson (1975) as well as Hall and Heyde (1980) employ it for some martingale convergence theorems while Giné and Zinn (1990) employs it for bootstrapping central limit theorems. Convergence in probability of random probability laws is important in studying the CLT for exchangeable random variables for the following reasons:

- via de Finetti's theorem, the laws for the conditional i.i.d. random variables are random probability measures;
- convergence in probability is equivalent to every sequence having a subsubsequence which converges almost surely;
- since, as random variables, random probability laws are uniformly bounded, hence uniformly integrable, it is easy to prove a theorem in the spirit of Theorem 4.5.4 in Chung (19974) and use de Finetti's theorem to obtain a convergence theorem for exchangeable random variables (see Lemma 3.2).

2. Results

This section discusses and places in context our main results. Their proofs appear in subsequent sections. Subsection 2.1. presents necessary and sufficient conditions for the partial sums of exchangeable random variables with symmetric i.i.d. mixands, normalized by a sequence of constants, to converge to a variance mixture of normals. When the normal mixands of the limit all have the same variance, the Klass and Teicher theorem for symmetric i.i.d. mixands (Theorem 1.3 follows immediately. Although these theorems do not assume any moment conditions on the random variables, the underlying mixing probability must be known.

Subsection 2.2. presents the fact that validity of the empirical CLT implies that of the regular CLT. Example 5.1 confirms the failure of the converse. Subsection 2.3. presents the conditions that characterize the empirical CLT and Subsection 2.4. does the same for the regular CLT.

Assumptions and Notation (unless specified otherwise): Assume $S_n = X_1 + \cdots + X_n$, $V_n^2 = X_1^2 + \cdots + X_n^2$, $X^{(1)} = \max\{|X_1|, \ldots, |X_n|\}$ *where* $X_1, X_2, \ldots$ *is a sequence of exchangeable random variables with symmetric mixands which are nondegenerate* Θ-*a.s. Let* $X_j(u), S_n(u), V_n(u)$, *and* $X^{(1)}(u)$ *denote the corresponding quantities for the mixands. Also,* Z *will always be a standard normal random variable.*

2.1. Regular CLT with the Limit Being a Variance Mixture of Normals Consider the situation when the partial sums S_n are normalized by a sequence of constants b_n to converge to a mixture of normals. Normal limits arise as a special case when the normal mixands in the limit all have the same variance.

DEFINITION 2.1 *For each u let the canonical norming constants $\tilde{b}_n(u)$ defined for $X_1(u)$ be given by*

$$\tilde{b}_n(u) = \sup\{t : nE^u X_1^2(u)I(|X_1(u)| < t) \geq t^2\}.$$

Via de Finetti's theorem,

$$\mathcal{L}\left(\frac{S_n}{b_n}\right) = \int \mathcal{L}\left(\frac{S_n(u)}{\tilde{b}_n(u)} \cdot \frac{\tilde{b}_n(u)}{b_n}\right) \Theta(du).$$

Heuristically, if $X_1(u)$ is in the DOA of a standard normal, then $\mathcal{L}\left(\dfrac{S_n(u)}{\tilde{b}_n(u)}\right)$ converges weakly to $N(0,1)$. If, in addition, $\tilde{b}_n(u)/b_n$ settles down to a $\sigma(u)$ in some fashion, then conceivably $\mathcal{L}\left(\dfrac{S_n}{b_n}\right)$ should converge to a mixture of normals. However, since $X_1(u)$ may not be in any DOA, we will use a truncated version of $X_1(u)$ with varying truncation levels and the norming constants $b_n(u)$ for this truncated version of $X_1(u)$ appearing in (1.3).

Theorem 2.1 gives necessary conditions for convergence to a mixture of normals, albeit under a restrictive assumption (2.5) on $\sigma(u)$ that is not necessary for the conclusion of the theorem. An example following the theorem demonstrates this. It is not surprising that the interplay of $b_n(u)$ and b_n as well as the fact that the Lévy measure is zero in the usual CLT for i.i.d. random variables should manifest themselves in some way.

THEOREM 2.1 *Suppose $\exists b_n \uparrow \infty$ such that*

$$\mathcal{L}\left(\frac{S_n}{b_n}\right) \to \mathcal{L}(W) = \int N(0, \sigma^2(u)) \Theta(du)$$

for $\sigma(u) > 0$ a.s.Θ and $\forall C_1 > 0 \; \exists C_2 > 0$ such that

$$\int \sigma(u) \exp\left(C_2 k \ln k - \frac{C_1 k^2}{\sigma^2(u)}\right) \Theta(du) \to 0 \quad as \; k \to \infty. \qquad (2.5)$$

Then
(i) *for any $\varepsilon > 0$*

$$nP^u(|X_1(u)| > \varepsilon b_n) \overset{\Theta}{\to} 0; \qquad (2.6)$$

(ii) *there exists $\varepsilon_n \downarrow 0$, satisfying (2.6) with ε_n in place of ε, such that upon letting $b_n(u)$ be as in (1.3) using this ε_n*

$$\frac{b_n(u)}{b_n} \overset{D}{\to} \tau(u), \qquad (2.7)$$

where $\tau(u)$ satisfies

$$\int N(0, \tau^2(u))\,\Theta(du) = \int N(0, \sigma^2(u))\,\Theta(du).$$

Condition (2.5) is automatically satisfied if $\sigma^2(u) = \sigma^2\ \forall u$, but is not necessary otherwise as seen by the following example.

EXAMPLE 2.1 *Let* $\mathbf{U} = (0,1)$, Θ *be Lebesgue measure on* $\mathbf{U}$, *and* $\sigma(u) = u^{-1}$. *Then*

$$\int \sigma(u) \exp\left\{C_2 k \ln k - \frac{k^2}{\sigma^2(u)}\right\} \Theta(du) = \int_0^1 \frac{1}{u} \exp\{C_2 k \ln k - u^2 k^2\}\,\Theta(du)$$

$$= k^{C_2 k} \int_0^1 \frac{du}{u e^{u^2 k^2}}.$$

Since $\int_0^1 \dfrac{du}{u e^{u^2 k^2}} = +\infty$, $\forall k \geq 1$, *regardless of the value of* $C_2 > 0$,

$$\int \sigma(u) \exp\left\{C_2 k \ln k - \frac{k^2}{\sigma^2(u)}\right\} \Theta(du) \not\to 0.$$

However, if $\mathcal{L}(X_1(u)) = N(0, u^{-2})$, *then*

$$\mathcal{L}\left(\frac{S_n(u)}{\sqrt{n}}\right) = N(0, u^{-2})$$

and consequently,

$$\mathcal{L}\left(\frac{S_n}{\sqrt{n}}\right) = \int_0^1 \mathcal{L}\left(\frac{S_n(u)}{\sqrt{n}}\right) du = \int_0^1 N(0, u^{-2})\,du,$$

a variance mixture of normals. □

The following theorem providing sufficient conditions contains a stronger condition than (2.7), namely (2.9).

THEOREM 2.2 *Suppose there exist* $b_n \uparrow \infty$ *and* $\varepsilon_n \downarrow 0$ *such that*

$$n P^u(|X_1(u)| > \varepsilon_n b_n) \overset{\Theta}{\to} 0 \tag{2.8}$$

and

$$\frac{b_n(u)}{b_n} \overset{\Theta}{\to} \sigma(u), \tag{2.9}$$

where $b_n(u)$ is defined as in (1.3) by using the same ε_n as in (2.8). Then

$$\mathcal{L}\left(\frac{S_n}{b_n}\right) \to \mathcal{L}(W) = \int N(0, \sigma^2(u))\,\Theta(du).$$

REMARK 2.1 *Condition 2.9 is not necessary, as demonstrated by Example 2.2, in which* $\dfrac{b_n(u)}{b_n} \overset{\Theta}{\nrightarrow} \sigma(u)$ *but* $\dfrac{b_n(u)}{b_n} \overset{D}{\to} \sigma(u).$

EXAMPLE 2.2 *Let the probability space* $(\mathbf{U}, \Theta)$ *be given by* $\mathbf{U} = \{0,1\}$ *with* $\Theta\{0\} = \Theta\{1\} = \dfrac{1}{2}.$ *Define* $\sigma_n(u)$ *and* $\sigma(u)$ *by*

$$\sigma_n(0) = 1, \quad \sigma_n(1) = 2, \quad \sigma(0) = 2, \quad \sigma(1) = 1.$$

Then $\sigma_n \overset{D}{=} \sigma$ *for all* n, *but* $\sigma_n(u) \overset{\Theta}{\nrightarrow} \sigma(u)$ *since* $\Theta \left\{ u : |\sigma_n(u) - \sigma(u)| > \dfrac{1}{2} \right\} \equiv 1.$

Let $\{X_k(0),\ k \in \mathbf{N}\}$ *be i.i.d. Rademacher random variables and* $\{X_k(1),\ k \in \mathbf{N}\}$ *be i.i.d. random variables distributed as twice Rademachers. By direct calculation with* $b_n = \sqrt{n}$ *and* $\varepsilon_n \downarrow 0$ *such that* $\varepsilon_n b_n \uparrow \infty$, *the norming constants* $b_n(u)$, *given by (1.3), are* $b_n(0) = \sqrt{n}$, $b_n(1) = 2\sqrt{n}$. *(In this example* $\tilde{b}_n(u) = b_n(u)$.) *Then, as* $n \to \infty$,

$$\mathcal{L}\left(\frac{S_n}{b_n}\right) = \int_{\mathbf{U}} \mathcal{L}\left(\frac{S_n(u)}{b_n}\right)\Theta(du) = \frac{1}{2}\mathcal{L}\left(\frac{S_n(0)}{b_n}\right) + \frac{1}{2}\mathcal{L}\left(\frac{S_n(1)}{b_n}\right)$$

$$\to \frac{1}{2}N(0,1) + \frac{1}{2}N(0,4) \equiv \mathcal{L}(W).$$

Set $\sigma_n(u) = \dfrac{b_n(u)}{b_n}$. *Then* $\sigma_n(0) = 1$ *and* $\sigma_n(1) = 2$. *Hence, by the first part of the example,*

$$\frac{b_n(u)}{b_n} \overset{D}{\to} \sigma(u) \qquad but \qquad \frac{b_n(u)}{b_n} \overset{\Theta}{\nrightarrow} \sigma(u). \qquad \square$$

REMARK 2.2 *If* $\sigma(u) \equiv 1$, *i.e., the limit is normal, then (2.5) always holds. In this situation, Theorem 2.1 together with Theorem 2.2 are equivalent to the Klass and Teicher theorem for symmetric mixands, except for their final statement concerning the slow variation of* $\dfrac{b_n}{\sqrt{n}}$.

Furthermore, it is remarkable that $\mathcal{L}(S_n/b_n) \to N(0,1)$, *implies* $\mathcal{L}\left(S_n(u)/\tilde{b}_n(u)\right) \overset{\Theta}{\to} N(0,1)$, *even though it may happen that* $X_1(u) \notin DOA$ *of any law for almost all* u. *(See Example 3 of Klass and Teicher (1987).)*

REMARK 2.3 *In Example 2.2,* $\mathcal{L}\left(\dfrac{S_n}{b_n}\right) \to \mathcal{L}(W)$ *with* $\sigma(u) \not\equiv 1$. *Let*

$$\eta_n = \frac{S_n(u)}{\tilde{b}_n} = \frac{S_n(u)}{b_n}, \quad \xi_n = \sigma_n, \quad \eta \sim N(0,1), \quad \text{and} \quad \xi = \sigma.$$

Via characteristic functions and the continuous mapping theorem,

$$\sigma_n \overset{\Theta}{\to} \sigma \Leftrightarrow N(0,\sigma_n^2) \overset{\Theta}{\to} N(0,\sigma^2).$$

Therefore, by Lemma 3.5 in Section 3.1.,

$$\mathcal{L}\left(\frac{S_n(u)}{b_n}\right) \overset{\Theta}{\nrightarrow} N(0,\sigma^2(u)). \tag{2.10}$$

2.2. The Relation Between the Empirical CLT and the Regular CLT for Exchangeable Random Variables   For exchangeable random variables,

$$\mathcal{L}\left(\frac{S_n}{b_n}\right) = \int \mathcal{L}\left(\frac{S_n(u)}{b_n}\right) \Theta(du) \quad \text{and} \quad \mathcal{L}\left(\frac{S_n}{V_n}\right) = \int \mathcal{L}\left(\frac{S_n(u)}{V_n(u)}\right) \Theta(du).$$

Applying Theorem 1.1 to the mixands, one sees that for each fixed u,

$$\mathcal{L}\left(\frac{S_n(u)}{V_n(u)}\right) \to N(0,1) \quad \Leftrightarrow \quad \mathcal{L}\left(\frac{S_n(u)}{\tilde{b}_n(u)}\right) \to N(0,1).$$

However, there may not exist b_n such that $\dfrac{\tilde{b}_n(u)}{b_n} \xrightarrow{\mathcal{D}} 1$, as Theorem 2.1 seems to require.

In fact $V_n(u)$, as it varies for different u, adapts itself to $S_n(u)$ whereas b_n, lacking this flexibility, requires all $S_n(u)$ to behave more or less the same. This kind of "self-adaptability" of self-normalization makes it more robust than constant normalization. Consequently, the CLT for exchangeable random variables with i.i.d. symmetric mixands does imply the empirical CLT, as formalized in Theorem 2.3.

THEOREM 2.3 *If there exists b_n such that $\dfrac{S_n}{b_n} \xrightarrow{\mathcal{D}} Z$, then $\dfrac{S_n}{V_n} \xrightarrow{\mathcal{D}} Z$.*

REMARK 2.4 *Example 5.1 shows that the Empirical CLT can hold while simultaneously for any b_n, either the weak limit of $\mathcal{L}(S_n/b_n)$ is degenerate at 0 with positive probability or $\{\mathcal{L}(S_n/b_n),\ n \in \mathbf{N}\}$ is not tight. Thus, the converse to Theorem 2.3 fails dramatically.*

2.3. Characterization of the Empirical CLT   Since the CLT and Empirical CLT are not equivalent for exchangeable random variables, we first identify which of the relationships in Theorem 1.4, if any, are equivalent to the Empirical CLT. Perhaps not surprisingly, the "negligibility" of the largest magnitude order statistic to both the the partial sum and the self-normalizers catches the essence of the empirically normalized asymptotic normality for exchangeable random variables with symmetric i.i.d. mixands.

THEOREM 2.4 *The following are equivalent:*

(i) $\mathcal{L}\left(\dfrac{S_n}{V_n}\right) \to \mathcal{L}(Z);$

(ii) $\mathcal{L}\left(\dfrac{X_n^{(1)}}{S_n}\right) \to \mathcal{L}(0);$

(iii) $\mathcal{L}\left(\dfrac{X_n^{(1)}}{V_n}\right) \to \mathcal{L}(0).$

REMARK 2.5 *Theorem 2.4 does not require knowledge of the underlying mixing probability measure, which facilitates its application.*

The next theorem relies on the explicit use of the mixing probability and exposes the behavior of the conditional i.i.d. random variables $X_1(u), X_2(u), \ldots$. In particular, notice that their partial sums $S_n(u)$ are normalized by $\tilde{b}_n(u)$, which depend on u, so that weak convergence occurs in Θ-probability.

THEOREM 2.5 *Let $\tilde{b}_n(u)$ be as in Defintion 2.1. Then the following are equivalent:*

(i) $\mathcal{L}\left(\dfrac{S_n}{V_n}\right) \to \mathcal{L}(Z);$

(ii) $\mathcal{L}\left(\dfrac{S_n(u)}{V_n(u)}\right) \overset{\Theta}{\to} \mathcal{L}(Z);$

(iii) $\mathcal{L}\left(\dfrac{X_n^{(1)}(u)}{S_n(u)}\right) \overset{\Theta}{\to} \mathcal{L}(0);$

(iv) $\mathcal{L}\left(\dfrac{X_n^{(1)}(u)}{V_n(u)}\right) \overset{\Theta}{\to} \mathcal{L}(0);$

(v) $\mathcal{L}\left(\dfrac{S_n(u)}{\tilde{b}_n(u)}\right) \overset{\Theta}{\to} \mathcal{L}(Z);$

(vi) $\mathcal{L}\left(\dfrac{V_n(u)}{\tilde{b}_n(u)}\right) \overset{\Theta}{\to} \mathcal{L}(1);$

(vii) *CCC holds in Θ-probability for $S_n(u)$ with norming $\tilde{b}_n(u)$, i.e.,*
$\forall \varepsilon > 0$,

$$nP^u\left\{|X_1(u)| > \varepsilon\tilde{b}_n(u)\right\} \overset{\Theta}{\to} 0 \quad \text{and}$$

$$\frac{n}{\tilde{b}_n^2(u)}E^u\left(X_1^2(u)I(|X_1(u)| \le \varepsilon\tilde{b}_n(u))\right) \overset{\Theta}{\to} 1.$$

2.4. Characterization of the Regular CLT Weak convergence of the laws of the constant-normalized partial sums S_n/b_n to a normal holds precisely when the self-normalizers $V_n(u)$ behave asymptotically like constant normalizers b_n, as evidenced by the following theorem.

THEOREM 2.6 *Let $\tilde{b}_n(u)$ be as in Definition 2.1. The following are equivalent:*

(i) $\mathcal{L}\left(\dfrac{S_n}{b_n}\right) \to \mathcal{L}(Z)$ *for some b_n;*

(ii) $\mathcal{L}\left(\dfrac{S_n(u)}{b_n}\right) \overset{\Theta}{\to} \mathcal{L}(Z)$ *for some b_n;*

(iii) $\mathcal{L}\left(\dfrac{V_n(u)}{b_n}\right) \overset{\Theta}{\to} \mathcal{L}(1)$ *for some b_n;*

(iv) $\lim\limits_{x \to \infty} \dfrac{x^2 P^u(|X_1(u)| > x)}{E^u X_1^2(u)I(|X_1(u)| \le x)} \overset{\Theta}{=} 0$ *and* $\lim\limits_{n \to \infty} \dfrac{\tilde{b}_n(u)}{b_n} \overset{\Theta}{=} 1;$

(v) *CCC holds in Θ-probability for $S_n(u)$ with norming b_n, i.e.,*

$$nP^u\left\{|X_1(u)| > \varepsilon b_n\right\} \overset{\Theta}{\to} 0 \quad \text{and} \quad \frac{n}{b_n^2}E^u\left(X_1^2(u)I(|X_1(u)| \le \varepsilon b_n)\right) \overset{\Theta}{\to} 1.$$

REMARK 2.6 *Theorem 2.6 does not generalize to the case when the limit is a mixture of normals. See Remark 2.3.*

REMARK 2.7 *Since the second part of condition (iv) in Theorem 2.6 requires $\tilde{b}_n(u)$ to behave more or less the same as b_n, one might expect to see the first part of condition (iv) as part of Theorem 2.5. Indeed,*

PROPOSITION 2.1

$$\lim_{x \to \infty} \frac{x^2 P^u(|X_1(u)| > x)}{E^u X_1^2(u) I(|X_1(u)| \le x)} \overset{\Theta}{=} 0 \tag{2.11}$$

implies

$$\mathcal{L}\left(\frac{S_n}{V_n}\right) \to N(0,1).$$

However, the converse implication is still an open question.

2.5. Organization of the Remaining Sections   The sections that follow contain the proofs of the results discussed above. In particular, Section 4.1. proves Theorem 2.1 and Section 4.2. contains the proof of Theorem 2.2. Section 5.1. proves Theorem 2.3, Section 5.2. proves both Theorem 2.4 and Theorem 2.5. Finally, Section 6. proves Theorem 2.6 and Proposition 2.1. Since many of the proofs use random measures and also a central limit theorem in probability, the ideas needed are consolidated in Section 3..

3. Preliminaries

This section establishes several results about random measures and central limit theorems in probability that facilitate the proofs of our main results.

3.1. Random Measures

DEFINITION 3.1 *Let $(U, \mathcal{U}, \Theta)$ and $(\Omega, \mathcal{F}, P)$ be probability spaces. A measure α : $(U \times \Omega, \mathcal{U} \times \mathcal{F}, \Theta \times P) \to \mathcal{P}(\Omega)$, where $\mathcal{P}(\Omega)$ is the space of all probability measures on Ω, is a random measure on Ω if*
  (i) *$u \mapsto \alpha(u, A)$ is a random variable on U for each $A \in \mathcal{F}$;*
  (ii) *$A \mapsto \alpha(u, A) \in \mathcal{P}(\Omega)$ for each $u \in U$.*

Let α_n be random measures on Ω. Define

$$\bar{\alpha}_n(A) = E\alpha_n(\cdot, A) = \int \alpha_n(u, A) \, \Theta(du). \tag{3.12}$$

Then $\bar{\alpha}_n(\cdot)$ are probability measures on Ω.

LEMMA 3.1 (LEMMA 7.14, ALDOUS (1983)) *Let α_n be random measures with $\{\bar{\alpha}_n\}$ as in (3.12). If $\{\bar{\alpha}_n\}$ are tight on $\mathcal{P}(\Omega)$, then $\{\mathcal{L}(\alpha_n)\}$ are tight on $\mathcal{P}(\mathcal{P}(\Omega))$.*

COROLLARY 3.1 *Let $\sigma_n(u) = \dfrac{b_n(u)}{b_n}$ and let random measures α_n on $\mathbf{R}$ be defined for any Borel set C in $\mathbf{R}$ by*

$$\alpha_n(u, C) = P^u\{\omega : \sigma_n(u)Z(\omega) \in C\} \equiv \mathcal{L}(\sigma_n(u)Z(\omega))(C).$$

Suppose

$$\lim_{n\to\infty} \int \mathcal{L}\left(\frac{b_n(u)}{b_n}Z\right)\Theta(du) = \mathcal{L}(W) = \int N(0, \sigma^2(u))\,\Theta(du). \qquad (3.13)$$

Then $\{\mathcal{L}(\sigma_n(u))\}$ are tight, i.e., $\forall \varepsilon > 0 \;\exists$ compact $K_\varepsilon \subset \mathbf{R}$ such that

$$\Theta\{u : \sigma_n(u) \in K_\varepsilon^c\} < \varepsilon \quad \forall n.$$

Proof of Corollary 3.1.

Since $\lim\limits_{n\to\infty} \bar\alpha_n = \lim\limits_{n\to\infty} \int \mathcal{L}\left(\dfrac{b_n(u)}{b_n}Z\right)\Theta(du) = \mathcal{L}(W)$, we see that $\{\bar\alpha_n\}$ is tight. By the Lemma 3.1, $\{\mathcal{L}(\alpha_n)\}$ is tight. Suppose $\{\mathcal{L}(\sigma_n(u))\}$ is not tight. Then $\exists \varepsilon_0 > 0$ such that $\forall k \in \mathbf{N} \;\exists n_k$ with

$$\Theta\{u : \sigma_{n_k}(u) \geq k\} \geq \varepsilon_0.$$

Consequently, there exists u such that $\mathcal{L}(\sigma_{n_k}(u)Z(\omega)) = N(0, \sigma_{n_k}^2(u))$ has variance at least k^2.

Let $\{\beta_i : i = 1, 2, \ldots\}$ be a dense subset in $\mathcal{P}(\mathbf{R})$. (The existence of such a dense subset is guaranteed by the fact that $\mathcal{P}(\mathbf{R})$ is Polish for the weak topology.) Define the ball

$$B(\beta, \delta) = \{\gamma \in \mathcal{P}(\Omega) : \rho(\gamma, \beta) < \delta\},$$

where ρ is the Levy distance on $\mathcal{P}(\mathbf{R})$. Let $K_{\varepsilon_0} \subset \mathcal{P}(\mathbf{R})$ be compact with

$$\Theta\{u : \alpha_n(u, \cdot) \in K_{\varepsilon_0}\} < \varepsilon_0, \quad \forall n.$$

Let $\delta_0 < 1/8$. Then, since K_{ε_0} is compact, there exist $\beta_i \in \mathcal{P}(\mathbf{R})$, $i = 1, \ldots, m$ such that

$$\bigcup_{i=1}^{m} B(\beta_i, \delta_0) \supset K_{\varepsilon_0}.$$

Pick $x_0 > 0$ such that $\beta_i(-\infty, x] \geq 1 - \delta_0$, $\forall x \geq x_0$, $i = 1, \ldots, m$. Notice that for any $x > 0$

$$\frac{e^{-x^2/2}}{1 + x} \leq \int_x^\infty e^{-t^2/2}\,dt \leq \frac{1}{x}e^{-x^2/2}. \qquad (3.14)$$

Let $t = \dfrac{s}{k}$ and then $y = kx$ to get

$$\frac{1}{\sqrt{2\pi}} \cdot \frac{e^{-y^2/2k^2}}{1 + \frac{y}{k}} \leq \frac{1}{\sqrt{2\pi}} \cdot \int_y^\infty \frac{1}{k}e^{-s^2/2k^2}\,ds \equiv 1 - \Phi_k(y).$$

Since for any fixed y, $e^{-y^2/2k^2}/(1+\frac{y}{k}) \to 1$ as $k \to \infty$, choose k_0 such that for any $k \geq k_0$, $e^{-x_0^2/2k^2}/(1+x_0/k) \geq 2/3$. Therefore,

$$N(0,k^2)(-\infty, x_0] = \Phi_k(x_0) \leq 1 - \frac{1}{\sqrt{2\pi}} \cdot \frac{2}{3} \leq \frac{3}{4}, \quad \forall k \geq k_0.$$

Then, for $k \geq k_0$, any u satisfying $\sigma_{n_k}(u) \geq k$ also satisfies $\rho(\beta_i, \alpha_{n_k}(u)) \geq \frac{1}{8} > \delta_0$, since $\alpha_{n_k}(u) \sim N(0, \sigma_{n_k}^2(u))$ with variance at least k^2.

Therefore, for $k \geq k_0$,

$$\Theta\{u : \alpha_{n_k}(u, \cdot) \in K_{\varepsilon_0}^c\} \geq \Theta\left\{u : \alpha_{n_k}(u, \cdot) \in \bigcap_{i=1}^{m} B^c(\beta_i, \delta_0)\right\}$$

$$= \Theta\{u : \rho(\alpha_{n_k}(u, \cdot), \beta_i) \geq \delta_0, \ i = 1, \ldots, m\}$$

$$\geq \Theta\{u : \sigma_{n_k}(u) \geq k\} \geq \varepsilon_0,$$

which is a contradiction. $\quad\square$

LEMMA 3.2 *For random probability measures α_n and α on $\mathbf{R}$ define probability measures $\bar{\alpha}_n$ and $\bar{\alpha}$ by*

$$\bar{\alpha}_n(A) = \int_{\mathcal{U}} \alpha_n(u, A)\, \Theta(du) \quad and \quad \bar{\alpha}(A) = \int_{\mathcal{U}} \alpha(u, A)\, \Theta(du),$$

where $A \subset \mathbf{R}$ is measurable. If for any fixed A, $\alpha_n(u, A) \overset{\Theta}{\to} \alpha(u, A)$, then $\bar{\alpha}_n(A) \to \bar{\alpha}(A)$.

Proof of Lemma 3.2.

Fix A to be any measurable set on $\mathbf{R}$. Since α_n are probability measures,

$$|\alpha_n(u, A)| \leq 1 \quad \forall n.$$

So $\{\alpha_n(\cdot, A)\}$ is uniformly integrable. Therefore, by Theorem 4.5.4 in Chung (1974),

$$\int \alpha_n(u, A)\, \Theta(du) \to \int \alpha(u, A)\, \Theta(du),$$

i.e., $\bar{\alpha}_n(A) \to \bar{\alpha}(A)$. $\quad\square$

LEMMA 3.3 *Let α_n and β_n be random measures. If for each fixed u*

$$\alpha_n(u, \cdot) \to \alpha(\cdot) \text{ is equivalent to } \beta_n(u, \cdot) \to \beta(\cdot), \tag{3.15}$$

then

$$\alpha_n(u, \cdot) \overset{\Theta}{\to} \alpha(\cdot) \text{ is equivalent to } \beta_n(u, \cdot) \overset{\Theta}{\to} \beta(\cdot). \tag{3.16}$$

Proof of Lemma 3.3.

$$\alpha_n(u, \cdot) \stackrel{\Theta}{\to} \alpha(\cdot) \Leftrightarrow \forall(n') \; \exists(n'') \subset (n') \text{ and } A \text{ with } \Theta(A) = 1 \text{ such that}$$
$$\alpha_{n''}(u, \cdot) \to \alpha(\cdot), \; \forall u \in A$$
$$\Leftrightarrow \forall(n') \; \exists(n'') \subset (n') \text{ and } A \text{ with } \Theta(A) = 1 \text{ such that}$$
$$\beta_{n''}(u, \cdot) \to \beta(\cdot), \; \forall u \in A$$
$$\Leftrightarrow \beta_n(u, \cdot) \stackrel{\Theta}{\to} \beta(\cdot). \qquad \qquad \square$$

LEMMA 3.4 *Let* $\eta_n(u), \eta(u), \xi_n(u) : (\mathcal{U} \times \Omega) \to \mathbf{R}$ *and* $\xi(u) : \mathcal{U} \to \mathbf{R}$ *be random variables and* $\mathcal{L}(\eta_n(u, \cdot)) \stackrel{\Theta}{\to} \mathcal{L}(\eta(u, \cdot))$. *Then*

$$\mathcal{L}(\xi_n(u, \cdot)) \stackrel{\Theta}{\to} \xi(u) \quad \Rightarrow \quad \mathcal{L}(\eta_n(u, \cdot)\xi_n(u, \cdot)) \stackrel{\Theta}{\to} \mathcal{L}(\eta(u, \cdot)\xi(u)). \qquad (3.17)$$

Proof of Lemma 3.4.

Suppose $\mathcal{L}(\eta_n(u, \cdot)) \stackrel{\Theta}{\to} \mathcal{L}(\eta(u, \cdot))$ and $\mathcal{L}(\xi_n(u, \cdot)) \stackrel{\Theta}{\to} \xi(u)$. Then for any subsequence n' there exists a further subsequence n'' such that Θ-a.s.

$$\mathcal{L}(\eta_{n''}(u, \cdot)) \to \mathcal{L}(\eta(u, \cdot)) \text{ and } \mathcal{L}(\xi_{n''}(u, \cdot)) \to \xi(u).$$

Notice that $\xi(u)$ is like a constant with respect to $\mathcal{L}(\cdot)$, hence, for any n' there exists n'' such that Θ-a.s.

$$\mathcal{L}(\eta_{n''}(u, \cdot)\xi_{n''}(u, \cdot)) \to \mathcal{L}(\eta(u, \cdot)\xi(u)),$$

which is equivalent to

$$\mathcal{L}(\eta_n(u, \cdot)\xi_n(u, \cdot)) \stackrel{\Theta}{\to} \mathcal{L}(\eta(u, \cdot)\xi(u)). \qquad \qquad \square$$

A partial converse to Lemma 3.4 holds under some restrictions.

LEMMA 3.5 *Let* $\eta_n(u), \eta(u) : (\mathcal{U} \times \Omega) \to \mathbf{R}$ *and* $\xi_n(u), \xi(u) : \mathcal{U} \to \mathbf{R}$ *be random variables and let*

$$\mathcal{L}(\eta_n(u, \cdot)) \stackrel{\Theta}{\to} \mathcal{L}(\eta(u, \cdot)). \qquad (3.18)$$

If, in addition, η *is such that*

$$\mathcal{L}(\eta(u, \cdot)\xi_n(u)) \stackrel{\Theta}{\to} \mathcal{L}(\eta(u, \cdot)\xi(u)) \quad \Rightarrow \quad \xi_n(u) \stackrel{\Theta}{\to} \xi(u), \qquad (3.19)$$

then

$$\mathcal{L}(\eta_n(u, \cdot)\xi_n(u)) \stackrel{\Theta}{\to} \mathcal{L}(\eta(u, \cdot)\xi(u)) \quad \Rightarrow \quad \xi_n(u) \stackrel{\Theta}{\to} \xi(u). \qquad (3.20)$$

REMARK 3.1 *The seemingly contrived condition (3.19) is actually fairly common. For example, if* $N(0, \sigma_n^2(u)) \stackrel{\Theta}{\to} N(0, \sigma^2(u))$, *then* $\sigma_n(u) \stackrel{\Theta}{\to} \sigma(u)$. *(This can be verified by using characteristic functions. In fact,* $\sigma_n(u)$ *converges to* $\sigma(u)$ *pointwise in this case.)*

Proof of Lemma 3.5.

Suppose that

$$\mathcal{L}(\eta_n(u, \cdot)) \xrightarrow{\Theta} \mathcal{L}(\eta(u, \cdot)) \tag{3.21}$$

and

$$\mathcal{L}(\eta_n(u, \cdot)\xi_n(u)) \xrightarrow{\Theta} \mathcal{L}(\eta(u, \cdot)\xi(u)). \tag{3.22}$$

We need only to show that

$$\mathcal{L}(\eta(u, \cdot)\xi_n(u)) \xrightarrow{\Theta} \mathcal{L}(\eta(u, \cdot)\xi(u)). \tag{3.23}$$

Let $\varepsilon > 0$. By considering a.s. convergent subsubsequences it is easy to see that (3.21) implies

$$\mathcal{L}(\eta_n(u, \cdot)\xi_m(u)) \xrightarrow{\Theta} \mathcal{L}(\eta(u, \cdot)\xi_m(u)) \quad \text{as } n \to \infty. \tag{3.24}$$

Furthermore, (3.21), an application of the triangle inequality, and the definition of convergence in θ-probability yield

$$\Theta\left\{u : d(\mathcal{L}(\eta_n(u, \cdot)\xi_m(u)), \mathcal{L}(\eta_m(u, \cdot)\xi_m(u)) > \frac{\varepsilon}{3}\right\} \to 0. \tag{3.25}$$

So, by (3.22), (3.24) and (3.25),

$$\Theta\left\{u : d\left(\mathcal{L}\left(\eta(u, \cdot)\xi_m(u)\right), \mathcal{L}\left(\eta(u, \cdot)\xi(u)\right)\right) > \varepsilon\right\}$$
$$\leq \Theta\left\{u : d\left(\mathcal{L}(\eta(u, \cdot)\xi_m(u)), \mathcal{L}(\eta_n(u, \cdot)\xi_m(u))\right) > \frac{\varepsilon}{3}\right\}$$
$$+ \Theta\left\{u : d\left(\mathcal{L}(\eta_n(u, \cdot)\xi_m(u)), \mathcal{L}(\eta_m(u, \cdot)\xi_m(u))\right) > \frac{\varepsilon}{3}\right\}$$
$$+ \Theta\left\{u : d\left(\mathcal{L}(\eta_m(u, \cdot)\xi_m(u)), \mathcal{L}(\eta(u, \cdot)\xi(u))\right) > \frac{\varepsilon}{3}\right\}$$
$$\to 0. \qquad \qquad \square$$

3.2. CLT in Probability We also require a CLT in probability for use in later sections.

DEFINITION 3.2 *Let* $\{\eta_{nk}(u, \omega) : 1 \leq k \leq k_n, n = 1, 2, \cdots\}$ *be a row-wise independent triangular array for each* u. *We say that* $\{\eta_{nk}(u, \omega)\}$ *is u.a.n. in* Θ-*probability if for each* $\varepsilon > 0$ *and* $\delta > 0$,

$$\lim_{n \to \infty} \Theta\{u : \max_k P^u\{|\eta_{nk}(u)| > \varepsilon\} > \delta\} = 0.$$

LEMMA 3.6 *Let* $\{\eta_{nk}(u, \omega) : 1 \leq k \leq k_n, n = 1, 2, \cdots\}$ *be an array and let* $a_n(u)$ *be a sequence of random variables in* u. *Then*

$$\mathcal{L}(S_n(u) - a_n(u)) \xrightarrow{\Theta} \mathcal{L}(Z) \sim N(0, 1) \tag{3.26}$$

and $\{\eta_{nk}\}$ is u.a.n. in Θ-probability if and only if for each $\varepsilon > 0$ the following three conditions hold:

$$\sum_{k=1}^{k_n} P^u \left(|\eta_{nk}(u)| > \varepsilon \right) \; \overset{\Theta}{\to} \; 0, \tag{3.27}$$

$$\sum_{k=1}^{k_n} E^u \left(\eta_{nk}(u) I \left(|\eta_{nk}(u)| < \varepsilon \right) \right) - a_n(u) \; \overset{\Theta}{\to} \; 0, \; and \tag{3.28}$$

$$\sum_{k=1}^{k_n} Var^u \left(\eta_{nk}(u) I \left(|\eta_{nk}(u)| < \varepsilon \right) \right) \; \overset{\Theta}{\to} \; 1. \tag{3.29}$$

Proof of Lemma 3.6.

Suppose (3.26) holds and $\{\eta_{nk}\}$ is u.a.n. in Θ-probability. Since the latter is equivalent to

$$\max_{k \leq k_n} E^u \frac{|\eta_{nk}|}{1 + |\eta_{nk}|} \; \overset{\Theta}{\to} \; 0,$$

upon letting $d(\cdot, \cdot)$ be the Lévy distance, the random vectors of $\mathbf{R}^2$

$$\left(\max_{k \leq k_n} E^u \frac{|\eta_{nk}|}{1 + |\eta_{nk}|}, \; d \left(\mathcal{L}(S_n(u) - a_n(u)), N(0,1) \right) \right) \overset{\Theta}{\to} 0.$$

So for any subsequence n' there exists a further subsequence $n'' \subset n'$ such that

$$\left(\max_{k \leq k_{n''}} E^u \frac{|\eta_{n''k}|}{1 + |\eta_{n''k}|}, \; d \left(\mathcal{L}(S_{n''}(u) - a_{n''}(u)), N(0,1) \right) \right) \to 0 \quad \Theta\text{-a.s.}$$

Hence, for any n' there exists $n'' \subset n'$ and A with $\Theta(A) = 1$ such that $\forall u \in A$

(i) $\max_{k \leq k_{n''}} E^u \dfrac{|\eta_{n''k}|}{1 + |\eta_{n''k}|} \to 0$, which is equivalent to u.a.n. along the subsequence n'';

(ii) $\mathcal{L} \left(S_{n''}(u) - a_{n''}(u) \right) \to N(0,1)$.

Therefore, by the usual Central Limit Theorem (e.g., Araujo and Giné (1980)), we have that for any subsequence n', there exists a further subsequence n'' and a set A with $\Theta(A) = 1$ such that for each $u \in A$

$$\sum_{k=1}^{k_{n''}} P^u \left(|\eta_{n''k}(u)| > \varepsilon \right) \; \to \; 0, \tag{3.30}$$

$$\sum_{k=1}^{k_{n''}} E^u \left(\eta_{n''k}(u) I \left(|\eta_{n''k}(u)| < \varepsilon \right) \right) - a_{n''}(u) \; \to \; 0, \tag{3.31}$$

$$\sum_{k=1}^{k_{n''}} Var^u \left(\eta_{n''k}(u) I \left(|\eta_{n''k}(u)| < \varepsilon \right) \right) \; \to \; 1. \tag{3.32}$$

Thus, these same conditions hold for the full sequence in Θ probability, i.e., (3.27), (3.28) and (3.29) hold.

Since (3.27) implies u.a.n. in Θ-probability, reversing the above argument immediately yields the converse. $\qquad \square$

4. CLT with Variance Mixtures of Normals as Limits

4.1. Necessary Conditions The proof of Theorem 2.1 consists of several lemmas, the first of which is an inequality in Araujo and Giné (1980):

LEMMA 4.1 (LEMMA 5.1, P.123, ARAUJO AND GINÉ (1980))
Let $\{X_i\}_{i=1}^n$ be independent symmetric random variables, A a symmetric Borel set, and $S = \sum_{i=1}^n X_i$, $S_A = \sum_{i=1}^n X_i I_A(X_i)$. Then for every convex symmetric Borel set K,

$$2P(S \notin K) \geq P(S_A \notin K).$$

We first establish that under the assumptions of Theorem 2.1 the Levy measures for the conditional i.i.d. random variables $X_1(u)$ converge to 0 in Θ-probability. Specifically,

LEMMA 4.2 *Suppose $\exists b_n \uparrow \infty$ such that*

$$\mathcal{L}\left(\frac{S_n}{b_n}\right) \to \mathcal{L}(W) = \int N(0, \sigma^2(u))\,\Theta(du) \tag{4.33}$$

for $\sigma(u) > 0$ a.s. Θ and $\forall C_1 > 0 \; \exists C_2 > 0$ such that

$$\int \sigma(u) \exp\left(C_2 k \ln k - \frac{C_1 k^2}{\sigma^2(u)}\right) \Theta(du) \to 0 \quad \text{as } k \to \infty. \tag{4.34}$$

Then

$$nP^u(|X_1(u)| > \varepsilon b_n) \overset{\Theta}{\to} 0. \tag{4.35}$$

Proof of Lemma 4.2.
The proof is similar to that of Lemma 1 in Klass and Teicher (1987). Let $\varepsilon > 0$ and $k > 0$. Then

$$\begin{aligned} S_n(u) &= \sum_{j=1}^n X_j(u)I(|X_j(u)| > \varepsilon b_n) + \sum_{j=1}^n X_j(u)I(|X_j(u)| \leq \varepsilon b_n) \\ &\equiv L_n(u) + M_n(u). \end{aligned}$$

The fact that each $X_j(u)$ is symmetric leads to symmetry of $L_n(u)$ and $M_n(u)$. Let $q > 0$. By Lemma 4.1 with $A = [-\varepsilon b_n, \varepsilon b_n]^c$ and $K = [-q, q]$,

$$P(|S_n(u)| > q) \geq P(L_n(u) > q). \tag{4.36}$$

Consequently, (4.33), inequality (3.14) and (4.36) imply that

$$\begin{aligned} \int \sqrt{\frac{2}{\pi}} \frac{\sigma(u)}{k\varepsilon} e^{-k^2\varepsilon^2/2\sigma^2(u)}\,\Theta(du) &\geq \lim_{n\to\infty} P(|S_n| > k\varepsilon b_n) \\ &= \lim_{n\to\infty} \int P^u(|S_n(u)| > k\varepsilon b_n)\,\Theta(du) \\ &\geq \limsup_{n\to\infty} \int P^u\left(\sum_{j=1}^n X_j(u)I(|X_j(u)| > \varepsilon b_n) > k\varepsilon b_n\right) \Theta(du). \end{aligned} \tag{4.37}$$

Let

$$\tau = \inf \left\{ 1 \leq h \leq n : \sum_{j=1}^{h} I(|X_j(u)| > \varepsilon b_n) \geq k \right\}$$

with $\tau = \infty$ if no such h exists. Hence, by the definition of τ, the independence of $X_1(u), \ldots, X_n(u)$, and the symmetry of $X_j(u)$,

$$P^u \left(\sum_{j=1}^{n} X_j(u) I(|X_j(u)| > \varepsilon b_n) > k\varepsilon b_n \right)$$

$$\geq P^u \left(\sum_{j=1}^{n} X_j(u) I(|X_j(u)| > \varepsilon b_n) > k\varepsilon b_n, \tau \leq n \right)$$

$$= \sum_{i=1}^{n} P^u \left(\left(\sum_{j=1}^{i} + \sum_{j=i+1}^{n} \right) X_j(u) I(|X_j(u)| > \varepsilon b_n) > k\varepsilon b_n, \tau = i \right)$$

$$\geq \sum_{i=1}^{n} P^u \left(\sum_{j=1}^{i} X_j(u) I(|X_j(u)| > \varepsilon b_n) > k\varepsilon b_n, \tau = i \right) \cdot \frac{1}{2} \qquad (4.38)$$

$$= \frac{1}{2} P^u \left(\sum_{j=1}^{\tau} X_j(u) I(|X_j(u)| > \varepsilon b_n) > k\varepsilon b_n, \tau \leq n \right)$$

$$\geq \frac{1}{2^{k+1}} P^u (\tau \leq n) \quad \text{using the definition of } \tau.$$

Let $\delta > 0$ and $Y_{n1}, \ldots, Y_{nn}$ be i.i.d. with

$$P^u(Y_{n1} = 1) = \delta/n = 1 - P^u(Y_{n1} = 0).$$

Let $A_n = \{u : nP^u(|X_1(u)| > \varepsilon b_n) > \delta\}$. Then, by the Poisson approximation to the binomial,

$$P^u((\tau \leq n) \cap A_n) \geq P^u \left(\sum_{j=1}^{n} I(|X_j(u)| > \varepsilon b_n) \geq k, A_n \right)$$

$$\geq P^u \left(\sum_{j=1}^{n} Y_{nj} \geq k, A_n \right) \qquad (4.39)$$

$$= \frac{\delta^k e^{-\delta}}{k!} + \mathcal{O}\left(\frac{1}{n} \right).$$

So, (4.37), (4.38), and (4.39) combine to yield

$$\int \sqrt{\frac{2}{\pi}} \frac{\sigma(u)}{k\varepsilon} e^{-k^2 \varepsilon^2 / 2\sigma^2(u)} \, \Theta(du)$$

$$\geq \quad \limsup_{n \to \infty} \frac{1}{2^{k+1}} \int_{A_n} P^u(\tau \leq n) \, \Theta(du) \qquad (4.40)$$

$$\geq \quad \frac{\delta^k e^{-\delta}}{2^{k+1}k!} \limsup_{n \to \infty} \Theta(A_n).$$

Using Stirling's formula, $k! \sim \sqrt{2\pi}k^{k+1/2}e^{-k}$ and (4.40), for k large,

$$\limsup_{n \to \infty} \Theta(A_n)$$

$$\leq \quad \frac{2\sqrt{2\pi}k^{k+1/2}e^{-k}}{(\delta/2)^k e^{-\delta}} \sqrt{\frac{2}{\pi}} \frac{1}{k\varepsilon} \int \sigma(u) \exp\left\{-\frac{k^2 \varepsilon^2}{2\sigma^2(u)}\right\} \Theta(du)$$

$$= \quad C_\delta \int \sigma(u) \exp\left\{(k+\frac{1}{2})\ln k - k - k\ln(\delta/2) - \ln k - \frac{k^2\varepsilon^2}{2\sigma^2(u)}\right\} \Theta(du)$$

$$\leq \quad C_\delta \int \sigma(u) \exp\left\{C_1 k \ln k - C_2 \frac{k^2}{\sigma^2(u)}\right\} \Theta(du)$$

$$\to \quad 0 \text{ as } k \to \infty, \text{ by condition (4.34).}$$

Since this behavior is valid for all $\delta > 0$, (4.35) holds. $\qquad \square$

In the presence of (4.34), validity of the CLT for a sequence $X_1, X_2, \ldots$ implies validity of the CLT for certain truncated terms from that sequence. Substantiation of this fact makes use of the following lemma whose proof is routine.

LEMMA 4.3 *Let η_n and ξ_n be sequences of random variables and assume $\xi_n \xrightarrow{D} \xi$. Then $P(\eta_n \neq \xi_n) \to 0$ implies $\eta_n \xrightarrow{D} \xi$.*

LEMMA 4.4 *Suppose $\exists b_n \uparrow \infty$ such that*

$$\mathcal{L}\left(\frac{S_n}{b_n}\right) \to \mathcal{L}(W) = \int N(0, \sigma^2(u)) \, \Theta(du) \qquad (4.41)$$

for $\sigma(u) > 0$ a.s.Θ and $\forall C_1 > 0 \, \exists C_2 > 0$ such that

$$\int \sigma(u) \exp\left(C_2 k \ln k - \frac{C_1 k^2}{\sigma^2(u)}\right) \Theta(du) \to 0 \quad \text{as } k \to \infty. \qquad (4.42)$$

Then there exists $\varepsilon_n \downarrow 0$ such that if $T_n \equiv \sum_{j=1}^n X_j I(|X_j| \leq \varepsilon_n b_n)$, then

$$\mathcal{L}\left(\frac{T_n}{b_n}\right) \to \mathcal{L}(W). \qquad (4.43)$$

Proof of Lemma 4.4
The proof follows that of Theorem 1 in Klass and Teicher (1987). For any $\varepsilon > 0$ and $\delta > 0$ let $A_n(\varepsilon, \delta) = \{u : nP^u(|X_1(u)| > \varepsilon b_n) > \delta\}$. By Lemma 4.2,

$nP^u(|X_1(u)| > \varepsilon b_n) \xrightarrow{\Theta} 0$, which implies that $\lim_{n\to\infty} \Theta\{A_n(\varepsilon, \delta)\} = 0$. Thus, there exists $\varepsilon_n \downarrow 0$ such that

$$\lim_{n\to\infty} \Theta\{A_n(\varepsilon_n, \delta)\} = 0. \qquad (4.44)$$

For each u, let $T_n(u) \equiv \sum_{j=1}^{n} X_j(u)I(|X_j(u)| \leq \varepsilon_n b_n)$. Thus,

$$
\begin{aligned}
P(S_n \neq T_n) &\leq P\left(\bigcup_{i=1}^{n}\{|X_i| > \varepsilon_n b_n\}\right) \\
&= \int P^u\left(\bigcup_{i=1}^{n}\{|X_i(u)| > \varepsilon_n b_n\}\right) \Theta(du) \\
&\leq \int_{A_n(\varepsilon_n,\delta)} \Theta(du) + \int_{A_n^c(\varepsilon_n,\delta)} nP^u(|X_1(u)| > \varepsilon_n b_n)\,\Theta(du) \\
&\leq \Theta(A_n(\varepsilon_n,\delta)) + \delta \\
&\to \delta \quad \text{as } n\to\infty \text{ by } (4.44) \\
&\to 0 \quad \text{as } \delta\to 0.
\end{aligned}
$$

Consequently, by Lemma 4.3, $\mathcal{L}\left(\dfrac{T_n}{b_n}\right) \to \mathcal{L}(W)$ since $\mathcal{L}(W)$. $\qquad\square$

REMARK 4.1 *Theorem 1 in Klass and Teicher (1987) is a special case of Lemma 4.4 when $\mathcal{L}(W) = N(0,1)$.*

LEMMA 4.5 *Suppose $\exists b_n \uparrow \infty$ such that*

$$\mathcal{L}\left(\frac{S_n}{b_n}\right) \to \mathcal{L}(W) = \int N(0, \sigma^2(u))\,\Theta(du) \qquad (4.45)$$

for $\sigma(u) > 0$ a.s.Θ and $\forall C_1 > 0$ $\exists C_2 > 0$ such that

$$\int \sigma(u)\exp\left(C_2 k \ln k - \frac{C_1 k^2}{\sigma^2(u)}\right)\Theta(du) \to 0 \quad \text{as } k\to\infty. \qquad (4.46)$$

Let $\varepsilon_n \downarrow 0$ be as in Lemma 4.4 and $b_n(u)$ be as in (1.3). Then

$$\lim_{n\to\infty} \int P^u\left(\frac{b_n(u)}{b_n} Z \leq x\right)\Theta(du) = P(W \leq x), \quad \forall x.$$

Proof of Lemma 4.5.

Let $T_n(u) \equiv \sum_{j=1}^{n} X_j(u)I(|X_j(u)| \leq \varepsilon_n b_n)$. Since Lemma 4.4 shows that

$$\lim_{n\to\infty} \int P^u\left(\frac{T_n(u)}{b_n} \leq x\right)\Theta(du) = P(W \leq x), \quad \forall x,$$

it suffices to show that

$$\lim_{n\to\infty} \int \left| P^u\left(\frac{b_n(u)}{b_n}Z \le x\right) - P^u\left(\frac{T_n(u)}{b_n} \le x\right) \right| \Theta(du) = 0, \quad \forall x.$$

To this end, let

$$A_n = \left\{ u : \frac{b_n(u)}{b_n} < \sqrt{\varepsilon_n} \right\}.$$

Via a string of routine inequalities,

$$\left| P^u\left(\frac{b_n(u)}{b_n}Z \le x\right) - P^u\left(\frac{T_n(u)}{b_n} \le x\right) \right|$$

$$\le \quad P^u\left(\left|\frac{b_n(u)}{b_n}Z\right| \ge \varepsilon_n^{1/4}\right) + P^u\left(\left|\frac{T_n(u)}{b_n}\right| \ge \varepsilon_n^{1/4}\right) + 2P^u(|x| < \varepsilon_n^{1/4}).$$

Therefore,

$$\int_{A_n} \left| P^u\left(\frac{b_n(u)}{b_n}Z \le x\right) - P^u\left(\frac{T_n(u)}{b_n} \le x\right) \right| \Theta(du)$$

$$\le \quad \int_{A_n} P^u\left(\left|\frac{b_n(u)}{b_n}Z\right| \ge \varepsilon_n^{1/4}\right) \Theta(du) + \int_{A_n} P^u\left(\left|\frac{T_n(u)}{b_n}\right| \ge \varepsilon_n^{1/4}\right) \Theta(du)$$

$$+2 \int_{A_n} P^u(|x| < \varepsilon_n^{1/4}) \Theta(du)$$

$$\equiv \quad I_{n1} + I_{n2} + 2I_{n3}.$$

By the definition of the set A_n and the continuity of the standard normal distribution,

$$I_{n1} \le \int_{A_n} P^u(|Z| \ge \varepsilon_n^{-1/4}) \Theta(du) \le P(|Z| \ge \varepsilon_n^{-1/4}) \to 0$$

since $\varepsilon_n \to 0$. The definition of $b_n(u)$ and Chebychev's inequality together with the definition of A_n give

$$I_{n2} \le \int_{A_n} \frac{b_n^2(u)}{b_n^2 \varepsilon_n^{1/2}} \Theta(du) \le \int_{A_n} \varepsilon_n^{1/2} \Theta(du) \to 0.$$

Applying Lemma 4.4 and the fact that $I_{n2} \to 0$ yields

$$\limsup_{n\to\infty} I_{n3} \quad \le \quad \limsup_{n\to\infty} \left\{ \int_{A_n} P^u\left(|x| < \varepsilon_n^{1/4}, \left|\frac{T_n(u)}{b_n}\right| \le \varepsilon_n^{1/4}\right) \Theta(du) \right.$$

$$\left. + \int_{A_n} P^u\left(\left|\frac{T_n(u)}{b_n}\right| \ge \varepsilon_n^{1/4}\right) \Theta(du) \right\}$$

$$\le \quad \limsup_{n\to\infty} \int_{A_n} P^u\left(\left|\frac{T_n(u)}{b_n} - x\right| \le 2\varepsilon_n^{1/4}\right) \Theta(du) + \limsup_{n\to\infty} I_{n2}$$

$$\le \quad \limsup_{n\to\infty} P(x - 2\varepsilon_n^{1/4} \le W \le x + 2\varepsilon_n^{1/4})$$

$$= \quad 0,$$

using the continuity of a variance mixture of nondegenerate normals.

On the other hand, since for any fixed u, the regular CLT holds for $T_n(u)$ with norming $b_n(u)$, i.e.,

$$\mathcal{L}\left(\frac{T_n(u)}{b_n(u)}\right) \to N(0,1),$$

the Berry-Esseen Theorem implies

$$\int_{A_n^c} \left| P^u\left(\frac{b_n(u)}{b_n}Z \le x\right) - P^u\left(\frac{T_n(u)}{b_n} \le x\right)\right| \Theta(du)$$

$$= \int_{A_n^c} \left| P^u\left(Z \le \frac{b_n}{b_n(u)}x\right) - P^u\left(\frac{T_n(u)}{b_n(u)} \le \frac{b_n}{b_n(u)}x\right)\right| \Theta(du)$$

$$\le C\int_{A_n^c} \frac{nE^u|X_1(u)|^3 I(|X_1(u)| \le \varepsilon_n b_n)}{b_n^3(u)} \Theta(du)$$

$$\le C\int_{A_n^c} \frac{\varepsilon_n b_n \cdot b_n^2(u)}{b_n^3(u)} \Theta(du)$$

$$\le C\sqrt{\varepsilon_n} \to 0.$$

Therefore,

$$\lim_{n\to\infty} \int \left| P^u\left(\frac{b_n(u)}{b_n}Z \le x\right) - P^u\left(\frac{T_n(u)}{b_n} \le x\right)\right| \Theta(du) = 0, \quad \forall x.$$

Thus,

$$\lim_{n\to\infty} \int \mathcal{L}\left(\frac{b_n(u)}{b_n}Z\right)\Theta(du) = \mathcal{L}(W). \qquad \square$$

LEMMA 4.6 *Suppose* $\exists b_n \uparrow \infty$ *such that*

$$\mathcal{L}\left(\frac{S_n}{b_n}\right) \to \mathcal{L}(W) = \int N(0,\sigma^2(u))\,\Theta(du)$$

for $\sigma(u) > 0$ *a.s.Θ and* $\forall C_1 > 0$ $\exists C_2 > 0$ *such that*

$$\int \sigma(u)\exp\left(C_2 k\ln k - \frac{C_1 k^2}{\sigma^2(u)}\right)\Theta(du) \to 0 \quad \text{as } k \to \infty.$$

Let ε_n *be as in Lemma 4.4 and let* $b_n(u)$ *be as in (1.3). Then there exists* $\tau(u)$ *satisfying*

$$\int N(0,\tau^2(u))\,\Theta(du) = \int N(0,\sigma^2(u))\,\Theta(du) \tag{4.47}$$

such that

$$\frac{b_n(u)}{b_n} \overset{\mathcal{D}}{\to} \tau(u).$$

Proof of Lemma 4.6. By Corollary 3.1, $\left\{\dfrac{b_n(u)}{b_n}\right\}$ is tight. So for every sequence n' there exists a further subsequence n'' and $\tau(u)$ such that

$$\frac{b_{n''}(u)}{b_{n''}} \xrightarrow{\ \mathcal{D}\ } \tau(u),$$

where $\tau(u)$ has a proper probability law. Also, by Lemma 4.5 and a characteristic function argument

$$\int \mathcal{L}\left(\tau(u)Z\right)\Theta(du) = \lim_{n''\to\infty}\int \mathcal{L}\left(\frac{b_{n''}(u)}{b_{n''}}Z\right)\Theta(du) = \int \mathcal{L}\left(\sigma(u)Z\right)\Theta(du).$$

Since n' is arbitrary, the above relation holds with the full sequence n in place of n''. Therefore,

$$\frac{b_n(u)}{b_n} \xrightarrow{\ \mathcal{D}\ } \tau(u)$$

and (4.47) holds.   □

Lemma 4.2 and Lemma 4.6 now give Theorem 2.1.

4.2. Sufficient Conditions
Proof of Theorem 2.2.

By assumption, there exist $b_n \uparrow \infty$ and $\varepsilon_n \downarrow 0$ such that

$$nP^u(|X_1(u)| > \varepsilon_n b_n) \xrightarrow{\ \Theta\ } 0. \tag{4.48}$$

Define, for b_n and ε_n given above, T_n and $T_n(u)$ as in Lemma 4.4 and let $b_n(u)$ be as in (1.3). By the usual CLT,

$$\mathcal{L}\left(\frac{T_n(u)}{b_n(u)}\right) \to N(0,1) \quad \forall u. \tag{4.49}$$

Then, since $\dfrac{b_n(u)}{b_n} \xrightarrow{\ \Theta\ } \sigma(u),$

$$\mathcal{L}\left(\frac{T_n(u)}{b_n}\right) = \mathcal{L}\left(\frac{T_n(u)}{b_n(u)}\cdot\frac{b_n(u)}{b_n}\right) \xrightarrow{\ \Theta\ } N(0,\sigma^2(u)) \tag{4.50}$$

by (4.49) and Lemma 3.4. Therefore, Lemma 3.2 and (4.50) lead to

$$\mathcal{L}\left(\frac{T_n}{b_n}\right) = \int \mathcal{L}\left(\frac{T_n(u)}{b_n}\right)\Theta(du) \to \int N(0,\sigma^2(u))\Theta(du). \tag{4.51}$$

Condition (4.48) and the proof of Lemma 4.4 imply that $P(S_n \neq T_n) \to 0$. Hence

$$\mathcal{L}\left(\frac{S_n}{b_n}\right) \to \int N(0,\sigma^2(u))\Theta(du),$$

by Lemma 4.3 and (4.51).   □

5. Empirical CLTs

This section will first prove that the regular CLT implies the empirical CLT. Then the two characterizations of the empirical CLT, namely Theorem 2.4 and Theorem 2.5, will be verified.

5.1. Validity of the Regular CLT Implies that of the Empirical CLT

Proof of Theorem 2.3.

Since, by assumption,

$$\frac{S_n}{b_n} = \frac{S_n}{V_n} \cdot \frac{V_n}{b_n} \xrightarrow{\mathcal{D}} Z,$$

it suffices to show that

$$K_n^2 \equiv \frac{V_n^2}{b_n^2} \xrightarrow{\mathcal{P}} 1. \tag{5.52}$$

By Theorem 1.3, $\dfrac{S_n}{b_n} \xrightarrow{\mathcal{D}} Z$ implies that for some $\varepsilon_n \downarrow 0$,

$$nP^u(|X_1(u)| > \varepsilon_n b_n) \xrightarrow{\Theta} 0 \tag{5.53}$$

and

$$\frac{b_n(u)}{b_n} \xrightarrow{\Theta} 1, \tag{5.54}$$

where $b_n(u)$ is as in (1.3) with ε_n as in (5.53). Thus,

$$\left(nP^u\left(|X_1(u)| > \varepsilon_n b_n\right), \frac{b_n(u)}{b_n}\right) \xrightarrow{\Theta} (0,1),$$

which implies that for any subsequence $\{n'\}$ there exists a further subsequence $\{n''\} \subset \{n'\}$ and a set A with $\Theta(A) = 1$ such that

$$\left(n''P^u\left(|X_1(u)| > \varepsilon_{n''}b_{n''}\right), \frac{b_{n''}(u)}{b_{n''}}\right) \to (0,1) \quad \forall u \in A.$$

Therefore, $\forall u \in A$,

$$n''P^u\left(|X_1(u)| > \varepsilon_{n''}b_{n''}\right) \to 0 \tag{5.55}$$

$$\frac{n''}{b_{n''}^2}E^u X_1^2(u)I(|X_1(u)| \le \varepsilon_{n''}b_{n''}) \to 1. \tag{5.56}$$

We first establish that (5.55) and (5.56) imply

$$K_{n''}^2(u) \equiv \frac{V_{n''}^2(u)}{b_{n''}^2} \xrightarrow{\mathcal{P}} 1 \quad \forall u \in A. \tag{5.57}$$

$$P^u\left(\left|K^2_{n''}(u)-1\right|>\varepsilon\right)$$

$$\leq P^u\left(\left|\frac{1}{b^2_{n''}}\sum_{i=1}^{n''}X^2_i(u)I(|X_i(u)|>\varepsilon_{n''}b_{n''})\right|>\frac{\varepsilon}{2}\right)$$

$$+P^u\left(\left|\frac{1}{b^2_{n''}}\sum_{i=1}^{n''}X^2_i(u)I(|X_i(u)|\leq\varepsilon_{n''}b_{n''})-1\right|>\frac{\varepsilon}{2}\right)$$

$$\leq P^u\left(\max_{1\leq i\leq n''}\frac{X^2_i(u)}{b^2_{n''}}>\varepsilon_{n''}\right)$$

$$+P^u\left(\left|\frac{1}{b^2_{n''}}\sum_{i=1}^{n''}X^2_i(u)I(|X_i(u)|\leq\varepsilon_{n''}b_{n''})-1\right|>\frac{\varepsilon}{2}\right)$$

$$\equiv I_{n''}(u)+II_{n''}(u).$$

For $u\in A$, $I_{n''}(u)\to 0$ by (5.55).

$$II_{n''}(u)\ \leq\ P^u\left(\frac{1}{b^2_{n''}}\left|\sum_{i=1}^{n''}X^2_i(u)I(|X_i(u)|\leq\varepsilon_{n''}b_{n''})\right.\right.$$

$$\left.\left.-n''E^uX^2_1(u)I(|X_1(u)|\leq\varepsilon_{n''}b_{n''})\right|>\frac{\varepsilon}{4}\right)$$

$$+P^u\left(\left|\frac{n''E^uX^2_1(u)I(X_1(u)|\leq\varepsilon_{n''}b_{n''})}{b^2_{n''}}-1\right|>\frac{\varepsilon}{4}\right).$$

The second term above converges to 0 by (5.56). The first term, by Chebychev's inequality, is bounded above by

$$\frac{16}{\varepsilon^2}\cdot\frac{1}{b^4_{n''}}n''E^uX^4_1(u)I(|X_1(u)|\leq\varepsilon_{n''}b_{n''})$$

$$\leq\frac{16}{\varepsilon^2}\cdot\varepsilon^2_{n''}\cdot\frac{n''}{b^2_{n''}}E^uX^2_1(u)I(|X_1(u)|\leq\varepsilon_{n''}b_{n''})$$

$$\to 0\quad\text{by (5.56) and the fact that }\varepsilon_{n''}\downarrow 0.$$

Therefore, (5.57) is valid which implies that

$$E^u\frac{|K_{n''}(u)-1|}{|K_{n''}(u)-1|+1}\to 0\quad\forall u\in A.$$

Invoking de Finetti's theorem and dominated convergence, implies that

$$E\frac{|K_{n''}-1|}{|K_{n''}-1|+1}=\int E^u\frac{|K_{n''}(u)-1|}{|K_{n''}(u)-1|+1}\Theta(du)\to 0,$$

or, $K_{n''}\overset{P}{\to}1$. So, for any $\{n'\}$ there exists $\{n''\}$ such that $K_{n''}\overset{P}{\to}1$. Hence $K_n\overset{P}{\to}1$, which finishes the proof of Theorem 2.3. $\square$

5.2.   Characterization of the Empirical CLT   The proof of Theorem 2.4 will use the following proposition from Griffin and Mason (1991) which does not require rowwise independence of the triangular array under consideration.

PROPOSITION 5.1 (PROPOSITION 3.1 OF GRIFFIN AND MASON (1991))
Let ε_i be a sequence of i.i.d. Bernoulli random variables with success probability $1/2$. Let $\{m\}$ denote a sequence of positive integers and let ξ_{mi}, $1 \leq i \leq m$, be a triangular array of random variables independent of the sequence ε_i. Let $\Delta_m = \xi_{m1}^2 + \cdots + \xi_{mm}^2$ and assume that

$$\Delta_m \leq 1 \quad and \quad \Delta_m \xrightarrow{P} 1. \tag{5.58}$$

Set

$$Z_m = \sum_{i=1}^{m} \varepsilon_i \xi_{mi} \quad and \quad \xi_m^* = \max_{1 \leq i \leq m} |\xi_{mi}|.$$

Then

$$Z_m \to Z \quad if\ and\ only\ if \quad \xi_m^* \xrightarrow{P} 0.$$

Proof of Theorem 2.4.
   (i) $\Leftrightarrow$ (ii): This is a direct application of Proposition 5.1 with

$$\xi_{mi} = \frac{|X_i|}{V_m} I(|V_m| > 0).$$

The non-degeneracy of $\mathcal{L}(X)$ guarantees that (5.58) holds.
   (iii) $\Rightarrow$ (ii): First, by de Finetti's theorem, for any $\varepsilon > 0$, as $n \to \infty$,

$$P\left(|X_n^{(1)}/V_n| > \varepsilon\right) = \int P^u\left(|X_n^{(1)}(u)/V_n(u)| > \varepsilon\right) \Theta(du) \to 0.$$

In particular, $P^u\left(|X_n^{(1)}(u)/V_n(u)| > \varepsilon\right)$ converges in L^1 to 0, hence converges in Θ probability to 0, namely,

$$\mathcal{L}\left(\frac{X_n^{(1)}(u)}{V_n(u)}\right) \xrightarrow{\Theta} \mathcal{L}(0).$$

Now let $\alpha_n(u, \cdot) = \mathcal{L}\left(\dfrac{X_n^{(1)}(u)}{V_n(u)}\right)$, $\alpha(\cdot) = \mathcal{L}(0)$, $\beta_n(u, \cdot) = \mathcal{L}\left(\dfrac{X_n^{(1)}(u)}{S_n(u)}\right)$, and finally let $\beta(\cdot) = \mathcal{L}(0)$. Theorem 1.4 indicates that for each fixed u

$$\alpha_n(\cdot, u) \equiv \mathcal{L}\left(\frac{X_n^{(1)}(u)}{V_n(u)}\right) \to \alpha(\cdot) \equiv \mathcal{L}(0) \Leftrightarrow$$

$$\beta_n(u, \cdot) \equiv \mathcal{L}\left(\frac{X_n^{(1)}(u)}{S_n(u)}\right) \to \beta(\cdot) \equiv \mathcal{L}(0).$$

Then, by Lemma 3.3,

$$\mathcal{L}\left(\frac{X_n^{(1)}(u)}{S_n(u)}\right) \overset{\Theta}{\to} \mathcal{L}(0).$$

Therefore, via de Finetti's theorem and the dominated convergence theorem,

$$\mathcal{L}\left(\frac{X_n^{(1)}}{S_n}\right) = \int \mathcal{L}\left(\frac{X_n^{(1)}(u)}{S_n(u)}\right) \Theta(du) \to 0,$$

which gives (ii).

(ii) $\Rightarrow$ (iii): The argument is exactly the same as (iii) $\Rightarrow$ (ii). $\qquad\square$

Proof of Theorem 2.5.

(i) $\Leftrightarrow$ (iii) $\Leftrightarrow$ (iv): These implications are embedded in the proof of Theorem 2.4.

(ii) $\Leftrightarrow$ (iii): Let $\alpha_n(u, \cdot) = \mathcal{L}\left(\frac{S_n(u)}{V_n(u)}\right)$, $\beta_n(u, \cdot) = \mathcal{L}\left(\frac{X_n^{(1)}(u)}{S_n(u)}\right)$,

$\alpha(\cdot) = \mathcal{L}(Z)$, and $\beta(\cdot) = \mathcal{L}(0)$. Theorem 1.4 indicates that for each fixed u

$$\mathcal{L}\left(\frac{S_n(u)}{V_n(u)}\right) \to \mathcal{L}(Z) \Leftrightarrow \mathcal{L}\left(\frac{X_n^{(1)}(u)}{S_n(u)}\right) \to \mathcal{L}(0).$$

Then, by Lemma 3.3,

$$\mathcal{L}\left(\frac{S_n(u)}{V_n(u)}\right) \overset{\Theta}{\to} \mathcal{L}(0) \Leftrightarrow \mathcal{L}\left(\frac{X_n^{(1)}(u)}{S_n(u)}\right) \overset{\Theta}{\to} \mathcal{L}(0).$$

(iv) $\Leftrightarrow$ (v) $\Leftrightarrow$ (vi): These are similar to (ii) $\Leftrightarrow$ (iii).

(v) $\Leftrightarrow$ (vii): This is an application of Lemma 3.6 with $\eta_{nk} = X_k(u)/\tilde{b}_n(u)$, $k = 1, 2, \ldots, n$, and $a_n(u) = 0$. $\qquad\square$

The next example shows the empirical CLT does not imply the Regular CLT for exchangeable random variables even when each mixand is in the DOA of the normal.

EXAMPLE 5.1 *For each $k \in \mathbf{N}$, let $X_1(k)$ be such that*

$$G_k(t) \equiv P(|X_1(k)| > t) = c_k t^{-2}(\ln t)^k, \quad c_k = (2e/k)^k, \quad t > e^{k/2}.$$

For each k,

$$U_k(t) \equiv EX^2(k)I(|X(k)| \le t) = 2c_k \int_{e^{k/2}}^t \frac{(\ln s)^k}{s}\, ds \to \infty \text{ as } t \to \infty.$$

Hence, $EX^2(k) = \infty$. But, by L'Hospital's rule

$$\lim_{t\to\infty} \frac{t^2 G_k(t)}{U_k(t)} = \lim_{t\to\infty} \frac{t^2 c_k t^{-2}(\ln t)^k}{2c_k \int_{e^{k/2}}^t \frac{(\ln s)^k}{s}\, ds} = \lim_{t\to\infty} \frac{k(\ln t)^{k-1}/t}{2(\ln t)^k/t} = 0.$$

Thus, $X(k)$ is in the DOA of the normal for each k.

Since $U_k(t)$ is a continuous function of t, the canonical norming constant

$$\tilde{b}_n(k) = \sup\left\{t > 0 : nU_k(t) \geq t^2\right\}$$
$$= \sup\left\{t > 0 : 2nc_k \int_{e^{k/2}}^t (\ln s)^k/s\, ds \geq t^2\right\},$$

is the solution to $nU_k(t) = t^2$, i.e.,

$$\tilde{b}_n^2(k) = n \cdot \frac{2c_k}{k+1}\left[\left(\ln \tilde{b}_n(k)\right)^{k+1} - \left(\frac{k}{2}\right)^{k+1}\right]. \tag{5.59}$$

Moreover, as canonical norming constants for convergence to a normal distribution, $\tilde{b}_n(k)$ has the form $\tilde{b}_n(k) = \sqrt{n}L_k(n)$, where $L_k(\cdot)$ is a slowly varying function. Consequently, after finitely many iterations

$$\frac{\ln\ln\ldots\ln \tilde{b}_n(k)}{\ln n} \to 0, \quad as\ n \to \infty. \tag{5.60}$$

For each $k \in \mathbf{N}$, define $b_n(k)$ by

$$b_n^2(k) = nd_k(\ln n)^{k+1}, \quad where\ d_k = \frac{c_k}{(k+1)2^k}.$$

Using the definition of $b_n(k)$, (5.59) and (5.60), routine computations lead to

$$\lim_{n\to\infty} \frac{\tilde{b}_n^2(k)}{b_n^2(k)} = 1 \quad for\ each\ k.$$

Thus, by the convergence of types theorem,

$$\mathcal{L}\left(\frac{S_n(k)}{b_n(k)}\right) \to N(0,1), \quad for\ each\ k. \tag{5.61}$$

Now, let

$$\mathcal{L}(X) = \sum_{k=1}^{\infty} 2^{-k}\mathcal{L}(X(k)).$$

Then, by (5.61) and Theorem 2.5,

$$\mathcal{L}\left(\frac{S_n}{V_n}\right) \to N(0,1).$$

However, there does not exist b_n such that for some W nondegenerate

$$\mathcal{L}\left(\frac{S_n}{b_n}\right) = \sum_{k=1}^{\infty} 2^{-k}\mathcal{L}\left(\frac{S_n(k)}{b_n(k)} \cdot \frac{b_n(k)}{b_n}\right) \to \mathcal{L}(W)$$

because

- if b_n is such that $\dfrac{b_n(k)}{b_n} \to 0$ for all k, then the limit of S_n/b_n is degenerate at 0;

- if b_n is such that $\dfrac{b_n(k)}{b_n} \to \infty$ for all k, then $\left\{ \mathcal{L}\left(\dfrac{S_n}{b_n} \right) \right\}$ is not tight;

- if there exists k_0 such that $b_n(k_0) \asymp b_n$, i.e.,

$$0 < \liminf_{n \to \infty} \frac{b_n(k_0)}{b_n} \le \limsup_{n \to \infty} \frac{b_n(k_0)}{b_n} < \infty,$$

then

$$\lim_{n \to \infty} \frac{b_n(k)}{b_n} = \infty, \quad \forall k > k_0$$

and

$$\lim_{n \to \infty} \frac{b_n(k)}{b_n} = 0, \quad \forall k < k_0,$$

which leads to lack of tightness of $\left\{ \mathcal{L}\left(\dfrac{S_n}{b_n} \right) \right\}$. $\qquad\square$

6. Regular CLT

Proof of Theorem 2.6.
(i) $\Rightarrow$ (ii): Let $\varepsilon > 0$. By Theorem 1.3,

$$nP^u\left(|X_1(u)| > \varepsilon b_n \right) \overset{\Theta}{\to} 0.$$

Since $\mathcal{L}(X_1(u))$ is symmetric,

$$\frac{1}{b_n} n E^u X_1(u) I(|X_1(u)| \le \varepsilon b_n) = 0.$$

Also, let $\varepsilon_n \downarrow 0$ be as in the definition of $b_n(u)$. Then by Theorem 1.3 and the definition of $b_n(u)$,

$$\frac{n}{b_n^2} E^u \left(X_1^2(u) I(|X_1(u)| \le \varepsilon b_n) \right)$$
$$= \frac{n}{b_n^2} E^u X_1^2(u) I(|X_1(u)| \le \varepsilon_n b_n) + \frac{n}{b_n^2} E^u X_1^2(u) I(\varepsilon_n b_n < |X_1(u)| \le \varepsilon b_n)$$
$$\le \frac{b_n^2(u)}{b_n^2} + \varepsilon^2 n P^u(|X_1(u)| > \varepsilon_n b_n) \overset{\Theta}{\to} 1.$$

Then appealing to Lemma 3.6, $\mathcal{L}\left(\dfrac{S_n(u)}{b_n} \right) \overset{\Theta}{\to} \mathcal{L}(Z)$.

(ii) $\Rightarrow$ (i): Use Lemma 3.2.
(ii) $\Leftrightarrow$ (iii): Let $\alpha_n(\cdot, u) \equiv \mathcal{L}\left(\dfrac{S_n(u)}{b_n} \right)$, $\alpha(\cdot) \equiv \mathcal{L}(Z)$, $\beta_n(\cdot, u) \equiv \mathcal{L}\left(\dfrac{V_n(u)}{b_n} \right)$, and $\beta(\cdot) \equiv \mathcal{L}(1)$. By Theorem 1.4, (3.15) is valid so by Lemma 3.3, (3.16) is valid as well.

(ii) $\Leftrightarrow$ (v): Apply Lemma 3.6 with $\eta_{nk} = X_k(u)/b_n$, $k = 1, 2, \ldots n$, and $a_n(u) = 0$.

(iv) $\Rightarrow$ (v): Define

$$U(x) = E^u X_1^2(u) I(|X_1(u)| \le x) \quad \text{and} \quad G(x) = P^u(|X_1(u)| > x). \tag{6.62}$$

First we will establish that $U(x)$ is slow varying in Θ probability, i.e.,

$$\forall t > 0: \quad \frac{U(tx)}{U(x)} \overset{\Theta}{\to} 1 \text{ as } x \to \infty. \tag{6.63}$$

Rewrite (iv) as

$$\Theta \left\{ u: \frac{x^2 G(x)}{U(x)} > \varepsilon \right\} \to 0 \text{ as } x \to \infty. \tag{6.64}$$

In order to verify (6.63), let $t > 0$. Then

$$\Theta \left\{ u: \left| \frac{U(tx)}{U(x)} - 1 \right| > \varepsilon \right\} \le \Theta \left\{ u: \left| \frac{U(tx) - U(x)}{U(x)} \right| > \varepsilon \right\}$$

$$\le \Theta \left\{ u: \frac{1}{U((t \wedge 1)x)} E^u X_1^2(u) I((t \wedge 1)x < |X_1(u)| \le (t \vee 1)x) > \varepsilon \right\}$$

$$\le \Theta \left\{ u: \frac{1}{U((t \wedge 1)x)} (t \vee 1)^2 x^2 P^u((t \wedge 1)x < |X_1(u)| \le (t \vee 1)x) > \varepsilon \right\}$$

$$\le \Theta \left\{ \frac{(t \vee 1)^2 x^2}{(t \wedge 1)^2 U((t \wedge 1)x)} G((t \wedge 1)x) > \varepsilon \right\}$$

$$\to \quad 0 \quad \text{as } x \to \infty \text{ by (6.64).}$$

Notice that (6.63) holds uniformly for all $t > 0$ in sets bounded away from 0.

Next we show that

$$\frac{U(\tilde{b}_n(u))}{U(b_n)} \overset{\Theta}{\to} 1 \quad \text{as } n \to \infty. \tag{6.65}$$

Notice first that by (iv), $\forall \delta > 0$,

$$E_n(\delta) = \left\{ u: \left| \frac{\tilde{b}_n(u)}{b_n} - 1 \right| > \delta \right\}$$

satisfies $\Theta(E_n(\delta)) \to 0$ as $n \to \infty$. Then

$$\Theta \left\{ \left| \frac{U(\tilde{b}_n(u))}{U(b_n)} - 1 \right| > \varepsilon \right\}$$

$$= \Theta \left\{ \frac{1}{U(b_n)} \left| U(\tilde{b}_n(u)) - U(b_n) \right| > \varepsilon \right\}$$

$$\le \Theta \left\{ \frac{1}{U(b_n)} E^u X_1^2(u) I\{b_n < |X_1(u)| \le \tilde{b}_n(u)\} > \varepsilon/2 \right\}$$

$$+\Theta\left\{\frac{1}{U(b_n)}E^u X_1^2(u)I\{\tilde{b}_n(u) < |X_1(u)| \leq b_n\} > \varepsilon/2\right\}$$

$$\leq \quad \Theta\left\{\frac{\tilde{b}_n^2(u)}{U(b_n)}P^u(|X_1(u)| > b_n) > \varepsilon/2\right\}$$

$$+\Theta\left\{\left\{\frac{b_n^2}{U(b_n)}P^u(|X_1(u)| > \tilde{b}_n(u)) > \varepsilon/2\right\} \cap E_n^c(\delta)\right\}$$

$$+\Theta\left\{\left\{\frac{b_n^2}{U(b_n)}P^u(|X_1(u)| > \tilde{b}_n(u)) > \varepsilon/2\right\} \cap E_n(\delta)\right\}$$

$$\leq \quad \Theta\left\{\frac{\tilde{b}_n^2(u)}{b_n^2} \cdot \frac{b_n^2 P^u(|X_1(u)| > b_n)}{U(b_n)} > \varepsilon/2\right\}$$

$$+\Theta\left\{\frac{U((1+\delta)b_n)}{U(b_n)} \cdot \frac{(1+\delta)^2 b_n^2 P^u(|X_1(u)| > (1+\delta)b_n)}{(1+\delta)^2 U((1+\delta)b_n)} > \varepsilon/2\right\}$$

$$+\Theta\{E_n(\delta)\}$$

$$\rightarrow \quad 0 \text{ by (iv), (6.63) and (6.64)}.$$

Therefore, (6.65) holds.

We are now ready to show (v). By definition, for n large,

$$\tilde{b}_n^2(u) = nE^u X_1^2(u)I(|X_1(u)| \leq \tilde{b}_n(u)) = nU(\tilde{b}_n(u)).$$

So, for any $\varepsilon > 0$ and for n large, by (iv), (6.63), (6.64) and (6.65),

$$nP^u(|X_1(u)| > \varepsilon b_n) = \frac{\tilde{b}_n^2(u)}{\varepsilon^2 b_n^2} \cdot \frac{\varepsilon^2 b_n^2 G(\varepsilon b_n)}{U(\tilde{b}_n(u))}$$

$$= \frac{\tilde{b}_n^2(u)}{\varepsilon^2 b_n^2} \cdot \frac{\varepsilon^2 b_n^2 G(\varepsilon b_n)}{U(\varepsilon b_n)} \cdot \frac{U(\varepsilon b_n)}{U(b_n)} \cdot \frac{U(b_n)}{U(\tilde{b}_n(u))}$$

$$\overset{\Theta}{\rightarrow} \quad 0 \text{ as } n \rightarrow \infty;$$

and

$$\frac{n}{b_n^2}E^u X_1^2(u)I(|X_1(u)| \leq \varepsilon b_n) = \frac{\tilde{b}_n^2(u)}{b_n^2} \cdot \frac{U(\varepsilon b_n)}{U(b_n)} \cdot \frac{U(b_n)}{U(\tilde{b}_n(u))}$$

$$\overset{\Theta}{\rightarrow} \quad 1 \text{ as } n \rightarrow \infty,$$

as desired.

(v) $\Rightarrow$ (iv): First let $x = b_n$. Then

$$\frac{x^2 P^u(|X_1(u)| > x)}{E^u X_1^2(u)I(|X_1(u)| \leq x)} = \frac{nG(b_n)}{\frac{n}{b_n^2}U(b_n)} \overset{\Theta}{\rightarrow} 0 \text{ as } x \rightarrow \infty. \tag{6.66}$$

For arbitrary x, there exists n' such that $b_{n'} \leq x < b_{n'+1}$. Then

$$\frac{b_{n'}^2}{b_{n'+1}^2} \cdot \frac{b_{n'+1}^2 G(b_{n'+1})}{U(b_{n'+1})} \leq \frac{x^2 P^u(|X_1(u)| > x)}{E^u X_1^2(u)I(|X_1(u)| \leq x)} \leq \frac{b_{n'+1}^2}{b_{n'}^2} \cdot \frac{b_{n'}^2 G(b_{n'})}{U(b_{n'})}. \tag{6.67}$$

By Theorem 1.3, $\dfrac{b_n}{\sqrt{n}}$ is slow varying. Using the form of a slowly varying function (e.g. the Corollary to Theorem 1 in Feller (1971) (VIII.9, Vol. 2)), it can be easily deduced that

$$\frac{b_n^2}{b_{n+1}^2} \to 1. \tag{6.68}$$

Then by (6.66) and (6.67)

$$\frac{x^2 P^u(|X_1(u)| > x)}{E^u X_1^2(u) I(|X_1(u)| \le x)} \overset{\Theta}{\to} 0.$$

To substantiate the second part of (iv), first use the equivalence of (v) and (ii) to obtain the validity of

$$\mathcal{L}\left(\frac{S_n(u)}{b_n}\right) \overset{\Theta}{\to} \mathcal{L}(Z),$$

which implies (i). Theorem 2.3 can be invoked so that $S_n/V_n \overset{D}{\to} Z$ also holds. Now by Theorem 2.5,

$$\mathcal{L}\left(\frac{S_n(u)}{\tilde{b}_n(u)}\right) \overset{\Theta}{\to} \mathcal{L}(Z).$$

Thus, $\forall (n') \; \exists (n'')$ and A with $\Theta(A) = 1$ such that $\forall u \in A$ both

$$\mathcal{L}\left(\frac{S_{n''}(u)}{b_{n''}}\right) \to \mathcal{L}(Z) \text{ and } \mathcal{L}\left(\frac{S_{n''}(u)}{\tilde{b}_{n''}(u)}\right) \to \mathcal{L}(Z).$$

By the usual convergence of types theorem, $\forall u \in A$, $\dfrac{\tilde{b}_{n''}(u)}{b_{n''}} \to 1$. Therefore, $\dfrac{\tilde{b}_n(u)}{b_n} \overset{\Theta}{\to} 1$, as desired. $\qquad\square$

Proof of Proposition 2.1.

By the proof of $(vi) \Rightarrow (v)$ in Theorem 2.6, we know that

$$\frac{x^2 G(x)}{U(x)} \overset{\Theta}{\to} 0 \quad \Rightarrow \quad \frac{U(tx)}{U(x)} \overset{\Theta}{\to} 1. \tag{6.69}$$

Since $\tilde{b}_n(u) \to \infty$ for any u, we have, in turn,

$$\frac{U(t\tilde{b}_n(u))}{U(\tilde{b}_n(u))} \overset{\Theta}{\to} 1.$$

Notice that, by definition, for n large, $\tilde{b}_n^2(u) = nU(\tilde{b}_n(u))$. Therefore, for any $\varepsilon > 0$ and n large,

$$
\begin{aligned}
nG(\varepsilon\tilde{b}_n(u)) &= \frac{\tilde{b}_n^2(u)G(\varepsilon\tilde{b}_n(u))}{U(\tilde{b}_n(u))} \\
&= \frac{\varepsilon^2 \tilde{b}_n^2(u)G(\varepsilon\tilde{b}_n(u))}{U(\varepsilon\tilde{b}_n(u))} \cdot \frac{U(\varepsilon\tilde{b}_n(u))}{U(\tilde{b}_n(u))} \overset{\Theta}{\to} 0
\end{aligned}
$$

and

$$\frac{nU(\varepsilon\tilde{b}_n(u))}{\tilde{b}_n^2(u)} = \frac{\tilde{b}_n^2(u)}{U(\tilde{b}_n(u))} \cdot \frac{U(\varepsilon\tilde{b}_n(u))}{\tilde{b}_n^2(u)} \xrightarrow{\Theta} 1.$$

Thus, by (vii) of Theorem 2.5, $\mathcal{L}\left(\dfrac{S_n}{V_n}\right) \to N(0,1)$. $\qquad\square$

Acknowledgement. We are indebted to Evarist Giné for his comments and suggestions which allowed the presentation to be improved and led to shortening of several of the original proofs.

References

[1] Aldous, D.J. (1983). *Exchangeability and Related Topics.* Lecture Notes in Mathematics. Springer-Verlag.

[2] Araujo, A. and Giné, E. (1980). *The Central Limit Theorem for Real and Banach Valued Random Variables.* John Wiley & Sons, Inc.

[3] Brown, B.M. and Eagleson, G.K. (1971). Martingale convergence to infinitely divisible laws with finite variances. *Trans. Amer. Math. Soc.* **162** 449–453.

[4] Chung, K.L. (1974). *A Course in Probability Theory.* Academic Press, Inc.

[5] De Finetti, B. (1937). La Prévision, ses lois logiques, ses sources subjectives. *Annales de l'Institut Henri Poincaré* **7** 1–68.

[6] Eagleson, G.K. (1975) Martingale convergence to mixtures of infinitely divisible laws. *Ann. Probab.* **3** 557–562.

[7] Feller, W. (1971). *An Introduction to Probability Theory and Its Applications.* Vol. II, John Wiley & and Sons, Inc.

[8] Giné, E., Götze, F., and Mason, D.M. (1996). When is the student *t*-statistic asymptotically standard normal? To appear in *Ann. Probab.*

[9] Giné, E. and Zinn, J (1990). Bootstrapping general empirical measures. *Ann. Probab.* **18** 851–869.

[10] Griffin, P.S. and Mason, D.M. (1991). On the asymptotic normality of self-normalized sums. *Math. Proc. Camb. Phil. Soc.* **109** 597–610.

[11] Hall, P. and Heyde, C.C. (1980). *Martingale limit theory and its applications.* Academic Press, Inc.

[12] Klass, M. and Teicher, H. (1987). The central limit theorem for exchangeable random variables without moments. *Ann. Probab.* **15** 138–153.

[13] Zhang, G. (1995). *Regular and Empirical Central Limit Theory for Exchangeable Random Variables.* Ph.D. dissertation, Tufts University, Medford, MA.

Marjorie G. Hahn
Department of Mathematics
Tufts University
Medford, MA 02155, USA
mhahn@diamond.tufts.edu

Gang Zhang
301 Waverly Ave.
Watertown, MA 02172
Medford, MA 02155, USA
gzhang@world.std.com

Progress in Probability, Vol. 43
© 1998 Birkhäuser Verlag Basel/Switzerland

Laws of Large Numbers and Continuity of Processes

BERNHARD HEINKEL

Lai [7] has shown that for a sequence $(X_k)_{k \geq 1}$ of independent copies of a real-valued, centered r.v. X, the strong law of large numbers (SLLN) holds if and only if that sequence is a.s. Abel convergent to 0. That surprising equivalence between Cesaro and Abel convergence remains true in other situations, for instance when the independent r.v. X_k are symmetric, but not necessarily identically distributed (see Martikainen [8], Mikosch and Norvaisa [9]). Stated in a slightly different way, these two convergence results assert that for a sequence (X_k) of independent r.v. which are either centered and identically distributed or symmetric, the SLLN is equivalent to the a.s. paths-continuity of the following process $(\zeta(t),\ t \in [0, 1])$:

$$\zeta(t) = \begin{cases} (1-t) \sum_{k \geq 1} t^k X_k & \forall t \in [0, 1[\,, \\ 0 \text{ a.s.} & \text{if } t = 1. \end{cases}$$

In the sequel, ζ will be called the first Lai process associated to the sequence (X_k).

The purpose of the present short note is to give new examples of such relations between laws of large numbers and regularity properties of suitable processes.

From now, $(X_k)_{k \geq 1}$ will be a sequence of real-valued r.v. defined on $(\Omega, \mathcal{F}, \mathcal{P})$, which are independent, symmetrically distributed and which fulfil the classical necessary condition for the SLLN:

$$X_k/k \quad \to \quad 0 \ a.s. \tag{0.1}$$

An auxiliary sequence $(\epsilon_k)_{k \geq 1}$ of independent Rademacher r.v. defined on another probability space $(\Omega', \mathcal{F}', \mathcal{P}')$ will also be involved.

1. Kolmogorov's SLLN Revisited

The most famous SLLN in the non-i.i.d. case is probably Kolmogorov's result [5] which can be stated as follows in the symmetric setting:

$$\sum_{k \geq 1} \frac{X_k^2}{k^2} \quad \text{converges a.s.} \implies (X_k) \in \text{SLLN}.$$

Godbole's characterization [3] of cotype 2 spaces gives a partial converse to Kolmogorov's result:

$$(X_k) \in \text{SLLN} \implies \frac{1}{n^2} \sum_{1 \leq k \leq n} X_k^2 \to 0 \text{ a.s.} \tag{1.1}$$

This implication leads to a natural question: "How far is property (1.1) from the SLLN?" Obviously (1.1) implies the weak law of large numbers (WLLN). The following process ζ' will be needed for answering more completely to the question:

DEFINITION 1.1.   The Lai-Rademacher process $(\zeta'(t),\ t \in [0,1])$ associated to (X_k) is defined as:

$$\zeta'(t) = \begin{cases} (1-t)\sum_{k\geq 1} t^k \epsilon_k X_k & \forall t \in [0,1[\ , \\ 0 \text{ a.s.} & \text{if } t = 1. \end{cases}$$

REMARK   Assumption (0.1) ensures that ζ' is well defined (at least a.s.).

That process ζ' is related to Kolmogorov's SLLN as follows:

THEOREM 1.2.   *The following are equivalent:*

(i) $P\left(\omega : \dfrac{1}{n^2} \sum_{1\leq k\leq n} X_k^2(\omega) \to 0\right) = 1.$

(ii) $P\left(\omega : (\zeta'(t,\omega,.), t \in [0,1])\ has\ a\ continuous\ covariance\right) = 1.$

Proof. Suppose that (i) holds. Choose an ω such that:

$$\frac{1}{n^2} \sum_{1\leq k\leq n} X_k^2(\omega) \;\to\; 0 \quad ;$$

define $x_k = X_k(\omega)$ and also: $0 \leq s < t < 1$,

$$\tau^2(s,t) = \int_{\Omega'} (\zeta'(t,\omega,\omega') - \zeta'(s,\omega,\omega'))^2\ d\mathrm{P}'(\omega')$$
$$= \sum_{k\geq 1} \big\{\ (1-t)t^k - (1-s)s^k\ \big\}^2 x_k^2\ .$$

If $0 \leq s < t < t_0 < 1$, as $x_k/k \to 0$, there exists a constant $c > 0$ such that:

$$\big\{\ (1-t)t^k - (1-s)s^k\ \big\}^2 x_k^2\ \leq\ c\ t_0^{2k}k^2\ ;$$

the majorizing sequence being summable, Lebesgue's dominated convergence theorem implies that τ is continuous on $[0,1[^2$.

By the choice of ω, the sequence (x_k^2/k) converges to 0 in the Cesaro mean and therefore also in the Abel mean. So the obvious inequality:

$$\tau^2(1,t)\ =\ \sum_{k\geq 1}(1-t)^2 t^{2k} x_k^2\ \leq\ (1-t)\sum_{k\geq 1} t^k\ \frac{x_k^2}{k}\ ,$$

leads to: $\lim_{t\uparrow 1^-}\ \tau^2(1,t)\ =\ 0$. So τ is continuous on $[0,1]^2$.

Suppose conversely that (ii) holds. Choose an ω such that the process $(\zeta'(t, \omega, .), \; t \in [0, 1])$ has a continuous covariance. The notations being the same as above, one has:

$$\lim_{t \uparrow 1^-} \tau^2(1, t) = 0 \implies \lim_{n \to +\infty} 2^{-2n} \sum_{2^n + 1 \leq k \leq 2^{n+1}} (1 - 2^{-n})^{2k} x_k^2 = 0 ,$$

which obviously implies: $1/n^2 \sum_{1 \leq k \leq n} x_k^2 \to 0.$

So the implication (ii) $\Rightarrow$ (i) holds.

A more complete answer to the question "How far is (1.1) from the SLLN?" can be given with the help of the second Lai process ξ :

DEFINITION 1.3.   The second Lai process $(\xi(t), \; t \in [0, 1])$ associated to $(X_k)_{k \geq 1}$ is defined as :

$$\xi(t) = \begin{cases} (1 - t)^2 \sum_{n \geq 3} t^n \sum_{0 < i < n} X_i X_{n-i} & \forall t \in [0, 1[, \\ 0 \text{ a.s.} & \text{if } t = 1. \end{cases}$$

REMARK   According to (0.1) the process ξ is well defined (at least a.s.).

From the equivalence between the a.s. paths-continuity of ζ and the SLLN, as well as from Godbole's result, one derives the following statement :

THEOREM 1.4. *The SLLN holds for* $(X_k)_{k \geq 1}$ *if and only if the following two properties are fulfilled :*

(i) $1/n^2 \sum_{1 \leq k \leq n} X_k^2 \to 0$ *a.s.*

(ii) *The process* $(\xi(t), \; t \in [0, 1])$ *has a.s. continuous paths.*

COMMENT 1.5   From the definition of ξ it appears that Theorem 1.4 is a criterion for $(X_k) \in$ SLLN in terms of a law of large numbers for quadratic forms built with the X_k. Such laws of large numbers have been studied by Cuzick, Gine and Zinn [2] ; it would be interesting to look if their techniques of proof can be used for showing the a.s. paths-continuity of ξ.

COMMENT 1.6   The process ξ has a more complicated definition than ζ, so its specific usefulness has to be justified : ξ has often a much more regular covariance than ζ has. Suppose for example that the X_k are independent copies of a square integrable, symmetrically distributed r.v. X. Then the covariance of the corresponding ξ is very smooth :

$$\rho^2(s, t) = E(\xi(t) - \xi(s))^2 \leq CE(X^2) |t - s|^2 ; \tag{1.2}$$

on the other hand :

$$\beta^2(1, t) = E(\zeta(1) - \zeta(t))^2 \geq \frac{1}{2} E(X^2) (1 - t) . \tag{1.3}$$

By (1.2), ξ belongs to a class of very regular processes, the ones satisfying Hahn's sufficient condition for the a.s. paths-continuity ([4] Theorem 2.3). By (1.3), ζ is far from belonging to that class of regular processes.

2. Appendix: WLLN and Regularity of the First Lai Process

As it was recalled above, in the symmetrical setting the SLLN is equivalent to the a.s. paths-continuity of the first Lai process ζ. The WLLN also can be characterized by a regularity property of ζ :

THEOREM 2.1. *Suppose that in addition to* (0.1) *one has :*

$$\forall k \geq 1 \ , \ \ |X_k| \ \leq \ k \ \ a.s. \tag{2.1}$$

Then the following are equivalent :

(i) $(X_k) \in$ WLLN;

(ii) *The process* ($\zeta(t), \ t \in [0,1]$) *has a continuous covariance.*

COMMENT 2.2   It is well known that under (0.1), by truncating the X_k at the level k if necessary, it is possible to reduce to the case (2.1); so that hypothesis isn't a loss of generality.

Sketch of the proof of Theorem 2.1

The proof follows the same lines as the one of Theorem 1.2. The only change to do is to define that time x_k^2 as $E(X_k^2)$ and then to notice that the key property $1/n^2 \ \sum_{1 \leq k \leq n} x_k^2 \to 0$ holds when $(X_k) \in$ SLLN. This latter fact of course follows from the well known equivalence

$$(X_k) \in \text{SLLN} \ \Longleftrightarrow \ \frac{1}{n^2} \ E\Big(\sum_{1 \leq k \leq n} X_k \Big)^2 \ \to \ 0 \ ,$$

which holds in the symmetric case, under restriction (2.1) (see [6] Lemma 2.3 or [1] Lemma 3.1).

Acknowledgement. I am indebted to the referee for suggesting a shortened proof of Theorem 1.2.

References

[1]   de Acosta, A. *Inequalities for B-valued random vectors with applications to the strong law of large numbers*, Ann. Probab. 9 (1981), p. 157–161.

[2]   Cuzick, J., Giné, E., Zinn, J. *Laws of large numbers for quadratic forms, maxima of products and truncated sums of i.i.d. random variables*, Ann. of Probab. 23 (1995), p. 292–333.

[3]   Godbole, A.P. *Strong laws of large numbers and laws of the iterated logarithm in Banach spaces*, Ph. D. Thesis of the Michigan State University, 1984.

[4]   Hahn, M.G. *Conditions for sample continuity and the central limit theorem*, Ann. Probab. 5 (1977) p. 351–360.

[5]   Kolmogorov, A. *Sur la loi forte des grands nombres*, C. R. Acad. Sci. Paris 191 (1930), p. 910–912.

[6]   Kuelbs, J., Zinn, J. *Some stability results for vector valued random variables*, Ann. Probab. 7 (1979), p. 75–84.

[7] Lai, T.L. *Summability methods for independent, identically distributed random variables*, Proc. Amer. Math. Soc. 45 (1974), p. 253–261.

[8] Martikainen, A.I. *Regular methods of summing random terms*, Theory Probab. Appl. 30 (1985), p. 9–18.

[9] Mikosch, T., Norvaisa, R. *Limit theorems for methods of summation of independent randomvariables II*, Lithanian Math. J. 27 (1987), p. 128–144.

Bernhard Heinkel
Département de Mathématique
7, rue René Descartes
67084 Strasbourg Cedex (France)
heinkel@math.u-strasbg.fr

Progress in Probability, Vol. 43
© 1998 Birkhäuser Verlag Basel/Switzerland

Convergence in Law of Random Elements and Random Sets

JØRGEN HOFFMANN-JØRGENSEN

1. Introduction

Throughout this paper, we let T denote a fixed topological space (not necessarily Hausdorff) and we let $(\Omega, \mathcal{F}, P)$ and $(\Upsilon, \mathcal{A}, Q)$ denote two fixed probability spaces. We let (E, E^*, E_*) denote the P-expectation, the upper and the lower P-expectation. Similarly, we let $(\mathbb{E}, \mathbb{E}^*, \mathbb{E}_*)$ denote the Q-expectation, the upper and the lower Q-expectation. Recall that a T-*valued random element* on $(\Omega, \mathcal{F}, P)$ or $(\Upsilon, \mathcal{A}, Q)$ is simply a map from Ω or Υ into T. We let $C(T)$ denote the set of all bounded continuous functions $\phi : T \to \mathbf{R}$ and we let $\mathcal{B}a(T)$ and $\mathcal{B}(T)$ denote *the Baire* and *Borel σ-algebra*; that is, $\mathcal{B}a(T)$ is the smallest σ-algebra making all functions in $C(T)$ measurable and $\mathcal{B}(T)$ is the smallest σ-algebra containing all closed (or all open) subsets of T. A random element Z on $(\Upsilon, \mathcal{A}, Q)$ is called *Baire (Borel) Q-measurable* if $Z^{-1}(B)$ is Q-measurable for every Baire (Borel) set $B \subseteq T$. Similarly, if μ is a probability measure on some σ-algebra on T, we say that μ is *Baireian (Borelian)* if every Baire (Borel) subset of T is μ-measurable.

Let Z be a T-valued random element on $(\Upsilon, \mathcal{A}, Q)$ and let $(X_\pi \mid \pi \in \Pi)$ be Π-net of T-valued random elements on $(\Omega, \mathcal{F}, P)$ (see Section 2). According to the current definition (see [[3]; (7.4) p. 149]) of convergence in law (which I shall rename to "convergence in Baire law"), we say that (X_π) is *convergent in Baire law* to Z and we write $X_\pi \to^\sim Z$ if

$$Z \text{ is Baire } Q\text{-measurable and } \lim_{\pi\uparrow\Pi} E^*\phi(X_\pi) = \mathbb{E}\phi(Z) \quad \forall \phi \in C(T) \qquad (1.1)$$

and if μ is probability measure on some σ-algebra $\mathcal{B}$ on T, we say that (X_π) *converge in Baire law* to μ and we write $X_\pi \to^\sim \mu$ if $X_\pi \to^\sim \mathrm{id}_\mu$ where $\mathrm{id}_\mu(t) := t$ is the identity map on T considered as an T-valued random element on the probability space $(T, \mathcal{B}, \mu)$; that is, $X_\pi \to^\sim \mu$ if and only if

$$\mu \text{ is Baireian and } \lim_{\pi\uparrow\Pi} E^*\phi(X_\pi) = \int_T \phi(t)\mu(dt) \quad \forall \phi \in C(T) \qquad (1.2)$$

If T is metrizable or more generally if T is perfectly normal, then the notion of convergence in Baire law behaves nicely and satisfies most of the calculus of the classical convergence in law for sequences of measurable random variables with values in a separable metric space. For instance, if S and T are separable metric spaces, $f : T \to S$ is a function with continuity set $C(f)$, $c \in S$ is a given point and $X_n : \Omega \to T$, $Z : \Upsilon \to T$ and $Y : \Omega \to S$ are Borel measurable functions, then we have the following two important rules:

$$X_\pi \to^\sim Z \text{ and } Z \in C(f) \ Q\text{-a.s.} \quad \Rightarrow \quad f(X_\pi) \to^\sim f(Z) \qquad (1.3)$$

$$X_\pi \to^\sim X \quad \text{and} \quad Y_\pi \to^\sim c \quad \Rightarrow \quad (X_\pi, Y_\pi) \to^\sim (X, c) \tag{1.4}$$

However, (1.3) and (1.4) fail when S and T are general topological spaces.

Observe that Baire convergence in law coincides for any two topologies on T with the same set of continuous real-valued functions. Thus, if the topology on T is not completely regular, we may expect peculiar properties of convergence in Baire law. At the first glance, it seems that class of completely regular spaces will serve for the most purposes. But in the context of nets of random sets (for instance, sets of maximum or zero points of stochastic processes), *the upper Fell topology* (see Section 5) present itself in a natural way. However, the upper Fell topology has the unpleasant property that every continuous real-valued function is constant. In particular, we have that any net of random elements converges in Baire law to any random element. Thus, we need a more restrictive definition of convergence in law in order to avoid trivialities. As a starting point, we note that even if $C(T)$ trivializes, we have still have sufficiently many upper semicontinuous functions to determine the topology (for instance, the indicators of closed sets).

Let $\text{Usc}(T)$ denote the set of all upper semicontinuous functions f from T into the extended real line $\overline{\mathbf{R}} := [-\infty, \infty]$ such that $\sup_{t \in T} f(t) < \infty$ and let $\text{Lsc}(T)$ denote the set of all lower semicontinuous functions $g : T \to \overline{\mathbf{R}}$ such that $\inf_{t \in T} g(t) > -\infty$; that is, $g \in \text{Lsc}(T)$ if and only if $(-g) \in \text{Usc}(T)$. If $(X_\pi \mid \pi \in \Pi)$ is a Π-net of T-valued random elements on the probability space $(\Omega, \mathcal{F}, P)$ and Z is a T-valued random element on $(\Upsilon, \mathcal{A}, Q)$, we say that (X_π) *converge in Borel law* to Z and we write $X_\pi \twoheadrightarrow^\sim Z$ if

$$\limsup_{\pi \uparrow \Pi} \mathbb{E}^* f(X_\pi) \le \mathbb{E}_* f(Z) \quad \forall f \in \text{Usc}(T) \tag{1.5}$$

If μ is probability measure on some σ-algebra $\mathcal{B}$ on T, we say that (X_π) *converge in Borel law* to μ and we write $X_\pi \twoheadrightarrow^\sim \mu$ if $X_\pi \twoheadrightarrow^\sim \text{id}_\mu$ where $\text{id}_\mu(t) := t$ is the identity map on T considered as an T-valued random element on the probability space $(T, \mathcal{B}, \mu)$. Hence, if $\int_* d\mu$ denotes the lower μ-integral, then $X_\pi \twoheadrightarrow^\sim \mu$ if and only if

$$\limsup_{\pi \uparrow \Pi} \mathbb{E}^* f(X_\pi) \le \int_* f(t) \mu(dt) \quad \forall f \in \text{Usc}(T) \tag{1.6}$$

The objective of this paper is to study the notions of convergence in law introduced in (1.1) and (1.5) for nets of random elements in general and nets of random sets in particular. In Section 2, you will find the notation used throughout this paper. Section 3 contains a careful study of the so-called probability extents which plays a cardinal role in the theory of convergence in law. Section 4 is devoted study of convergence in Baire and Borel law including the appropriate portmanteau lemmas and general versions of the formulas (1.3) and (1.4). In Section 5, the general theory of Section 4 is applied to the study of convergence of random sets in general with applications to law convergence of statistical estimators in particular. Before we proceed with the general theory, let me give an example illustrating some of the peculiarities of convergence in law which seems to have been overlooked in the past.

1.1. Example (The notation used in this example may be found in the Section 2). Let $(\Omega, \mathcal{F}, P) = ([0,1], \mathcal{B}[0,1], \lambda)$ be the unit interval with its Lebesgue measure and let $(\Upsilon, \mathcal{A}, Q)$ be a arbitrary probability space. Let $\widehat{T} = [0,1]^{[0,1]}$ be the set of all functions $t : [0,1] \rightarrow [0,1]$ equipped with the product topology. Then $\widehat{T}$ is a compact Hausdorff space and if $n \geq 1$, $\alpha \in [0,1]$ and $\psi(v) = (\zeta(v), \&bgr;(v))$: $\Upsilon \rightarrow [0,1]^2$ are given, we shall consider the $\widehat{T}$-valued random elements Y_α, Y_ψ, X_n and X_n^α given by

$$Y_\alpha(\omega, s) = \alpha 1_D(\omega, s) \;,\;\; Y_\psi(\omega, s) = Y_{\beta(v)}(\zeta(v), s)$$
$$\xi_n(\omega, s) = (\omega + \frac{1-\alpha}{2n}) \wedge (\frac{1}{3}\omega + \frac{2}{3} - \frac{1-\alpha}{2n})$$
$$X_n(\omega, s) = (1 - 2n|\omega - s|)^+ \;,\;\; X_n^\alpha(s, \omega) = X_n(\xi_n(\omega, s), s)$$

where $D = \{(\omega, s) \mid \omega = s\}$ is the diagonal in the unit square. Let $\mathbf{0} \in \widehat{T}$ denote the function which is identically equal to zero, let D denote the set of all functions $t \in \widehat{T}$ such that $t(s) > 0$ for at most one $s \in [0,1]$ and set

$$\mathcal{X} = (X_n) \;,\;\; C_n = X_n([0,1]) = \{X_n(\omega) \mid \omega \in [0,1]\} \;,\;\; C^n = \bigcup_{i=n}^{\infty} C_i$$

Let T denote a fixed subset of $\widehat{T}$ such that $T \supseteq C^1$. Then D is compact and T is a completely regular Hausdorff space such that X_n and X_n^α are T-valued random elements. Note that $Y_0(\omega) = \mathbf{0}$ and $X_n(\omega, s) = X_n^1(\omega, s)$ for all $(\omega, s) \in [0,1]^2$. Moreover, the reader easily verifies that we have

(1) $\xi_n : [0,1]^2 \rightarrow [0,1]$ and $X_n^\alpha : [0,1] \rightarrow T$ are continuous and X_n^α is injective

(2) $\lim_{n\to\infty} X_n^\alpha(\omega, s) = Y_\alpha(\omega, s) \quad \forall \omega, s, \alpha \in [0,1]$

(3) $C_1, C_2, \ldots$ are disjoint and compact and $\operatorname{cl} C^n = T \cap (D \cup C^n)$

Baire and Borel measurability: Let $Y : [0,1] \rightarrow T$ be a given function. If Y is continuous, then Y is Borel and Baire measurable and since Y is Baire measurable if and only if $\phi \circ Y$ is measurable for all $\phi \in C(T)$, we have that any pointwise limit of a sequence of Baire measurable functions is Baire measurable. It is a common mistake to claim that this also holds for Borel measurability. By (1) and (2), we see that Y_α is the pointwise limit in $\widehat{T}$ of a sequence of continuous functions. Thus, $Y_\alpha : [0,1] \rightarrow \widehat{T}$ is Baire measurable. However, if $\alpha > 0$ and $A \subseteq [0,1]$, then Y_α is injective and $Y_\alpha(A) \subseteq \operatorname{cl}_{\widehat{T}} Y_\alpha(A) \subseteq Y_\alpha(A) \cup \{\mathbf{0}\}$. Hence, $A = Y_\alpha^{-1}(Y_\alpha(A))$ and $Y_\alpha(A) \in \mathcal{B}(\widehat{T})$ for all $A \subseteq [0,1]$. Thus, if $\alpha > 0$, then Y_α is neither Borel measurable nor Borel P-measurable.

Baire convergence in law: Let $Z : \Upsilon \rightarrow T$ be a given T-valued random element. Note that $X_n(\cdot, s) \rightarrow 0$ P-a.s. for all $s \in [0,1]$. Hence, if $X_n \rightarrow^\sim Z$, then $Z(\cdot, s) = 0$ Q-a.s. for all $s \in [0,1]$. Using Bockstein's theorem, it follows easily that the converse implication holds whenever T is compact. Note that $C^n \downarrow \emptyset$ and $I^{\mathcal{X}}(1_{C^n}) = 1$ for all $n \geq 1$ where $I^{\mathcal{X}}$ is given by (3.14). So by Lemma 4.1 we have

(4) If $X_n \rightarrow^\sim Z$, then $Z(\cdot, s) = 0$ Q-a.s. for all $s \in [0,1]$ and the converse implication holds if T is compact (for instance, if $T = \widehat{T}$)

(5) If $C^n \in \mathcal{F}_\circ(T)$ for all $n \geq 1$ (for instance, if $T = C^1$), then (X_n) does *not* converge in Baire law on T to any T-valued random element

In particular, we see that (X_n) may converge in Baire law on $\widehat{T}$ to a T-valued random element without converging in Baire law on T to any T-valued random element.

Borel convergence in law: Let $Z : \Upsilon \to T$ and $Y : \Upsilon \to T$ be given T-valued random elements. By Corollary 4.8, we have that $X_n \twoheadrightarrow^\sim Z$ on T if and only if $X_n \twoheadrightarrow^\sim Z$ on $\widehat{T}$. Thus, in this respect, Borel convergence in law behaves more decent than Baire convergence in law. However, Borel convergence in law have peculiarities which are not present for Baire convergence in law. By Corollary 4.8, we have that $X_n \to^\sim Y$ and $X_n \to^\sim Z$ implies that Z and Y are Baire Q-measurable maps such that Q_Y and Q_Z coincides on the Baire σ-algebra. It is tempting to believe that Borel convergence of (X_n) to Z and Y implies that Z and Y are Borel Q-measurable maps such that Q_Y and Q_Z coincides on the Borel σ-algebra. None of these statements are true. Recall that axiom RM states that the Lebesgue measure on $[0,1]$ admits a σ-additive extension to the σ-algebra $2^{[0,1]}$ of all subsets of $[0,1]$ and that RM is independent of the usual axioms of set theory (including the axiom of choice) and implies the negation of the special continuum hypothesis (i.e., $RM \Rightarrow \aleph_1 < 2^{\aleph_0}$). Let $Z : \Upsilon \to T$ be a T-valued random element and let $\alpha, \beta \in [0,1]$ be given numbers such that $Y_\alpha(\omega) \in T$ and $Y_\beta(\omega) \in T$ for all $\omega \in [0,1]$. Then Y_α and Y_β are T-valued random elements and using the results of Section 4, it can be shown that we have

(6) If $X_n \twoheadrightarrow^\sim Z$ and $Z \neq \mathbf{0}$ Q-a.s., then there exists a function $\psi(v) = (\zeta(v), \alpha(v)) : \Upsilon \to [0,1]^2$ such that $Z = Y_\psi$ Q-a.s., $\alpha(v) > 0$ for all $v \in \Upsilon$ and ζ is Q-measurable and uniformly distributed

(7) $X_n \twoheadrightarrow^\sim Y_\alpha$ and $X_n \twoheadrightarrow^\sim Y_\beta$ but if $\alpha > 0$, then Y_α is *not* Borel P-measurable. Moreover, if $\alpha \neq \beta$, then there exists a set $B \in \mathcal{B}(T) \cap \mathcal{M}(P_{Y_\alpha}) \cap \mathcal{M}(P_{Y_\beta})$ such that $P(Y_\alpha \in B) = 1$ and $P(Y_\beta \in B) = 0$; that is the distributions of Y_α and Y_β are mutually singular

(8) If $X_n \twoheadrightarrow^\sim \mu$ for some Borel probability measure μ with $\mu(T \cap \{\mathbf{0}\}) = 0$, then axiom RM holds

(9) If axiom RM holds, then P_{Y_α} admits a σ-additive extension μ to 2^T such that $X_n \twoheadrightarrow^\sim \mu$

Comparing (4) and (6), we see that Baire convergence in law does *not* imply Borel convergence in law (not even when T is a compact Hausdorff space). Moreover, (7) shows that we have may have Borel convergence to many different non-Borel measurable random elements with mutually singular distributions. Finally, (8+9) show that the simple question: "Does (X_n) converge in Borel law to some Borel probability" may not be decidable within the usual axioms of set theory (including the axiom of choice). This means that convergence in Baire and Borel law should be treated with care and Section 4 is devoted to a study to what extend the classical calculus for convergence in law of measurable random variables with values in a separable metric space carries over to Baire and Borel convergence for nets of random elements with values in an arbitrary topological space.

2. Notation and Preliminary Facts

In this section, I shall fix the notation used throughout this paper.

Posets: Recall that a *poset* is a partially ordered set $(\Gamma, \leq)$ and that a *net* is an *upwards directed* poset $(\Gamma, \leq)$ (i.e., $\forall \alpha, \beta \in \Gamma \; \exists \gamma \in \Gamma$ so that $\gamma \geq \alpha$ and $\gamma \geq \beta$). If Γ is net, then a Γ-*net* on the set S is an indexed family $(s_\gamma \mid \gamma \in \Gamma) \subseteq S$. Let Λ and Γ be posets. Then we write $\Lambda \looparrowright \Gamma$ if there exists a function $\sigma : \Lambda \to \Gamma$ such that σ is *increasing* (i.e., $\lambda \leq \mu \Rightarrow \sigma(\lambda) \leq \sigma(\mu)$) and *cofinal* (i.e., $\forall \gamma \in \Gamma \; \exists \lambda \in \Lambda : \sigma(\lambda) \geq \gamma$) and we write $\Lambda \approx \Gamma$ if $\Lambda \looparrowright \Gamma$ and $\Gamma \looparrowright \Lambda$. We let $\mathbf{N} = \{1, 2, \ldots\}$ denote the set of positive integers with its usual ordering and we say that the poset Γ is *countably cofinal* if $\mathbf{N} \looparrowright \Gamma$. A poset Γ is called *finitely founded* if the initial intervals $\{\gamma \in \Gamma \mid \gamma \leq \alpha\}$ are finite for all $\alpha \in \Gamma$ and Γ is a *Frechet net* if Γ is a net without maximal elements. If S and L are given sets, we let 2^S denote the set of all subsets of S and we let L^S denote the set of all functions $f : S \to L$. We shall always consider 2^S and its non-empty subsets (called *pavings*) as posets with respect to inclusion $\subseteq$. We let $\overline{\mathbf{R}} := [-\infty, \infty]$ denote the extended real line and we shall always consider $\overline{\mathbf{R}}^S$ and its non-empty subsets (called *function spaces*) as posets under *the pointwise ordering*: $f \leq g \iff f(s) \leq g(s) \;\; \forall s \in S$. In particular, if $2^{(S)}$ denotes the set of all finite non-empty subsets of S, then $2^{(S)}$ is a finitely founded net and if S is infinite, then $2^{(S)}$ is a finitely founded Frechet net. Moreover, it is well-known and easily verified that $2^{(\Gamma)} \looparrowright \Gamma$ for every net Γ and that $\Gamma \approx 2^{(\Gamma)}$ for every finitely founded net Γ.

Function spaces: If $f, g \in \overline{\mathbf{R}}^T$ are given functions and $A \subseteq T$, we set $\sup_A f := \sup_{t \in A} f(t)$, $\inf_A f := \inf_{t \in A} f(t)$ and $\|f - g\|_A := \sup_{t \in A} |f(t) - g(t)|$ with the conventions $\sup \emptyset := -\infty$, $\inf \emptyset := -\infty$ and $\infty - \infty := 0$. Then $\|f - g\|_T$ is a pseudo-metric on $\overline{\mathbf{R}}^T$ and we let $\overline{\Phi}$ denote the closure of $\Phi \subseteq \overline{\mathbf{R}}^T$ with respect to pseudo-metric $\| \cdot \|_T$. If $\Phi \subseteq \overline{\mathbf{R}}^T$ is a function space and Γ is a non-empty set, we let $\Phi_{\wedge \Gamma}$ (resp. $\Phi_{\vee \Gamma}$) denote the set of all functions $f \in \overline{\mathbf{R}}^T$ of the form $f = \inf_{\psi \in \Psi} \psi$ (resp. $f = \sup_{\psi \in \Psi} \psi$) for some non-empty set $\Psi \subseteq \Phi$ with card $\Psi \leq$ card Γ. In particular, we set $\Phi_\wedge := \Phi_{\wedge \Phi}$ and $\Phi_\vee := \Phi_{\vee \Phi}$ and we let $\Phi_\downarrow$ (resp. $\Phi_\uparrow$) denote the set of infimums (resp. supremums) of all downwards directed (resp. upwards directed) non-empty subsets of Φ. Similarly, if Γ is a net, we let $\Phi_{\downarrow \Gamma}$ denote the set of all functions $h \in \overline{\mathbf{R}}^T$ for which there exists a net $(\phi_\gamma \mid \gamma \in \Gamma) \subseteq \Phi$ such that $\phi_\gamma \downarrow h$ (i.e., $h = \inf_{\gamma \in \Gamma} \phi_\gamma$ and $\phi_\gamma \leq \phi_\beta$ for all $\beta \leq \gamma$) and we define $\Phi_{\uparrow \Gamma}$ and $\phi_\gamma \uparrow h$ similarly. Let $\mathcal{E} \subseteq 2^T$ be a paving on T. Then we let $U(\mathcal{E})$ denote the set of all $f \in \overline{\mathbf{R}}^T$ such that $\sup_T f < \infty$ and $\{f \geq a\} \in \mathcal{E}$ for all $a \in \mathbf{R}$.

Topology: I shall use *compact* (*Lindelöf*) in the wide sense: "every open cover has a finite (countable) subcover" without assuming the Hausdorff property and we say that T is

- *regular* if for every closed set $F \subseteq T$ and every $t \in T \setminus F$ there exist disjoint open sets U and V with $t \in U$ and $F \subseteq V$

- an *Urysohn space* if for all $t_0, t_1 \in T$ with $\mathrm{cl}(t_0) \cap \mathrm{cl}(t_1) = \emptyset$ there exists $\phi \in C(T)$ such that $\phi(t_0) = 0$ and $\phi(t_1) = 1$

- *completely regular* if for every closed set $F \subseteq T$ and every $t \in T \setminus F$ there exists $\phi \in C(T)$ with $\phi(t) = 0$ and $\phi(u) = 1$ for all $u \in F$
- *normal* if for every two disjoint closed sets $F_0, F_1 \subseteq T$ there exists $\phi \in C(T)$ with $\phi(t) = 0$ for all $t \in F_0$ and $\phi(t) = 1$ for all $u \in F_1$
- *perfectly normal* if T is normal and every open set is a countable union of closed sets
- *pseudo-metrizable* if the topology is induced by a pseudo-metric
- *countably paracompact* if for every sequence $\{F_n\}$ of closed subsets of T with $F_n \downarrow \emptyset$, there exist open sets G_n such that $G_n \downarrow \emptyset$ and $G_n \supseteq F_n$ for all $n \geq 1$

If "xxx" is topological property, we say that T is *hereditarily* "xxx" if every sub-space of T has the property "xxx". Recall that a set $F \subseteq T$ is called *functionally closed* if there exists $\phi \in C(T)$ such that $F = \phi^{-1}(0)$ and that $G \subseteq T$ is *function-ally open* if the complement $T \setminus G$ is functionally closed. We let $\mathcal{F}(T)$, $\mathcal{G}(T)$, $\mathcal{F}_o(T)$, $\mathcal{G}_o(T)$ and $\mathcal{K}(T)$ denote the sets of all closed, all open, all functionally closed, all functionally open and all closed compact subsets of T. If ρ denotes *the regulariza-tion* topology on T (i.e., the weakest topology on T making all functions $\phi \in C(T)$ continuous), then $C(T, \rho) = C(T)$ and $Ba(T, \rho) = Ba(T)$, $Usc(T, \rho) = C_\wedge(T)$ and $Lsc(T, \rho) = C_\vee(T)$. We let $Usc_o(T)$ and $Lsc_o(T)$ denote the countable analogues of $Usc(T, \rho)$ and $Lsc(T, \rho)$:

$$Usc_o(T) := C_{\wedge \mathbf{N}}(T) = C_{\downarrow \mathbf{N}}(T) \ , \quad Lsc_o(T) := C_{\vee \mathbf{N}}(T) = C_{\uparrow \mathbf{N}}(T) \tag{2.1}$$

We let $\operatorname{cl} A$ and $\operatorname{int} A$ denote the closure and interior of $A \subseteq T$ and if $(A_\gamma \mid \gamma \in \Gamma)$ is a Γ-net of subsets of T, we define *the upper topological limit* of (A_γ) as usual:

$$\operatorname{up lim} A_\gamma := \bigcap_{\gamma \in \Gamma} \operatorname{cl} \left(\bigcup_{\beta \geq \gamma} A_\beta \right) \tag{2.2}$$

Let $f : T \to S$ be a given function with values in the topological space S. If $A, B \subseteq T$, we say that f is *continuous at A along B* if $f(t_\sigma) \to f(t)$ for any $t \in A$ and any net $\{t_\sigma\} \subseteq B$ such that $t_\sigma \to t$ or equivalently if for every $t \in A$ and every open neighborhood U of $f(t)$ there exists an open neighborhood V of t such that $f(B \cap V) \subseteq U$.

Increasing functionals: Let $\mathbf{D} \subseteq \overline{\mathbf{R}}^T$ be a function space and let $I : \mathbf{D} \to \overline{\mathbf{R}}$ be a given functional. Then we set $-\mathbf{D} := \{-f \mid f \in \mathbf{D}\}$ and the functional $I^\circ : (-\mathbf{D}) \to \overline{\mathbf{R}}$ given by $I^\circ(f) := -I(-f)$ for all $f \in -\mathbf{D}$ is called *the dual functional*. We say that I is *increasing* if $I(f) \leq I(g)$ for all $f, g \in \mathbf{D}$ with $f \leq g$. Note that $I^{\circ\circ} = I$ and that I is increasing if and only if I° is increasing. Let $(\Gamma, \leq)$ be a net, let $\mathcal{L} \subseteq 2^T$ be a paving and let $\Phi, \Psi, \mathbf{D} \subseteq \overline{\mathbf{R}}^T$ be function spaces such that $\Phi, \Psi \subseteq \mathbf{D}$. If $I : \mathbf{D} \to \overline{\mathbf{R}}$ is an increasing functional, we say that I is

- *Γ-smooth at Ψ along Γ* if $I(\phi_\gamma) \downarrow I(\psi)$ whenever $(\phi_\gamma \mid \gamma \in \Gamma) \subseteq \Phi$ is a net and $\psi \in \Psi$ is a function such that $\phi_\gamma \downarrow \psi$
- *Φ-regular at Ψ* if $I(\psi) = \inf \{I(\phi) \mid \phi \in \Phi , \ \phi \geq \psi\} \ \forall \psi \in \Psi$
- *$\mathcal{L}$-tight* if $1_L \in \mathbf{D}$ for all $L \in \mathcal{L} \cup \{T\}$ and $I(1_T) = \sup\{I(1_L) \mid L \in \mathcal{L}\}$

- $\mathcal{L}$-*tight along* Φ if $1_T \in \Phi$ and the functional $I^\Phi : \overline{\mathbf{R}}^T \to \overline{\mathbf{R}}$ given by $I^\Phi(f) :=$ $\inf\{I(\phi) \mid \phi \in \Phi , \phi \geq f\}$ is $\mathcal{L}$-tight; that is, if $1_T \in \Phi$ and

$$\forall a < I(1_T) \; \exists L \in \mathcal{L} \text{ so that } I(\phi) \geq a \quad \forall \phi \in \Phi \text{ with } \phi \geq 1_L$$

If $\mathcal{G}$ and $\mathcal{H}$ are pavings on T such that $1_A \in \mathbf{D}$ for all $A \in \mathcal{G} \cup \mathcal{H}$, we say that I is Γ-*smooth at* $\mathcal{G}$ *along* $\mathcal{H}$ if I is Γ-smooth at $\{1_G \mid G \in \mathcal{G}\}$ along $\{1_H \mid H \in \mathcal{H}\}$ and we define $\mathcal{H}$-*regularity at* $\mathcal{G}$ and $\mathcal{L}$-*tightness along* $\mathcal{H}$ similarly. If $\mathcal{D} \subseteq 2^T$ is a paving and $\upsilon : \mathcal{D} \to [0, \infty]$ is a set function, we say that υ is Γ-*smooth at* $\mathcal{G}$ *along* $\mathcal{H}$ or $\mathcal{H}$-*regular at* $\mathcal{G}$ or $\mathcal{L}$-*tight along* $\mathcal{G}$ if the functional $1_D \frown \upsilon(D)$ is so. We use the terminology σ-*smooth* if I is $\mathbf{N}$-smooth and we use the terminology τ-*smooth* if I is Γ-smooth for every net Γ.

Measure theory: I shall use the terminology of [[2]; Chap. 1]. In particular, if $\mu : \mathcal{G} \to [0, \infty]$ is a set function, then μ_* and μ^* denote *the inner* and *outer* μ-*measures* and if $\nu : 2^T \to [0, \infty]$ is an everywhere defined set function, then $\mathcal{M}(\nu)$ denote the algebra of all μ-*measurable* sets in the sense of Carathéodory (see [[2]; 1.22 p. 23]). Let $(S, \mathcal{B}, \mu)$ be a probability space. Then we set $\mathcal{M}(\mu) := \mathcal{M}(\mu^*)$ and $\bar{\mu}$ denotes the completion of μ; that is, $\bar{\mu}$ is the restriction of μ^* to the σ-algebra $\mathcal{M}(\mu)$. Let g be a map from S into the set L. Then we let $\mathcal{M}_\mu(g)$ denote the paving of all sets $A \subseteq M$ such that $g^{-1}(A) \in \mathcal{M}(\mu)$ and we define *the maximal* μ-*distribution of* g to be the set function $\mu_g(A) := \bar{\mu}(g^{-1}(A))$ for all $A \in \mathcal{M}_\mu(g)$. Note that $\mathcal{M}_\mu(g)$ is the largest σ-algebra on M making g μ-measurable and that μ_g is a measure on $\mathcal{M}_\mu(g)$ such that $\mathcal{M}_\mu(g) = \mathcal{M}(\mu_g)$ and

$$\int_* h \, d\mu_g \leq \int_* (h \circ g) d\mu \leq \int^* (h \circ g) d\mu \leq \int^* h \, d\mu_g \quad \forall h : S \to \overline{\mathbf{R}} \qquad (2.3)$$

If μ is a probability measure on some σ-algebra on T, we say that μ is *Radon* if μ is Borelian and

$$\mu(B) = \sup\{\mu(K) \mid K \in \mathcal{K}(T) , \; K \subseteq B\} \quad \forall B \in \mathcal{B}(T) \qquad (2.4)$$

and we say that μ is *Radonian* if μ is Baireian and $\bar{\mu}$ coincides with a Radon probability on the Baire σ-algebra.

A.s. continuity points: Let Z be a T-valued random element on the probability space $(\Upsilon, \mathcal{A}, Q)$. If ζ is a function from T into the topological space, we let $C(\zeta)$ denote the set of all continuity points of ζ and we say that ζ is *continuous* Q_Z-*a.s.* if $T \setminus C(\zeta)$ is a Q_Z-null set or equivalently if the set $\{Z \notin C(\zeta)\}$ is a Q-null set. If $h \in \overline{\mathbf{R}}^T$ is a given function, we let $\bar{h}$ and h° denote *the upper* and *lower semicontinuous envelopes* of h and we let $\mathrm{Usc}[Z]$ denote the set of all functions $f \in \overline{\mathbf{R}}^T$ satisfying one of the following two equivalent conditions:

$$\sup_T f < \infty \text{ and } \mathbb{E}_* f(Z) = \inf\{\mathbb{E}_* h(Z) \mid h \in \mathrm{Usc}(T) , \; h \geq f\} \qquad (2.5)$$

$$\sup_T f < \infty \text{ and } \mathbb{E}_* f(Z) = \mathbb{E}_* \bar{f}(Z) \qquad (2.6)$$

We let $\mathrm{Lsc}[Z]$ denote the set of all functions $g \in \overline{\mathbf{R}}^T$ such that $(-g) \in \mathrm{Usc}[Z]$; that is, $g \in \mathrm{Lsc}[Z]$ if and only if g satisfies of the following two equivalent conditions

$$\inf_T g > -\infty \quad \text{and} \quad \mathbb{E}^* g(Z) = \sup\{\mathbb{E}^* h(Z) \mid h \in \mathrm{Lsc}(T) , h \le g\} \qquad (2.7)$$

$$\inf_T g > -\infty \quad \text{and} \quad \mathbb{E}^* g(Z) = \mathbb{E}^* g^\circ(Z) \qquad (2.8)$$

We set $C[Z] := \mathrm{Usc}[Z] \cap \mathrm{Lsc}[Z]$ and $C_\circ[Z] := \mathrm{Usc}_\circ[Z] \cap \mathrm{Lsc}_\circ[Z]$ where:

$$f \in \mathrm{Usc}_\circ[Z] \;\Leftrightarrow\; \sup_T f < \infty \quad \text{and} \quad \mathbb{E}_* f(Z) = \inf_{\phi \in C(T) , \phi \ge f} \mathbb{E}_* \phi(Z) \qquad (2.9)$$

$$g \in \mathrm{Lsc}_\circ[Z] \;\Leftrightarrow\; \inf_T g > -\infty \quad \text{and} \quad \mathbb{E}^* g(Z) = \sup_{\phi \in C(T) , \phi \le f} \mathbb{E}^* \phi(Z) \qquad (2.10)$$

3.   Probability Extents

Let $I : \overline{\mathbf{R}}^T \to \overline{\mathbf{R}}$ be a given functional with dual functional I°. Then I is called a probability functional, if I is increasing and $I(a\,f + b\,1_T) = a\,I(f) + b$ for all $0 \le a < \infty$, all $b \in \mathbf{R}$ and all $f \in \overline{\mathbf{R}}^T$ with the convention $0 \cdot (\pm\infty) := 0$. It is easily verified that I is a probability functional if and only if I° is a probability functional. The pair (I, I°) is called *probability extent* on T if I is increasing $I(1_T) = I^\circ(1_T) = 1$ and

$$I(af) = aI(f) , \quad I(f \dotplus g) \le I(f) \dotplus I(g) \quad \forall f, g \in \overline{\mathbf{R}}^T \;\forall 0 \le a < \infty \qquad (3.1)$$

where $\dotplus$ denotes the addition on $\overline{\mathbf{R}}$ with the convention $\infty \dotplus (-\infty) := \infty$. Let (I, I°) be a probability extent. Then it follows easily that I and I° are probability functionals such that $I^\circ \le I$ and

$$I^\circ(f) \dotplus I^\circ(g) \le I^\circ(f + g) \le I^\circ(f \dotplus g) \le I^\circ(f) \dotplus I(g) \quad \forall f, g \in \overline{\mathbf{R}}^T \qquad (3.2)$$

$$I^\circ(f) + I(g) \le I(f + g) \le I(f \dotplus g) \le I(f) \dotplus I(g) \quad \forall f, g \in \overline{\mathbf{R}}^T \qquad (3.3)$$

where $+$ denotes the addition on $\overline{\mathbf{R}}$ with the convention $\infty + (-\infty) := -\infty$. If (I, I°) is a probability extent, we let $L(I, I^\circ)$ and $L^1(I, I^\circ)$ denote the following function spaces:

$$L(I, I^\circ) = \{f \in \overline{\mathbf{R}}^T \mid I(f) = I^\circ(f)\} \qquad (3.4)$$

$$L^1(I, I^\circ) = \{f \in \mathbf{R}^T \cap L(I, I^\circ) \mid I(f) \ne \pm\infty\} \qquad (3.5)$$

Then $L^1(I, I^\circ)$ is a linear subspace of $\mathbf{R}^T$ containing 1_T and I is linear on $L^1(I, I^\circ)$. The probability extents plays cardinal role in the theory of convergence law as demonstrated by the following typically examples:

If $\mathcal{B}$ is a σ-algebra on T and μ is a probability measure on $(T, \mathcal{B})$, we define the probability extent (I^μ, I_μ) as follows:

$$I^\mu(f) = \int^* f d\mu \quad \text{and} \quad I_\mu(f) = \int_* f d\mu \qquad (3.6)$$

It is well-known that we have

$$L^1(I^\mu, I_\mu) = L^1(\mu) \tag{3.7}$$

$$I^\mu(f) = \inf_{c \in \mathbf{R}} I^\mu(f \vee c) \quad \text{and} \quad I_\mu(f) = \sup_{c \in \mathbf{R}} I_\mu(f \wedge c) \quad \forall f \in \overline{\mathbf{R}}^T \tag{3.8}$$

$$I^\mu(f) = \sup_{c \in \mathbf{R}} I^\mu(f \wedge c) \quad \Leftrightarrow \quad \text{either } I^\mu(f^+) < \infty \text{ or } I_\mu(f^-) < \infty \tag{3.9}$$

$$I_\mu(f) = \inf_{c \in \mathbf{R}} I_\mu(f \vee c) \quad \Leftrightarrow \quad \text{either } I_\mu(f^+) < \infty \text{ or } I^\mu(f^-) < \infty \tag{3.10}$$

$$I_\mu \text{ is } \sigma\text{-smooth at } \overline{\mathbf{R}}^T \text{ along } \{f \in \overline{\mathbf{R}}^T \mid I_\mu(f) < \infty\} \tag{3.11}$$

If Z is a T-valued random element on $(\Upsilon, \mathcal{A}, Q)$, we define the probability extent (I^Z, I_Z) as follows:

$$I^Z(f) := \mathbb{E}^* f(Z) \quad \text{and} \quad I_Z(f) := \mathbb{E}_* f(Z) \quad \forall f \in \overline{\mathbf{R}}^T \tag{3.12}$$

Then (I^Z, I_Z) satisfies (3.8)–(3.11) and we have

$$L^1(I^Z, I_Z) = L^1(Q_Z) = \{h \in \mathbf{R}^T \mid h(Z) \in L^1(Q)\} \tag{3.13}$$

If $\mathcal{X} = (X_\pi \mid \pi \in \Pi)$ is a Π-net of T-valued random elements on $(\Omega, \mathcal{F}, P)$, we define the probability extent $(I^\mathcal{X}, I_\mathcal{X})$ as follows:

$$I^\mathcal{X}(f) = \limsup_{\pi \uparrow \Pi} E^* f(X_\pi) \quad \text{and} \quad I_\mathcal{X}(f) = \liminf_{\pi \uparrow \Pi} E_* f(X_\pi) \quad \forall f \in \overline{\mathbf{R}}^T \tag{3.14}$$

It is easily verified that we have

$$h \in L(I^\mathcal{X}, I_\mathcal{X}) \text{ if and only if } \lim_{\pi \uparrow \Pi} E^* h(X_\pi) = \lim_{\pi \uparrow \Pi} E_* h(X_\pi) \tag{3.15}$$

$$X_\pi \to^{\sim} Z \quad \Leftrightarrow \quad I^\mathcal{X}(\phi) = I_\mathcal{X}(\phi) = I^Z(\phi) = I_Z(\phi) \quad \forall \phi \in C(T) \tag{3.16}$$

$$X_\pi \twoheadrightarrow^{\sim} Z \quad \Leftrightarrow \quad I^\mathcal{X}(f) \leq I_Z(f) \quad \forall f \in \mathrm{Usc}(T) \tag{3.17}$$

Finally, let (I, I°) be an arbitrary probability extent and let $\Phi \subseteq \mathbf{R}^T$ be a linear space containing 1_T. Then it is easily verified that the prescription:

$$I^\Phi(f) = \inf_{\phi \in \Phi, \, \phi \geq f} I(\phi) \quad \text{and} \quad I_\Phi(f) = \sup_{\phi \in \Phi, \, \phi \leq f} I^\circ(\phi) \quad \forall f \in \overline{\mathbf{R}}^T \tag{3.18}$$

defines a probability extent (I^Φ, I_Φ) satisfying

$$I_\Phi(f) \leq I^\circ(f) \leq I(f) \leq I^\Phi(f) \quad \forall f \in \overline{\mathbf{R}}^T \tag{3.19}$$

$$I_\Phi(\phi) = I^\circ(\phi) \quad \text{and} \quad I^\Phi(\phi) = I(\phi) \quad \forall \phi \in \Phi \tag{3.20}$$

$$L(I^\Phi, I_\Phi) \subseteq L(I, I^\circ) \quad \text{and} \quad L^1(I^\Phi, I_\Phi) \subseteq L^1(I, I^\circ) \tag{3.21}$$

3.1. Lemma *Let $I : \overline{\mathbf{R}}^T \to \overline{\mathbf{R}}$ be an increasing functional, let $\Phi, \Psi, \Xi \subseteq \overline{\mathbf{R}}^T$ be given function spaces and let Γ and Λ be two nets. Then we have*

(1) *If I is Γ-smooth at Ψ along Φ and $\Gamma \looparrowright \Lambda$, then I is Λ-smooth at Ψ along Φ and I is Φ-regular at $\Psi \cap \Phi_{\downarrow\Gamma}$*

(2) *If Γ is a countably cofinal Frechet net, then $\Gamma \approx \mathbf{N}$ and I is σ-smooth at Ψ along Φ if and only if I is Γ-smooth at Ψ along Φ*

(3) *If I is Φ-regular at Ψ and I is Ξ-regular at Φ, then I is Ξ-regular at Ψ*

(4) *I is τ-smooth at Ψ along Φ $\iff$ I is $2^{(\Phi)}$-smooth at Ψ along Φ*

Proof. (1)–(3) are easy and I shall leave the details to the reader.

(4): Suppose that I is $2^{(\Phi)}$-smooth at Ψ along Φ. Let Δ be a given net and let $\psi \in \Psi$ and $(\phi_\delta \mid \delta \in \Delta) \subseteq \Phi$ be given such that $\phi_\delta \downarrow \psi$. Let $\delta^* \in \Gamma$ be a given point and set $\Pi = 2^{(\Phi)}$, $\Delta^* = \{\delta \in \Delta \mid \delta \geq \delta^*\}$ and $\Phi^* = \{\phi_\delta \mid \delta \in \Delta^*\}$. If $\pi \in \Pi$, we chose a finite (possibly empty) set $\eta(\pi) \subseteq \Delta^*$ such that

$$\operatorname{card}\eta(\pi) = \operatorname{card}(\pi \cap \Phi^*) \ \text{ and } \ \pi \cap \Phi^* = \{\phi_\delta \mid \delta \in \eta(\pi)\}$$

Let Π_n denote the set of all $\pi \in \Pi$ such that π has at most n elements for $n = 1, 2 \ldots$. If $\pi \in \Pi_1$, then $\eta(\pi)$ has at most one element and we set $\tau(\pi) = \delta^*$ if $\eta(\pi) = \emptyset$ and we let $\tau(\pi)$ denote the unique element in $\eta(\pi)$ if $\eta(\pi) \neq \emptyset$. Let $n \geq 1$ be given and suppose that $\sigma_1, \ldots, \sigma_n$ has been constructed such that

(i) $\sigma_1 = \tau$ and σ_i is an increasing map from Π_i into Δ^* for all $i = 1, \ldots, n$

(ii) $\sigma_i(\xi) = \sigma_j(\xi) \quad \forall \xi \in \Pi_i \ \forall 1 \leq i \leq j \leq n$

(iii) $\sigma_i(\xi) \geq \delta \quad \forall \delta \in \eta(\xi) \ \forall i = 1, \ldots, n$

Set $\sigma_{n+1}(\xi) = \sigma_n(\xi)$ for all $\xi \in \Pi_n$ and let $\pi \in \Pi_{n+1} \setminus \Pi_n$ be given. Since Δ is upwards directed, $\eta(\pi)$ is finite and $\{\xi \mid \xi \not\subseteq \pi\}$ is a finite subset of Π_n, we may chose $\sigma_{n+1}(\pi) \in \Delta^*$ such that $\sigma_{n+1}(\pi) \geq \beta$ for all $\beta \in \eta(\pi)$ and $\sigma_{n+1}(\pi) \geq \sigma_n(\xi)$ for all $\xi \not\subseteq \pi$. Then $\{\sigma_1, \ldots, \sigma_{n+1}\}$ satisfies (i)–(iii). So by induction in n, there exists a sequence (σ_i) satisfying (i)–(iii) for all $n \geq 1$. By (i) and (ii), we see that $\sigma(\pi) := \sigma_n(\pi)$ is a well-defined increasing map from Π into Δ^* such that $\sigma(\pi) = \tau(\pi)$ for all $\pi \in \Pi_1$. In particular, we have $\phi_{\sigma(\xi)} \leq \phi_{\sigma(\pi)}$ for all $\xi \subseteq \pi$. Let $\delta \in \Delta$ be given. Then there exists $\beta \in \Delta$ such that $\beta \geq \delta$ and $\beta \geq \delta^*$. Hence, if $\pi = \{\phi_\beta\}$, then $\phi_{\sigma(\pi)} = \phi_\beta \leq \phi_\delta$. Hence, $\phi_{\sigma(\pi)} \downarrow \psi$ and $\inf_{\pi \in \Pi} I(\phi_{\sigma(\pi)}) = \inf_{\delta \in \Delta} I(\phi_\delta)$. So by $2^{(\Phi)}$-smoothness of I, we conclude that $I(\phi_\delta) \downarrow I(\psi)$; that is, I is τ-smooth at Ψ along Φ. The converse implication is evident. $\qquad \square$

3.2. Lemma *Let $\Psi, \Phi, \Phi^* \subseteq \overline{\mathbf{R}}^T$ be given function spaces and let $I : \overline{\mathbf{R}}^T \to \overline{\mathbf{R}}$ be an increasing functional. Let Γ be a net and let $J : \Psi \to \overline{\mathbf{R}}$ be function satisfying*

(1) $\lim_{\gamma \uparrow \Gamma} I(\phi_\gamma) \leq J(\psi) \quad \forall \psi \in \Psi \ \forall (\phi_\gamma \mid \gamma \in \Gamma) \subseteq \Phi$ *with* $\phi_\gamma \downarrow \psi$

Then we have

(2) $I(\psi) \leq \lim_{\gamma \uparrow \Gamma} I(\phi_\gamma^*) \leq J(\psi) \quad \forall \psi \in \Psi \ \forall (\phi_\gamma^* \mid \lambda \in \Lambda) \subseteq \Phi^*$ *with* $\phi_\gamma^* \downarrow \psi$

(3) $I(\psi) \leq J(\psi) \quad \forall \psi \in \Psi \cap (\Phi_{\downarrow\Gamma} \cup \Phi_{\downarrow\Gamma}^*)$

in each of the following five cases:

(I) *For every $\psi \in \Psi$ and every net $(\phi_\gamma^* \mid \gamma \in \Gamma) \subseteq \Phi^*$ with $\phi_\gamma^* \downarrow \psi$, there exists a net $(\phi_\gamma \mid \gamma \in \Gamma) \subseteq \Phi$ such that $\phi_\gamma \downarrow \psi$ and $\phi_\gamma \geq \phi_\gamma^*$ for all $\gamma \in \Gamma$*

(II) *Γ is a finitely founded Frechet net, $\Phi^* \subseteq \overline{\Phi}$ and $a\,1_T + \phi \in \overline{\Phi}$ for all $\phi \in \Phi$ and all $0 < a < 1$*

(III) *Γ is a finitely founded Frechet net, $\Phi^* \subseteq \overline{\Phi}_{\wedge\Gamma}$ and $a\,1_T + \phi \in \overline{\Phi}$ and $\phi \wedge \psi \in \overline{\Phi}$ for all $\phi, \psi \in \Phi$ and all $0 < a < 1$*

(IV) *Γ is a finitely founded Frechet net, $\Phi^* \subseteq \overline{\Phi}_{\wedge\Gamma}$ and Φ is an algebra of bounded functions containing 1_T*

(V) *Γ is finitely founded, $\Phi^* \subseteq \overline{\Phi}_{\wedge\Gamma}$ and $\phi \wedge \psi \in \Phi$ for all $\phi, \psi \in \Phi$*

Proof. Note that (3) follows from (1), (2) and monotonicity of I.

Case I: Evident!

Case II: Let me show that Case II implies Case I for every function space Ψ. So let $(\phi_\gamma^*) \subseteq \Phi^*$ be a given net such that $\phi_\gamma^* \downarrow \psi$ for some function $\psi \in \overline{\mathbf{R}}^T$. By the assumption, it follows easily that $a\,1_T + \phi^* \in \overline{\Phi}$ for all $\phi^* \in \Phi^*$ and all $0 < a < 1$. Let $\gamma \in \Gamma$ be given and let $|\gamma|$ denote the number of elements in the set $\{\alpha \in \Gamma \mid \alpha \leq \gamma\}$. Then $1 \leq |\gamma| < \infty$ and since $\phi_\gamma^* + a\,1_T \in \overline{\Phi}$ for all $0 < a < 1$, there exists $\phi_\gamma \in \Phi$ such that

$$\left\| \phi_\gamma - \left(\phi_\gamma^* + 3 \cdot 2^{-|\gamma|-2} \right) \right\|_T \leq 2^{-|\gamma|-2}$$

Let $\beta, \gamma \in \Gamma$ be given such that $\beta \leq \gamma$ and $\beta \neq \gamma$. Then $\gamma \not\leq \beta$ and so we have that $|\gamma| \geq 1 + |\beta|$ and since (ϕ_γ^*) is decreasing, we have

$$\phi_\gamma \leq \phi_\gamma^* + 2^{-|\gamma|} \leq \phi_\beta^* + 2^{-|\gamma|} \leq \phi_\beta + 2^{-|\beta|-2} - 3 \cdot 2^{-|\beta|-2} + 2^{-|\gamma|}$$
$$\leq \phi_\beta - 2^{-|\beta|-1} + 2^{-|\beta|-1} = \phi_\beta$$

Hence, (ϕ_γ) is decreasing and $\|\phi_\gamma - \phi_\gamma^*\|_T \leq 2^{-|\gamma|}$ for all $\gamma \in \Gamma$. Since Γ is a Frechet net, we have that $2^{-|\gamma|} \to 0$ and since $\phi_\gamma^* \downarrow \psi$, we see that $\phi_\gamma \downarrow \psi$ and that $\phi_\gamma^* \leq \phi_\gamma - 2^{-|\gamma|-1} \leq \phi_\gamma$ for all $\gamma \in \Gamma$. Thus, Case II follows from Case I.

Case III: Let me show that Case III implies Case I for every function space Ψ. So let $(\phi_\gamma^*) \subseteq \Phi^*$ be a given net such that $\phi_\gamma^* \downarrow \psi$ for some function $\psi \in \overline{\mathbf{R}}^T$. By the assumption, its follows easily that $f \wedge g \in \overline{\Phi}$ and $a\,1_T + f \in \overline{\Phi}$ for all $0 < a < 1$ and all $f, g \in \overline{\Phi}$. Since $\phi_\gamma^* \in \overline{\Phi}_{\wedge\Gamma}$, there exist functions $(\phi_{\alpha,\gamma} \mid \alpha \in \Gamma) \subseteq \overline{\Phi}$ such that $\phi_\gamma^* = \inf_{\alpha \in \Gamma} \phi_{\alpha\gamma}$ for all $\gamma \in \Gamma$. Set

$$\zeta_\gamma = \inf\{\phi_{\alpha\beta} \mid \alpha \leq \gamma \,, \ \beta \leq \gamma\} \quad \forall \gamma \in \Gamma$$

Then (ζ_γ) is decreasing and since $\phi_\gamma^* \downarrow \psi$, we have

$$\inf_{\gamma \in \Gamma} \zeta_\gamma = \inf_{\gamma \in \Gamma} \inf_{\alpha \in \Gamma} \phi_{\alpha\gamma} = \inf_{\gamma \in \Gamma} \phi_\gamma^* = \psi$$
$$\zeta_\gamma = \inf_{\alpha \leq \gamma} \inf_{\beta \leq \gamma} \phi_{\mu,\kappa(\nu)} \geq \inf_{\alpha \leq \gamma} \phi_\alpha^* = \phi_\gamma^* \quad \forall \gamma \in \Gamma$$

Thus, $\zeta_\gamma \downarrow \psi$ and since Γ is finitely founded, we see that $a\,1_T + \zeta_\gamma \in \overline{\Phi}$ for all $\gamma \in \Gamma$ and all $0 < a < 1$. Hence, there exist functions $\phi_\gamma \in \Phi$ satisfying

$$\left\|\phi_\gamma - (\zeta_\gamma + 3 \cdot 2^{-|\gamma|-2})\right\|_T \le 2^{-|\gamma|-2} \qquad \forall \gamma \in \Gamma$$

Thus, in exactly the same manner as in the proof of Case II, we conclude that (ϕ_γ) satisfies the hypothesis of Case I.

Case IV: By Stone-Weierstrass' theorem, we see that the hypothesis of Case III holds. So Case IV follows from Case III.

Case V: Since $\phi_\gamma^* \in \Phi_{\wedge\Gamma}$, there exist functions $(\phi_{\alpha\gamma} \mid \alpha \in \Gamma) \subseteq \Gamma$ such that $\phi_\gamma^* = \inf_{\alpha\in\Gamma} \phi_{\alpha\gamma}$. If we define ζ_γ as in the proof of Case III, we see that the hypothesis of Case I holds with $\phi_\gamma := \zeta_\gamma$. $\qquad\Box$

3.3. Lemma *Let $\mathcal{K}, \mathcal{L} \subseteq 2^T$ be given pavings, let $I, J : \overline{\mathbf{R}}^T \to \overline{\mathbf{R}}$ be two probability functionals satisfying*

(1) $I\left(\sum_{i=1}^n 1_{K_i}\right) \le \sum_{i=1}^n I(1_{K_i})$ $\forall K_1, \ldots, K_n \in \mathcal{K}$ with $K_1 \supseteq \cdots \supseteq K_n$

(2) $J\left(\sum_{i=1}^n 1_{L_i}\right) \ge \sum_{i=1}^n J(1_{L_i})$ $\forall L_1, \ldots, L_n \in \mathcal{L}$ with $L_1 \supseteq \cdots \supseteq L_n$

If $f, g \in \overline{\mathbf{R}}^T$ are given functions satisfying

(4) $f \le g$ and for all $x, y \in \mathbf{R}$ with $\inf_T f < x < y < \sup_T f$ we have $\{f \ge y\} \in \mathcal{K}$, $\{g \ge x\} \in \mathcal{L}$ and $I(1_{\{f\ge y\}}) \le J(1_{\{g\ge x\}})$

then we have

(5) $I((f \wedge c) \vee a) \le J((g \wedge c) \vee a)$ $\forall a, c \in \mathbf{R}$

Moreover, if Γ is a net satisfying the following condition:

(5) $\lim_{\gamma\uparrow\Gamma} I(1_{K_\gamma}) \le J(1_L)$ $\forall L \in \mathcal{L}\ \forall (K_\gamma \mid \gamma \in \Gamma) \subseteq \mathcal{K}$ with $K_\gamma \downarrow L$

and $(\phi_\gamma \mid \gamma \in \Gamma) \subseteq U(\mathcal{K})$ is a given net such that $\phi_\gamma \downarrow \psi$ for some $\psi \in U(\mathcal{L})$, then we have

(6) $I(\psi) \le \lim_{\gamma\uparrow\Gamma} I(\phi_\gamma) \le \lim_{a\downarrow-\infty} J(\psi \vee a) = \inf_{a\in\mathbf{R}} J(\psi \vee a)$

(7) $I(f) \le \lim_{a\downarrow-\infty} J(f \vee c) = \inf_{a\in\mathbf{R}} J(f \vee a)$ $\forall f \in U(\mathcal{L}) \cap U_{\downarrow\Gamma}(\mathcal{K})$

Remark: If I satisfies (1), we say that I is *subrectilinear* on $\mathcal{K}$, if J satisfies (2), we say that J is *subrectilinear* on $\mathcal{L}$ and if I is sub and superrectlinear on $\mathcal{K}$, we say that I is *rectilinear* on $\mathcal{K}$. We say that I *sub/super/rectilinear* if I sub/super/rectilinear on 2^T. If (I, I°) is a probability extent, then I is subrectilinear and I° is superrectilinear. If μ is a probability measure and Z is a T-valued random element, then it is well-known that I^μ, I_μ, I^Z and I_Z are rectilinear.

Proof. (4): Let $a, c \in \mathbf{R}$ be given and set $f_0 = (f \wedge c) \vee a$ and $g_0 = (g \wedge c) \vee a$. If $c \le a$, then $f_0 = g_0 = a\,1_T$ and so $I(f_0) = J(g_0) = a$. So suppose that $c > a$ and set $a^* = \inf_T f_0$ and $b^* = \sup_T f_0$. Then $a \le a^* \le b^* \le c < \infty$ and if $a^* = b^*$, then $f_0 = a^* 1_T \le g_0$ and so $I(f_0) = a \le J(g_0)$. So suppose that $a^* < b^*$ and set $d = b^* - a^*$. Then $0 < d < \infty$ and $\inf_T f \le a^* < b^* \le \sup_T f$. Let $n \ge 3$ and $2 \le i \le n - 1$ be given integers and set $K_{in} = \{f \ge a^* + \frac{i}{n} d\}$ and $L_{in} = \{g \ge a^* + \frac{i-1}{n} d\}$. Applying (3.3.3) with $x = a^* + \frac{i-1}{n} d$ and $y = a^* + \frac{i}{n} d$, we see that $K_{in} \in \mathcal{K}$, $L_{in} \in \mathcal{L}$ and $I(1_{K_{in}}) \leq J(1_{L_{i,n}})$ for all $2 \leq i \leq n-1$ and since $g_0 \geq f_0 \geq a^*$, we have

$$a^* + \frac{d}{n}\sum_{i=2}^{n-1} 1_{G_{i,n}} \leq \frac{2d}{n} + g_0 \;,\;\; f_0 \leq a^* + \frac{2d}{n} + \frac{d}{n}\sum_{i=2}^{n-1} 1_{F_{in}}$$

Moreover, since I and J are probability functionals, then by (1) and (2) we have

$$I(f_0) \leq a^* + \frac{2d}{n} + \frac{d}{n}\sum_{i=2}^{n-1} I(1_{F_{in}}) \leq a^* + \frac{2d}{n} + \frac{d}{n}\sum_{i=2}^{n-1} J(1_{G_{in}})$$

$$\leq \frac{2d}{n} + J\left(a^* + \frac{d}{n}\sum_{i=2}^{n-1} 1_{G_{in}}\right) \leq \frac{4d}{n} + J(g_0)$$

Letting $n \to \infty$, we see that $I(f_0) \leq J(g_0)$ and so (4) follows.

(6 + 7): Suppose that (5) holds and let $(\phi_\gamma \mid \gamma \in \Gamma) \subseteq U(\mathcal{K})$ be a given net. Let $a \in \mathbf{R}$ be a given number and let $\psi \in U(\mathcal{L})$ be a given function such that $\phi_\gamma \downarrow \psi$ and set $\phi_\gamma^a = \phi_\gamma \vee a$ and $\psi^a - \psi \vee a$. Since ϕ_γ is bounded from above, there exist $\gamma^* \in \Gamma$ and $0 < b < \infty$ such that $a \leq \phi_\gamma^a \leq a + b$ for all $\gamma \geq \gamma^*$. Let $\gamma \geq \gamma^*$ be given and set $K_{in}(\gamma) = \{\phi_\gamma \geq a + \frac{i}{n}b\}$ and $L_{in} = \{\psi \geq a + \frac{i}{n}b\}$ for all $i, n \geq 1$. Since $K_{in}(\gamma) \in \mathcal{K}$ and $L_{in} \in \mathcal{L}$ for all $i, n \geq 1$, then in the same manner as above we obtain the following inequalities:

$$I(\phi_\gamma) \leq I(\phi_\gamma^a) \leq a + \frac{b}{n} + \frac{b}{n}I\left(\sum_{i=1}^{n} 1_{K_{in}(\gamma)}\right) \leq a + \frac{b}{n} + \frac{b}{n}\sum_{i=1}^{n} I\left(1_{K_{in}(\gamma)}\right)$$

$$a + \frac{b}{n} + \frac{b}{n}\sum_{i=1}^{n} J\left(1_{L_{in}}\right) \leq \frac{b}{n} + J\left(a + \frac{b}{n}\sum_{i=1}^{n} 1_{L_{in}}\right) \leq \frac{b}{n} + J(\psi^a)$$

and since $\phi_\gamma^a \downarrow \psi^a$, then for every fixed $1 \leq i \leq n$ we have that $K_{in}(\gamma) \downarrow L_{in}$. So by (5), we see that $\lim_{\gamma \uparrow \Gamma} I(\phi_\gamma) \leq \frac{b}{n} + J(\psi^a)$ for all $n \geq 1$ and all $a \in \mathbf{R}$. Letting $n \to \infty$, we obtain (6) and (7) is an immediate consequence of (6). $\square$

3.4. Lemma   *Let $\Phi, \Psi \subseteq \mathbf{D} \subseteq \overline{\mathbf{R}}^T$ be given function spaces such that $(\phi - \psi)^+ \in \Phi$ for all $\phi \in \Phi$ and all $\psi \in \Psi$ with the convention $(\infty - \infty)^+ := 0$. Let Γ be a net and let $I : \mathbf{D} \to \overline{\mathbf{R}}$ be an increasing functional such that I is Γ-smooth at 0 along Φ and $I(\phi + \psi) \leq I(\phi) \dot{+} I(\psi)$ for all $\phi \in \Phi^+$ and all $\psi \in \Psi$. Let $f \in \overline{\mathbf{R}}^T$ be a given function and let $(\phi_\gamma \mid \gamma \in \Gamma) \subseteq \Phi$ be a given net such that $\phi_\gamma \downarrow f$. Then we have*

(1) $\lim_{\gamma \uparrow \Gamma} I(\phi_\gamma) = \inf_{\gamma \in \Gamma} I(\phi_\gamma) \leq \inf \{I(\psi) \mid \psi \in \Psi \,,\, \psi \geq f\}$

In particular, we see that I is Γ-smooth at Ξ along Φ, where Ξ is the set of all functions $f \in \mathbf{D}$ such that I is Ψ-regular at f.

Proof. Let $(\phi_\gamma) \subseteq \Phi$ be given such that $\phi_\gamma \downarrow f$ and let c denote the last infimum in (1). If $c = \infty$, then there is nothing to prove. So suppose that $c < \infty$ and let $a > c$ be given. Then there exists $\psi \in \Psi$ such that $\psi \geq f$ and $I(\psi) < a$.

Set $\zeta_\gamma = (\phi_\gamma - \psi)^+$ for all $\gamma \in \Gamma$. Then $\zeta_\gamma \in \Phi$ and since $\phi_\gamma \leq \zeta_\gamma \dotplus \psi$, we have $I(\phi_\gamma) \leq I(\zeta_\gamma) \dotplus I(\psi)$. Moreover, since $\phi_\gamma \downarrow f \leq \psi$, we see that $\zeta_\gamma \downarrow 0$. Hence, by Γ-smoothness of I we have

$$\lim_{\gamma \uparrow \Gamma} I(\phi_\gamma) \leq I(\psi) \dotplus \lim_{\gamma \uparrow \Gamma} I(\zeta_\gamma) = I(\psi) < a$$

Letting $a \downarrow c$, we see that (1) follows. □

3.5. Theorem *Let (I, I°) be a probability extent on T, let $\Phi \subseteq \mathbf{R}^T$ be a linear space containing 1_T and let (I^Φ, I_Φ) denote the probability extent given by (3.18). Let $\mathcal{L} \subseteq 2^T$ be a paving on T and set $\mathcal{L}^* = \{K \subseteq T \mid K \cap L \in \mathcal{L} \ \forall L \in \mathcal{L}\}$. Suppose that the following three conditions hold:*

(1) $\mathcal{L}$ *is $(\cap c)$-stable and $\{\phi \geq a\} \in \mathcal{L}^* \ \forall a \in \mathbf{R} \ \forall \phi \in \Phi$*

(2) $I^\Phi(1_{L_1 \cup L_2}) \geq I^\Phi(1_{L_1}) + I^\Phi(1_{L_2}) \quad \forall L_1, L_2 \in \mathcal{L}$ *with $L_1 \cap L_2 = \emptyset$*

(3) I^Φ *is σ-smooth at $\emptyset$ along $\mathcal{L}$*

and let μ denotes the restriction of ν to $\mathcal{M}(\nu)$ where ν is the increasing set function given by

$$\nu(A) = \sup\{I^\Phi(1_L) \mid L \in \mathcal{L}, \ L \subseteq A\} \quad \forall A \subseteq T$$

Then $\mathcal{M}(\nu)$ is a σ-algebra and μ is a measure on $(T, \mathcal{M}(\nu))$ such that $\mu(T) \leq 1$ and

(4) $\mu(T) = 1$ *if and only if I is $\mathcal{L}$-tight along Φ*

(5) $\nu(L) = \mu(L) = I^\Phi(1_L) \ \forall L \in \mathcal{L}$ *and $\nu(A) = \mu_*(A) \leq I^\Phi(1_A) \ \forall A \subseteq T$*

(6) $\emptyset \in \mathcal{L} \cup \{T\} \subseteq \mathcal{L}^*$ *and $\sigma(\Phi) \subseteq \sigma(\mathcal{L}^*) \subseteq \mathcal{M}(\nu)$*

Moreover, if $\mu(T) = 1$ and $h, g \in \overline{\mathbf{R}}^T$ are given functions satisfying

(7) $h \leq g$ *and for all $\varepsilon > 0$ and all $x, y \in \mathbf{R}$ with $\inf_T h < x < y < \sup_T h$, there exists $\phi \in \Phi$ such that $\phi \geq 1_{\{h \geq y\}}$ and $I(\phi) < \varepsilon + \nu(g \geq x)$*

then we have

(8) $I(h \wedge c) \leq I^\Phi(h \wedge c) \leq \int_* (g \wedge c) d\mu \leq \int_* g \, d\mu \quad \forall c \in \mathbf{R}$

(9) $I(f) \leq I^\Phi(f) \leq \int_* f \, d\mu \ \forall f \in U(\mathcal{L})$

(10) *If Φ is an algebra of bounded functions, then we have*
    (a) $\Phi \subseteq L^1(\mu) \cap L^1(I, I^\circ) \cap L^1(I^\Phi, I_\Phi)$
    (b) $I^\circ(\phi) = I(\phi) = I_\Phi(\phi) = I^\Phi(\phi) = \int_T \phi \, d\mu \quad \forall \phi \in \Phi$

Proof. (4)–(6): By [[2]; 1.23 p. 22], we have that $\mathcal{M}(\nu)$ is an algebra and that μ is a finitely additive content on $\mathcal{M}(\nu)$. Hence, (3.5.4) follows from the definition of μ. Since I^Φ is increasing, we have that $\nu(L) = I^\Phi(1_L)$ for all $L \in \mathcal{L}$ and since $\mathcal{L}$ is $(\cap c)$-stable and $0 \in \Phi$, we see that $\emptyset \in \mathcal{L}$, $T \in \mathcal{L}^*$ and $\mathcal{L} \subseteq \mathcal{L}^*$. Hence, the first part of (6) follows and by (1), we have that $\sigma(\Phi) \subseteq \sigma(\mathcal{L}^*)$. By (3), it follows easily that ν is superadditive on 2^T. Let $L \in \mathcal{L}$, $L^* \in \mathcal{L}^*$ and $0 < \varepsilon < 1$ be given. Since $K := L \cap L^* \in \mathcal{L}$, there exist $\phi \in \Phi$ such that $\phi \geq 1_K$ and $I(\phi) < \varepsilon + \nu(K)$. Setting $E = L \cap \{\phi \leq 1 - \varepsilon\}$, we see that $E \subseteq L \setminus L^*$ and by (1) we have that $E \in \mathcal{L}$. Hence, there exist $\psi \in \Phi$ such that $\psi \geq 1_E$ and $I(\psi) \leq \varepsilon + \nu(E)$. Set $\zeta = \phi + \psi$ and let $t \in L$ be given. If $t \in E$, then $\zeta(t) \geq \psi(t) \geq 1_E(t) = 1$ and if $t \notin E$, then

$\zeta(t) \geq \phi(t) \geq 1 - \varepsilon$. Hence, $\zeta \geq (1 - \varepsilon) 1_L$ and since $\zeta \in \Phi$ and $E \subseteq L \setminus L^*$, we have

$$(1 - \varepsilon)\nu(L) \leq I(\zeta) \leq I(\phi) + I(\psi) \leq 2\varepsilon + \nu(K) + \nu(E) \leq 2\varepsilon + \nu(K) + \nu(L \setminus L^*)$$

Letting $\varepsilon \downarrow 0$, we see that $\nu(L) \leq \nu(L \cap L^*) + \nu(L \setminus L^*)$ for all $L \in \mathcal{L}$ and all $L^* \in \mathcal{L}^*$. Let $A \subseteq T$ and $L^* \in \mathcal{L}^*$ be given. Taking supremum over all $L \in \mathcal{L}$ contained in the set A, we see that $\nu(A) \leq \nu(A \cap L^*) + \nu(A \setminus L^*)$ and converse inequality follows from superadditivity of ν. Hence, $\mathcal{L} \subseteq \mathcal{L}^* \subseteq \mathcal{M}(\nu)$ and so (5) follows monotonicity of I^Φ. Moreover, I claim that we have

(i) $\lim_{n \to \infty} \nu(A_n) = \nu(A) \quad \forall A_1, A_2, \dots \subseteq T$ with $A_n \downarrow A$

So let $\varepsilon > 0$ and $A_1, A_2, \dots \subseteq T$ be given such that $A_n \downarrow A$. Then there exist sets $L_n \in \mathcal{L}$ such that $L_n \subseteq A_n$ and $\nu(A_n) \leq 2^{-n}\varepsilon + \nu(L_n)$ for all $n \geq 1$. Since $L_n \in \mathcal{M}(\nu)$, we have that $\nu(A_n) = \nu(L_n) + \nu(A_n \setminus L_n)$ and so we conclude that $\nu(A_n \setminus L_n) \leq 2^{-n}\varepsilon$ for all $n \geq 1$. If we set $L^n = L_1 \cap \cdots \cap L_n$, then $L^n \in \mathcal{L} \subseteq \mathcal{M}(\nu)$ and $L_n \setminus L_i \subseteq A_i \setminus L_i$ for all $n \geq i \geq 1$. Hence, by additivity of ν on the algebra $\mathcal{M}(\nu)$ we have

$$\nu(A_n) \leq 2^{-n}\varepsilon + \nu(L^n) + \nu(L_n \setminus L^n) \leq 2^{-n}\varepsilon + \nu(L^n) + \sum_{i=1}^n \nu(L_n \setminus L_i)$$

$$\leq 2^{-n}\varepsilon + \nu(L^n) + \sum_{i=1}^n \nu(A_i \setminus L_i) \leq 2^{-n}\varepsilon + \nu(L^n) + \varepsilon \sum_{i=1}^n 2^{-i} = \varepsilon + \nu(L^n)$$

Setting $L = \cap_{n=1}^\infty L_n$, we see that $L \in \mathcal{L}$ and that $L^n \setminus L \downarrow \emptyset$. Hence, by the same procedure as above, there a decreasing sequence $(L_n') \subseteq \mathcal{L}$ such that $L_n' \subseteq L_n \setminus L$ and $\nu(L_n \setminus L) \leq \varepsilon + \nu(L_n')$ for all $n \geq 1$. In particular, we have that $L_n' \downarrow \emptyset$. So by σ-smoothness of I^Φ we see that $\nu(L_n') \to 0$ and since $L \in \mathcal{L} \subseteq \mathcal{L}^*$, then by additivity of ν on the algebra $\mathcal{M}(\nu)$ we have

$$\nu(A) \leq \lim_{n \to \infty} \nu(A_n) \leq \varepsilon + \lim_{n \to \infty} \nu(L^n) \leq \varepsilon + \nu(L) + \lim_{n \to \infty} \nu(L^n \setminus L)$$
$$\leq 2\varepsilon + \nu(L) + \lim_{n \to \infty} \nu(L_n') \leq 2\varepsilon + \nu(A)$$

Letting $\varepsilon \downarrow 0$, we obtain (i).

Let $M_1, M_2, \dots$ be mutually disjoint sets in $\mathcal{M}(\nu)$ and set $M = \cup_{n=1}^\infty M_n$. Let $A \subseteq T$ be given. By [[2]; (1.23.3) p. 23] and (i), we have that

$$\nu(A \setminus M) + \nu(A \cap M) = \lim_{n \to \infty} \nu(A \setminus (M_1 \cup \cdots \cup M_n)) + \sum_{i=1}^\infty \nu(A \cap M_i)$$

$$= \lim_{n \to \infty} \left\{ \nu(A \setminus (M_1 \cup \cdots \cup M_n)) + \sum_{i=1}^{n-1} \nu(A \cap M_i) \right\} = \nu(A)$$

Hence, we see that $M \in \mathcal{M}(\nu)$. Thus, $\mathcal{M}(\nu)$ is a σ-algebra containing $\sigma(\mathcal{L}^*)$ and μ is a measure on $\mathcal{M}(\nu)$. Thus, (4)–(6) are proved.

(8): By (7), we see that (h, g) satisfies (3) in Lemma 3.3 with $I := I^\Phi$ and $J := I_\mu$ and since μ is a probability measure, I^Φ is subrectilinear on 2^T and I_μ is rectilinear on 2^T, we see that (8) follows from Lemma 3.3 (4) and (3.10).

(9): Let $f \in U(\mathcal{L})$ be given. Since f is bounded from above and (f, f) satisfies (7), we see that (9) follows from (8).

(10): Suppose that Φ is an algebra of bounded function and let $\phi \in \Phi$, $\varepsilon > 0$ and $-\infty < x < y < \infty$ be given. Then the function $\zeta(t) = (\frac{\phi(t)-x}{y-x})^+ \wedge 1$ belongs to $\overline{\Phi}$ and so there exists $\psi_0 \in \Phi$ such that $||\zeta - \psi_0||_T < \varepsilon$. Hence, we see that $\psi := \psi_0 + \varepsilon 1_T \in \Phi$ and since $1_{\{\phi \geq y\}} \leq \zeta \leq 1_{\{\phi \geq x\}}$, we see that $1_{\{\phi \geq y\}} \leq \zeta \leq \psi \leq 1_{\{\phi \geq x\}} + 2\varepsilon 1_T$. Thus, we conclude that (ϕ, ϕ) satisfies (7) and since ϕ is bounded, then by (8) we have $I(\phi) \leq I^{\Phi}(\phi) \leq I_\mu(\phi)$. Applying, this to $-\phi$, we see that (10) follows. $\square$

3.6. Definition (Completions of probability extents)   Let (I, I°) be a probability extent on T, let $\mathcal{L} \subseteq 2^T$ be a given paving on T. If I is $\mathcal{L}$-tight along $C(T)$ and satisfies the conditions (1)–(3) in Theorem 3.5 hold with $\Phi = C(T)$, we say that (I, I°) is $\mathcal{L}$-*complete* and the probability measure μ defined in Theorem 3.5 will be called *the $\mathcal{L}$-completion*. If (I, I°) is $\mathcal{L}$-complete, then Theorem 3.5 shows that the Borel completion is a Baireian probability measure such that I^μ is $C(T)$-regular at $\mathcal{L}$ and $\int_T \phi \, d\mu = I(\phi) = I^\circ(\phi)$ for all $\phi \in C(T)$. Moreover, the reader easily verifies that (I, I°) is $\mathcal{L}$-complete if and only if the following four conditions hold:

(1) $\mathcal{L}$ is ($\cap c$)-stable and $F \cap L \in \mathcal{L}$ $\forall F \in \mathcal{F}_\circ(T)$ $\forall L \in \mathcal{L}$

(2) I is $\mathcal{L}$-tight along $C(T)$ and $C(T) \subseteq L^1(I, I^\circ)$
     For all $\varepsilon > 0$ and all $L_1, L_2 \in \mathcal{L}$ with $L_1 \cap L_2 = \emptyset$ there exist $\phi_1, \phi_2 \in C(T)$
     such that

(3) $\phi_1 \geq 1_{L_1}$, $\phi_2 \geq 1_{L_2}$ and $I(\phi_1 \wedge \phi_2) \leq \varepsilon$

(4) For all $\varepsilon > 0$ and all sequences $L_1, L_2, \ldots \in \mathcal{L}$ with $L_n \downarrow \emptyset$ there exist
     $\phi_1, \phi_2, \ldots \in C(T)$ such that
        (a) $\phi_n \geq 1_{L_n}$ $\forall n \geq 1$ and $\inf_{n \geq 1} I(\phi_n) \leq \varepsilon$

Let (I, I°) be an arbitrary probability extent, let $L_1, L_2 \subseteq T$ be given sets and let $\phi : T \to [0, 1]$ be a function such that $\phi(t) = 1$ for all $t \in L_1$ and $\phi(t) = 0$ for all $t \in L_2$. Then $\phi_1 := (3\phi_1 - 1)^+ \geq 1_{L_1}$, $\phi_2 := (1 - 3\phi)^+ \geq 1_{L_2}$ and $\phi_1 \wedge \phi_2 = 0$ and so we see that (ϕ_1, ϕ_2) satisfies (3a). Hence, if $\mathcal{L}$ is a ($\cap f$)-stable paving and $\mathcal{L}^*$ is defined as in Theorem 3.5, then we have

(5) If $T \in \mathcal{L}$, then I is $\mathcal{L}$-tight along $C(T)$ and $\mathcal{L}^* = \mathcal{L}$

(6) If $\mathcal{L} \subseteq \mathcal{F}_\circ(T)$, then (3) holds. If T is normal and $\mathcal{L} \subseteq \mathcal{F}(T)$, then (3) holds.
     If $\mathcal{L} \subseteq \mathcal{K}(T)$, then (4) holds. If T is an Urysohn space and $\mathcal{L} \subseteq \mathcal{K}(T)$, then
     (3+4) hold

(7) The pavings $\mathcal{F}_\circ(T)$, $\mathcal{F}(T)$ and $\mathcal{K}(T)$ satisfy (1) and $\mathcal{F}_\circ^*(T) = \mathcal{F}_\circ(T)$, $\mathcal{F}^*(T) = \mathcal{F}(T)$ and $\mathcal{K}^*(T) \supseteq \mathcal{F}(T)$

We say that (I, I°) is *Baire complete* or *Borel complete* or *Radon complete* if (I, I°) is $\mathcal{F}_\circ(T)$-complete or $\mathcal{F}(T)$-complete or $\mathcal{K}(T)$-complete and if so then the respective completions will be called *the Baire completion, the Borel completion* or *the Radon completion*. Note that the Baire completion is a Baireian probability measure, the Borel completion is a Borelian probability measure and that the Radon completion is a Radon probability measure. Let "xxx" stand for either "$\mathcal{L}$" or "Baire" or "Borel" or "Radon". If ν is a probability measure on some σ-algebra on T, Z is a random element on $(\Upsilon, \mathcal{A}, Q)$ and $\mathcal{X} = (X_\pi \mid \pi \in \Pi)$ is Π-net of random elements on $(\Omega, \mathcal{F}, P)$, then we say that ν or Z or (X_π) is *"xxx" complete* with *"xxx" completion* μ if the probability extents (I^ν, I_ν) or (I^Z, I_Z) or $(I^\mathcal{X}, I_\mathcal{X})$ is "xxx" complete with "xxx" completion μ.

3.7.   Theorem (Completeness)   *Let (I, I°) be a probability extent on T, let $\mathcal{L}, \mathcal{K} \subseteq 2^T$ be two pavings on T satisfying Definition 3.6 (1) and let $\mathcal{L}_\mu$ denote the set of all $L \in \mathcal{M}(\mu)$ such that I^μ is $C(T)$-regular at 1_L whenever μ is a probability measure on T. Then we have*

(1) *(I, I°) is $\mathcal{L}$-complete if and only there exists a probability measure μ on some σ-algebra $\mathcal{B}$ on T satisfying*
    (a) *I^μ is $C(T)$-regular at $\mathcal{L}$ and I^μ is $\mathcal{L}$-tight along $C(T)$*
    (b) *$\mathcal{L} \subseteq \mathcal{M}(\mu)$ and $I(\phi) \leq \int_* \phi\, d\mu$ $\forall \phi \in C(T)$*
  and if so, then $\bar{\mu}$ is an extension of the $\mathcal{L}$-completion of (I, I°)

(2) *Suppose that (I, I°) is $\mathcal{L}$-complete with $\mathcal{L}$-completion μ. Then $\mathcal{L}_\mu$ is a $(\cup f, \cap c)$-stable paving containing $\mathcal{L} \cup \mathcal{F}_\circ(T)$ and satisfying $(1+2)$ in Definition 3.6 and if I is $\mathcal{K}$-tight along $C(T)$ and $\mathcal{K} \subseteq \mathcal{L}_\mu$, then I is $\mathcal{K}$-complete and μ is an extension of the $\mathcal{K}$-completion κ of (I, I°) and if $\mathcal{L} \subseteq \mathcal{M}(\kappa)$, then $\mu = \kappa$*

(3) *(I, I°) is Baire complete if and only if $C(T) \subseteq L^1(I, I^\circ)$ and I is σ-smooth at 0 along $C(T)$*

(4) *If (I, I°) is $\mathcal{L}$-complete, then (I, I°) is Baire complete and the $\mathcal{L}$-completion of (I, I°) is an extension of the Baire completion*

(5) *(I, I°) is Borel complete if and only if there exists a Borelian probability measure μ such that I^μ is $C(T)$-regular at $\mathcal{F}(T)$ and $I(\phi) \leq \int_T \phi\, d\mu$ for all $\phi \in C(T)$ and if so then $\bar{\mu}$ is an extension of the Borel completion of (I, I°)*

(6) *If (I, I°) is Radon complete, then I is $\mathcal{K}(T)$-tight along $C(T)$ and $C(T) \subseteq L^1(I, I^\circ)$ and the converse implication holds if T is an Urysohn space*

  Moreover, we have that (I, I°) is Borel complete in either of the following six cases

(I) *T is perfectly normal and (I, I°) is Baire complete*

(II) *T is normal and countably paracompact and (I, I°) is Baire complete*

(III) *(I, I°) is Baire complete and I_μ is $C(T)$-regular at $\mathcal{F}(T)$ where μ is the Baire completion of (I, I°)*

(IV) *T is completely regular, $C(T) \subseteq L^1(I, I^\circ)$ and I is τ-smooth at 0 along $C(T)$*

(V) *T is completely regular and (I, I°) is Radon complete*

(VI) *$C(T) \subseteq L^1(I, I^\circ)$, I° is $C(T)$-regular at $\mathcal{F}(T)$ and I° is σ-smooth at $\emptyset$ along $\mathcal{F}(T)$*

In particular, if Z is a given T-valued random element on $(\Upsilon, \mathcal{A}, Q)$, then we have

(7) *Z is $\mathcal{L}$-complete if and only if Q_Z is $\mathcal{L}$-complete*

(8) *Z is Baire complete if and only if Z is Baire Q-measurable*

(9) *If Z is Baire Q-measurable and I_Z is $C(T)$-regular at $\mathcal{F}(T)$, then Z is Borel Q-measurable and Borel complete*

(10) *If Z is Radon complete, then Q_Z is Radonian*

(11) *If T is an Urysohn space and Q_Z is Radonian, then Z is Radon complete*

Proof. Throughout the proof, we let (J^*, J_*) be given by (3.18) with $\Phi = C(T)$ and if μ is a probability measure, we define (J^Z, J_Z) and (J^μ, J_μ) similarly with (I, I°) replaced by (I^Z, I_Z) and (I^μ, I_μ), respectively.

(1): If (I, I°) is $\mathcal{L}$-complete, then the $\mathcal{L}$-completion satisfies (1.a+b). So suppose that μ is a probability measure satisfying (1.a+b). Applying (1.b) on $\pm\phi$, we see that $I^\mu(\phi) \leq I^\circ(\phi) \leq I(\phi) \leq I_\mu(\phi)$ for all $\phi \in C(T)$. So μ is Baireian, $C(T) \subseteq L^1(I, I^\circ)$ and $I(\phi) = I^\mu(\phi)$ for all $\phi \in C(T)$. Hence, we see that $J^* = J^\mu$ and by (1.a) we have that $J^\mu(1_L) = \bar{\mu}(L)$ for all $L \in \mathcal{L}$. Thus, we see that $(I, I^\circ, \mathcal{L}, C(T))$ satisfies Theorem 3.5 (1)–(3) and since I^μ is $\mathcal{L}$-tight along $C(T)$, then so is I. Thus, (I, I°) is $\mathcal{L}$-complete and if λ denotes the $\mathcal{L}$-completion, then $\lambda(L) = \bar{\mu}(L)$ for all $L \in \mathcal{L}$. So by the definition of λ, we have that $\lambda_*(A) \leq \mu_*(A)$ for all $A \subseteq T$. Applying this on the complement $T \setminus A$, we see that $\lambda_*(A) \leq \mu_*(A) \leq \mu^*(A) \leq \lambda^*(A)$ for all $A \subseteq T$. In particular, we see that $\mathcal{M}(\lambda) \subseteq \mathcal{M}(\mu)$ and that $\bar{\mu}(B) = \lambda(B)$ for all $B \in \mathcal{M}(\lambda)$. Thus, $\bar{\mu}$ is an extension of λ.

(2): Let $L_1, L_2 \in \mathcal{L}_\mu$ and $\varepsilon > 0$ be given. Since μ is Baireian and $L_1, L_2 \in \mathcal{M}(\mu)$, there exist $\phi_1, \phi_2 \in C(T)$ such that $\phi_i \geq 1_{L_i}$ and $I^\mu(\phi_i - 1_{L_i}) < \varepsilon$ for $i = 1, 2$ and since $0 \leq \phi_1 \wedge \phi_2 - 1_{L_1 \cap L_2} \leq (\phi_1 - 1_{L_1}) + (\phi_2 - 1_{L_2})$ and $0 \leq \phi_1 \vee \phi_2 - 1_{L_1 \cup L_2} \leq (\phi_1 - 1_{L_1}) + (\phi_2 - 1_{L_2})$, we see that $L_1 \cap L_2$ and $L_1 \cup L_2 \in \mathcal{L}_\mu$. Thus, $\mathcal{L}_\mu$ is $(\cap f, \cup f)$-stable. So let $\varepsilon > 0$ and $(L_n) \subseteq \mathcal{L}_\mu$ be given and set $L = \cap_{n=1}^\infty L_n$. Then there exists an integer $n \geq 1$ such that $\bar{\mu}(L^n) < \varepsilon + \bar{\mu}(L)$ where $L^n = L_1 \cap \cdots \cap L_n$ and since $L^n \in \mathcal{L}_\mu$, there exists $\phi \in C(T)$ such that $\phi \geq 1_{L^n} \geq 1_L$ and $I^\mu(\phi) < \varepsilon + \bar{\mu}(L^n) < 2\varepsilon + \bar{\mu}(L)$. Thus, $L \in \mathcal{L}_\mu$ and so $\mathcal{L}$ is $(\cup f, \cap c)$-stable. By the definition of μ, we have that $\mathcal{L} \subseteq \mathcal{L}_\mu$ and since μ is Baireian, we have that $\mathcal{F}_\circ(T) \subseteq \mathcal{L}_\mu$. In particular, we see that $\mathcal{L}_\mu$ satisfies Definition 3.6 (1+2). So suppose that I is $\mathcal{K}$-tight along $C(T)$ and that $\mathcal{K} \subseteq \mathcal{L}_\mu$. By the definition of $\mathcal{L}_\mu$, we see that $(\mu, \mathcal{K})$ satisfies (1.a+b). So by (1) we have that (I, I°) is $\mathcal{K}$-complete and that μ is an extension of κ. Moreover, if $\mathcal{L} \subseteq \mathcal{M}(\kappa)$, then $\kappa(L) = \bar{\mu}(L)$ for all $L \in \mathcal{L}$. So by the definition of μ, we have that $\mu_*(A) \leq \kappa_*(A)$ for all $A \subseteq T$. Hence, in the same manner as in the proof of (1a) we conclude that κ is an extension of μ; that is, $\mu = \kappa$.

(3): Suppose that $C(T) \subseteq L^1(I, I^\circ)$ and that I is σ-smooth at $\emptyset$ along $C(T)$. By Definition 3.6 (5)–(7), we see that $\mathcal{F}_\circ(T)$ satisfies Definition 3.6 (1)–(3) and by Case IV in Lemma 3.2, we have that I is σ-smooth at $\emptyset$ along $\text{Usc}_\circ(T)$. Thus, $\mathcal{F}_\circ(T)$ satisfies Definition 3.6 (1)–(4) and so (I, I°) is Baire complete. The converse implication follows from Theorem 3.5 (11) and the definition of Baire completeness,

(4) follows from (2) and (5) follows directly from (1) and Theorem 3.5.

(6): If (I, I°) is Radon complete, then by Definition 3.6 (2), we have that $C(T) \subseteq L^1(I, I^\circ)$ and that I is $\mathcal{K}(T)$-tight along $C(T)$ and if T is an Urysohn space, then the converse implication follows from Definition 3.6 (6+7).

So let me show that (I, I°) is Borel complete in each of the six cases in the theorem

Case I: Since T is perfectly normal, we have that $\mathcal{F}_\circ(T) = \mathcal{F}(T)$. Thus, (I, I°) is Borel complete.

Case II: By Definition 3.6 (5)–(7), we have that $\mathcal{F}(T)$ satisfies Definition 3.6 (1)–(3). So let (L_n) be a sequence of closed sets such that $L_n \downarrow \emptyset$. Since T is normal and countably paracompact, there exist $(\phi_n) \subseteq C(T)$ such that $\phi_n \downarrow 0$ and $\phi_n \geq 1_{L_n}$ for all $n \geq 1$. So by (3) we see that $\mathcal{F}(T)$ satisfies Definition 3.6 (4).

Case III: Let $F \in \mathcal{F}(T)$ and $\varepsilon > 0$ be given. Since μ is Baireian and I_μ is $C(T)$-regular at F, there exists $\phi \in C(T)$ such that $\phi \geq 1_F$ and $\mu^*(F) \leq I^\mu(\phi) < \varepsilon + \mu_*(F)$. Letting $\varepsilon \downarrow 0$, we conclude that $\mathcal{F}(T) \subseteq \mathcal{M}(\mu)$. Hence, $(\mu, \mathcal{F}(T))$ satisfies (1.a+b) and so Case III follows from (1).

Case IV: By Definition 3.6 (5+7), we have that $\mathcal{F}(T)$ satisfies Definition 3.6 (1+2). Let $\varepsilon > 0$ be given and let $L_1, L_2 \in \mathcal{F}(T)$ be disjoint closed sets. Since T is completely regular, there exist nets $(\phi_\gamma) \subseteq C(T)$ and $(\psi_\gamma) \subseteq C(T)$ such that $\phi_\gamma \downarrow 1_{L_1}$ and $\psi_\gamma \downarrow 1_{L_2}$. Since $L_1 \cap L_2 = \emptyset$, we have that, $\phi_\gamma \wedge \psi_\gamma \downarrow 0$ and since I is τ-smooth at 0 along $C(T)$ there exists a γ such that $I(\phi_\gamma \wedge \psi_\gamma) < \varepsilon$. Hence, $\mathcal{F}(T)$ satisfies Definition 3.6 (3). Let $\varepsilon > 0$ and $(L_n) \subseteq \mathcal{F}(T)$ be given such that $L_n \downarrow \emptyset$ and set $\Lambda = \{\phi \in C(T) \mid \exists n \geq 1 : \phi \geq 1_{L_n}\}$. Then it follows easily that Λ is downwards directed and since T is completely regular, we have that $\inf_{\phi \in \Lambda} \phi = 0$. So by τ-smoothness of I there exist $k \geq 1$ and $\phi \in C(T)$ such that $\phi \geq 1_{L_k}$ and $I(\phi) < \varepsilon$. Setting $\phi_n = 1_T$ for $n < k$ and $\phi_n = \phi$ for $n \geq k$, we see that $\mathcal{F}(T)$ satisfies Definition 3.6 (4).

Case V: Let μ be the Radon completion of (I, I°). Then $I(\phi) = I^\mu(\phi)$ for all $\phi \in C(T)$ and since μ is a Radon measure, we have that I^μ is τ-smooth at 0 along $C(T)$. Thus, Case V follows from Case IV.

Case VI: By assumption, we have that $I^\circ(1_F) = J^*(1_F)$ for all $F \in \mathcal{F}(T)$. Hence, $(I, I^\circ, \mathcal{F}(T), C(T))$ satisfies Theorem 3.5 (1)–(3) and so (I, I°) is Borel complete.

(7)–(10): Follows easily from (3) and case VI.

(11): Suppose that T is an Urysohn space and that μ is a Radon probability which coincides with Q_Z on the Baire σ-algebra. Let $\varepsilon > 0$ and $K \in \mathcal{K}(T)$ be given. Since μ is a Radon measure, there exists a closed compact set $C \subseteq T \setminus K$ such that $\mu(T \setminus K) < \varepsilon + \mu(C)$ and since T is an Urysohn space, there exists a continuous function $\phi : T \to [0, 1]$ such that $1_K \leq \phi \leq 1_{T \setminus C}$. Hence, we have $\int_T \phi \, d\mu \leq \mu(T \setminus C) < \varepsilon + \mu(K)$. Thus, I^μ is $C(T)$-regular at $\mathcal{K}(T)$ and since μ is a Radon measure, we see that $(\mu, \mathcal{K}(T))$ satisfies (1.a+b). So by (1) we conclude that Z is Radon complete. $\qquad\square$

3.8. Lemma *Let Z be a T-valued random element on $(\Upsilon, \mathcal{A}, Q)$ and set $\mathcal{F}_Z := \{F \subseteq T \mid 1_F \in \mathrm{Usc}_\circ[Z]\}$. Let ρ denote the regularization topology on T and let $f \in \overline{\mathbf{R}}^T$ be a given function such that $\sup_T f < \infty$. If $C(f)$ denotes the set of continuity points of f, then we have*

(1) $\mathrm{Usc}(T) \subseteq \mathrm{Usc}[Z]$, $\mathrm{Usc}_\circ(T) \subseteq \mathrm{Usc}_\circ[Z]$ and $\mathcal{F}_\circ(T) \subseteq \mathcal{F}_Z$

(2) *If f is upper semicontinuous Q_Z-a.s., then $f \in \mathrm{Usc}[Z]$ and the converse implication holds if Z is Borel Q-measurable and $\mathbb{E}_* f(Z) > -\infty$*

(3) *$f \in \mathrm{Usc}_\circ[Z] \iff I_Z$ is $C(T)$-regular at $f \iff I_Z$ is $\mathrm{Usc}_\circ[Z]$-regular at f*

(4) *I_Z is $C(T)$-regular at $\mathcal{F}(T) \iff \mathcal{F}(T) \subseteq \mathcal{F}_Z$*

(5) *If $f_1, f_2, \ldots \in \mathrm{Usc}_\circ[Z]$ and $f_n \downarrow f$, then $f \in \mathrm{Usc}_\circ[Z]$*

(6) *If T is normal, $\sup\{ Q_*(Z \in F) \mid F \in \mathcal{F}(T) ,\ F \subseteq C(f) \} = 1$ and f is bounded, then $f \in C_\circ[Z]$*

(7) *If T is perfectly normal, then $\mathrm{Usc}(T) \subseteq \mathrm{Usc}_\circ[Z]$*

(8) *If $\Psi \subseteq \mathrm{Usc}(T, \rho)$ and I_Z is τ-smooth at Ψ along $C(T)$, then $\Psi \subseteq \mathrm{Usc}_\circ[Z]$*

Moreover, if Z is Baire Q-measurable, then we have

(9) *$\mathcal{F}_Z$ is $(\cup f, \cap c)$-stable and $U(\mathcal{F}_Z) \subseteq \mathrm{Usc}_\circ[Z]$*

(10) *If $f \in \mathrm{Usc}_\circ[Z]$ and $\mathbb{E}_* f(Z) > -\infty$, then $f \in U(\mathcal{F}_Z)$*

(11) *If f is bounded, then $f \in C_\circ[Z]$ if and only if for every $\varepsilon >$ there exist $\psi_1, \psi_2 \in C(T)$ such that $\psi_1 \le f \le \psi_2$ and $Q_Z(\psi_1 \ne \psi_2) < \varepsilon$*

(12) *If T is normal, $\sup\{ Q^*(Z \in F) \mid F \in \mathcal{F}(T) ,\ F \subseteq C(f) \} = 1$ and f is bounded, then $f \in C_\circ[Z]$*

(13) *I_Z is $C(T)$-regular at $\mathcal{F}(T) \iff \mathcal{F}(T) \subseteq \mathcal{F}_Z \iff \mathrm{Usc}(T) \subseteq \mathrm{Usc}_\circ[Z]$*

Proof. Throughout the proof we let $\bar{h}$ and h° denote the upper and lower semicontinuous envelopes of h whenever $h \in \overline{\mathbf{R}}^T$.

(1): The first inclusion is evident and the last two inclusions follows from (3.11).

(2): Suppose that f is upper semicontinuous Q_Z-a.s. Then $f = \bar{f}\ Q_Z$-a.s. and so by (2.6) we conclude that $f \in \mathrm{Usc}[Z]$. So suppose that Z is Borel Q-measurable, $\mathbb{E}_* f(Z) > -\infty$ and $f \in \mathrm{Usc}[Z]$. Then $\bar{f} \ge f$ and $-\infty < \mathbb{E}\bar{f}(Z) = \mathbb{E}_* f(Z) < \infty$. Hence, we see that $\bar{f} = f\ Q_Z$-a.s. and so f is upper semicontinuous Q_Z-a.s.

(3+4): Evident!

(5): Let $a > \mathbb{E}_* f(Z)$ be given. By (3.11) there exists an integer $n \ge 1$ such that $\mathbb{E}_* f_n(Z) < a$ and since $f_n \in \mathrm{Usc}_\circ[Z]$, then exists $\phi \in C(T)$ such that $\phi \ge f_n \ge f$ and $\mathbb{E}_* \phi(Z) < a$. Hence, we see that $f \in \mathrm{Usc}_\circ[Z]$.

(6+12): Since f is bounded, we may assume that $0 \le f \le 1$. Let $F \in \mathcal{F}(T)$ be a given closed set such that $F \subseteq C(f)$. Then $\bar{f}(t) = f(t) = f^\circ(t)$ for all $t \in F$ and since T is normal, we may apply [[4]; Cor.5 (5.5)] on the quadruples $(f^\circ, \bar{f}, F, T)$ and $(f^\circ, \bar{f}, T, F)$ to obtain continuous functions $\psi_1, \psi_2 : T \to [0, 1]$ satisfying

$$\bar{f}(t) \le \psi_1(t) \quad \forall t \in F \ \text{ and } \ 0 \le \psi_1(t) \le f^\circ(t) \quad \forall t \in T$$
$$\bar{f}(t) \le \psi_2(t) \le 1 \quad \forall t \in T \ \text{ and } \ \psi_2(t) \le f^\circ(t) \quad \forall t \in F$$

Hence, we have $\psi_1(t) \le f^\circ(t) \le f(t) \le \bar{f}(t) \le \psi_2(t)$ for all $t \in T$ and $\psi_1(t) = \psi_2(t)$ for all $t \in F$ and so we obtain the following inequalities:

$$\mathbb{E}^* \psi_1(Z) \le \mathbb{E}^* f(Z) \le \mathbb{E}^* \psi_1(Z) + \mathbb{E}^* \{\psi_2(Z) - \psi_1(Z)\}$$
$$\mathbb{E}_* \psi_2(Z) \ge \mathbb{E}_* f(Z) \ge \mathbb{E}_* \psi_2(Z) - \mathbb{E}^* \{\psi_2(Z) - \psi_1(Z)\}$$

Since $0 \le \psi_2 - \psi_1 \le 1_{T \setminus F}$, we see that (6) and (12) follow.

(7): If T is perfectly normal, then $\mathrm{Usc}(T) = \mathrm{Usc}_\circ(T) \subseteq \mathrm{Usc}_\circ[Z]$.

(8): Follows from (3) and (1).

(9): Since Z is Baire Q-measurable, then by Theorem 3.7 (2), we have that $\mathcal{F}_Z$ is $(\cup f, \cap c)$-stable. Let (J^Z, J_Z) be given by (3.18) with $(I, I^\circ, \Phi) = (I^Z, I_Z, C(T))$. Since Z is Baire Q-measurable, we have that $J^Z(1_F) \leq I_Z(1_F)$ for all $F \in \mathcal{F}_Z$ and since J^Z is subrectilinear and I_Z is rectilinear, then by Lemma 3.3 (4) and (3.11) we conclude that $J^Z(h) \leq I_Z(h)$ for all $h \in U(\mathcal{F}_Z)$; that is, $U(\mathcal{F}_Z) \subseteq \mathrm{Usc}_\circ[Z]$.

(10): Let $f \in \mathrm{Usc}_\circ[Z]$ be given such that $\mathbb{E}_* f(Z) > -\infty$. Since Z is Baire Q-measurable, there exists $\phi_n \in C(T)$ such that $\phi_n \geq f$ and $\mathbb{E}^*\{\phi_n(Z) - f(Z)\} \leq 2^{-n}$ for all $n \geq 1$. Let $c \in \mathbf{R}$ be given and set $\zeta_n(x) = (nx - nc + 1)^+ \wedge 1$ for all $x \in \overline{\mathbf{R}}$. Then ζ_n is increasing and $|\zeta_n(x) - \zeta_n(y)| \leq n|x - y|$ for all $x, y \in \overline{\mathbf{R}}$ with the convention $\infty - \infty := 0$. Setting $\psi_n = \zeta_n \circ \phi_n$ and $f_n = \zeta_n \circ f$, we see that $\psi_n \in C(T)$, $0 \leq \psi_n - f_n \leq n(\phi_n - f)$ and $f_n \downarrow 1_F$ where $F = \{f \geq c\}$. Hence, we have

$$\mathbb{E}_* \psi_n(Z) \leq \mathbb{E}_* f_n(Z) + \mathbb{E}^*\{\psi_n(Z) - f_n(Z)\} \leq \mathbb{E}_* f_n(Z) + n2^{-n}$$

Letting $n \to \infty$ and applying (3.11), we see that $F \in \mathcal{F}_Z$; that is, $f \in U(\mathcal{F}_Z)$.

(11): Suppose that $f \in C_\circ[Z]$ and let $\varepsilon > 0$ be given. Since f is bounded, we may assume that $0 \leq f \leq 1$. Since Z is Baire Q-measurable, there exist functions $\xi_n, \eta_n \in C(T)$ such that $0 \leq \xi_n \leq f \leq \eta_n \leq 1$ and $\mathbb{E}\{\eta_n(Z) - \xi_n(Z)\} \leq 2^{-n}$ for all $n \geq 1$. Set $d(u, v) = \sum_{n=1}^\infty 2^{-n}|\xi_n(u) - \xi_n(v)| + \sum_{n=1}^\infty 2^{-n}|\eta_n(u) - \eta_n(v)|$ for all $u, v \in T$. Then d is a separable pseudo-metric on T such that the d-topology is weaker than the given topology on T and ξ_n and η_n are d-continuous for all $n \geq 1$. Set $\xi = \sup_{n \geq 1} \xi_n$ and $\eta = \inf_{n \geq 1} \eta_n$. Then $\xi \in \mathrm{Lsc}(T, d)$, $\eta \in \mathrm{Usc}(T, d)$, $\xi \leq f \leq \eta$ and $\mathbb{E}\{\eta(Z) - \xi(Z)\} = 0$. Hence, the set $L := \{\xi = \eta\}$ belongs to $\mathcal{B}a(T, d)$ and $\bar{Q}(Z \in L) = 1$. Thus, there exists $F \in \mathcal{F}(T, d)$ such that $F \subseteq L$ and $\bar{Q}(Z \in F) > 1 - \varepsilon$ and since f is d-continuous at every $t \in L$, then in the same manner as in the proof of (6+12), we can find function $\psi_1, \psi_2 \in C(T, d)$ such that $\psi_1(t) \leq \xi(t) \leq f(t) \leq \eta(t) \leq \psi_2(t)$ for all $t \in T$ and $\psi_1(t) = \psi_2(t)$ for all $t \in F$. Thus, the pair (ψ_1, ψ_2) satisfies the hypothesis of (12). The converse implication is evident.

(13): Follows from (9) and (4). $\qquad\qquad\qquad\qquad\qquad\qquad\qquad\qquad\square$

4. The Calculus of Convergence in Law

With the provision of Section 3, we are now ready for the study of the calculus of convergence in Baire and Borel law. We shall also study the notion of convergence in probability for T-valued random elements. Let X be a T-valued random element, let $(X_\pi \mid \pi \in \Pi)$ be a net of T-valued random elements and let d be a pseudo-metric on T. According to classical definition, we say that (X_π) converge in d-*probability* to X and we write $X_\pi \to^d X$ if $P^*(d(X_\pi, X) > \varepsilon) \to 0$ for all $\varepsilon > 0$. However, there exist a real valued random element X and a metric δ inducing the Euclidian topology on $\mathbf{R}$ such that $P^*(\delta(X, X + \frac{1}{n}) > 1) = 1$ for all $n \geq 1$. Since $X + \frac{1}{n} \to^d X$ when $d(x, y) = |x - y|$ is the usual Euclidian metric, we see that the classical definition is *not* topological invariant and depends on the particular metric. This observation calls for a new definition of convergence in probability. As for convergence in law, we have a Baire and a Borel version: We say that $X_\pi \to X$

in *Baire probability* and we write $X_\pi \to^{ba} X$ if

$$\lim_{\pi\uparrow\Pi} E^* \phi(X, X_\pi) = 0 \quad \forall \phi \in C^+(T \times T) \text{ so that } \phi(t,t) = 0 \ \forall t \in T \quad (4.1)$$

and we say that $X_\pi \to X$ in *Borel probability* and we write $X_\pi \to^{bo} X$ if

$$\lim_{\pi\uparrow\Pi} P^*((X, X_\pi) \in F) = 0 \quad \forall F \in \mathcal{F}(T \times T) \text{ with } P^*((X, X) \in F) = 0 \quad (4.2)$$

By Markov's inequality, it follows easily that $X_\pi \to^{bo} X \Rightarrow X_\pi \to^{ba} X \Rightarrow X_\pi \to^d X$ whenever d is a continuous pseudo-metric. In Corollary 4.6, we shall see the converse implications hold whenever d induces the topology on T and the limit X is Baire measurable. I shall extend the convergence notion $\to^d$ in the following way: Let $\Theta = (\theta_\gamma \mid \gamma \in \Gamma)$ be a Γ-net of set valued functions $\theta_\gamma : T \to 2^T$. Then we say that (X_π) *converges in Θ-probability* to X and we write $X_\pi \to^\Theta X$ if $P^*(X_\pi \notin \theta_\gamma(X)) \to 0$ for all $\gamma \in \Gamma$ or equivalently if $P^*((X, X_\pi) \notin G_\gamma) \to 0$ for all $\gamma \in \Gamma$ where $G_\gamma := \{(t, u) \mid u \in \theta_\gamma(t)\}$ is *the graph* of the set valued function θ_γ. If $\theta : T \to 2^T$ is a set valued function with graph G, then $\theta(t) = G(t)$ where $G(t) = \{u \in T \mid (t, u)\}$ is the t-section of G. Conversely if $G \subseteq T \times T$, then $t \curvearrowright G(t)$ is a set valued function with graph G. In particular, we shall consider so-called Γ-grids. Let $D \subseteq T$ be a given subset of T and let Γ be a net. Then a *decreasing Γ-grid on D* is a Γ-net $\Theta = (\theta_\gamma \mid \gamma \in \Gamma)$ of set valued functions $\theta_\gamma : T \to 2^T$ satisfying

$$\theta_\gamma(t) \subseteq \theta_\beta(t) \quad \forall t \in T \ \forall \gamma \geq \beta \quad (4.3)$$

$$\forall t \in D \ \forall G \in \mathcal{N}_t \ \exists \gamma \in \Gamma \ \exists V \in \mathcal{N}_t \text{ so that } \theta_\gamma(v) \subseteq G \quad \forall v \in V \quad (4.4)$$

where $\mathcal{N}_t$ denotes the set of all open neighborhoods of t. If $\Theta = (\theta_\gamma \mid \gamma \in \Gamma)$ is a decreasing Γ-grid on D such that the graph of θ_γ is an open subset of $T \times T$ containing $\{(t, t) \mid t \in D\}$ for all $\gamma \in \Gamma$, we say that Θ is a *decreasing Γ-base on D*. Let me provide some examples of decreasing Γ-grids and bases:

(1): Let d is a pseudo-metric inducing the topology on T and let $\theta_n(t)$ be the open d-ball with center t and radius $\frac{1}{n}$. Then $\Theta = (\theta_n \mid n \in \mathbf{N})$ is a decreasing $\mathbf{N}$-base on T such that $X_\pi \to^\Theta X$ if and only if $X_\pi \to^d X$. (2): Let $(U_\gamma \mid \gamma \in \Gamma)$ be a decreasing uniformity base inducing the topology on T and set $\Theta = (U_\gamma(t) \mid \gamma \in \Gamma)$. Then Θ is a decreasing Γ-base for T such that $X_\pi \to^\Theta X$ if and only if $P^*((X, X_\pi) \notin U_\gamma) \to 0$ for all $\gamma \in \Gamma$. (3): Let $T = \mathbf{R}$ with its right Sorgenfrey topology and set $\theta_n(t) = [t, t + \frac{1}{n}[$, then $\Theta = (\theta_n \mid n \in \mathbf{N})$ is a decreasing $\mathbf{N}$-grid on T such that $X_\pi \to^\Theta X$ if and only if $P_*(X \leq X_\pi < X + \varepsilon) \to 1$ for all $\varepsilon > 0$. However, Θ is *not* a decreasing $\mathbf{N}$-base for T. (4): Let T be a regular topological space and let $(U_\gamma \mid \gamma \in \Gamma)$ be a decreasing family of subsets of $T \times T$ such that for every open set $G \supseteq \{(t, t) \mid t \in T\}$ there exists $\gamma \in \Gamma$ with $U_\gamma \subseteq G$. Then $\Theta = \{U_\gamma(t) \mid \gamma \in \Gamma\}$ is a decreasing Γ-grid on T such that $X_\pi \to^\Theta X$ if and only if $P^*((X, X_\pi) \notin U_\gamma) \to 0$ for all $\gamma \in \Gamma$ and if U_γ is open and contains the diagonal $\{(t, t) \mid t \in T\}$, then Θ is a decreasing Γ-base for T. (5): Let $c \in T$ be a given point and let $(V_\gamma \mid \gamma \in \Gamma)$ be a decreasing open neighborhood base at c. If $\theta_\gamma(t) = V_\gamma$ for all $\gamma \in \Gamma$ and all $t \in T$, then $\Theta = (\theta_\gamma \mid \gamma \in \Gamma)$ is a decreasing Γ-grid at $D := \text{cl}(c)$ and a decreasing Γ-base at $D := \{c\}$ such that $X_\pi \to^\Theta X$ if and only if $P^*(X_\pi \notin V_\gamma) \to 0$ for all $\gamma \in \Gamma$.

As a starting point for our study of convergence in law, I shall establish the appropriate portmanteau lemmas for Baire/Borel convergence in law/probability.

4.1. Lemma (The Baire-law portmanteau lemma)   *Let* $\mathcal{X} = (X_\pi \mid \pi \in \Pi)$ *and* $(Y_\pi \mid \pi \in \Pi)$ *be two nets of T-valued random elements on* $(\Omega, \mathcal{F}, P)$ *such that* $\lim_{\pi \uparrow \Pi} E^* |\phi(X_\pi) - \phi(Y_\pi)| = 0$ *for all* $\phi \in C(T)$ *and let Z be a T-valued random element on* $(\Upsilon, \mathcal{A}, Q)$. *Let* $\mathbf{F}, \mathbf{G} \subseteq \overline{\mathbf{R}}^T$ *be given function spaces satisfying*

(1) $\mathbf{F} \subseteq \mathrm{Usc}_\circ[Z]$, $Q_*(Z \in F) = \inf\{\mathbb{E}_* f(Z) | f \in \mathbf{F}, f \ge 1_F\}$ $\forall F \in \mathcal{F}_\circ(T)$

(2) $\mathbf{G} \subseteq \mathrm{Lsc}_\circ[Z]$, $Q^*(Z \in G) = \sup\{\mathbb{E}^* g(Z) | g \in \mathbf{G}, g \le 1_G\}$ $\forall G \in \mathcal{G}_\circ(T)$

Then the following five statements are equivalent

(3) $X_\pi \to^\sim Z$

(4) $Y_\pi \to^\sim Z$

(5) $X_\pi \to^\sim Q_Z$

(6) $\limsup_{\pi \uparrow \Pi} E^* f(X_\pi) \le \mathbb{E}_* f(Z)$ $\forall f \in \mathbf{F}$

(7) $\liminf_{\pi \uparrow \Pi} E_* g(X_\pi) \ge \mathbb{E}^* g(Z)$ $\forall g \in \mathbf{G}$

and we have

(8) $X_\pi \twoheadrightarrow^\sim Q_Z$ $\Rightarrow$ $X_\pi \twoheadrightarrow^\sim Z$ $\Rightarrow X_\pi \to^\sim Z$ $\Leftrightarrow$ $X_\pi \to^\sim Q_Z$

(9) *If I_Z is $C(T)$-regular at $\mathcal{F}(T)$, then:* $X_\pi \to^\sim Z$ $\Leftrightarrow$ $X_\pi \twoheadrightarrow^\sim Z$

Remarks: (a): Note that $\mathbf{F}$ satisfies (1) if and only if $\mathbf{F} \subseteq \mathrm{Usc}_\circ[Z]$ and I_Z is $\mathbf{F}$-regular at $\mathcal{F}_\circ(T)$. Hence, if $\mathbf{F} \subseteq \mathrm{Usc}_\circ[Z]$ is a function space such that $1_F \in \mathbf{F}_{\downarrow \mathbf{N}}$ for all $F \in \mathcal{F}_\circ(T)$, then by (3.11) and Lemma 3.1 (1) we see that $\mathbf{F}$ satisfies (1). Similarly, if $\mathbf{G} \subseteq \mathrm{Lsc}_\circ[Z]$ is a function space such that $1_G \in \mathbf{G}_{\uparrow \mathbf{N}}$ for all $G \in \mathcal{G}_\circ(T)$, then $\mathbf{G}$ satisfies (2).

(b): Recall that Lemma 3.8 gives a series of criteria for I_Z to be $C(T)$-regular at $\mathcal{F}(T)$ and thus for equivalence of Baire and Borel convergence in law. In particular, we have that Baire and Borel convergence in law are equivalent in the following cases: (i): T is perfectly normal. (ii): T is pseudo-metrizable. (iii): T is completely regular and I_Z is τ-smooth at $\mathcal{F}(T)$ along $C(T)$. Thus, convergence in Baire and Borel law coincide in most cases but Example 1.1 shows the two concepts differs when the space T is "sufficiently large".

Proof. (3) $\Leftrightarrow$ (4) $\Leftrightarrow$ (5): Evident!

(3) $\Leftrightarrow$ (6): Suppose that (3) holds. Then $I^{\mathcal{X}}(\phi) \le I_Z(\phi)$ for all $\phi \in C(T)$ and by Lemma 3.8 (1+3), we have that I_Z is $C(T)$-regular at $\mathcal{F}_\circ(T)$ and so $I^{\mathcal{X}}(1_F) \le I_Z(1_F)$ for all $F \in \mathcal{F}_\circ(T)$. Thus, by Lemma 3.3 (4) and (3.10) we have that $I^{\mathcal{X}}(f) \le I_Z(f)$ for all $f \in \mathrm{Usc}_\circ[Z]$ and so (6) holds. Conversely, suppose that (6) holds. By (1) and (6), we see that $I^{\mathcal{X}}(1_F) \le I_Z(1_F)$ for all $F \in \mathcal{F}_\circ(T)$. So by Lemma 3.3 (4), we have that $I^{\mathcal{X}}(\phi) \le I_Z(\phi)$ for all $\phi \in C(T)$. Applying this to $\pm\phi$, we see that $I^{\mathcal{X}}(\phi) \le I_Z(\phi) \le I^Z(\phi) \le I_{\mathcal{X}}(\phi)$ for all $\phi \in C(T)$. Thus, (3) follows.

(3) $\Leftrightarrow$ (4.1.7): Since $\{h \in \overline{\mathbf{R}}^T \mid -h \in \mathbf{G}\}$ satisfies (1), this follows from the equivalence of (3) and (6).

(8): The first implication follows from (2.3). So suppose that $X_\pi \twoheadrightarrow^\sim Z$. Then $\mathrm{Usc}_\circ(T)$ satisfies (1) and (6) and so $X_\pi \to^\sim Z$. The last equivalence follows directly from the definition of convergence in Baire law.

(9): By (8), we have that the implication "$\Rightarrow$" holds in general and if I_Z is $C(T)$-regular at $\mathcal{F}(T)$, then the converse implication follows from Lemma 3.8 (13) and the equivalence of (3) and (6) with $\mathbf{F} = \mathrm{Usc}_\circ[Z]$. $\qquad\square$

4.2.   Lemma (The Borel-law portmanteau lemma)   *Let $\mathcal{X} = (X_\pi \mid \pi \in \Pi)$ be a net of T-valued random elements on $(\Omega, \mathcal{F}, P)$ and let Z be a T-valued random element on $(\Upsilon, \mathcal{A}, Q)$. Let $\mathbf{F}, \mathbf{G} \subseteq \overline{\mathbf{R}}^T$ be given function spaces satisfying*

(1) $\mathbf{F} \subseteq \mathrm{Usc}[Z]$, $Q_*(Z \in F) = \inf\{\mathbb{E}_* f(Z) | f \in \mathbf{F}, f \geq 1_F\}$ $\forall F \in \mathcal{F}(T)$

(2) $\mathbf{G} \subseteq \mathrm{Lsc}[Z]$, $Q^*(Z \in G) = \sup\{\mathbb{E}^* g(Z) | g \in \mathbf{G}, g \leq 1_G\}$ $\forall G \in \mathcal{G}(T)$

Then the following three statements are equivalent

(3) $X_\pi \twoheadrightarrow^\sim Z$

(4) $\limsup_{\pi \uparrow \Pi} E^* f(X_\pi) \leq \mathbb{E}_* f(Z)$ $\forall f \in \mathbf{F}$

(5) $\liminf_{\pi \uparrow \Pi} E_* g(X_\pi) \geq \mathbb{E}^* g(Z)$ $\forall g \in \mathbf{G}$

and if I_Z is Π-smooth at $\mathcal{F}(T)$ along $\mathcal{F}(T)$, then (3)–(5) are equivalent to either of the following two statements:

(6) $\limsup_{\pi \uparrow \Pi} P^*(A_\pi) \leq Q_*(Z \in \mathrm{uplim} X_\pi(A_\pi))$ $\forall \{A_\pi\} \subseteq 2^\Omega$

(7) $\limsup_{\pi \uparrow \Pi} P^*(X_\pi \in F) \leq Q_*(Z \in \mathrm{uplim}(F \cap X_\pi(\Omega)))$ $\forall F \in \mathcal{F}(T)$

In particular, if I_Z is Π-smooth at $\mathcal{F}(T)$ along $\mathcal{F}(T)$ and $\{A_\pi \mid \pi \in \Pi\}$ is a net of subsets of T, then we have

(8) $X_\pi \twoheadrightarrow^\sim Z$ and $\limsup_{\pi \uparrow \Pi} P^*(X_\pi \in A_\pi) = 1$ $\Rightarrow$ $Z \in \mathrm{uplim} X_\pi(A_\pi)$ Q-a.s.

Remarks: (a): Note that $\mathbf{F}$ satisfies (1) if and only if $\mathbf{F} \subseteq \mathrm{Usc}[Z]$ and I_Z is $\mathbf{F}$-regular at $\mathcal{F}(T)$. Hence, if $\mathbf{F} \subseteq \mathrm{Usc}[Z]$ is a function space containing 1_F for all $F \in \mathcal{F}(T)$, then $\mathbf{F}$ satisfies (1). Similarly, if $\mathbf{G} \subseteq \mathrm{Lsc}[Z]$ is a function space containing 1_G for all $G \in \mathcal{G}(T)$, then $\mathbf{G}$ satisfies (2).

(b): Suppose that Π is a countably cofinal Frechet net. By (3.11) and (2), we have that I_Z is Π-smooth at $\overline{\mathbf{R}}^T$ along $\{h \mid \mathbb{E}_* h(Z) < \infty\}$ and so (3)–(7) are equivalent and (8) holds.

Proof. Evidently, we have that (3) implies (4). So suppose that (4) holds. By (1) and (4), we have that $I^{\mathcal{X}}(1_F) \leq I_Z(1_F)$ for all $F \in \mathcal{F}(T)$. So by Lemma 3.3 (4) and (3.10), we have that $I^{\mathcal{X}}(f) \leq I_Z(f)$ for all $f \in \mathrm{Usc}(T)$. Hence, (3) and (4) are equivalent and since $\{h \in \overline{\mathbf{R}}^T \mid -h \in \mathbf{G}\}$ satisfies (3), we see that (3) and (5) are equivalent.

Setting $A_\pi = X_\pi^{-1}(F)$, we see that (6) implies (7) and that (7) implies (4) with $\mathbf{F} := \{1_F \mid F \in \mathcal{F}(T)\}$. So let $\{A_\pi\}$ be a given net of subsets of Ω and suppose that I_Z is Π-smooth at $\mathcal{F}(T)$ along $\mathcal{F}(T)$ and that (3) holds. Setting $F = \mathrm{up}\lim A_\pi$ and $F_\pi = \mathrm{cl}(\cup_{\gamma \geq \pi} A_\gamma)$ for all $\pi \in \Pi$, we see that $\{F_\pi\}$ is a net of closed subsets of T such that $F_\pi \downarrow F$ and $A_\pi \subseteq X_\pi^{-1}(F_\gamma)$ for all $\pi \geq \gamma$. Hence, by (3) we have

$$\limsup_{\pi \uparrow \Pi} P^*(A_\pi) \leq \limsup_{\pi \uparrow \Pi} P^*(X_\pi \in F_\gamma) \leq Q_*(Z \in F_\gamma)$$

for all $\gamma \in \Pi$. Letting $\gamma \uparrow \Pi$, we see that (6) follows from Π-smoothness of I_Z. Thus, (3)–(7) are equivalent whenever I_Z is Π-smooth at $\mathcal{F}(T)$ along $\mathcal{F}(T)$ and (8) follows directly from the equivalence of (6) and (3). $\qquad\square$

4.3.   Lemma (The probability portmanteau lemma)   *Let $(X_\pi \mid \pi \in \Pi)$ be a net of T-valued random elements on $(\Omega, \mathcal{F}, P)$ and let X be a T-valued random element on $(\Omega, \mathcal{F}, P)$. Set $U_\pi = (X, X_\pi)$, $U = (X, X)$ and let $\mathcal{F}_U$ be the paving of all sets $F \in \mathcal{F}(T \times T)$ with $P^*(U \in F) = 0$. Then we have*

(1) $X_\pi \to^{ba} X \;\Leftrightarrow\; \limsup_{\pi \uparrow \Pi} E^* f(U_\pi) \le E^* f(U) \;\forall f \in \mathrm{Usc}_\circ[U]$

(2) $X_\pi \to^{bo} X$ *if and only if* $\limsup_{\pi \uparrow \Pi} E^* f(X, X_\pi) \le E^* g(X)$ *for all* $f \in \mathrm{Usc}(T \times T)$ *and* $g \in \mathrm{Lsc}(T)$ *with* $f(X, X) \le g(X)$ *P-a.s.*

(3) *If $X_\pi \to^{bo} X$, then $X_\pi \to^{ba} X$*

(4) *(X_π) converge in Baire (Borel) probability to X if and only if (U_π) converge in Baire (Borel) probability to U*

(5) *If $U_\pi \to^\sim U$, then X is Baire P-measurable, $X_\pi \to^\sim X$ and $X_\pi \to^{ba} X$*

(6) *If X is Baire P-measurable and $X_\pi \to^{ba} X$, then $U_\pi \to^\sim U$ and $X_\pi \to^\sim X$*

(7) *If $U_\pi \twoheadrightarrow^\sim U$, then X is Borel P-measurable, $X_\pi \twoheadrightarrow^\sim X$ and $X_\pi \to^{bo} X$*

(8) *If X is Borel P-measurable, I_X is $\mathcal{G}(T)$-regular at $\mathcal{F}(T)$ and $X_\pi \to^{bo} X$, then $U_\pi \twoheadrightarrow^\sim U$ and $X_\pi \twoheadrightarrow^\sim X$*

(9) *If either $T \times T$ is perfectly normal or I^U is $\mathrm{Usc}_\circ[U]$-regular at $\mathcal{F}_U$, then*
$$m\,(a)\quad X_\pi \to^{ba} X \;\Leftrightarrow\; X_\pi \to^{bo} X$$

Remark: By (9) and (3.11), we have that Baire and Borel convergence in probability are equivalent in the following cases: (i): $T \times T$ is perfectly normal. (ii): T is pseudo-metrizable. (iii): T is completely regular and I^U is τ-smooth at $\mathcal{F}(T)$ along $C(T)$. Thus, Baire and Borel convergence in probability coincide in most cases but it can be shown that the two concepts differs when the space T is "sufficiently large".

Proof. (1): Suppose that $X_\pi \to^{ba} X$ and let $f \in \mathrm{Usc}_\circ[U]$ and $a \in \mathbf{R}$ be given such that $a > E^* f(U)$. Then there exists $\phi \in C(T \times T)$ such that $\phi \ge f$ and $E^* \phi(U) < a$. Setting $\psi(t, t') = |\phi(t, t') - \phi(t, t)|$, we see that $\psi \in C^+(T \times T)$ and $\psi(t, t) = 0$ for all $t \in T$ and since $f(t, t') \le \phi(t, t') \le \psi(t, t') + \phi(t, t)$ for all (t, t'), then by (4.1) we have $\limsup E^* f(U_\pi) \le E^* \phi(U) < a$. Letting $a \downarrow E^* f(U)$, we see that $\limsup E^* f(U_\pi) \le E^* f(U)$. The converse implication is evident.

(2): Suppose that $X_\pi \to^{bo} X$ and let $f \in \mathrm{Usc}(T \times T)$ and $g \in \mathrm{Lsc}(T)$ be given such that $f(X, X) \le g(X)$ P-a.s. Let $\varepsilon > 0$ be given and set $F = \{(t, u) \mid f(t, u) \ge g(t) + \varepsilon\}$. Then $F \in \mathcal{F}(T \times T)$ and $P^*((X, X) \in F) = 0$ and if $0 < c < \infty$ is chosen such that $f \le c$ and $g \ge -c$, then we have $f \le \varepsilon 1_{T \setminus F} + 2c 1_F + g(t)$. Hence, by (4.2) we have $\limsup E^* f(U_\pi) \le \varepsilon + E^* g(X)$. Thus, the "only if" part of (2) follows and the converse implication is evident.

(3)–(6) are easy consequences of the definitions and $(1 + 2)$.

(7): Suppose that $U_\pi \twoheadrightarrow^\sim U$. Then $X_\pi \twoheadrightarrow^\sim X$ and by (4.3.2), we see that $X_\pi \to^{bo} X$. So let $F \in \mathcal{F}(T)$ be given. Then $P^*(X \in F) = P^*(U_\pi \in T \times F)$

and so we have $P^*(X \in F) \le P_*(U \in T \times F) = P_*(X \in F)$. Hence, we see that X is Borel P-measurable.

(8): Let $F \in \mathcal{F}(T \times T)$ be given and set $\hat{F} = \{t \mid (t,t) \in F\}$. Then $\hat{F}$ is a closed subset of T and by (2) and Borel P-measurability of X, we have that $\limsup P^*(U_\pi \in F) \le P_*(X \in G)$ for every open set $G \supseteq \hat{F}$. So by $\mathcal{G}(T)$-regularity of I_X, we have $\limsup P^*(U_\pi \in F) \le P_*(X \in \hat{F}) = P_*(U \in F)$. Hence (8) follows from Lemma 4.2.

(9): By (3), we have that Borel convergence in probability implies Baire convergence in probability. So suppose that $X_\pi \to^{ba} X$. If $T \times T$ is perfectly normal, then $\mathcal{F}_U \subseteq \mathcal{F}_\circ(T \times T)$ and so by (4.3.1) we conclude that $\limsup P^*(U_\pi \in F) = 0$ for all $F \in \mathcal{F}_U$; that is, $X_\pi \to^{bo} X$. So suppose that I^U is $\mathrm{Usc}_\circ[Z]$-regular at $\mathcal{F}_U$ and let $F \in \mathcal{F}_U$ be given. Since $P^*(U \in F) = 0$, there exists $f \in \mathrm{Usc}_\circ[Z]$ such that $f \ge 1_F$ and $E^* f(U) < \varepsilon$. So by (1), we see that $\limsup P^*(U_\pi \in F) < \varepsilon$. Letting $\varepsilon \downarrow 0$, we conclude that $X_\pi \to^{bo} X$. $\qquad\square$

4.4. Lemma *Let $(X_\pi \mid \pi \in \Pi)$ be a Π-net of T-valued random elements on $(\Omega, \mathcal{F}, P)$, let X be a T-valued random element on $(\Omega, \mathcal{F}, P)$ and let $\Theta = (\theta_\gamma \mid \gamma \in \Gamma)$ be a Γ-net of set valued functions $\theta_\gamma : T \to 2^T$ with graphs $G_\gamma := \{(t, u) \mid u \in \theta_\gamma(t)\}$. If G_γ is open and $(X, X) \in G_\gamma$ P-a.s. for all $\gamma \in \Gamma$, then we have*

(1) $X_\pi \to^{bo} X \quad \Rightarrow \quad X_\pi \to^\Theta X$

Let S be a topological space, let $(Y_\pi \mid \pi \in \Pi)$ be a Π-net of S-valued random elements on $(\Omega, \mathcal{F}, P)$, let Y be a S-valued random element on $(\Omega, \mathcal{F}, P)$ and let $\Xi = (\xi_\lambda \mid \lambda \in \Lambda)$ be a Λ-net of set valued functions $\xi_\lambda : S \to 2^S$. If we set $\Sigma = (\sigma_{\lambda\gamma} \mid (\lambda, \gamma) \in \Lambda \times \Gamma)$ where $\sigma_{\lambda\gamma}(s,t) = \xi_\lambda(s) \times \theta_\gamma(t)$, then we have

(2) $Y_\pi \to^\Xi Y$ and $X_\pi \to^\Theta X \quad \Rightarrow \quad (Y_\pi, X_\pi) \to^\Sigma (Y, X)$

(3) $X_\pi \to^\Theta X$ and $Y \in \xi_\lambda(Y)$ a.s. $\forall \lambda \in \Lambda \quad \Rightarrow \quad (Y, X_\pi) \to^\Sigma (Y, X)$

(4) *If Θ is a decreasing Γ-grid on $D \subseteq T$ and Ξ is a decreasing Λ-grid on $E \subseteq S$, then Σ is a decreasing Γ-grid on $E \times D$*

Suppose that Θ is a decreasing Γ-grid on the set $D \subseteq T$ and that $X_\pi \to^\Theta X$. If η and η° are given by

$$\eta(A) = \limsup_{\pi \uparrow \Pi} P^*((Y_\pi, X) \in A) \ , \quad \eta^\circ(A) = \inf\{\eta(G) \mid G \in \mathcal{G}(T), \ G \supseteq A\}$$

for all $A \subseteq S \times T$, then we have the following:

(A): Let $\mu : 2^{S \times T} \to [0, \infty]$ be an increasing set function such that $\eta(F) \le \mu(F \cap (S \times D))$ for all $F \in \mathcal{F}(S \times T)$. Then $P^*(X \in C) = 0$ for every closed set $C \subseteq T \setminus D$ and if μ is Γ-smooth at $\mathcal{F}(T)$ along $\mathcal{F}(T)$, then we have

(5) $\limsup_{\pi \uparrow \Pi} P^*((Y_\pi, X_\pi) \in F) \le \mu(F) \quad \forall F \in \mathcal{F}(S \times T)$

(B): Suppose that D is closed, $X \in D$ P-a.s. and that η is Γ-smooth at $\emptyset$ along $\mathcal{F}(T)$. Then we have

(6) $\limsup_{\pi \uparrow \Pi} P^*((Y_\pi, X_\pi) \in F) \le \eta^\circ(F \cap (S \times D)) \le \eta^\circ(F) \quad \forall F \in \mathcal{F}(S \times T)$

Proof. (1)–(4) are easy and shall leave the details to the reader. So suppose that Θ is a decreasing Γ-grid on D and that $X_\pi \to^\Theta X$ and let $F \in \mathcal{F}(S \times T)$ be given.

If we set

$$H_\gamma = \{(s,t) \in S \times T \mid F \cap (\{s\} \times \theta_\gamma(t)) \neq \emptyset\} \ , \quad F_\gamma = \mathrm{cl}H_\gamma \ , \quad L = \bigcap_{\gamma \in \Gamma} F_\gamma$$

then by (4.3), we see that $F_\gamma \downarrow L$. Let $(s,t) \in (S \times D) \setminus F$ be given. Then there exist open neighborhoods U and G of s and t such that $(U \times G) \cap F = \emptyset$ and by (4.4), there exists $\gamma \in \Gamma$ and $V \in \mathcal{N}_t$ such that $V \subseteq G$ and $\theta_\gamma(y) \subseteq G$ for all $y \in V$. Hence, we see that $F \cap (\{x\} \cap \theta_\gamma(y)) = \emptyset$ for all $(x,y) \in U \times V$; that is, $(U \times V) \cap H_\gamma = \emptyset$ and so $(s,t) \in (S \times T) \setminus F_\gamma \subseteq (S \times T) \setminus L$. Thus, we conclude that $L \cap (S \times D) \subseteq F$. Moreover, if $(s,t) \in F$ and $(s,u) \notin H_\gamma$, then $F \cap (\{s\} \times \theta_\gamma(u)) = \emptyset$. Hence, $t \notin \theta_\gamma(u)$ and so we have

$$P^*((Y_\pi, X_\pi) \in F) \leq P^*((Y_\pi, X) \in H_\gamma) + P^*((Y_\pi, X) \notin H_\gamma \ , \ (Y_\pi, X_\pi) \in F)$$
$$\leq P^*((Y_\pi, X) \in F_\gamma) + P^*(X_\pi \notin \theta_\gamma(X))$$

and since $X_\pi \to^\Theta X$, we conclude that

(i) $F_\gamma \downarrow L$, $L \cap (S \times D) \subseteq F$ and $\limsup_{\pi \uparrow \Pi} P^*((Y_\pi, X_\pi) \in F) \leq \inf_\gamma \eta(F_\gamma)$

(A): Let C be a closed subset of $T \setminus D$. Then $P^*(X \in C) = \eta(S \times C) \leq \mu((S \times C) \cap (S \times D)) = \mu(\emptyset) = 0$ and if μ is Γ-smooth at $\mathcal{F}(T)$ along $\mathcal{F}(T)$, then (5) follows from (i).

(B): Let $F \in \mathcal{F}(S \times T)$ be given and let (F_γ) and L be defined as above. Let $G \in \mathcal{G}(T)$ be given such that $G \supseteq F \cap (S \times D)$ and set $F_\gamma^* = F_\gamma \cap (S \times D)$. Then $F_\gamma^* \setminus G \downarrow \emptyset$ and $F_\gamma^* \setminus G \in \mathcal{F}(T)$. Hence, there exists $\gamma \in \Gamma$ such that $\eta(F_\gamma^* \setminus G) < \varepsilon$ and since $F_\gamma \subseteq (F_\gamma^* \setminus G) \cup G \cup S \times (T \setminus D)$ and η is subadditive, then by (i) we have

$$\limsup_{n \to \infty} P^*((Y_\pi, X_\pi) \in F) \leq \eta(F_\gamma) \leq \eta(F_\gamma^* \setminus G) + \eta(G) + \eta(S \times (T \setminus D))$$
$$\leq \varepsilon + \eta(G) + P^*(X \notin D) = \varepsilon + \eta(G)$$

Letting $\varepsilon \downarrow 0$ and taking infimum over G, we see that (6) follows.   □

4.5. Theorem *Let $\Theta = (\theta_\gamma \mid \gamma \in \Gamma)$ be decreasing Γ-grid on the set $D \subseteq T$, let $\mathcal{X} = (X_\pi \mid \pi \in \Pi)$ be a net of T-valued random elements on $(\Omega, \mathcal{F}, P)$ and let X be a T-valued random element on $(\Omega, \mathcal{F}, P)$ such that $X_\pi \to^\Theta X$.*

(A): Suppose that $X \in D$ P-a.s. and that I^X is Γ-smooth at $\mathcal{F}(T)$ along $\mathcal{F}(T)$. Then we have

(1) $X_\pi \to^{bo} X$ and $\limsup_{\pi \uparrow \Pi} E^* f(X, X_\pi) \leq E^* f(X, X) \quad \forall f \in \mathrm{Usc}(T \times T)$

Moreover, if S is a topological space and Y is an S-valued random element such that $I^{(Y,X)}$ is Γ-smooth at $\mathcal{F}(S \times T)$ along $\mathcal{F}(S \times T)$, then $(Y, X_\pi) \to^{bo} (Y, X)$.

(B): Let S be a topological space, let $(Y_\pi \mid \pi \in \Pi)$ be a net of random elements on $(\Omega, \mathcal{F}, P)$, set $\mathcal{U} = ((Y_\pi, X) \mid \pi \in \Pi)$ and let Z be an $(S \times T)$-valued random element on $(\Upsilon, \mathcal{A}, Q)$. Then we have

(2) *If $Z \in S \times D$ Q-a.s. and I_Z is Γ-smooth at $\mathcal{F}(S \times T)$ along $\mathcal{F}(S \times T)$, then*
   (a) $(Y_\pi, X) \twoheadrightarrow^\sim Z \quad \Rightarrow \quad (Y_\pi, X_\pi) \twoheadrightarrow^\sim Z$

(3) *If D is closed, $X \in D$ P-a.s. and $I^\mathcal{U}$ is Γ-smooth at $\emptyset$ along $\mathcal{F}(S \times T)$, then*
   (a) $(Y_\pi, X) \to^\sim Z \quad \Rightarrow \quad (Y_\pi, X_\pi) \to^\sim Z$

Remarks: (a): Let $\Theta = (\theta_\gamma \mid \gamma \in \Gamma)$ be a decreasing Γ-base on T. By Lemma 4.4 (1), we have that $X_\pi \to^{bo} X$ implies $X_\pi \to^\Theta X$ and by (1) we have that the converse implication holds whenever I^X is Γ-smooth at $\mathcal{F}(T)$ along $\mathcal{F}(T)$.

(b): Recall that I^X is Γ-smooth at $\mathcal{F}(T)$ along $\mathcal{F}(T)$ in either of the following seven cases: (i): T is completely regular and I^X is τ-smooth at $\mathcal{F}(T)$ along $C(T)$. (ii): T is hereditarily Lindelöf and I^X is σ-smooth at $\mathcal{F}(T)$ along $\mathcal{F}(T)$. (iii): T is separable and pseudo-metrizable and I^X is σ-smooth at $\mathcal{F}(T)$ along $\mathcal{F}(T)$. (iv): T is hereditarily Lindelöf and X is Borel P-measurable. (v): T is separable and pseudo-metrizable and X is Baire P-measurable. (vi): Γ is countably cofinal and I^X is σ-smooth at $\mathcal{F}(T)$ along $\mathcal{F}(T)$, (vii): Γ is countably cofinal and X is Borel P-measurable.

(c): Recall that I_Z is Γ-smooth at $\mathcal{F}(S \times T)$ along $\mathcal{F}(S \times T)$ in either of the following four cases: (i): S and T are completely regular and I_Z is τ-smooth at $\mathcal{F}(S \times T)$ along $C(S \times T)$; (ii): $S \times T$ is hereditarily Lindelöf. (iii): S and T is separable and pseudo-metrizable. (iv): Γ is countably cofinal.

(d): Suppose that $(Y_\pi, X) \to^\sim Z$. Then $I^{\mathcal{U}}(\phi) \leq I_Z(\phi)$ for all $\phi \in C(S \times T)$. Hence, we have that $I^{\mathcal{U}}$ is Γ-smooth at $\emptyset$ along $\mathcal{F}(S \times T)$ in either of the following five cases: (i): S and T are completely regular and $I^{\mathcal{U}}$ is τ-smooth at 0 along $C(S \times T)$. (ii): S and T are completely regular and I_Z is τ-smooth at 0 along $C(S \times T)$. (iii): $S \times T$ is perfectly normal and Γ is countably cofinal. (iv): $S \times T$ is Lindelöf and perfectly normal. (v): S and T are separable and pseudo-metrizable.

Proof. (A): Set $S = T$, $Y_\pi = X$ and $\rho(A) = P^*((X, X) \in A)$ for all $A \subseteq T \times T$ and let η be given as in Lemma 4.4 with this choice of (Y_π). Then $\eta(A) = \rho(A)$ for all $A \subseteq T \times T$ and since $X \in D$ P-a.s., we have that $\rho(A) = \rho(A \cap (T \times D))$ for all $A \subseteq T \times T$. Moreover, since $\rho(A) = I^X(1_{\zeta^{-1}(A)})$ where $\zeta(t) = (t, t)$, we see that ρ is Γ-smooth at $\mathcal{F}(T \times T)$ along $\mathcal{F}(T \times T)$. Hence, if $\mathcal{V} = ((X, X_\pi) \mid \pi \in \Pi)$ and $V = (X, X)$, then by Lemma 4.4 (5) we have that $I^{\mathcal{V}}(1_F) \leq \rho(1_F) = I^V(1_F)$ for all $F \in \mathcal{F}(T \times T)$. In particular, we have $X_\pi \to^{bo} X$ and since $I^{\mathcal{V}}$ is subrectilinear and I^V is rectilinear, we see that (1) follows from Lemma 3.3 (4). So suppose that Y is an S-valued random element such that $I^{(Y,X)}$ is Γ-smooth at $\mathcal{F}(S \times T)$ along $\mathcal{F}(S \times T)$. If we set $\Sigma = (\{s\} \times \theta_\gamma(t) \mid \gamma \in \Gamma)$ and $\xi_1(s) = \{s\}$ for all $s \in S$, then it follows easily that $\Xi = (\xi_1)$ is a decreasing $\{1\}$-grid on S. Hence, by Lemma 4.4 (3+4) we have that Σ is a decreasing Γ-grid on $S \times D$ such that $(Y, X_\pi) \to^\Sigma (Y, X)$. So by (1) we conclude that $(Y, X_\pi) \to^{bo} (Y, X)$.

(2): Suppose that $(Y_\pi, X) \twoheadrightarrow^\sim Z$. Since $Z \in S \times D$ Q-a.s., we see that $\mu(A) := Q_*(Z \in A)$ satisfies the assumptions of Lemma 4.4 (A). Hence, by Lemma 4.4 (5) and Lemma 4.2 we conclude that $(Y_\pi, X_\pi) \twoheadrightarrow^\sim Z$.

(3): Suppose that $(Y_\pi, X) \to^\sim Z$ and let η and η° be defined as in Lemma 4.4. Let $F \in \mathcal{F}_\circ(S \times T)$ and $\varepsilon > 0$ be given. Then Z is Baire Q-measurable and there exists $\phi \in C(S \times T)$ such that $\phi \geq 1_F$ and $E\phi(Z) < \varepsilon + \bar{Q}(Z \in F)$. Set $G = \{\phi > 1 - \varepsilon\}$ and $H = \{\phi \geq 1 - \varepsilon\}$. Then $G \in \mathcal{G}_\circ(S \times T)$, $H \in \mathcal{F}_\circ(S \times T)$, $F \subseteq G \subseteq H$ and $1_H \leq \frac{1}{1-\varepsilon} \phi$. So by Lemma 4.4 (6) and Lemma 4.1 we have

$$\limsup_{\pi \uparrow \Pi} P^*((Y_\pi, X_\pi) \in F) \leq \eta^\circ(F) \leq \eta(G) \leq \eta(H) \leq \bar{Q}(Z \in H)$$

$$\leq \frac{1}{1-\varepsilon} E\phi(Z) \leq \frac{\varepsilon}{1-\varepsilon} + \frac{1}{1-\varepsilon} \bar{Q}(Z \in F)$$

Letting $\varepsilon \downarrow 0$ and applying Lemma 4.1, we see that $(Y_\pi, X_\pi) \to^\sim (Y, X)$. $\qquad\square$

4.6. Corollary *Let $\mathcal{X} = (X_\pi \mid \pi \in \Pi)$ be a net of T-valued random elements on $(\Omega, \mathcal{F}, P)$ and let X be a T-valued random element on $(\Omega, \mathcal{F}, P)$. Let d be a pseudo-metric inducing the topology on T. Then we have*

(1) $X_\pi \to^{ba} X \quad \Leftrightarrow \quad X_\pi \to^{bo} X \quad \Rightarrow \quad X_\pi \to^d X$

and if I^X is σ-smooth at $\mathcal{F}(T)$ along $\mathcal{F}(T)$ (for instance, if X is Baire P-measurable), then we have

(2) $X_\pi \to^{ba} X \quad \Leftrightarrow \quad X_\pi \to^{bo} X \quad \Leftrightarrow \quad X_\pi \to^d X$

Let S be a topological space, let $(Y_\pi \mid \pi \in \Pi)$ be a net of S-valued random elements on $(\Omega, \mathcal{F}, P)$ and let Z be an $S \times T$-valued random element on $(\Upsilon, \mathcal{A}, Q)$. Then we have

(3) $X_\pi \to^d X$ *and* $(Y_\pi, X) \twoheadrightarrow^\sim Z \quad \Rightarrow \quad (Y_\pi, X_\pi) \twoheadrightarrow^\sim Z$

(4) $X_\pi \to^d X$ *and* $(Y_\pi, X) \to^\sim Z \quad \Rightarrow \quad (Y_\pi, X_\pi) \to^\sim Z$

Proof. Set $\theta_n(t) = \{u \mid d(t, u) < \frac{1}{n}\}$. Then $\Theta = (\theta_n \mid n \in \mathbf{N})$ is a decreasing $\mathbf{N}$-base on $D := T$ such that $X_\pi \to^\Theta X$ if and only if $X_\pi \to^d X$. Hence, (1) and (2) follow from Lemma 4.4 (1), Theorem 4.5 (1) and Lemma 4.3 (9). Moreover, since I_Z is σ-smooth at $2^{S \times T}$ along $2^{S \times T}$, we see that (3) follows from Theorem 4.5 (2). Finally, if $(Y_\pi, X) \to^\sim Z$ and $\mathcal{U} = ((Y_\pi, X) \mid \pi \in \Pi)$, then by Lemma 4.1 we have that $I^\mathcal{U}(1_F) \le I_Z(1_F)$ for all $F \in \mathcal{F}_\circ(T) = \mathcal{F}(T)$ and since, I_Z is σ-smooth at $\emptyset$ along 2^T, we see that (4) follows from Lemma 4.5 (3). $\qquad\square$

4.7. Corollary *Let $(d_\gamma \mid \gamma \in \Gamma)$ be an increasing family of pseudo-metrics inducing the topology on T, let $\mathcal{X} = (X_\pi \mid \pi \in \Pi)$ be a net of T-valued random elements on $(\Omega, \mathcal{F}, P)$ and let X be a T-valued random element on $(\Omega, \mathcal{F}, P)$. Then we have*

(1) $X_\pi \to^{bo} X \quad \Rightarrow \quad X_\pi \to^{ba} X \quad \Rightarrow \quad X_\pi \to^{d_\gamma} X \quad \forall \gamma \in \Gamma$

and if I^X is Γ-smooth and σ-smooth at $\mathcal{F}(T)$ along $\mathcal{F}(T)$, then we have

(2) $X_\pi \to^{ba} X \quad \Leftrightarrow \quad X_\pi \to^{bo} X \quad \Leftrightarrow \quad X_\pi \to^{d_\gamma} X \quad \forall \gamma \in \Gamma$

Let S be a topological space, let $(Y_\pi \mid \pi \in \Pi)$ be a net of S-valued random elements on $(\Omega, \mathcal{F}, P)$, set $\mathcal{U} = ((Y_\pi, X) \mid \pi \in \Pi)$ and let Z be an $S \times T$-valued random element on $(\Upsilon, \mathcal{A}, Q)$. If I_Z is Γ-smooth and σ-smooth at $\mathcal{F}(S \times T)$ along $\mathcal{F}(S \times T)$, then we have

(3) $(Y_\pi, X) \twoheadrightarrow^\sim Z$ *and* $X_\pi \to^{d_\gamma} X \quad \forall \gamma \in \Gamma \quad \Rightarrow \quad (Y_\pi, X_\pi) \twoheadrightarrow^\sim Z$

and if $I^\mathcal{U}$ is Γ-smooth and σ-smooth at $\emptyset$ along $\mathcal{F}(S \times T)$, then we have

(4) $(Y_\pi, X) \to^\sim Z$ *and* $X_\pi \to^{d_\gamma} X \quad \forall \gamma \in \Gamma \quad \Rightarrow \quad (Y_\pi, X_\pi) \to^\sim Z$

Remark: In the Remarks (b)–(d) to Theorem 4.5, you will find a series of criteria for Γ-smoothness and σ-smoothness of I^X, I_Z and $I^\mathcal{U}$.

Proof. Set $\theta_{\gamma, n}(t) := \{u \mid d_\gamma(t, u) < \frac{1}{n}\}$ for all $t \in T$ and all $(\gamma, n) \in \Gamma \times \mathbf{N}$. Then it is easily verified that $(\theta_{\gamma, n} \mid n \in \mathbf{N})$ is a decreasing $(\Gamma \times \mathbf{N})$-base on $D := T$ such that $X_\pi \to^\Theta X$ if and only if $X_\pi \to^{d_\gamma} X$ for all $\gamma \in \Gamma$ and since an increasing functional I on the topological space M is $(\Gamma \times \mathbf{N})$-smooth at $\mathcal{F}(M)$ along $\mathcal{F}(M)$ if and only if I is Γ-smooth and σ-smooth at $\mathcal{F}(M)$ along $\mathcal{F}(M)$, we see that the corollary follows from Theorem 4.5. $\qquad\square$

4.8. Corollary   *Let $(X_\pi \mid \pi \in \Pi)$ be a net of T-valued random elements on $(\Omega, \mathcal{F}, P)$, let $(\Sigma, \mathcal{B}, R)$ be a probability space and let $Z : \Upsilon \to T$ and $Y : \Sigma \to T$ be given random elements. Then we have*

(1) *If $X_\pi \to^\sim Z$, then $X_\pi \to^\sim Y$ if and only if Y is Baire R-measurable and $R_Y(B) = Q_Z(B)$ for all $B \in \mathcal{B}a(T)$*

(2) *If $X_\pi \twoheadrightarrow^\sim Z$ and $R_*(Y \in F) \geq Q_*(Z \in F)$ for all $F \in \mathcal{F}(T)$, then $X_\pi \twoheadrightarrow^\sim Y$*

(3) *If $X_\pi \twoheadrightarrow^\sim Z$ and $(Q_Z)_*(F) \geq Q_*(Z \in F)$ for all $F \in \mathcal{F}(T)$, then $X_\pi \twoheadrightarrow^\sim Q_Z$*

Moreover, if $T_0 \subseteq T$ is a given set such that $X_\pi(\omega) \in T_0$ and $Z(\upsilon) \in T_0$ for all $\omega \in \Omega$, all $\pi \in \Pi$ and all $\upsilon \in \Upsilon$, then we have

(4) *$X_\pi \twoheadrightarrow^\sim Z$ in T_0 $\iff$ $X_\pi \twoheadrightarrow^\sim Z$ in T*

(5) *If $X_\pi \to^\sim Z$ in T_0, then $X_\pi \to^\sim Z$ in T and the converse implication holds if T is normal and T_0 is closed*

Remark: Note that (4) states that Borel convergence in law is preserved under imbedding in a larger topological space or restriction to a subspace of T. Similarly, (5) states that Baire convergence in law is preserved under imbedding in a larger topological space, and if T is normal, then Baire convergence in law is preserved under restrictions to a closed subspaces of T. However, Example 1.1 shows that Baire convergence is *not* preserved under restrictions to arbitrary subspaces.

Proof. (1)–(4) follow easily from the portmanteau lemmas 4.1 and 4.2. and since the restriction of every $\phi \in C(T)$ to T_0 belongs to $C(T_0)$, we see that the first part of (5) holds. Moreover, if T is normal and T_0 is closed, then every $\psi \in C(T_0)$ is the restriction of a function $\phi \in C(T)$. Thus, the second part of (5) follows. $\qquad\square$

4.9. Corollary   *Let $\mathcal{X} = (X_\pi \mid \pi \in \Pi)$ be a net of T-valued random elements on $(\Omega, \mathcal{F}, P)$ and let $c \in T$ be a given point. Let $\delta_c(A) := 1_A(c)$ for $A \subseteq T$ be the Dirac measure at c and let $(V_\gamma \mid \gamma \in \Gamma)$ be a decreasing neighborhood base at c where Γ is a given net. Then we have*

(1) *$X_\pi \to^\sim c \Leftrightarrow X_\pi \to^\sim \delta_c \Leftrightarrow X_\pi \to^{ba} c \Leftrightarrow \lim_{\pi \uparrow \Pi} E^* |\phi(X_\pi) - \phi(c)| = 0 \, \forall \phi \in C(T)$*

(2) *$X_\pi \twoheadrightarrow^\sim c \Leftrightarrow X_\pi \twoheadrightarrow^\sim \delta_c \Leftrightarrow X_\pi \to^{bo} c \Leftrightarrow \lim_{\pi \uparrow \Pi} P(X_\pi \in V_\gamma) = 0 \, \forall \gamma \in \Gamma$*

Let S be a topological space, let $\mathcal{Y} = (Y_\pi \mid \pi \in \Pi)$ be a net of S-valued random elements on $(\Omega, \mathcal{F}, P)$ and let Z be an S-valued random element on $(\Upsilon, \mathcal{A}, Q)$. If I_Z is Γ-smooth at $\mathcal{F}(S)$ along $\mathcal{F}(S)$, then we have

(3) *$Y_\pi \twoheadrightarrow^\sim Z$ and $X_\pi \to^{bo} c$ $\Rightarrow$ $(Y_\pi, X_\pi) \twoheadrightarrow^\sim (Y, c)$*

and if $I^{\mathcal{Y}}$ is Γ-smooth at $\emptyset$ along $\mathcal{F}(S)$, then we have

(4) *$Y_\pi \to^\sim Z$ and $X_\pi \to^{bo} c$ $\Rightarrow$ $(Y_\pi, X_\pi) \to^\sim (Y, c)$*

Remarks: (a): Note that (3) and (4) show that Borel and Baire convergence in law satisfies formula (1.4) in the introduction under the appropriate smoothness conditions. Also, observe that Theorem 4.5 (2+3), Corollary 4.6 (3+4) and Corollary 4.7 (3+4) are extensions of formula (1.4) to the case where c is replaced by a random element X.

(b): Since I_Z is σ-smooth at 2^S along 2^S, we see that (3) holds in either of the following four cases: (i): S is hereditarily Lindelöf; (ii): S is separable and pseudo-metrizable; (iii): Z is Borel measurable and Q_Z is Γ-smooth at $\mathcal{F}(S)$ along $\mathcal{F}(S)$; (iv): c admits countable neighborhood base.

(c): If $Y_\pi \to^\sim Z$, then by Lemma 4.1 we have that $I^{\mathcal{Y}}(1_F) \leq I_Z(1_F)$ for all $F \in \mathcal{F}_\circ(T)$. Hence, we see that (4) holds in either of the following five cases: (i): S is completely regular and I_Z is τ-smooth at 0 along $C(T)$; (ii): S is perfectly normal and c admits a countable neighborhood base; (iii): S and T are pseudo-metrizable; (iv): S is Lindelöf and perfectly normal; (v): S is separable and pseudo-metrizable.

Proof. (1) and (2) follow easily from the definitions and Corollary 4.8 (3). Set $\mathcal{U} = ((Y_\pi, c) \mid \pi \in \Pi)$ and $\theta_\gamma(t) = V_\gamma$ for all $\gamma \in \Gamma$ and all $t \in T$. Then it follows easily that $\Theta = (\theta_\gamma \mid \gamma \in \Gamma)$ is a decreasing Γ-grid on $D := \mathrm{cl}(c)$ such that $X_\pi \to^\Theta c$ if and only if $X_\pi \to^{bo} c$. and since $I^{\mathcal{U}}$ is Γ-smooth at $\emptyset$ along $\mathcal{F}(S \times T)$ if and only if $I^{\mathcal{Y}}$ is Γ-smooth at $\emptyset$ along $\mathcal{F}(T)$, we see that Corollary 4.9 follows Theorem 4.5. $\square$

4.10. Corollary *Let Γ be a net and let $(T_\gamma \mid \gamma \in \Gamma)$ be a Γ-net of topological spaces. Let $(\wp_\gamma \mid \gamma \in \Gamma)$ be a net of continuous functions $\wp_\gamma : T \to T_\gamma$ and set*

$$\mathbf{F} = \{f \circ \wp_\gamma \mid \gamma \in \Gamma \,,\, f \in \mathrm{Usc}(T_\gamma)\} \,\,,\,\, \mathbf{F}_\circ = \{f \circ \wp_\gamma \mid \gamma \in \Gamma \,,\, f \in \mathrm{Usc}_\circ(T_\gamma)\}$$

Let $(X_\pi \mid \pi \in \Pi)$ be a net of T-valued random elements on $(\Omega, \mathcal{F}, P)$ and let Z be a T-valued random element on $(\Upsilon, \mathcal{A}, Q)$. If I_Z is $\mathbf{F}$-regular at $\mathcal{F}(T)$, then we have

(1) $X_\pi \to^\sim Z \iff \wp_\gamma(X_\pi) \to^\sim \wp_\gamma(Z) \quad \forall \gamma \in \Gamma$

and if I_Z is $\mathbf{F}_\circ$-regular at $\mathcal{F}_\circ(T)$, then we have

(2) $X_\pi \to^\sim Z \iff \wp_\gamma(X_\pi) \to^\sim \wp_\gamma(Z) \quad \forall \gamma \in \Gamma$

Remarks: (a): Set $\mathcal{G}_\gamma = \wp_\gamma^{-1}(\mathcal{G}(T_\gamma))$ and suppose that $(\mathcal{G}_\gamma)$ is increasing and that $\cup_{\gamma \in \Gamma} \mathcal{G}_\gamma$ is a base for the topology on T. Then it follows easily that $1_F \in \mathbf{F}_{\downarrow\Gamma}$ for all $F \in \mathcal{F}(T)$. Hence, if I_Z is Γ-smooth at $\mathcal{F}(T)$ along $\mathcal{F}(T)$, then by Lemma 3.1 (1) we have that I_Z is $\mathbf{F}$-regular at $\mathcal{F}(T)$ and so (1) holds.

(b): Suppose that $T = \prod_{i \in I} T_i$ is the product of the topological spaces (T_i) with its product topology. Then (1+2) show that Baire/Borel convergence in law is equivalent to Baire/Borel convergence in law of the finite dimensional marginals (under the appropriate regularity condition of the limit).

(c): Let (S, d) be a pseudo-metric space and let M be a given set. If $A \subseteq M$, we let d_A denote the pseudo-metric on S^M given by $d_A(f, g) := \sup_{x \in A} d(f(x), g(x))$. Let $\mathcal{K} \subseteq 2^M$ be an upwards directed paving on M and let $T := S^M$ be equipped with *the topology of uniform convergence on $\mathcal{K}$*; i.e., the topology $u(\mathcal{K})$ induced by the pseudo-metric $\{d_K \mid K \in \mathcal{K}\}$. Then (1+2) show that Baire/Borel convergence in law is equivalent to Baire/Borel convergence in law of the restrictions to K for all $K \in \mathcal{K}$ (under the appropriate regularity condition of the limit).

Proof. Since $\wp_\gamma$ is continuous, we see that $\mathbf{F} \subseteq \mathrm{Usc}(T) \subseteq \mathrm{Usc}[Z]$ and $\mathbf{F}_\circ \subseteq \mathrm{Usc}_\circ(T) \subseteq \mathrm{Usc}_\circ[Z]$. Hence, the corollary follows from Lemmas 4.1 and 4.2. $\square$

4.11. Theorem *Let $(X_\pi \mid \pi \in \Pi)$ be a net of T-valued random elements on $(\Omega, \mathcal{F}, P)$, let X be a T-valued random element on $(\Omega, \mathcal{F}, P)$ and let Z be a T-valued random element on $(\Upsilon, \mathcal{A}, Q)$. Let S be a topological space, let $\zeta : T \to S$ be a given function and set $\zeta_2(t,u) = (\zeta(t), \zeta(u))$ for all $(t,u) \in T \times T$. Then we have*

(1) *If $\psi \circ \zeta \in C_\circ[Z]$ for all $\psi \in C(S)$, then we have*

    (a) $X_\pi \to^\sim Z \;\Rightarrow\; \zeta(X_\pi) \to^\sim \zeta(Z)$

(2) *If $\psi \circ \zeta_2 \in C_\circ[(X,X)]$ for all $\psi \in C(S \times S)$, then we have*

    (a) $X_\pi \to^{ba} Z \;\Rightarrow\; \zeta(X_\pi) \to^{ba} \zeta(Z)$

(3) *If for every $\varepsilon > 0$ there exist $D, L \subseteq T$ such that ζ is continuous at D along L, $Q^*(Z \notin D) < \varepsilon$ and $\limsup_{\pi \uparrow \Pi} P^*(X_\pi \notin L) < \varepsilon$, then*

    (a) $X_\pi \twoheadrightarrow^\sim Z \;\Rightarrow\; \zeta(X_\pi) \twoheadrightarrow^\sim \zeta(Z)$

(4) *If for every $\varepsilon > 0$ there exist $D, L \subseteq T$ such that ζ is continuous at D along L, $P^*(X \notin L) < \varepsilon$, $\limsup_{\pi \uparrow \Pi} P^*(X_\pi \notin L) < \varepsilon$ and $X \in D$ P-a.s., then*

    (a) $X_\pi \to^{bo} X \;\Rightarrow\; \zeta(X_\pi) \to^{bo} \zeta(X)$

Remarks: (a): Observe that (1) is a Baire version of the formula (1.3) in the introduction and note that every continuous function $\zeta : T \to S$ satisfies the hypotheses of (1) and (2). Moreover, recall that Lemma 3.8 gives a series a criteria for the hypotheses of (1) or (2).

(b): If ζ is Q_Z-a.s. continuous, then ζ satisfies the hypotheses of (3) and (4) with $D = C(\zeta)$ and $L = T$. In particular, we see that Borel convergence in law satisfies the formula (1.3) in the introduction (even under weaker assumptions).

Proof. (1) and (2) follow easily from Lemma 4.1 and Lemma 4.3 (1).

(3): Let $H \in \mathcal{F}(S)$ and $\varepsilon > 0$ be given. Let $D, L \subseteq T$ be chosen according to assumption and set $F = \mathrm{cl}(\zeta^{-1}(H) \cap L)$. Since ζ is continuous at D along L, we have that $F \cap D \subseteq \zeta^{-1}(H)$. Hence, we have

$$\limsup_{\pi \uparrow \Pi} P^*(\zeta(X_\pi) \in H) \leq \limsup_{\pi \uparrow \Pi} P^*(X_\pi \in F) + \limsup_{\pi \uparrow \Pi} P^*(X_\pi \notin L)$$
$$\leq \varepsilon + Q_*(Z \in F) \leq 2\varepsilon + Q_*(Z \in F \cap D)$$
$$\leq 2\varepsilon + Q_*(\zeta(Z) \in H)$$

Letting $\varepsilon \downarrow 0$ and applying Lemma 4.2, we see that $\zeta(X_\pi) \twoheadrightarrow^\sim \zeta(Z)$.

(4): Set $Y = \zeta(X)$ and $Y_\pi = \zeta(X_\pi)$ and let $H \in \mathcal{F}(S \times S)$ be given such that $P^*((Y,Y) \in H) = 0$. Then $\hat{H} := \{s \in S \mid (s,s) \in H\}$ is a closed subset of T such that $P^*(Y \in \hat{H}) = 0$. Let $\varepsilon > 0$ be given and let $D, L \subseteq T$ be chosen according to assumption and set $F = \mathrm{cl}(\zeta_2^{-1}(H) \cap (L \times L))$. Since ζ_2 is continuous at $D \times D$ along $L \times L$, we have that $F \cap (D \times D) \subseteq \zeta_2^{-1}(H)$ and since $(X,X) \in D \times D$ P-a.s. and $P^*((X,X) \in \zeta_2^{-1}(H)) = P^*((Y,Y) \in H) = 0$, we conclude that $P^*((X,X) \in F) = 0$. So by (4.2) we have

$$\limsup_{\pi \uparrow \Pi} P^*((Y, Y_\pi) \in H) \leq \limsup_{\pi \uparrow \Pi} P^*((X, X_\pi) \in F) + \limsup_{\pi \uparrow \Pi} P^*((X, X_\pi) \notin L \times L)$$
$$\leq P^*(X \notin L) + \limsup_{\pi \uparrow \Pi} P^*(X_\pi \notin L) \leq 2\varepsilon$$

Letting $\varepsilon \downarrow 0$, we see that $\zeta(X_\pi) \to^{bo} \zeta(X)$. $\square$

4.12. Theorem (The continuity theorem) *Let $\mathcal{X} = (X_\pi \mid \pi \in \Pi)$ be a net of T-valued random elements on $(\Omega, \mathcal{F}, P)$, let Z be T-valued random element on $(\Upsilon, \mathcal{A}, Q)$ and let $\mathcal{L} \subseteq 2^T$ be a given paving satisfying (3.6.1). Then we have the following:*

(A): Suppose that I_Z is $C(T)$-regular at $\mathcal{F}(T)$ and $X_\pi \to^\sim Z$. Then Z is Borel measurable and Borel complete and $X_\pi \twoheadrightarrow^\sim Z$. Moreover, if Y is a T-valued random element on the probability space $(\Sigma, \mathcal{B}, R)$ such that $X_\pi \to^\sim Y$, then $\int^* f(Y)dR \le \mathbb{E}f(Z)$ for all $f \in \mathrm{Usc}(T)$.

(B): (X_π) converges in Baire law to some $\mathcal{L}$-complete random element Y if and only if (X_π) is $\mathcal{L}$-complete and if so, then the $\mathcal{L}$-completion μ of (X_π) equals the $\mathcal{L}$-completion of Y and μ is an $\mathcal{L}$-complete probability measure such that $X_\pi \to^\sim \mu$.

(C): (X_π) converges in Baire law some random element Y if and only if $I^\mathcal{X}$ is σ-smooth at 0 along $C(T)$ and the limits $\lim_{\pi \uparrow \Pi} E^* \phi(X_\pi)$ exist for all $\phi \in C(T)$.

(D): (X_π) converges in Borel law to some Borel complete random element if and only if (X_π) is Borel complete and if so, then the Borel completion μ is a Borelian probability measure such that μ is Borel complete, I_μ is $C(T)$-regular at $\mathcal{F}(T)$ and $X_\pi \twoheadrightarrow^\sim \mu$.

(E): Suppose that T is completely regular and that $I_\mathcal{X}$ is $\mathcal{K}(T)$-tight along $C(T)$. If $\Phi \subseteq C(T)$ is an algebra satisfying

(1) $\forall t_0 \ne t_1 \in T \ \forall \phi \in \Phi$ *so that* $\phi(t_0) = 0$ *and* $\phi(t_1) = 1$

(2) Q_Z *is a Radon measure and* $\limsup_{\pi \uparrow \Pi} E^* \phi(X_\pi) \le \mathbb{E}\phi(Z) \ \ \forall \phi \in \Phi$

then we have $X_\pi \twoheadrightarrow^\sim Q_Z$ and $X_\pi \twoheadrightarrow^\sim Z$.

Remarks: (a): Recall that Theorem 3.7 gives a series of criteria for Baire, Borel, Radon and $\mathcal{L}$-completeness of (X_π) and Y. The usual criteria for convergence in law appeals to some sort of $\mathcal{K}(T)$-tightness of $I^\mathcal{X}$. However, the only use of $\mathcal{K}(T)$-tightness is to establish the appropriate smoothness of (X_π) and the theorem above implies most of the usual criteria for convergence in law.

(b): Example 1.1 shows that (X_π) may converge in Borel law to many different probability measures. However, if (X_π) is Borel complete with Borel completion μ, then by (A) and (D) we see that $X_\pi \twoheadrightarrow^\sim \mu$ and that μ dominates any Baire limit of (X_π) on the set $\mathrm{Usc}(T)$.

Proof. (A): Since Z is Baire measurable, then by Theorem 3.7 (9) we have that Z is Borel measurable and Borel complete and by Lemma 4.1 (9), we have that $X_\pi \twoheadrightarrow^\sim Z$. So suppose that $X_\pi \to^\sim Y$ and let $F \in \mathcal{F}(T)$ be given. Then we have

$$R^*(Y \in F) \le \inf\left\{ \int_\Sigma \phi(Y)dR \ \middle| \ \phi \in C(T) \,, \ \phi \ge 1_F \right\}$$
$$= \inf\{\mathbb{E}\phi(Z) \mid \phi \in C(T) \,, \ \phi \ge 1_F\} = Q_*(Z \in F)$$

and so the last statement in (A) follows from Lemma 3.3 (4).

(B): Suppose that $X_\pi \to^\sim Y$ for some $\mathcal{L}$-complete random element Z. Then $I^\mathcal{X}(\phi) = I^Y(\phi)$ for all $\phi \in C(T)$. Hence, $I^\mathcal{X}$ is $\mathcal{L}$-complete with the same $\mathcal{L}$-completion as Y. Conversely, if (X_π) is $\mathcal{L}$-complete with $\mathcal{L}$-completion μ, then μ

is Baireian and $\mathcal{L}$-complete and by Theorem 3.5 (10) we have that $I^{\mathcal{X}}(\phi) = I_\mu(\phi)$ for all $\phi \in C(T)$; that is, $X_\pi \to^\sim \mu$.

(C+D): Follows from directly from (B) and Theorem 3.8 (3).

(E): Without loss of generality, we may assume that Φ is $\| \cdot \|_T$-closed and $1_T \in \Phi$. Let $\varepsilon > 0$ and $\psi \in C(T)$ be given and let $K \in \mathcal{K}(T)$ be chosen such that $Q_Z(T \setminus K) < \varepsilon$ and $I^{\mathcal{X}}(f) < \varepsilon$ for all $f \in C(T)$ with $f \leq 1_{T \setminus K}$. Since Φ is a $\| \cdot \|_T$-closed algebra containing 1_T and separating points in T, then by Stone-Weierstrass' theorem there exists $\phi \in \Phi$ such that $\|\phi - \psi\|_K < \varepsilon$ and $\|\phi\|_T \leq c$ where $c = \|\psi\|_T$. Then $|\phi - \psi| \leq \varepsilon 1_T + 2c\, 1_{T \setminus K}$ and so we have $I^{\mathcal{X}}(|\phi - \psi|) < 4c\varepsilon$ and $I^Z(|\phi - \psi|) < 4c\varepsilon$. Hence, by (1) we have

$$I^{\mathcal{X}}(\psi) \leq I^{\mathcal{X}}(\phi) + I^{\mathcal{X}}(|\phi - \psi|) \leq I_Z(\phi) + 4c\varepsilon$$
$$\leq I_Z(\psi) + I^Z(|\phi - \psi|) + 4c\varepsilon \leq I_Z(\psi) + 8c\varepsilon$$

Letting $\varepsilon \downarrow 0$, we see that $I^{\mathcal{X}}(\psi) \leq I_Z(\psi)$ for all $\psi \in C(T)$ and since T is completely regular and Q_Z is a Radon measure, we see that Theorem 3.7 (1+2) hold with $\mathcal{L} = \mathcal{F}(T)$, $\mu = Q_Z$ and $(I, I^\circ) = (I^{\mathcal{X}}, I_{\mathcal{X}})$. Hence, by Theorem 3.7 (1) we have that (X_π) is Borel complete and that Q_Z is an extension of the Borel completion of (X_π). So (E) follows from (D) and Lemma 4.1 (8).   □

5. Convergence of Random Sets

In this section, we shall consider convergence in law of random sets; that is, random elements with values in 2^T. To do this we need a topology on 2^T. Typically, we shall consider random sets which are the maximum points or the zeroes of real valued stochastic processes with time set T. With this in mind, the so-called upper Fell topologies present themselves in a natural way.

If $A, B \subseteq T$ are subsets of T, we set $\mathcal{U}(A, B) = \{E \in 2^T \mid E \cap B \subseteq A\}$ and if $\mathcal{K}$ is a paving on T, we define *the upper Fell topology* $\varpi^*(T, \mathcal{K})$ to be the topology generated by the sets $\{\mathcal{U}(G, K) \mid (G, K) \in \mathcal{G}(T) \times \mathcal{K}\}$. In particular, we set $\varpi^*(T) := \varpi^*(T, \{T\})$. The upper Fell topology has very few topological properties. However, it is compact for the trivial reason that 2^T is the only Fell-open set containing T. Note that $D \in \mathcal{U}(A, B) \Rightarrow E \in \mathcal{U}(A, B)$ for all $E \subseteq D$. Hence, the upper Fell topology is neither Hausdorff nor a T_1-space. Let Γ be a net and let $A \subseteq B \subseteq T$ be given sets, then we say that A is Γ-*net compact in* B if every Γ-net $(t_\gamma) \subseteq A$ has a cluster point belonging to B or equivalently if for every decreasing Γ-net $(F_\gamma) \subseteq \mathcal{F}(T)$ with $F_\gamma \cap B \downarrow \emptyset$ there exists $\beta \in \Gamma$ such that $F_\beta \cap A = \emptyset$. If $(A_\gamma \mid \gamma \in \Gamma)$ is a Γ-net of subsets of T, then the reader easily verifies the following simple properties of the upper Fell topology:

If $\mathcal{V} \subseteq 2^T$ is $\varpi^*(T, \mathcal{K})$-open and $D \in \mathcal{V}$, then $E \in \mathcal{V}$ for all $E \subseteq D$ (5.1)

If $\mathcal{H} \subseteq 2^T$ is $\varpi^*(T, \mathcal{K})$-closed and $D \in \mathcal{H}$, then $E \in \mathcal{H}$ for all $E \supseteq D$ (5.2)

If $f : 2^T \to \mathbf{R}$ is $\varpi^*(T, \mathcal{K})$-continuous, then f is constant   (5.3)

$A_\gamma \to A$ in $\varpi^*(T, \mathcal{K}) \Leftrightarrow A_\gamma \cap K \to A \cap K$ in $\varpi^*(K) \quad \forall K \in \mathcal{K}$ (5.4)

If T is regular and $A_\gamma \to A$ in $\varpi^*(T, \mathcal{K})$, then $\mathrm{cl}(A \cap K) \supseteq$ up $\lim(A_\gamma \cap K)$ for all $K \in \mathcal{K}$ (5.5)

If for every set $K \in \mathcal{K}$ there exists $\beta \in \Gamma$ such that $\cup_{\gamma \geq \beta}(A_\gamma \cap K)$ is Γ-net compact in K, then $A_\gamma \to A$ in $\varpi^*(T, \mathcal{K})$ for every set $A \subseteq T$ satisfying $A \supseteq K \cap \mathrm{up} \lim(A_\gamma \cap K)$ for all $K \in \mathcal{K}$ (5.6)

Note that (5.3) shows that Baire convergence in law on $(2^T, \varpi^*(T, \mathcal{K}))$ is trivial and worthless (every net of 2^T-valued random elements is Baire convergent in law to any 2^T-valued random element). However, Borel convergence in law turns out to be non-trivial and useful – in particular, in the context of the following statistical estimation method:

Let T be *the parameter space*, let *the state space* be a given topological space S and let the observations be a Π-net $(X_\pi \mid \pi \in \Pi)$ of S-valued random elements on $(\Omega, \mathcal{F}, P)$. If $J : S \times [0, \infty] \to 2^T$ is a given set valued map, then a net $(\theta_\pi \mid \pi \in \Pi)$ of T-valued random elements will be called a *net of approximate J-estimators for (X_π) with error terms (R_π)* if (R_π) is a Π-net of $[0, \infty]$-valued random elements satisfying

$$\limsup_{\pi \uparrow \Pi} P^*(\theta_\pi \notin J(X_\pi, R_\pi)) = 0 \quad \text{and} \quad R_\pi \to^{ba} 0 \tag{5.7}$$

In particular, if $S \subseteq \overline{\mathbf{R}}^T$, we define arg max and arg zero, as follows:

$$\arg\max(f, \alpha) := \left\{ t \in T \mid \sup_T f \leq f(t) + \alpha \right\} \quad \forall f \in S \; \forall 0 \leq \alpha \leq \infty \tag{5.8}$$

$$\arg\mathrm{zero}(f, \alpha) := \left\{ t \in T \mid |f(t)| \leq \alpha \right\} \quad \forall f \in S \; \forall 0 \leq \alpha \leq \infty \tag{5.9}$$

with the convention $\infty - \infty := \infty$. Then $\arg\max(f, 0)$ is the set of all maximum points of f and we may consider $\arg\max(f, \alpha)$ as the set of all approximative maximum points of f and a net (θ_π) of approximate arg max-estimators for (X_π) will be called a *net of approximative maximum estimators for (X_π)*. Similarly, $\arg\mathrm{zero}(f, 0)$ is the set of all zero points of f and we may consider $\arg\mathrm{zero}(f, \alpha)$ as the set of all approximative zero points of f and a net (θ_π) of approximate arg zero-estimators for (X_π) will be called a *net of approximative zero estimators for (X_π)*.

5.1. Lemma *Let S be a topological space and let $\mathcal{K} \subseteq 2^T$ be a given paving on T. Let $m : S \times [0, \infty] \to \overline{\mathbf{R}}$ and $R : S \times T \to \overline{\mathbf{R}}$ be given functions and set $M_A(s) = \sup_{t \in A} R(s, t)$ for all $A \subseteq T$ and*

(1) $J(s, \alpha) = \{ t \in T \mid m(s, \alpha) \leq R(s, t) \} \quad \forall s \in S \; \forall 0 \leq \alpha \leq \infty$

Then $J : S \times [0, \infty] \to (2^T, \varpi^*(T, \mathcal{K}))$ is continuous at (s, α) for all $(s, \alpha) \in S \times [0, \infty]$ with $m(s, \alpha) = -\infty$. Moreover, if S_0 denotes the set of all $s_0 \in S$ satisfying the following conditions:

(2) *m is lower semicontinuous at $(s_0, 0)$ and $R(s_0, \cdot)$ is upper semicontinuous*

(3) *$M_{K \cap F}$ is upper semicontinuous at s_0 for all $K \in \mathcal{K}$ and all $F \in \mathcal{F}(T)$*

(4) *For all $K \in \mathcal{K}$ there exists $c < m(s_0, 0)$ such that $\{ t \in K \mid \rho(s_0, t) \geq c \}$ is $\mathbf{N}$-net compact in K*

Then $J : S \times [0, \infty] \to (2^T, \varpi^*(T, \mathcal{K}))$ is continuous at $(s_0, 0)$ for all $s_0 \in S_0$.

Remark: Suppose that $S = \mathbf{R}^T$ with the topology $u(\mathcal{K})$ of uniform convergence on $\mathcal{K}$ (see Remark (c) to Corollary 4.9). Let $\text{Usc}(T, \mathcal{K})$ denote the set of all upper semicontinuous functions $s \in \mathbf{R}^T$ satisfying the following condition

(5) $\forall K \in \mathcal{K} \; \exists c < \sup_T s$ so that $\{t \in K \mid s(t) \geq c\}$ is $\mathbf{N}$-net compact in K

Setting $R(s,t) = s(t)$ and $m(s,\alpha) = \sup_T s - \alpha$ with the convention $\infty - \infty = -\infty$, we see that $\arg\max$ is given on the form (1) and that (3+4) hold every $s \in \text{Usc}(T, \mathcal{K})$. Moreover, if $T = \cup_{K \in \mathcal{K}} K$, then m is lower semicontinuous on $S \times [0, \infty]$ and so we conclude that $\arg\max$ is $\varpi^*(T, \mathcal{K})$-continuous at $(s, 0)$ for all $s \in \text{Usc}(T, \mathcal{K})$. Similarly, setting $R(s,t) = -|s(t)|$ and $m(s,\alpha) = -\alpha$, we see that $\arg\text{zero}$ is given on the form (1) and that $\arg\text{zero}$ is $\varpi^*(T, \mathcal{K})$-continuous at $(s, 0)$ for all $s \in \mathbf{R}^T$ with $-|s(\cdot)| \in \text{Usc}(T, \mathcal{K})$.

Proof. Let $(s, \alpha) \in S \times [0, \infty]$ be given such that $m(s, \alpha) = -\infty$. Then $J(s, \alpha) = T$, so by (5.1) we see that J is continuous at (s, α).

Let $s_0 \in S_0$, $K \in \mathcal{K}$ and $G \in \mathcal{G}(T)$ be given such that $J(s_0, 0) \cap K \subseteq G$ and let $-\infty < c < m(s_0, 0)$ be chosen according to (3). Let (c_n) be a strictly increasing sequence of real numbers such that $c_0 = c$ and $c_n \uparrow m(s_0, 0)$ and set $F_n = \{t \in T \setminus G \mid R(s_0, t) \geq c_n\}$ for all $n \geq 0$. Then $F_n \cap K \downarrow J(s_0, 0) \cap K \setminus G = \emptyset$ and by upper semicontinuity of $R(s_0, \cdot)$, we have that F_n is closed subset of F_0 for all $n \geq 1$. Hence, by $\mathbf{N}$-net compactness of F_0 in K, there exists an integer $k \geq 1$ such that $F_k \cap K = \emptyset$. In particular, we see that $M_{K \setminus G}(s_0) \leq c_k$. Since $c_k < m(s_0, 0)$, there exists $0 < \delta < \infty$ such that $c_k + 2\delta < m(s_0, 0)$. So by (1) and (2) there exist $\kappa > 0$ and an open neighborhood U of s_0 such that $m(s, \alpha) > c_k + 2\delta$ and $M_{K \setminus G}(s) < c_k + \delta$ for all $0 \leq \alpha \leq \kappa$ and all $s \in U$. Let $s \in U$, $0 \leq \alpha \leq \kappa$ and $t \in J(s, \alpha) \cap K$ be given. Then we have $c_k + 2\delta < m(s, \alpha) \leq \rho(s, t) + \delta$ and so $M_{K \setminus G}(s) < c_k + \delta < \rho(s, t)$. Hence, we see that $t \in G$; that is $J(s, \alpha) \cap K \subseteq G$ for all $s \in U$ and all $0 \leq \alpha \leq \kappa$. Thus, we conclude that J is $\varpi^*(T, \mathcal{K})$-continuous at $(s_0, 0)$. $\qquad \square$

5.2.   Theorem   *Let 2^T have its upper Fell topology $\varpi^*(T, \mathcal{K})$ where $\mathcal{K}$ is a given paving on T. Let $(M_\pi \mid \pi \in \Pi)$ and $(L_\pi \mid \pi \in \Pi)$ be two nets of 2^T-valued random elements on $(\Omega, \mathcal{F}, P)$ and let $M, L : \Upsilon \to 2^T$ be two set valued random elements. Then we have*

(1) $M_\pi \to^{bo} \emptyset \iff \lim_{\pi \uparrow \Pi} P_*(M_\pi \cap K = \emptyset) = 1 \quad \forall K \in \mathcal{K}$

(2) $L_\pi \twoheadrightarrow^\sim L$ *and* $\limsup_{\pi \uparrow \Pi} P^*(M_\pi \not\subseteq L_\pi) = Q^*(L \not\subseteq M) = 0 \implies M_\pi \twoheadrightarrow^\sim M$

Moreover, if $(\vartheta_\pi \mid \pi \in \Pi)$ is a net of T-valued random element on $(\Omega, \mathcal{F}, P)$ and Z is a T-valued random element on $(\Upsilon, \mathcal{A}, Q)$ satisfying

(3) $L(v) \subseteq \{Z(v)\}$ *for Q-a.a. $v \in \Upsilon$*

(4) *For all $\varepsilon > 0$ and all $G \in \mathcal{G}(T)$ there exists $K \in \mathcal{K}$ satisfying*

   (a) $Q^*(L \subseteq G) \leq \varepsilon + \liminf_{n \to \infty} P_*(L_\pi \cap K \subseteq G)$

   (b) $\limsup_{\pi \uparrow \Pi} P^*(\vartheta_\pi \notin G \, , \, L_\pi \cap K \subseteq G) < \varepsilon$

then $\vartheta_\pi \twoheadrightarrow^\sim Z$ and $L(\cdot) = \{Z(\cdot)\}$ Q-a.s.

Proof. (1): Let $\mathcal{K}_0$ denote the set of all finite unions of sets from $\mathcal{K}$. Then $P_*(M_\pi \cap K = \emptyset) \to 0$ for all $K \in \mathcal{K}$ if and only if $P_*(M_\pi \cap K = \emptyset) \to 0$ for all $K \in \mathcal{K}_0$ and since $\{\mathcal{U}(\emptyset, K) \mid K \in \mathcal{K}_0\}$ is a decreasing neighborhood base of $\emptyset$, we see that (1) follows from Corollary 4.9 (2).

(2): Let $\mathcal{H}$ be a given $\varpi^*(T, \mathcal{K})$-closed subset of 2^T. By (5.2) and the hypothesis we have

$$\limsup_{\pi \uparrow \Pi} P^*(M_\pi \in \mathcal{H}) \leq \limsup_{\pi \uparrow \Pi} P^*(L_\pi \in \mathcal{H}) + \limsup_{\pi \uparrow \Pi} P^*(M_\pi \not\subseteq L_\pi)$$
$$\leq Q_*(L \in \mathcal{H}) \leq Q_*(M \in \mathcal{H}) + Q^*(L \not\subseteq M) = Q_*(M \in \mathcal{H})$$

Hence, (2) follows from Lemma 4.2.

Suppose that (3+4) hold and let $\varepsilon > 0$ be given. If $K \in \mathcal{K}$ is chosen according to (4) with $G = \emptyset$, then by (4.a+b) we have

$$Q^*(L = \emptyset) \leq \varepsilon + \liminf_{\pi \uparrow \Pi} P_*(L_\pi \cap K = \emptyset) < 2\varepsilon$$

Letting $\varepsilon \downarrow 0$ and applying (3), we see that $L(\cdot) = \{Z(\cdot)\}$ Q-a.s. Let $\varepsilon > 0$ and $G \in \mathcal{G}(T)$ be given and let $K \in \mathcal{K}$ be chosen according to (4), then by (4.a+b) we have

$$Q^*(Z \in G) = Q^*(L \subseteq G) \leq \varepsilon + \liminf_{\pi \uparrow \Pi} P_*(L_\pi \cap K \subseteq G)$$
$$\leq \varepsilon + \liminf_{\pi \uparrow \Pi} P_*(\vartheta_\pi \in G) + \limsup_{\pi \uparrow \Pi} P^*(\vartheta_\pi \notin G,\ L_\pi \cap K \subseteq G)$$
$$\leq 2\varepsilon + \liminf_{\pi \uparrow \Pi} P_*(\vartheta_\pi \in G)$$

Letting $\varepsilon \downarrow 0$ and applying Lemma 4.2, we see that $\vartheta_\pi \twoheadrightarrow^\sim Z$. □

5.3. Corollary Let $\mathcal{K} \subseteq 2^T$ be a paving on T and let 2^T be equipped with the $\varpi^*(T, \mathcal{K})$-topology. Let S be a topological space and let $J : S \times [0, \infty] \to 2^T$ be a set valued map. Let $(X_\pi \mid \pi \in \Pi)$ be a Π-net of S-valued random elements on $(\Omega, \mathcal{F}, P)$ and let (ϑ_π) be a net of approximate J-estimators for (X_π) with error terms (R_π). Let $Y : \Upsilon \to T$ be a random element satisfying

(1) $X_\pi \twoheadrightarrow^\sim Y$ and J is $\varpi^*(T, \mathcal{K})$-continuous at $(Y(\upsilon), 0)$ for Q-a.a. $\upsilon \in \Upsilon$

Then $(X_\pi, R_\pi) \twoheadrightarrow^\sim (Y, 0)$ and $J(X_\pi, R_\pi) \twoheadrightarrow^\sim J(Y, 0)$ and if $Z : \Upsilon \to T$ is a random element such that $J(Y, 0) \subseteq \{Z\}$ Q-a.s. and

(2) $\forall \varepsilon > 0\ \exists K \in \mathcal{K}$ so that $\limsup_{\pi \uparrow \Pi} P^*(\vartheta_\pi \notin K) < \varepsilon$

then $\vartheta_\pi \twoheadrightarrow^\sim Z$ and $J(Y, 0) = \{Z\}$ Q-a.s.

Proof. By (1), Corollary 4.6 (2) and Corollary 4.9 (3), we have that $(X_\pi, R_\pi) \twoheadrightarrow^\sim (Y, 0)$ and since J continuous $Q_{(Y,0)}$-a.s., then by Theorem 4.11 (3) we see that $J(X_\pi, R_\pi) \twoheadrightarrow^\sim J(Y, 0)$ and the last part of the corollary follows easily from (2) and Theorem 5.2 with $L_\pi = J(X_\pi, R_\pi)$ and $L(\upsilon) = J(Y(\upsilon), 0)$. □

5.4. Example Corollary 5.3 offers a method of determining the asymptotic distribution of a net of approximate J-estimators. However, usually a net (θ_π) of approximate J-estimators converge to a constant (*the true parameter*) in probability and we want to find non-degenerated asymptotic distributions of some transforms of the estimators – typically of the form $\kappa_\pi^{-1}(\theta_\pi - \theta_0)$ where θ_0 is true parameter and (κ_π) is a specified net of positive numbers. To achieve this objective, we have find a new net (Y_π) and possibly a new set valued map $\tilde{J}$ such the transformed estimators is a net of approximative $\tilde{J}$-estimators for (Y_π). Let me illustrate the idea by the following example (see [[5]], [[1]] and [[2]; Vol.II Chap. 14]):

Let T be a linear topological space, let $\mathcal{K}$ be an upwards directed paving on T such that $\mathcal{K} \uparrow T$ and let $\mathbf{R}^T$ be equipped with the topology $u(\mathcal{K})$ of uniform convergence on $\mathcal{K}$ (see Remark (c) to Corollary 4.9). Let W be a linear space, let $(t, w) \curvearrowright t \bullet w$ be a bilinear map on $T \times W$ such that

$$q_K(w) := \sup_{t \in K} |t \bullet w| < \infty \qquad \forall w \in W \;\; \forall K \in \mathcal{K}$$

and let $u^\star(\mathcal{K})$ denote the (locally convex) topology on W induced by the seminorms $(q_K \mid K \in \mathcal{K})$. Let $(S, \mathcal{B})$ be a measurable space, let (ξ_n) be a sequence of independent, identically distributed S-valued random variables and let $h : S \times T \to \mathbf{R}$ be a given function satisfying the following Taylor expansion at the given parameter point $\theta_0 \in T$:

(1) $\;h(s, t + \theta_0) = h(s, \theta_0) + t \bullet D(s) - \frac{1}{2} t \bullet \Lambda t + r(s, t) \qquad \forall t \in T$

where $D : S \to W$ and $r : S \times T \to \mathbf{R}$ are a given functions and $\Lambda : T \to W$ is a given linear operator. Let (κ_n) be a given sequence of strictly positive numbers and set $\beta_n = n^{-1} \kappa_n^{-2}$ and

$$X_n(t) = \sum_{i=1}^n h(\xi_i, t) \;,\;\; U_n(t) = \frac{1}{n\kappa_n^2} \sum_{i=1}^n r(\xi_i, \kappa_n \cdot t) \;,\;\; D_n = \frac{1}{n\kappa_n} \sum_{i=1}^n D(\xi_i)$$
$$Y_n(t) = \beta_n \big(X_n(\kappa_n \cdot t + \theta_0) - X_n(\theta_0)\big) = t \bullet D_n - t \bullet \Lambda t + U_n(t)$$

Let (θ_n) be a sequence of T-valued random elements and let (R_n) be a sequence of $[0, \infty]$-valued random elements satisfying

(2) $\;\lim_{n \to \infty} P^*(\theta_n \notin \arg\max(X_n, R_n)) = 0$ $\;$ and $\;$ $\beta_n \cdot R_n \to^{ba} 0$

If we set $\vartheta_n = \kappa_n^{-1}(\theta_n - \theta_0)$ and $R_n^* = \beta_n R_n$, it follows easily that (ϑ_n) is a sequence of approximate maximum estimators for (Y_n) with error terms (R_n^*). Now suppose that $Y : \Upsilon \to \mathbf{R}^T$ is a random element satisfying

(3) $\;Y_\pi \twoheadrightarrow^\sim Y$ in $(\mathbf{R}^T, u(\mathcal{K}))$

(4) $\;Y \in \mathrm{Usc}(T, \mathcal{K})$ $\;Q$-a.s. (see the remark to Lemma 5.1)

Then Lemma 5.1 and Corollary 5.3 show that $(Y_\pi, R_\pi) \twoheadrightarrow^\sim (Y, 0)$ and that

$$\arg\max(Y_\pi, R_\pi) \twoheadrightarrow^\sim \arg\max(Y, 0) \;\text{ in }\; (2^T, \varpi^*(T, \mathcal{K}))$$

and if $Z : \Upsilon \to T$ is a random element with $\arg\max(Y, 0) \subseteq \{Z\}$ Q-a.s. and

(5) $\;\forall \varepsilon > 0 \; \exists K \in \mathcal{K}$ so that $\;\limsup_{n \to \infty} P^*(\vartheta_n \notin K) < \varepsilon$

then $\kappa_\pi^{-1}(\theta_\pi - \theta_0) \twoheadrightarrow^\sim Z$ and $Z(\upsilon)$ is the unique maximum point of $Y(\upsilon, \cdot)$ for Q-a.a. $\upsilon \in \Upsilon$. Note that (5.4.4) holds if Y has upper semicontinuous sample paths and every set $K \in \mathcal{K}$ is $\mathbf{N}$-net compact in itself. Recall that a linear operator $\Lambda : V \to W$ is *non-negative definite* if $t \bullet \Lambda u = u \bullet \Lambda t$ and $t \bullet \Lambda t \geq 0$ for all

$t, u \in T$ and observe that Corollary 4.6, Corollary 4.9 and Theorem 4.10 delivers the following:

(6) Suppose that V is a W-valued random element on $(\Upsilon, \mathcal{A}, Q)$ satisfying

   (a) $D_n \twoheadrightarrow^{\sim} V$ in $(W, u^{\star}(\mathcal{K}))$ and $\sup_{t \in K} |U_n(t)| \rightarrow^{ba} 0 \quad \forall K \in \mathcal{K}$

Then (3) holds with $Y(\upsilon, t) := t \bullet V(\upsilon) - \frac{1}{2} t \bullet \Lambda t$. Moreover, if Λ is non-negative definite and bijective with inverse operator Λ^{-1}, then

   (a) $\arg\max(Y(\upsilon, \cdot), 0) = \{\Lambda^{-1} V(\upsilon)\} \quad \forall \upsilon \in \Upsilon$

(7) Suppose that U is a $\mathbf{R}^T$-valued random element on $(\Upsilon, \mathcal{A}, Q)$ satisfying

   (a) $U_n \twoheadrightarrow^{\sim} U$ in $(\mathbf{R}^T, u(\mathcal{K}))$ and $D_n \rightarrow^{bo} 0$ in $(W, u^{\star}(\mathcal{K}))$

Then (3) holds with $Y(\upsilon, t) := U(\upsilon, t) - \frac{1}{2} t \bullet \Lambda t$

References

[1] Arcones, M.A. (1994): *Distributional convergence of M-estimators under unusual rates*, Stat. Probab. Lett. 20 No.1, p. 57–62

[2] Hoffmann-Jørgensen, J. (1994): *Probability with a view toward statistics*, Vol.I Chapman & Hall, New York and London

[3] Hoffmann-Jørgensen, J. (1991): *Stochastic processes on Polish spaces*, Var. Publ. Ser. No.39, Mat. Inst. Aarhus University

[4] Hoffmann-Jørgensen, J. (1982): *A general "in between theorem"*, Math. Scand. 50, p. 55–65

[5] Kim, J. & Pollard, D. (1990): *Cube root asymptotics*, Ann. Stat. 18 No.1, p. 191–219

Jørgen Hoffmann-Jørgensen
Institute of Mathematical Sciences
University of Aarhus
Ny Munkegade
Universitetsparken
DK-8000 Aarhus C, Denmark
hoff@mi.aau.dk

Progress in Probability, Vol. 43
© 1998 Birkhäuser Verlag Basel/Switzerland

Asymptotics of Spectral Projections of Some Random Matrices Approximating Integral Operators

VLADIMIR I. KOLTCHINSKII*

ABSTRACT. Given a probability space $(S, \mathcal{S}, P)$ and a symmetric measurable real valued kernel h on $S \times S$, which defines a compact integral operator H from $L_2(P)$ into $L_2(P)$, we study the asymptotic behavior of spectral projections of the random $n \times n$ matrices $\tilde{H}_n$ with entries $n^{-1}h(X_i, X_j), 1 \leq i, j \leq n$, and H_n, obtained from $\tilde{H}_n$ by deleting of its diagonal. Here $(X_1, \ldots, X_n)$ is a sample of independent random variables in $(S, \mathcal{S})$ with common distribution P. We show that if H is a Hilbert-Schmidt operator, then the spectral projections of H_n converge a.s. to the spectral projections of H (the convergence is understood in the sense of quadratic forms). Under slightly stronger assumptions, the convergence also holds for the spectral projections of $\tilde{H}_n$. Moreover, under much stronger assumptions on the kernel h, we show that the fluctuations of the spectral projections of $\tilde{H}_n$ or H_n are asymptotically Gaussian.

1. Introduction

Let $(S, \mathcal{S}, P)$ be a probability space. Consider a Hilbert-Schmidt integral operator $H : L_2(P) \mapsto L_2(P)$ with symmetric measurable kernel $h : S \times S \mapsto \mathbb{R}$:

$$Hg(x) := \int_S h(x, y)g(y)P(dy), \ g \in L_2(P).$$

It is well known that H has a spectrum, consisting of at most countable number of distinct real eigenvalues and the unique limit point 0 (in the case of infinite spectrum). Each nonzero eigenvalue has finite multiplicity, and there exists a finite dimensional (eigen)subspace of $L_2(P)$ (its dimension is equal to the multiplicity), consisting of all eigenfunctions, corresponding to this eigenvalue. The orthogonal projections on such eigenspaces are called spectral projections of the operator H.

Let $\{X_n\}_{n \geq 1}$ be a sequence of independent random variables in $(S, \mathcal{S})$ with common distribution P, defined on a probability space $(\Omega, \Sigma, \Pr)$. Consider the $n \times n$ random matrix $\tilde{H}_n$ with the entries $n^{-1}h(X_i, X_j)$, $1 \leq i, j \leq n$. It is easy to check that $\tilde{H}_n$ can be considered as a matrix of the operator (for which we use the same notation), defined by

$$\tilde{H}_n g(x) = \int_S h(x, y)g(y)P_n(dy), \ g \in L_2(P_n),$$

where P_n is the empirical measure based on a sample $(X_1, \ldots, X_n)$. Indeed, the map $g \mapsto n^{-1/2}\big(g(X_1(\omega)), \ldots, g(X_n(\omega))\big)$ defines a (random) isometry between

---

*)   Partially supported by a Sandia National Laboratories grant AU-4005

$L_2(P_n(\omega))$ and a subspace of $\mathbb{R}^n$. The matrix (in the canonical basis of $\mathbb{R}^n$) of the "image" of the operator $\tilde{H}_n$ under this isometry is exactly $\tilde{H}_n$. Clearly, the operator $\tilde{H}_n$ is a natural empirical version of H. Let H_n denote the matrix obtained from $\tilde{H}_n$, by changing its diagonal entries to 0. Note that the operator H and the distribution of the random operator H_n depend only on h up to equality a.s. $P \times P$, while $\tilde{H}_n$ depends also on the values of $h(x, x)$.

The main goal of this paper is to study how the spectral projections of the random matrices H_n and $\tilde{H}_n$ (considered as random operators from $\mathbb{R}^n$ into $\mathbb{R}^n$) are related to the spectral projections of H. We show that if H is a Hilbert-Schmidt operator, then the spectral projections of H_n converge (in a certain sense, described precisely below) a.s. to the spectral projections of H as $n \to \infty$. For $\tilde{H}_n$, the a.s. convergence also holds, but under stronger conditions. We also show that under much stronger conditions on the kernel h the fluctuations of the spectral projections of H_n and $\tilde{H}_n$ about the corresponding spectral projections of H are asymptotically Gaussian random operators.

It is worth noting that most of the previously known results about spectral properties of large random matrices dealt either with sample covariance matrices (see, e.g., Dauxois, Pousse and Romain (1982)), or with symmetric (hermitian) random matrices, having independent entries (as, e.g. in the famous Gaussian ensembles, introduced and studied by E. Wigner, see e.g. Mehta (1991) or Girko (1990)). These two classes of matrices are used in important problems of mathematical statistics and theoretical (especially, nuclear) physics. The random matrices discussed in this paper arise in a problem of Monte Carlo evaluation of the spectra of integral operators. They are also related to some problems of the asymptotic theory of U-statistics (see Giné and Zhang (1996)), but, maybe, most importantly, they provide a very simple model of a random perturbation of infinite dimensional operators (see, e.g., Simon (1979) for physically meaniful examples of random perturbations).

The matrices H_n and $\tilde{H}_n$ were introduced and the asymptotics of their spectra studied by Koltchinskii and Giné (1996). In the rest of the section, we briefly discuss some of their results. First, we introduce some notations. It is well known that any symmetric Hilbert-Schmidt kernel h has the following representation

$$h(x, y) = \sum_{i \in I} \lambda_i \phi_i(x) \phi_i(y), \tag{1.1}$$

where I is a countable or finite set, $\{\lambda_i, \ i \in I\}$ is the sequence of eigenvalues of h (with their multiplicities), $\{\phi_i, \ i \in I\}$ are the corresponding orthonormal eigenfunctions. The series (1.1) converges in the space $L_2(P \times P)$.

Regularly, the spectrum of H is defined as the set $\sigma(H)$ of all λ such that $H - \lambda I$ does not have a bounded inverse operator (here I is the identity operator). In the case of compact (in particular, Hilbert-Schmidt) operator, the spectrum is at most countable, and 0 is the only possible limit point of it. Any nonzero point of the spectrum is an eigenvalue of H. In this paper, though, it will be convenient to view the spectrum as a sequence $\{\lambda_i : i \in I\}$ of eigenvalues with their multiplicities rather than a set. Such a sequence will be considered either as a point of the space $c_0 = c_0(I)$ of all sequences tending to 0 with the sup-norm, or (which is especially natural in the case of Hilbert-Schmidt operators) as a point of the space $\ell_2 = \ell_2(I)$ of all square summable sequences with the standard ℓ_2-norm. In what follows we always assume that I is countable, extending the sequences, if necessary, by adding zeroes. This convention applies, in particular, to the spectra of finite matrices.

Two sequences $x := \{x_i : i \in I\}$ and $y := \{y_i : i \in I\}$ are called equivalent iff there is a bijection $\pi : I \mapsto I$ such that $y_i = x_{\pi(i)}$, $i \in I$. Often, it is convenient to identify equivalent sequences, i.e. to consider the spectrum of the operator H as a point of the quotient space of c_0 or ℓ_2 with respect to this equivalence. In such cases, we use the notation $\lambda(H)$ for the spectrum of H. To be more precise, such a spectrum will be referred to as the λ-spectrum of H.

Given two sequences $x, y \in \ell_2$, we define the following distance between x and y (in fact, rather between their equivalence classes):

$$\delta_2(x, y) := \inf_{\pi} \left[\sum (x_i - y_{\pi(i)})^2 \right]^{1/2},$$

where the infimum is taken over the set of all bijections π from $\mathbb{Z}$ onto $\mathbb{Z}$.

Given a $d \times d$ matrix A (which, of course, defines an operator on $\mathbb{R}^d$), denote by $\lambda^{\downarrow}(A)$ (resp. $\lambda^{\uparrow}(A)$) the vector of its eigenvalues (with their multiplicities) arranged in non-increasing (resp. non-decreasing) order.

The sign $\oplus$ will stand for the direct sum of vectors. Say, if

$$u = (u_1, \ldots, u_m) \in \mathbb{R}^m \text{ and } v = (v_1, \ldots, v_n) \in \mathbb{R}^n,$$

then

$$u \oplus v := (u_1, \ldots, u_m, v_1, \ldots, v_n) \in \mathbb{R}^{m+n}$$

with obvious extension to finite and infinite direct sums.

Often, it will be convenient to index the eigenvalues of H (and other compact operators), arranged in decreasing order, by all integers, the nonnegative eigenvalues being indexed by nonnegative integers, the nonpositive eigenvalues, respectively, by negative integers. If the positive (resp., the negative) part of the spectrum is finite, we extend it to the infinite sequence by adding zeroes. Thus, the vector of all eigenvalues of H (counted with their multiplicities) can be represented as

$$\lambda_0 \geq \lambda_1 \geq \lambda_2 \geq \ldots \geq 0 \geq \ldots \geq \lambda_{-2} \geq \lambda_{-1}.$$

The notation we use for such a spectrum is $\lambda^{\uparrow\downarrow}(H)$. More precisely, $\lambda^{\uparrow\downarrow}(H)$ is defined as follows:

- in the case when H has infinitely many positive and infinitely many negative eigenvalues, $\lambda^{\uparrow\downarrow}(H)$ is the direct sum of the sequence of positive eigenvalues, arranged in non-increasing order, and the sequence of negative eigenvalues, arranged in non-decreasing order;

- in the case when H has infinitely many positive and a finite number of negative eigenvalues, we first extend the vector of negative eigenvalues to the infinite sequence by adding zeroes. Then, $\lambda^{\uparrow\downarrow}(H)$ is the direct sum of the sequence of positive eigenvalues, arranged in non-increasing order, and the extended sequence of negative eigenvalues, arranged in non-decreasing order;

- similarly, in the case when H has a finite number of positive and infinitely many negative eigenvalues, we extend the vector of positive eigenvalues to the infinite sequence by adding zeroes. Then, $\lambda^{\uparrow\downarrow}(H)$ is the direct sum of the extended sequence of positive eigenvalues, arranged in non-increasing order, and the sequence of negative eigenvalues, arranged in non-decreasing order;

- finally, in the case when the numbers of positive and negative eigenvalues of H are finite (in particular, when H is finite dimensional), we extend both vectors of positive and negative eigenvalues to the infinite sequences by adding zeroes. Then, $\lambda^{\uparrow\downarrow}(H)$ is the direct sum of the extended sequences of positive eigenvalues, arranged in non-increasing order, and negative eigenvalues, arranged in non-decreasing order.

$\lambda^{\uparrow\downarrow}(H)$ will be considered as a vector in such spaces of two-sided sequences as $\ell^2(\mathbb{Z})$ or $c_0(\mathbb{Z})$ and referred to as the $\lambda^{\uparrow\downarrow}$-spectrum of H.

We denote by $\{\mu_r\}$ the ordered set of distinct nonzero eigenvalues of H,

$$\mu_0 > \mu_1 > \mu_2 > \ldots > 0 > \ldots > \mu_{-2} \geq \mu_{-1}.$$

Let m_r be the multiplicity of the eigenvalue μ_r, $r \in \mathbb{Z}$ (i.e. there are m_1 eigenvalues λ_i equal to μ_1, etc.).

With these notations and conventions, we formulate now the main results of the paper of Koltchinskii and Giné (1996). We start with their Strong Law of Large Numbers (SLLN) for the spectra.

1.1. THEOREM. *If* $\mathbb{E}h^2(X,Y) < +\infty$ *(i.e., H is a Hilbert-Schmidt operator), then*

$$\delta_2(\lambda(H_n), \lambda(H)) \to 0 \text{ as } n \to \infty \text{ a.s.}$$

If the sequence of the spectra $\lambda(H_n)$ is stochastically δ_2-bounded, then $\mathbb{E}h^2(X,Y) < +\infty$.

Koltchinskii and Giné (1996) also proved two versions of the central limit theorem (CLT). One of these versions gives the limit distributions of particular eigenvalues, another one demonstrates the convergence in the spaces of sequences.

Assume that the eigenfunctions of H are in $L_4(P)$: $\int_S \phi_i^4 dP < +\infty$, $i \in \mathbb{Z}$.

Let G_P denote the generalized P-Brownian bridge. This is the centered Gaussian process G_P indexed by functions $f \in L_2(P)$ with covariance

$$\mathbb{E}G_P(f)G_P(g) = \int_S fg \, dP - \int_S f \, dP \int_S g \, dP, \ f, g \in L_2(P).$$

The limit distribution of the spectra can be expressed in terms of the process G_P.

1.2. THEOREM. *Let h be a symmetric kernel in $L_2(P \times P)$ with the property that there exists a sequence $R_n \to \infty$ satisfying*

$$\sum_{|i|>R_n} \lambda_i^2 = o(n^{-1}) \tag{1.2}$$

and

$$\sum_{|i|\leq R_n,|j|\leq R_n} \int_S \phi_i^2 \phi_j^2 dP \sum_{|i|\leq R_n,|j|\leq R_n} (\lambda_i^2 + \lambda_j^2) \int_S \phi_i^2 \phi_j^2 dP = o(n). \tag{1.3}$$

Suppose that moreover

$$\sum_{i\in\mathbb{Z}} |\lambda_i|\phi_i^2 \in L_2(P). \tag{1.4}$$

Let Δ_r be the set of indices $i \in \mathbb{Z}$ such that $H\phi_i = \mu_r\phi_i$. Let $\Gamma_{P,r}$ be the Gaussian matrix

$$\Gamma_{P,r} := \mu_r \big(G_P(\phi_i\phi_j) : i,j \in \Delta_r \big), \ r \in \mathbb{Z}. \tag{1.5}$$

Then the finite dimensional distributions of the sequence

$$\left\{ n^{1/2}(\lambda^{\uparrow\downarrow}(H_n) - \lambda^{\uparrow\downarrow}(H)) \right\}_{n=1}^{\infty}$$

converge weakly to the finite dimensional distributions of $\oplus_{r\geq 0}\lambda^{\downarrow}(\Gamma_{P,r}) \oplus \oplus_{r<0}\lambda^{\uparrow}(\Gamma_{P,r})$.

If condition (1.4) is replaced by

$$\sum_{i\in\mathbb{Z}} \lambda_i \phi_i^2(x) \text{ converges to } h(x,x) \in L_2(P), \tag{1.4'}$$

then the statement holds with H_n replaced by $\tilde{H}_n$.

1.3. THEOREM. *Suppose that*

$$\sum_{i,j\in\mathbb{Z}} (\lambda_i^2 + \lambda_j^2) \int_S \phi_i^2 \phi_j^2 dP < +\infty, \tag{1.6}$$

and that, moreover, there exists a sequence $R_n \to \infty$ such that both

$$\sum_{|i|\leq R_n,|j|\leq R_n} \int_S \phi_i^2 \phi_j^2 dP = o(n) \tag{1.7}$$

and condition (1.2) hold. Suppose, in addition, that P-a.s.

$$h(x,x) = \sum_{i\in\mathbb{Z}} \lambda_i \phi_i^2(x). \tag{1.8}$$

Then the sequence

$$\left\{ n^{1/2} \big(\lambda^{\uparrow\downarrow}(\tilde{H}_n) - \lambda^{\uparrow\downarrow}(H) \big) \right\}_{n=1}^{\infty} \tag{1.9}$$

converges weakly in $\ell_2(\mathbb{Z})$ *to the random vector* $\oplus_{r\geq 0} \lambda^{\downarrow}(\Gamma_{P,r}) \oplus \oplus_{r<0} \lambda^{\uparrow}(\Gamma_{P,r})$.
If conditions (1.2), (1.6) and (1.7) hold, then the sequence of random vectors

$$\left\{ n^{1/2} \big(\lambda^{\uparrow\downarrow}(H_n) - \lambda^{\uparrow\downarrow}(H) \big) \right\}_{n=1}^{\infty}$$

converges weakly in $c_0(\mathbb{Z})$ *to the random vector* $\oplus_{r\geq 0} \lambda^{\downarrow}(\Gamma_{P,r}) \oplus \oplus_{r<0} \lambda^{\uparrow}(\Gamma_{P,r})$.

The conditions of Theorems 1.2 and 1.3 are rather restrictive. For many kernels these conditions do not hold and the rate of convergence of $\lambda(H_n)$ to $\lambda(H)$ is slower than $n^{-1/2}$. The following theorem of Koltchinskii and Giné (1996) describes the relationship between the spectral decomposition of the kernel h and the rate of convergence.

1.4. THEOREM. *Suppose that for some* $\alpha > 0$ *and* $\beta > 0$

$$\sum_{|i|\leq R, |j|\leq R} (\lambda_i^2 + \lambda_j^2) \int_S \phi_i^2 \phi_j^2 dP = O(R^\alpha) \text{ as } R \to \infty \tag{1.10}$$

and

$$\sum_{|i|>R} \lambda_i^2 = O(R^{-\beta}) \text{ as } R \to \infty. \tag{1.11}$$

Then

$$\mathbb{E}\delta_2^2(\lambda(H_n), \lambda(H)) = O(n^{-\frac{\beta}{\alpha+\beta}}) \text{ as } n \to \infty. \tag{1.12}$$

2. SLLN and CLT for Spectral Projections

The main goal of the paper is to study the limit behavior of spectral projections of the matrices $\tilde{H}_n$ and H_n as $n \to \infty$. We start with giving more precise definitions of the projections in question.

Given a compact linear operator A from a Hilbert space $\mathcal{H}$ into itself and $\varepsilon > 0$, consider a subset $\tilde{\Lambda}$ of the spectrum $\sigma(A)$ such that the diameter of $\tilde{\Lambda}$ is smaller than ε and $\tilde{\Lambda}$ is at a distance larger than ε from $\sigma(A) \setminus \tilde{\Lambda}$. Let us consider only the sets consisting either of positive, or of negative eigenvalues. Next we define Λ as a vector whose coordinates are the elements of the set $\tilde{\Lambda}$ each counted with its multiplicity, arranged in non-increasing order in the case of "positive" sets (resp., non-decreasing order for "negative" sets). We call Λ an ε-cluster of the spectrum of H. We use all integers (positive, negative, zero) to number such clusters. For $r \geq 0$ (resp. $r < 0$), an ε-cluster of an operator A is the r-th ε-cluster of C, $\Lambda_r^\varepsilon(C)$ if there are exactly r (resp. $-r-1$) ε-clusters of C whose components are larger (resp. smaller) than those of $\Lambda_r^\varepsilon(C)$. $\Lambda_r^\varepsilon(C)$ does not always exist.

Given $\varepsilon > 0$ and $r \in \mathbb{Z}$, denote by $P_r(C) = P_r^\varepsilon(C)$ the orthogonal projection on the linear subspace of $\mathcal{H}$, spanned by all eigenvectors, corresponding to the eigenvalues in the ε-cluster $\Lambda_r^\varepsilon(C)$.

Consider again the kernel h given by (1.1) with distinct eigenvalues $\{\mu_r\}$. Define

$$\delta_r = \frac{1}{2}\min\{\mu_i - \mu_{i+1} : 0 \le i \le r\} \text{ for } r > 0,$$

$$\delta_r = \frac{1}{2}\min\{\mu_{i-1} - \mu_i : r \le i \le -2\} \text{ for } r < 0. \qquad (2.1)$$

If $\varepsilon < 2\delta_r$, then $P_j(H) = P_j^\varepsilon(H)$ does not depend on ε (for all $0 \le j \le r$, $r > 0$, and $0 > j \ge r$, $r < 0$), and it is exactly the projection on the eigenspace of μ_j.

A difficulty one has to deal with in considering the asymptotic relationship between such operators as $P_r(\tilde{H}_n)$ and $P_r(H_n)$ on the one hand, and $P_r(H)$ on the other hand, is that these operators act in different spaces (the first two in $L_2(P_n)$, the last one in $L_2(P)$). Due to this reason, we are going to study, in fact, the convergence of the bilinear forms generated by these operators.

Given a function $f \in L_2(P)$, we denote by $\tilde{f} = \tilde{f}_\omega$ the restriction of f to the sample $(X_1(\omega), \ldots, X_n(\omega))$ (considered as a function in the space $L_2(P_n(\omega))$.)

Let $\mathcal{F}$ be a class of measurable functions on $(S, \mathcal{S})$. We assume that the reader is familiar with the notions of P-Glivenko-Cantelli (the notation $\mathcal{F} \in GC(P)$) and P-Donsker classes (the notation $\mathcal{F} \in CLT(P)$), see, e.g., Dudley (1984) or van der Vaart and Wellner (1996).

First we formulate the SLLN for spectral projections.

2.1. THEOREM. *Let h be a symmetric kernel such that $\mathbb{E}h^2(X, Y) < +\infty$ (so the operator H is Hilbert-Schmidt). Suppose that $\mathcal{F}$ is a class of measurable functions on $(S, \mathcal{S})$ with a square integrable envelope F, i.e. $|f(x)| \le F(x)$ for all $x \in S$, $f \in \mathcal{F}$ and $F \in L_2(P)$. Moreover, suppose that for all $i \in \mathbb{Z}$*

$$\mathcal{F}\phi_i := \{f\phi_i : f \in \mathcal{F}\} \in GC(P).$$

Then, for all $r \in \mathbb{Z}$ with $\mu_r \ne 0$ and for all $\varepsilon < \delta_r/2$,

$$\sup_{f,g \in \mathcal{F}} \left| \left\langle P_r^\varepsilon(H_n)\tilde{f}, \tilde{g} \right\rangle_{L_2(P_n)} - \left\langle P_r(H)f, g \right\rangle_{L_2(P)} \right| \to 0 \text{ as } n \to \infty \text{ a.s.} \qquad (2.2)$$

If, in addition, $\mathbb{E}|h(X, X)| < +\infty$, then

$$\sup_{f,g \in \mathcal{F}} \left| \left\langle P_r^\varepsilon(\tilde{H}_n)\tilde{f}, \tilde{g} \right\rangle_{L_2(P_n)} - \left\langle P_r(H)f, g \right\rangle_{L_2(P)} \right| \to 0 \text{ as } n \to \infty \text{ a.s.} \qquad (2.3)$$

Next we consider a version of the CLT for spectral projections. In fact, we show that the limit distribution of the processes

$$n^{1/2}\left[\left\langle P_r^\varepsilon(\tilde{H}_n)\tilde{f}, \tilde{g} \right\rangle_{L_2(P_n)} - \left\langle P_r(H)f, g \right\rangle_{L_2(P)} \right], \quad f, g \in \mathcal{F}$$

and

$$n^{1/2}\left[\left\langle P_r^\varepsilon(H_n)\tilde{f}, \tilde{g} \right\rangle_{L_2(P_n)} - \left\langle P_r(H)f, g \right\rangle_{L_2(P)} \right], \quad f, g \in \mathcal{F}$$

can be expressed in terms of P-Brownian bridge.

Recall that Δ_r is the set of all $i \in \mathbb{Z}$ such that $H\phi_i = \mu_r\phi_i$. Given $r \in \mathbb{Z}$, denote by $\Gamma_P^{(r)}$ the infinite random Gaussian matrix with the following entries:

$$\left(\Gamma_P^{(r)}\right)_{i,j} := \begin{cases} \frac{\lambda_j}{\mu_r - \lambda_j} G_P(\phi_i\phi_j) & \text{for } i \in \Delta_r, j \notin \Delta_r \\ \frac{\lambda_i}{\mu_r - \lambda_i} G_P(\phi_i\phi_j) & \text{for } i \notin \Delta_r, j \in \Delta_r \\ 0 & \text{otherwise,} \end{cases} \tag{2.4}$$

It's easy to see that under condition (2.6) below the relationship

$$\Gamma_P^{(r)} f := \sum_{i,j\in\mathbb{Z}} \left(\Gamma_P^{(r)}\right)_{i,j} \langle f, \phi_i \rangle_{L_2(P)} \phi_j \tag{2.5}$$

defines a Gaussian random Hilbert-Schmidt operator (which we also denote by $\Gamma_P^{(r)}$) on the space $L_2(P)$. Indeed, we have

$$\mathbb{E} \sum_{i,j\in\mathbb{Z}} \langle \Gamma_P^{(r)}\phi_i, \phi_j \rangle_{L_2(P)}^2 = \mathbb{E} \sum_{i,j\in\mathbb{Z}} \left(\Gamma_P^{(r)}\right)_{i,j}^2$$

$$= \sum_{i\in\Delta_r, j\notin\Delta_r} \left(\frac{\lambda_j}{\mu_r - \lambda_j}\right)^2 \mathbb{E}G_P^2(\phi_i\phi_j) + \sum_{i\notin\Delta_r, \in\Delta_r} \left(\frac{\lambda_i}{\mu_r - \lambda_i}\right)^2 \mathbb{E}G_P^2(\phi_i\phi_j)$$

$$\leq \sum_{i\in\Delta_r, j\notin\Delta_r} \left(\frac{\lambda_j}{\mu_r - \lambda_j}\right)^2 \int_S \phi_i^2\phi_j^2 dP + \sum_{i\notin\Delta_r, \in\Delta_r} \left(\frac{\lambda_i}{\mu_r - \lambda_i}\right)^2 \int_S \phi_i^2\phi_j^2 dP$$

$$\leq \frac{1}{2\delta_r^2} \sum_{i\in\Delta_r, j\notin\Delta_r} \lambda_j^2 \int_S \phi_i^2\phi_j^2 dP < +\infty,$$

which implies the statement.

In the next theorem we use the notion of weak convergence in the space $\ell^\infty(\mathcal{F} \times \mathcal{F})$ (see, e.g., van der Vaart and Wellner (1996)).

Denote

$$\zeta_n^2(R) := \zeta_n^2(\mathcal{F}; R) := \sup_{f\in\mathcal{F}} \sum_{|j|\leq R} \left[n^{1/2} \int_S f\phi_j d(P_n - P) \right]^2.$$

2.2. THEOREM. *Suppose there exists a sequence $R_n \to \infty$ such that conditions*

$$\sum_{|i|>R_n} \lambda_i^2 = o(n^{-1}) \tag{1.2}$$

and

$$\sum_{|i|\leq R_n, |j|\leq R_n} \int_S \phi_i^2\phi_j^2 dP \sum_{|i|\leq R_n, |j|\leq R_n} (\lambda_i^2 + \lambda_j^2) \int_S \phi_i^2\phi_j^2 dP = o(n) \tag{1.3}$$

hold. Let $r \in \mathbb{Z}$ be such that $\mu_r \neq 0$. Suppose that for all $i \in \Delta_r$

$$\sum_{j\notin\Delta_r} \lambda_j^2 \int_S \phi_i^2\phi_j^2 dP < +\infty. \tag{2.6}$$

Suppose also that $\mathcal{F}$ is a class of measurable functions on $(S, \mathcal{S})$ with a square integrable envelope F, that for all $i \in \mathbb{Z}$ $\mathcal{F}\phi_i \in CLT(P)$, and, moreover, that

$$\zeta_n^2(\mathcal{F}; R_n) = o_P(n) \text{ as } n \to \infty. \tag{2.7}$$

If condition

$$\sum_{i \in \mathbb{Z}} \lambda_i \phi_i^2(x) \text{ converges to } h(x, x) \in L_2(P) \tag{1.4'}$$

holds, then, for all $\varepsilon < \delta_r/2$, the sequence

$$n^{1/2}\left[\langle P_r^\varepsilon(\tilde{H}_n)\tilde{f}, \tilde{g}\rangle_{L_2(P_n)} - \langle P_r(H)f, g\rangle_{L_2(P)}\right], \ f, g \in \mathcal{F}$$

converges weakly in the space $\ell^\infty(\mathcal{F} \times \mathcal{F})$ to the continuous Gaussian process

$$G_P(fP_r(H)g) + G_P(gP_r(H)f) + \langle \Gamma_P^{(r)}f, g\rangle_{L_2(P)}, \ f, g \in \mathcal{F}.$$

If, instead of (1.4'), condition

$$\sum_{i \in \mathbb{Z}} |\lambda_i| \phi_i^2 \in L_2(P) \tag{1.4}$$

holds, then the conclusion holds with matrix $\tilde{H}_n$ replaced by H_n.

As in the case of asymptotics of the spectra, the conditions of the CLT for spectral projections are rather restrictive and do not hold for typical kernels, like Green functions and heat kernels of differential operators (see Koltchinskii and Giné (1996)). In such cases one can expect slower rates of convergence for the spectral projections than the CLT-rate $n^{-1/2}$. The following theorem is a result in this direction.

2.3. THEOREM. *Suppose that for some $\alpha > 0$ and $\beta > 0$*

$$\sum_{|i| \leq R, |j| \leq R} \int_S \phi_i^2 \phi_j^2 dP = O(R^\alpha) \text{ as } R \to \infty \tag{2.8}$$

and

$$\sum_{|i| > R} \lambda_i^2 = O(R^{-\beta}) \text{ as } R \to \infty. \tag{2.9}$$

Suppose also that $\mathcal{F}$ is a class of measurable functions on $(S, \mathcal{S})$ with an envelope F in $L_4(P)$, and, moreover, $\mathcal{F}\phi_i \in CLT(P)$ for all $i \in \mathbb{Z}$ and

$$\sup_{n \geq 1} \mathbb{E}\zeta_n^2(\mathcal{F}; R) = O(R^\alpha) \text{ as } R \to \infty. \tag{2.10}$$

Then, for all $\varepsilon < \delta_r/2$

$$\mathbb{E}\sup_{f, g \in \mathcal{F}}\left|\langle P_r^\varepsilon(H_n)\tilde{f}, \tilde{g}\rangle_{L_2(P_n)} - \langle P_r(H)f, g\rangle_{L_2(P)}\right| = O(n^{-\frac{\beta}{2(\alpha+\beta)}}) \text{ as } n \to \infty. \tag{2.11}$$

If, in addition to (2.8)–(2.10), the condition

$$\sum_{i \in \mathbb{Z}} \lambda_i \phi_i^2(x) \text{ converges to } h(x,x) \in L_2(P) \tag{1.4'}$$

holds, then

$$\mathbb{E} \sup_{f,g \in \mathcal{F}} \left| \langle P_r^\varepsilon(\tilde{H}_n)\tilde{f}, \tilde{g} \rangle_{L_2(P_n)} - \langle P_r(H)f, g \rangle_{L_2(P)} \right| = O(n^{-\frac{\beta}{2(\alpha+\beta)}}) \text{ as } n \to \infty. \tag{2.12}$$

2.4. EXAMPLE. Let $d_{Q,2}$ be the metric of the space $L_2(Q)$, Q being a probability measure on $(S, \mathcal{S})$. Given a class $\mathcal{G} \subset L_2(Q)$, $H_{d_{Q,2}}(\mathcal{G}; \varepsilon)$ denotes the ε-entropy of $\mathcal{G}$ with respect to the metric $d_{Q,2}$. Define

$$J(\delta; \mathcal{G}) := \sup_Q \int_0^\delta \sqrt{1 + H_{d_{Q,2}}(\mathcal{G}; \varepsilon \|G\|_{L_2(Q)})} \, d\varepsilon,$$

where the supremum is taken over all probability measures Q with finite support and G denotes a measurable envelope of $\mathcal{G}$. It is well known (see, e.g., van der Vaart and Wellner (1996), p. 239) that

$$\mathbb{E} \sup_{g \in \mathcal{G}} \left| n^{1/2} \int_S g \, d(P_n - P) \right|^2 \leq C \, J^2(1; \mathcal{G}) \, \|G\|_{L_2(P)}^2.$$

Now suppose that our class $\mathcal{F}$ satisfies the uniform entropy condition

$$\int_0^{+\infty} \sup_Q H_{d_{Q,2}}^{1/2}(\mathcal{F}; \varepsilon \|F\|_{L_2(Q)}) \, d\varepsilon < +\infty,$$

which implies that $J(1; \mathcal{F}) < +\infty$ (in particular, this holds for VC-subgraph classes of functions). Let $\phi = \phi_j$ for some $j \in \mathbb{Z}$. Given a probability measure Q (with finite support), define

$$\hat{Q}(A) = \frac{\int_A \phi^2 \, dQ}{\int_S \phi^2 \, dQ}, \quad A \in \mathcal{S}.$$

For an envelope $F|\phi|$ of the class $\mathcal{F}\phi$, we have

$$\|F|\phi|\|_{L_2(Q)} = \|F\|_{L_2(\hat{Q})} \|\phi\|_{L_2(Q)}.$$

We also have

$$d_{Q,2}(f\phi, g\phi) = d_{\hat{Q},2}(f,g) \, \|\phi\|_{L_2(Q)},$$

which implies

$$H_{d_{Q,2}}(\mathcal{F}\phi; \varepsilon) = H_{d_{\hat{Q},2}}\left(\mathcal{F}; \frac{\varepsilon}{\|\phi\|_{L_2(Q)}}\right), \quad \varepsilon > 0.$$

It easily follows that

$$J(1; \mathcal{F}\phi) = \sup_Q \int_0^1 \sqrt{1 + H_{d_{Q,2}}\big(\mathcal{F}\phi; \varepsilon \|F|\phi|\|_{L_2(Q)}\big)} d\varepsilon$$

$$\leq \sup_{\hat{Q}} \int_0^1 \sqrt{1 + H_{d_{\hat{Q},2}}\big(\mathcal{F}; \varepsilon \|F\|_{L_2(\hat{Q})}\big)} d\varepsilon$$

$$= J(1; \mathcal{F}).$$

Therefore we have the following bound

$$\mathbb{E}\sup_{f \in \mathcal{G}}\Big|n^{1/2}\int_S f\phi d(P_n - P)\Big|^2 \leq C\ J^2(1; \mathcal{F})\ \|F\phi\|^2_{L_2(P)},$$

and condition (2.10) in this case can be represented as follows:

$$\sum_{|j| \leq R} \|F\phi_j\|^2_{L_2(P)} = O(R^\alpha).$$

Another possibility to check conditions like (2.7) or (2.10) is to use entropies with bracketing (see, e.g., Dudley (1984) or van der Vaart and Wellner (1996)).

2.5. EXAMPLE. Let $S = G$ be a bounded open connected subset of $\mathbb{R}^d$, and let P be the uniform distribution in G. Consider a strictly elliptic differential operator

$$L := -\sum_{i,j=1}^d \frac{\partial}{\partial x_i}\Big\{a_{ij}(x)\frac{\partial}{\partial x_j}\Big\},$$

where $x \mapsto a(x) := \big(a_{ij}(x)\big)_{i,j=1}^d$ is a locally integrable function in G with values in the set of nonnegative real symmetric matrices. Moreover, for some positive constants $c < C$, we have $c \leq a(x) \leq C$, $x \in G$. We consider the operator L with Dirichlet boundary condition. Denote by h_t the heat kernel of L at $t > 0$, i.e. h_t is the kernel of the semigroup $e^{-Lt}, t > 0$ (see, Davies (1989)). Suppose that for some fixed $t > 0$ $h \equiv h_t$. Using the methods developed in Koltchinskii and Giné (1996) and the bounds for heat kernels obtained in the book of Davies (1989), one can check the conditions of Theorem 2.3. In this case, Theorem 2.3 gives the rate of convergence for spectral projections of the order $o(n^{-1/2+\delta})$ for any $\delta > 0$. If, instead of the heat kernel, one takes h to be the kernel of $(L + sI)^{-\gamma}$, for some $s > 0$ and $\gamma > 0$ (a "Green function" of L), then the rate given by Theorem 2.3 is going to be

$$O\big(n^{-\frac{1}{2}\frac{4\gamma-d}{4\gamma+d}}\big)$$

for each $\gamma > d/4$. Note that in this case Theorem 1.4 gives the same rate of convergence for the spectra (Koltchinskii and Giné (1996)). Similar results can be obtained for the kernels of functions of some other differential operators (like, e.g., Schrödinger operators, or Laplace-Beltrami operators on compact Riemannian manifolds).

The main ingredient of the proofs of the theorems is an isometric representation (developed in Koltchinskii and Giné (1996)) of the operators $\tilde{H}_n$ on the space $L_2(P)$ in the case of kernels of finite rank, which we consider now.

CONSTRUCTION OF ISOMETRY FOR KERNELS OF FINITE RANK. Suppose that for some fixed $R < +\infty$

$$h(x,y) := h_R(x,y) := \sum_{j=1}^{R} \lambda_j \phi_j(x)\phi_j(y).$$

Let $\mathcal{L} := \mathcal{L}_R$ be the linear span of functions $\{\phi_j : j = 1,\ldots,R\}$ in $L_2(P)$. Denote by $H := H_R$ the restriction of the integral operator with kernel h in $L_2(P)$ to the space $\mathcal{L}$.

Next we define a matrix

$$E_n := E_{n,R} := \left(\int_S \phi_i \phi_j d(P_n - P) : 1 \le i,j \le R \right),$$

and consider a linear operator from $\mathcal{L}$ into $\mathcal{L}$ whose matrix in the basis $\{\phi_j : j = 1,\ldots,R\}$ is E_n. We use the same notation E_n (or $E_{n,R}$) for this operator. Let $I = I_R$ denote the identity operator on $\mathcal{L}$. Given $f \in \mathcal{L}$, we have

$$\langle (I + E_n)f, f \rangle_{L_2(P)} = \sum_{i,j=1}^{R} \int_S \phi_i \phi_j dP_n \langle f, \phi_i \rangle_{L_2(P)} \langle f, \phi_j \rangle_{L_2(P)}$$

$$= \int_S \left(\sum_{j=1}^{R} \langle f, \phi_j \rangle_{L_2(P)} \phi_j \right)^2 dP_n \ge 0,$$

showing that $I + E_n$ is a nonnegatively definite operator. Let $A_n := A_{n,R} := (I + E_n)^{1/2}$. Given $f \in \mathcal{L}$, we define $\hat{f} := A_n f$, and show that the map $\tilde{f} \mapsto \hat{f}$ is an isometry of the linear space $\tilde{\mathcal{L}} := \{\tilde{f} : f \in \mathcal{L}\} \subset L_2(P_n)$ and the linear space $\hat{\mathcal{L}} := (I + E_n)\mathcal{L} \subset \mathcal{L} \subset L_2(P)$. Indeed,

$$\langle \hat{\phi}_i, \hat{\phi}_j \rangle_{L_2(P)} = \langle A_n \phi_i, A_n \phi_j \rangle_{L_2(P)} = \langle A_n^2 \phi_i, \phi_j \rangle_{L_2(P)}$$

$$= \langle (I + E_n)\phi_i, \phi_j \rangle_{L_2(P)} = \int_S \phi_i \phi_j dP_n = \langle \tilde{\phi}_i, \tilde{\phi}_j \rangle_{L_2(P_n)}.$$

Thus, the correspondence

$$\tilde{\phi}_j \leftrightarrow \hat{\phi}_j, \ j = 1,\ldots,R$$

defines an isometry of the linear spans of $\{\tilde{\phi}_j : j = 1,\ldots,R\}$ and $\{\hat{\phi}_j : j = 1,\ldots,R\}$, which are $\tilde{\mathcal{L}}$ and $\hat{\mathcal{L}}$, respectively.

The operator $\hat{H}_n := \hat{H}_{n,R} = A_n H A_n$ corresponds under the isometry to the operator $\tilde{H}_n = \tilde{H}_{n,R}$. Indeed, we have

$$\tilde{H}_n \tilde{f} = \sum_{j=1}^{R} \lambda_j \langle \tilde{f}, \tilde{\phi}_j \rangle_{L_2(P_n)} \tilde{\phi}_j$$

and

$$\hat{H}_n\hat{f} = A_n H A_n\hat{f} = A_n \sum_{j=1}^{R} \lambda_j \langle A_n\hat{f}, \phi_j \rangle_{L_2(P)} \phi_j$$

$$= \sum_{j=1}^{R} \lambda_j \langle \hat{f}, A_n\phi_j \rangle_{L_2(P)} A_n\phi_j = \sum_{j=1}^{R} \lambda_j \langle \hat{f}, \tilde{\phi}_j \rangle_{L_2(P)} \tilde{\phi}_j,$$

proving the statement, since, by the isometry,

$$\langle \hat{f}, \hat{\phi}_j \rangle_{L_2(P)} = \langle \tilde{f}, \tilde{\phi}_j \rangle_{L_2(P_n)}.$$

It follows from the above observations that for all r and for all $f, g \in \mathcal{L}$

$$\langle P_r(\tilde{H}_n)\tilde{f}, \tilde{g} \rangle_{L_2(P_n)} = \langle P_r(\hat{H}_n)\hat{f}, \hat{g} \rangle_{L_2(P)} = \langle A_n P_r(\hat{H}_n) A_n f, g \rangle_{L_2(P)}. \quad (2.13)$$

This relationship is crucial for the following investigation of the asymptotics of $P_r(\tilde{H}_n)$, since it allows (in the finite rank case) one to investigate instead the asymptotics of the operator $A_n P_r(\hat{H}_n) A_n$, which is a small random perturbation of the projection $P_r(H)$ in the Hilbert space $L_2(P)$. Now we are going to modify (2.13) slightly and to extend it to the case of any functions $\tilde{f}, \tilde{g} \in L_2(P_n)$ (not necessarily from $\tilde{\mathcal{L}}$.) To this end, we note that the operator $\tilde{H}_n$ is equal to 0 on the orthocomplement of the subspace $\tilde{\mathcal{L}}$. Thus

$$\langle P_r(\tilde{H}_n)\tilde{f}, \tilde{g} \rangle_{L_2(P_n)} = \langle P_r(\tilde{H}_n) P_{\tilde{\mathcal{L}}}\tilde{f}, P_{\tilde{\mathcal{L}}}\tilde{g} \rangle_{L_2(P_n)}, \quad \tilde{f}, \tilde{g} \in L_2(P_n), \quad (2.14)$$

where $P_{\tilde{\mathcal{L}}}$ denotes the orthoprojection on $\tilde{\mathcal{L}}$. Given $\tilde{f} \in L_2(P_n)$, $P_{\tilde{\mathcal{L}}}f$ is the point of $\tilde{\mathcal{L}}$ such that

$$\tilde{f} - P_{\tilde{\mathcal{L}}}\tilde{f} \perp \tilde{\phi}_j, \ j = 1, \ldots, R \text{ in } L_2(P_n). \quad (2.15)$$

Suppose that

$$P_{\tilde{\mathcal{L}}}\tilde{f} = \sum_{i=1}^{R} c_i \tilde{\phi}_i.$$

Then (2.15) yields

$$\sum_{i=1}^{R} c_i \langle \tilde{\phi}_i, \tilde{\phi}_j \rangle_{L_2(P_n)} = \langle \tilde{f}, \tilde{\phi}_j \rangle_{L_2(P_n)}, \ j = 1, \ldots, R. \quad (2.16)$$

Define

$$T_n f := T_{n,R} f := \sum_{j=1}^{R} \langle \tilde{f}, \tilde{\phi}_j \rangle_{L_2(P_n)} \phi_j \in \mathcal{L}.$$

Note that T_n coincides with $I + E_n$ on $\mathcal{L}$. Using (2.16) and the definition of the operator E_n, we get

$$T_n f = \sum_{j=1}^{R} \sum_{i=1}^{R} c_i \langle \tilde{\phi}_i, \tilde{\phi}_j \rangle_{L_2(P_n)} \phi_j = \sum_{j=1}^{R} \sum_{i=1}^{R} c_i \langle (I + E_n)\phi_i, \phi_j \rangle_{L_2(P)} \phi_j$$

$$= \sum_{j=1}^{R} \langle (I + E_n) \sum_{i=1}^{R} c_i \phi_i, \phi_j \rangle_{L_2(P)} \phi_j = (I + E_n) \sum_{i=1}^{R} c_i \phi_i.$$

Assuming now that $I + E_n$ is invertible (recall that it is nonnegatively definite), we find that

$$\sum_{i=1}^{R} c_i \phi_i = (I + E_n)^{-1} T_n f$$

and

$$P_{\tilde{\mathcal{L}}} \tilde{f} = \widetilde{(I + E_n)^{-1} T_n f}.$$

Thus, equations (2.13) and (2.14) imply for all $\tilde{f}, \tilde{g} \in L_2(P_n)$

$$\begin{aligned}
\left\langle P_r(\tilde{H}_n)\tilde{f}, \tilde{g} \right\rangle_{L_2(P_n)} &= \left\langle A_n P_r(\hat{H}_n) A_n (I + E_n)^{-1} T_n f, (I + E_n)^{-1} T_n g \right\rangle_{L_2(P)} \\
&= \left\langle A_n^{-1} P_r(\hat{H}_n) A_n^{-1} T_n f, T_n g \right\rangle_{L_2(P)}.
\end{aligned}$$

$$(2.17)$$

3. Proofs of the Main Results

We use the notations of Sections 1 and 2. For the sake of simplicity, we assume that *all eigenvalues of H are nonnegative.* This allows us to deal with one-sided sums, like e.g.

$$h(x, y) = \sum_{j=1}^{\infty} \lambda_j \phi_j(x)\phi_j(y).$$

The extension to the general case is trivial. In what follows, $\|\cdot\|$ will denote the operator norm and $\|\cdot\|_{HS}$ the Hilbert-Schmidt norm. The facts of perturbation theory used in the proofs can be found in the Appendix. Given $r \geq 1$, we fix $\varepsilon < \delta_r/2$. This ε will be skipped in notations like $P_r^\varepsilon(H) = P_r(H), P_r^\varepsilon(H_n) = P_r(H_n)$, etc.

PROOF OF THEOREM 2.1. Let $R < +\infty$ be a fixed number. Since for all $i, j = 1, \dots, R$ $\phi_i, \phi_j \in L_2(P)$, we have $\phi_i \phi_j \in L_1(P)$. By the strong law of large numbers,

$$\int_S \phi_i \phi_j d(P_n - P) \to 0 \text{ as } n \to \infty \text{ a.s. for all } i, j = 1, \dots, R,$$

which implies $\|E_{n,R}\| \to 0$ a.s. as $n \to \infty$. Since $\lambda(E_{n,R}) \subset [-1, +\infty)$ (recall that $I_R + E_{n,R}$ is nonnegatively definite) and

$$|(1 + \lambda)^{1/2} - 1| \leq |\lambda|, \ \lambda \geq -1$$

we have

$$\|A_{n,R} - I_R\| = \|(I_R + E_{n,R})^{1/2} - I_R\| \leq \|E_{n,R}\|,$$

and it follows that $A_{n,R} \to I_R$ as $n \to \infty$ in the operator norm a.s. Since the mapping $A \mapsto A^{-1}$ is continuous at $A = I_R$, we also have

$$A_{n,R}^{-1} \to I_R \text{ as } n \to \infty \text{ a.s..}$$

By continuity of the product of operators,

$$\hat{H}_{n,R} = A_{n,R} H_R A_{n,R} \to H_R \text{ as } n \to \infty \text{ a.s.},$$

and by continuity of orthogonal eigenprojections as functions of an operator (see Lemma A.1, iii)), we get (again, in the operator norm)

$$P_r(\hat{H}_{n,R}) \to P_r(H_R) \text{ as } n \to \infty \text{ a.s..}$$

Again, applying the continuity of the product,

$$A_{n,R}^{-1} P_r(\hat{H}_{n,R}) A_{n,R}^{-1} \to P_r(H_R) \text{ as } n \to \infty \text{ in the operator norm a.s..} \quad (3.1)$$

Recall the definition of $T_{n,R}$ (at the end of Section 2) and define

$$T_R f := \sum_{i=1}^{R} \langle f, \phi_i \rangle_{L_2(P)} \phi_i, \ f \in L_2(P).$$

Since $\mathcal{F}\phi_i \in GC(P)$, we have

$$\sup_{f \in \mathcal{F}} \|T_{n,R} f - T_R f\|_{L_2(P)}^2 = \sup_{f \in \mathcal{F}} \sum_{i=1}^{R} \left[\int_S f \phi_i d(P_n - P) \right]^2$$

$$\leq \sum_{i=1}^{R} \sup_{f \in \mathcal{F}} \left[\int_S f \phi_i d(P_n - P) \right]^2 \to 0 \text{ as } n \to \infty \text{ a.s.} \quad (3.2)$$

Suppose that R is large enough, so that $\Delta_r \subset \{1, \ldots, R\}$. Then $P_r(H_R) T_R f = P_r(H_R) f$. Now we use the basic isometry relationship (2.17) along with (3.1) and (3.2) to obtain (via a simple continuity argument)

$$\sup_{f,g \in \mathcal{F}} \left| \langle P_r(\tilde{H}_{n,R}) \tilde{f}, \tilde{g} \rangle_{L_2(P_n)} - \langle P_r(H_R) f, g \rangle_{L_2(P)} \right|$$

$$= \sup_{f,g \in \mathcal{F}} \left| \langle A_{n,R}^{-1} P_r(\hat{H}_{n,R}) A_{n,R}^{-1} T_{n,R} f, T_{n,R} g \rangle_{L_2(P)} \right. \quad (3.3)$$

$$\left. - \langle P_r(H_R) T_R f, T_R g \rangle_{L_2(P)} \right| \to 0 \text{ as } n \to \infty \text{ a.s..}$$

Next we replace $\tilde{H}_{n,R}$ by $H_{n,R}$. To do this, first note that, by isometry, $\lambda(\tilde{H}_{n,R}) = \lambda(\hat{H}_{n,R})$. Since $\hat{H}_{n,R}$ converges in the Hilbert-Schmidt norm to H_R a.s. as $n \to \infty$, we also have, by Lidskii's inequality (see the Appendix), that

$$\delta_2\big((\lambda(\tilde{H}_{n,R}), \lambda(H_R)\big) \to 0 \text{ as } n \to \infty \text{ a.s.} \quad (3.4)$$

Therefore, for $r = 1, \ldots, R$, the δ_r-numbers of $\tilde{H}_{n,R}$ (defined in Lemma A.1, in the Appendix) tend to δ_r as $n \to \infty$ a.s. Since

$$\mathbb{E}|h_R(X_i, X_i)| \leq \sum_{j=1}^{R} |\lambda_j| \, \mathbb{E}\phi_j^2(X_i) = \sum_{j=1}^{R} |\lambda_j| < +\infty,$$

we get

$$\|\tilde{H}_{n,R} - H_{n,R}\| = \frac{1}{n} \max_{1 \le i \le n} h_R(X_i, X_i) \to 0 \text{ as } n \to \infty \text{ a.s.,} \qquad (3.5)$$

which, by Lemma A.1, iii), implies that

$$\|P_r(\tilde{H}_{n,R}) - P_r(H_{n,R})\| \to 0 \text{ as } n \to \infty \text{ a.s..}$$

The class $\mathcal{F}$ has a square integrable envelope F. Thus

$$\limsup_{n \to \infty} \sup_{f \in \mathcal{F}} \|\tilde{f}\|_{L_2(P_n)}^2 = \limsup_{n \to \infty} \sup_{f \in \mathcal{F}} \int_S f^2 dP_n \le \lim_{n \to \infty} \int_S F^2 dP_n = \int_S F^2 dP \text{ a.s.} \qquad (3.6)$$

Therefore

$$\limsup_{n \to \infty} \sup_{f,g \in \mathcal{F}} \left| \left\langle P_r(\tilde{H}_{n,R})\tilde{f}, \tilde{g} \right\rangle_{L_2(P_n)} - \left\langle P_r(H_{n,R})\tilde{f}, \tilde{g} \right\rangle_{L_2(P_n)} \right|$$

$$\le \limsup_{n \to \infty} \|P_r(\tilde{H}_{n,R}) - P_r(H_{n,R})\| \sup_{f \in \mathcal{F}} \|\tilde{f}\|_{L_2(P_n)} \sup_{g \in \mathcal{F}} \|\tilde{g}\|_{L_2(P_n)} = 0 \text{ a.s.,}$$

which along with (3.3) implies

$$\sup_{f,g \in \mathcal{F}} \left| \left\langle P_r(H_{n,R})\tilde{f}, \tilde{g} \right\rangle_{L_2(P_n)} - \left\langle P_r(H_R)f, g \right\rangle_{L_2(P)} \right| \to 0 \text{ as } n \to \infty \text{ a.s.} \qquad (3.7)$$

Note that $P_r(H_R) = P_r(H)$ for a fixed r and large enough R (so that $\Delta_r \subset \{1, \ldots R\}$.) Thus

$$\sup_{f,g \in \mathcal{F}} \left| \left\langle P_r(H)f, g \right\rangle_{L_2(P)} - \left\langle P_r(H_R)f, g \right\rangle_{L_2(P)} \right| \to 0 \text{ as } R \to \infty \text{ a.s..} \qquad (3.8)$$

We also have

$$\sup_{f,g \in \mathcal{F}} \left| \left\langle P_r(H_n)f, \tilde{g} \right\rangle_{L_2(P)} - \left\langle P_r(H_{n,R})\tilde{f}, \tilde{g} \right\rangle_{L_2(P_n)} \right|$$

$$\le \|P_r(H_n) - P_r(H_{n,R})\| \sup_{f \in \mathcal{F}} \|\tilde{f}\|_{L_2(P_n)} \sup_{g \in \mathcal{F}} \|\tilde{g}\|_{L_2(P_n)}. \qquad (3.9)$$

By (3.5) and Lidskii's inequality we have

$$\delta_2\big((\lambda(\tilde{H}_{n,R}), \lambda(H_{n,R})\big) \to 0 \text{ as } n \to \infty \text{ a.s.,}$$

which, in view of (3.4), implies

$$\delta_2\big((\lambda(H_{n,R}), \lambda(H_R)\big) \to 0 \text{ as } n \to \infty \text{ a.s..}$$

In particular, this means that, for $r = 1, \ldots, R$, the δ_r-numbers of $H_{n,R}$ converge to δ_r as $n \to \infty$ a.s. Using the law of large numbers for U-statistics, we get

$$\|H_n - H_{n,R}\|_{HS}^2 = n^{-2} \sum_{1 \le i \ne j \le n} (h - h_R)^2(X_i, X_j) \to$$

$$\mathbb{E}(h - h_R)^2(X, Y) = \sum_{r=R+1}^{\infty} \lambda_r^2 \text{ a.s.} \qquad (3.10)$$

Since h is a Hilbert-Schmidt kernel, we also have $\sum_{r=R+1}^{\infty} \lambda_r^2 \to 0$ as $R \to \infty$. Application of Lemma A.1, iii) now yields

$$\lim_{R \to \infty} \limsup_{n \to \infty} \|P_r(H_n) - P_r(H_{n,R})\| = 0 \text{ a.s.}$$

Therefore, (3.6) and (3.9) imply

$$\lim_{R \to \infty} \limsup_{n \to \infty} \sup_{f,g \in \mathcal{F}} \left| \langle P_r(H_n)\tilde{f}, \tilde{g} \rangle_{L_2(P)} - \langle P_r(H_{n,R})\tilde{f}, \tilde{g} \rangle_{L_2(P_n)} \right|$$

$$\leq \lim_{R \to \infty} \limsup_{n \to \infty} \left| \|P_r(H_n) - P_r(H_{n,R})\| \limsup_{n \to \infty} \sup_{f \in \mathcal{F}} \|\tilde{f}\|_{L_2(P_n)} \right. \tag{3.11}$$

$$\limsup_{n \to \infty} \sup_{g \in \mathcal{F}} \|\tilde{g}\|_{L_2(P_n)} = 0 \text{ a.s.}$$

The first statement of the theorem now follows from (3.7), (3.8), (3.11), and the following bound:

$$\sup_{f,g \in \mathcal{F}} \left| \langle P_r(H_n)\tilde{f}, \tilde{g} \rangle_{L_2(P_n)} - \langle P_r(H)f, g \rangle_{L_2(P)} \right|$$

$$\leq \sup_{f,g \in \mathcal{F}} \left| \langle P_r(H_{n,R})\tilde{f}, \tilde{g} \rangle_{L_2(P_n)} - \langle P_r(H_R)f, g \rangle_{L_2(P)} \right|$$

$$+ \sup_{f,g \in \mathcal{F}} \left| \langle P_r(H_{n,R})\tilde{f}, \tilde{g} \rangle_{L_2(P_n)} - \langle P_r(H_n)\tilde{f}, \tilde{g} \rangle_{L_2(P_n)} \right|$$

$$+ \sup_{f,g \in \mathcal{F}} \left| \langle P_r(H_R)f, g \rangle_{L_2(P)} - \langle P_r(H)f, g \rangle_{L_2(P)} \right|.$$

To prove the second statement, note that under the condition $\mathbb{E}|h(X, X)| < +\infty$,

$$\|H_n - \tilde{H}_n\| = \frac{\max_{1 \leq i \leq n} |h(X_i, X_i)|}{n} \to 0 \text{ as } n \to \infty \text{ a.s.}$$

By Theorem 1.1,

$$\delta_2\big((\lambda(H_n), \lambda(H)\big) \to 0 \text{ as } n \to \infty \text{ a.s.},$$

which implies that, for $r = 1, \ldots, R$, the δ_r-numbers of H_n converge to δ_r as $n \to \infty$ a.s. Given this, it follows from Lemma A.1 iii) that

$$\|P_r(H_n) - P_r(\tilde{H}_n)\| \to 0 \text{ as } n \to \infty \text{ a.s.}$$

Taking into account (3.6), we have

$$\sup_{f,g \in \mathcal{F}} \left| \langle P_r(H_n)\tilde{f}, \tilde{g} \rangle_{L_2(P_n)} - \langle P_r(\tilde{H}_n)\tilde{f}, \tilde{g} \rangle_{L_2(P_n)} \right|$$

$$\leq \|P_r(H_n) - P_r(\tilde{H}_n)\| \sup_{f \in \mathcal{F}} \|\tilde{f}\|_{L_2(P_n)} \sup_{g \in \mathcal{F}} \|\tilde{g}\|_{L_2(P_n)} \to 0 \text{ as } n \to \infty \text{ a.s.},$$

and the second statement follows. $\qquad\square$

PROOF OF THEOREM 2.2. The proof consists of several steps. First, we use perturbation theory to represent the operator $A_{n,R}^{-1} P_r(\hat{H}_{n,R}) A_{n,R}^{-1}$ as a sum of $P_r(H_R)$, a linear function of the operator $E_{n,R}$ and a remainder term (for a fixed R.) Then we obtain bounds for the expectation of the Hilbert-Schmidt norm of the remainder (first for a fixed R and then for $R \to \infty$.) Next, we approximate h by a sequence of kernels of finite rank. We use the CLT for Hilbert-Schmidt operators to study the asymptotics of the linear term. This gives us the first statement of the theorem. Finally, we prove the second statement by comparing the quadratic forms generated by $P_r(\tilde{H}_{n,R_n})$ and $P_r(H_{n,R_n})$.

Throughout the first two steps of the proof, R is a fixed number, so we skip it in the notations like $E_{n,R}, H_{n,R}$, etc.

STEP 1. LINEARIZATION. Denote

$$S_n := S_{n,R} := A_n^{-1} P_r(\hat{H}_n) A_n^{-1} - P_r(\hat{H}_n) + \frac{E_n P_r(\hat{H}_n) + P_r(\hat{H}_n) E_n}{2}$$

and

$$\bar{S}_n := \bar{S}_{n,R} := -\frac{E_n \big(P_r(\hat{H}_n) - P_r(H)\big) + \big(P_r(\hat{H}_n) - P_r(H)\big) E_n}{2}.$$

Then

$$A_n^{-1} P_r(\hat{H}_n) A_n^{-1} = P_r(\hat{H}_n) - \frac{E_n P_r(H) + P_r(H) E_n}{2} + S_n + \bar{S}_n. \qquad (3.12)$$

First we bound $\|\bar{S}_n\|_{HS}$. We have

$$\|\bar{S}_n\|_{HS} \leq \frac{\|E_n\|_{HS}\big\|P_r(\hat{H}_n) - P_r(H)\big\|_{HS} + \big\|P_r(\hat{H}_n) - P_r(H)\big\|_{HS}\|E_n\|_{HS}}{2}$$

$$= \|E_n\|_{HS}\big\|P_r(\hat{H}_n) - P_r(H)\big\|_{HS}.$$

Assume that $\|\hat{H}_n - H\|_{HS} < \varepsilon$. Using Lemma A.1 iii), we get

$$\|\bar{S}_n\|_{HS} \leq \frac{2}{\delta_r}\|E_n\|_{HS}\|\hat{H}_n - H\|_{HS}. \qquad (3.13)$$

To bound $\|S_n\|_{HS}$, we use Lemma A.3 to the effect that

$$\|S_n\|_{HS} \leq 6\|E_n\|_{HS}\big[\|E_n P_r(\hat{H}_n)\|_{HS} + \|P_r(\hat{H}_n) E_n\|_{HS}\big], \qquad (3.14)$$

given that $\|E_n\| < 1/2$. Now

$$\|E_n P_r(\hat{H}_n)\|_{HS} \leq \|E_n P_r(H)\|_{HS} + \big\|E_n \big(P_r(\hat{H}_n) - P_r(H)\big)\big\|_{HS},$$

where the last term, quite similarly to (3.13), is bounded by the quantity

$$\frac{2}{\delta_r}\|E_n\|_{HS}\|\hat{H}_n - H\|_{HS}.$$

Thus

$$\|E_n P_r(\hat{H}_n)\|_{HS} \le \|E_n P_r(H)\|_{HS} + \frac{2}{\delta_r}\|E_n\|_{HS}\|\hat{H}_n - H\|_{HS},$$

and, similarly,

$$\|P_r(\hat{H}_n)E_n\|_{HS} \le \|P_r(H)E_n\|_{HS} + \frac{2}{\delta_r}\|E_n\|_{HS}\|\hat{H}_n - H\|_{HS}.$$

Given $\|E_n\| < 1/2$, (3.14) now implies

$$\|S_n\|_{HS} \le 6\|E_n\|_{HS}\big[\|E_n P_r(H)\|_{HS} + \|P_r(H)E_n\|_{HS}\big] + \frac{24}{\delta_r}\|E_n\|_{HS}^2\|\hat{H}_n - H\|_{HS}. \tag{3.15}$$

Next we use Lemma A.1 iv) to represent the operator $P_r(\hat{H}_n)$ as a sum of a linear term in $\hat{H}_n - H$ and a remainder of the second order. We get

$$P_r(\hat{H}_n) = P_r(H) + L_r(\hat{H}_n - H) + V_n,$$

where

$$L_r(B) := \begin{cases} \frac{1}{\mu_r - \lambda_j}\langle B\phi_i, \phi_j\rangle_{L_2(P)} & \text{for } i \in \Delta_r, j \notin \Delta_r \\ \frac{1}{\mu_r - \lambda_i}\langle B\phi_i, \phi_j\rangle_{L_2(P)} & \text{for } i \notin \Delta_r, j \in \Delta_r \\ 0 & \text{otherwise,} \end{cases}$$

and, for the operator V_n, the following bound holds:

$$\|V_n\|_{HS} \le 2\frac{\|\hat{H}_n - H\|_{HS}^2}{\delta_r^2}. \tag{3.16}$$

It follows from Lemma A.2, that

$$\left\|\hat{H}_n - H - \frac{E_n H + H E_n}{2}\right\|_{HS} \le 3\|E_n\|_{HS}\big[\|E_n H\|_{HS} + \|H E_n\|_{HS}\big].$$

Let

$$\bar{V}_n := L_r(\hat{H}_n - H) - L_r\Big(\frac{E_n H + H E_n}{2}\Big).$$

Then it is easy to check that

$$\|\bar{V}_n\|_{HS} \le \sup\Big\{\frac{1}{|\mu_r - \lambda_j|} : i \in \Delta_r, j \notin \Delta_r\Big\} \bigvee \sup\Big\{\frac{1}{|\mu_r - \lambda_i|} : i \notin \Delta_r, j \in \Delta_r\Big\}$$
$$\times \left\|\hat{H}_n - H - \frac{E_n H + H E_n}{2}\right\|_{HS} \le \frac{3}{2\delta_r}\|E_n\|_{HS}\big[\|E_n H\|_{HS} + \|H E_n\|_{HS}\big]. \tag{3.17}$$

Clearly, we have

$$P_r(\hat{H}_n) = P_r(H) + L_r\Big(\frac{E_n H + H E_n}{2}\Big) + V_n + \bar{V}_n. \tag{3.18}$$

Next we combine (3.12) and (3.18) to get

$$A_n^{-1} P_r(\hat{H}_n) A_n^{-1} =$$

$$= P_r(H) - \frac{E_n P_r(H) + P_r(H) E_n}{2} + L_r\Big(\frac{E_n H + H E_n}{2}\Big) + S_n + \bar{S}_n + V_n + \bar{V}_n. \quad (3.19)$$

An easy calculation shows that

$$\Big\langle L_r\Big(\frac{E_n H + H E_n}{2}\Big) - \frac{E_n P_r(H) + P_r(H) E_n}{2} \phi_i, \phi_j \Big\rangle_{L_2(P)}$$

$$:= L_{n,R}^{(r)} := \begin{cases} \frac{\lambda_j}{\mu_r - \lambda_j} \int_S \phi_i \phi_j \, d(P_n - P) & \text{for } i \in \Delta_r, j \notin \Delta_r \\ \frac{\lambda_i}{\mu_r - \lambda_i} \int_S \phi_i \phi_j \, d(P_n - P) & \text{for } i \notin \Delta_r, j \in \Delta_r \\ 0 & \text{otherwise.} \end{cases}$$

We use the same notation $L_{n,R}^{(r)}$ for the operator on $\mathcal{L}$, whose matrix in the basis $\{\phi_i : 1 \le i \le R\}$ is $L_{n,R}^{(r)}$.

STEP 2. BOUNDS FOR THE REMAINDER. Next we calculate expectations of the norms involved in the bounds (3.13) and (3.15)–(3.17). We easily get

$$\mathbb{E}\|E_n\|_{HS}^2 = \mathbb{E} \sum_{i,j=1}^R \Big[\int_S \phi_i \phi_j d(P_n - P)\Big]^2 = \sum_{i,j=1}^R \mathbb{E}\Big[\int_S \phi_i \phi_j d(P_n - P)\Big]^2$$

$$= n^{-1} \sum_{i,j=1}^R \Big[\int_S \phi_i^2 \phi_j^2 dP - \delta_{ij}\Big] \le n^{-1} \sum_{i,j=1}^R \int_S \phi_i^2 \phi_j^2 dP. \quad (3.20)$$

Similarly

$$\mathbb{E}\|P_r(H) E_n\|_{HS}^2 \le n^{-1} \sum_{i \in \Delta_r} \sum_{j=1}^R \int_S \phi_i^2 \phi_j^2 dP \quad (3.21)$$

and

$$\mathbb{E}\|E_n P_r(H)\|_{HS}^2 \le n^{-1} \sum_{i=1}^R \sum_{j \in \Delta_r} \int_S \phi_i^2 \phi_j^2 dP. \quad (3.22)$$

To get a bound for the expectation of $\|\hat{H}_n - H\|_{HS}^2$, we follow Koltchinskii and Giné (1996). Let $g_j, j = 1, \dots, R$ be orthonormal eigenfunctions of the random symmetric operator E_n, corresponding to the eigenvalues $\gamma_j, \ j = 1, \dots, R$. The matrix of A_n in the basis $\{g_j : j = 1, \dots, R\}$ is diagonal with entries $(1+\gamma_j)^{1/2}, \ j = 1, \dots, R$. Let

$$h_{ij} := \langle H g_i, g_j \rangle_{L_2(P)}, \ i, j = 1, \dots, R.$$

Then

$$\|\hat{H}_n - H\|_{HS}^2 = \|A_n H A_n - H\|_{HS}^2 = \sum_{1 \le i,j \le R} h_{ij}^2 \big[(1+\gamma_i)^{1/2}(1+\gamma_j)^{1/2} - 1\big]^2.$$

An elementary inequality

$$\left[(1+a)^{1/2}(1+b)^{1/2}-1\right]^2 \leq a^2+b^2, \ a,b \geq -1,$$

gives the bound

$$\|\hat{H}_n - H\|_{HS}^2 \leq \sum_{1 \leq i,j \leq R} h_{ij}^2(\gamma_i^2 + \gamma_j^2)$$

$$= \|E_n H\|_{HS}^2 + \|HE_n\|_{HS}^2 = 2\|E_n H\|_{HS}^2.$$

Since

$$\|E_n H\|_{HS}^2 = \sum_{i,j=1}^{R} \lambda_j^2 \Big[\int_S \phi_i \phi_j d(P_n - P)\Big]^2,$$

we now easily get

$$\mathbb{E}\|\hat{H}_n - H\|_{HS}^2 \leq 2 \sum_{i,j=1}^{R} \lambda_j^2 \mathbb{E}\Big[\int_S \phi_i \phi_j d(P_n - P)\Big]^2$$

$$= n^{-1} \sum_{i,j=1}^{R} (\lambda_i^2 + \lambda_j^2)\Big[\int_S \phi_i^2 \phi_j^2 dP - \delta_{ij}\Big] \leq n^{-1} \sum_{i,j=1}^{R} (\lambda_i^2 + \lambda_j^2) \int_S \phi_i^2 \phi_j^2 dP. \tag{3.23}$$

STEP 3. BOUNDS FOR LARGE R. Now we use the above representations and bounds in the case when $R := R_n \to \infty$, in such a way that conditions (1.2) and (1.3) hold. Under condition (1.3), bounds (3.20), (3.23) and Cauchy inequality imply that

$$\mathbb{E}\|E_{n,R_n}\|_{HS}\|\hat{H}_{n,R_n} - H_{R_n}\|_{HS} \leq \mathbb{E}^{1/2}\|E_{n,R_n}\|_{HS}^2 \, \mathbb{E}^{1/2}\|\hat{H}_{n,R_n} - H_{R_n}\|_{HS}^2$$

$$\leq \Big(n^{-1}\sum_{i,j=1}^{R_n}\int_S \phi_i^2 \phi_j^2 dP\Big)^{1/2}\Big(n^{-1}\sum_{i,j=1}^{R_n}\int_S (\lambda_i^2 + \lambda_j^2)\phi_i^2 \phi_j^2 dP\Big)^{1/2} = o(n^{-1/2}). \tag{3.24}$$

By (3.23) and condition (1.3), we also have

$$\|\hat{H}_{n,R_n} - H_{R_n}\|_{HS} = o_P(1).$$

Now (3.13) and (3.24) imply that

$$\|\bar{S}_{n,R_n}\|_{HS} = o_P(n^{-1/2}) \ \text{as} \ n \to \infty. \tag{3.25}$$

Next, again under condition (1.3), (3.20) and (3.22) yield

$$\mathbb{E}\|E_{n,R_n}\|_{HS}\|E_{n,R_n}P_r(H_{R_n})\|_{HS} \leq \mathbb{E}^{1/2}\|E_{n,R_n}\|_{HS}^2 \mathbb{E}^{1/2}\|E_{n,R_n}P_r(H_{R_n})\|_{HS}^2$$

$$\leq \Big(n^{-1}\sum_{i,j=1}^{R_n}\int_S \phi_i^2 \phi_j^2 dP\Big)^{1/2} \Big(n^{-1}\sum_{i=1}^{R_n}\sum_{j\in\Delta_r}\int_S \phi_i^2 \phi_j^2 dP\Big)^{1/2}$$

$$\leq \frac{1}{\mu_r^2}\Big(n^{-1}\sum_{i,j=1}^{R_n}\int_S \phi_i^2 \phi_j^2 dP\Big)^{1/2} \Big(n^{-1}\sum_{i=1}^{R_n}\sum_{j\in\Delta_r}\int_S \lambda_j^2 \phi_i^2 \phi_j^2 dP\Big)^{1/2}$$

$$\leq \frac{1}{\mu_r^2}\Big(n^{-1}\sum_{i,j=1}^{R_n}\int_S \phi_i^2 \phi_j^2 dP\Big)^{1/2} \Big(n^{-1}\sum_{i,j=1}^{R_n}\int_S (\lambda_i^2 + \lambda_j^2)\phi_i^2 \phi_j^2 dP\Big)^{1/2} = o(n^{-1/2}).$$

This implies that

$$\|E_{n,R_n}\|_{HS}\,\|E_{n,R_n}P_r(H_{R_n})\|_{HS} = o_P(n^{-1/2}) \text{ as } n \to \infty \qquad (3.26)$$

and, similarly,

$$\|E_{n,R_n}\|_{HS}\,\|P_r(H_{R_n})E_{n,R_n}\|_{HS} = o_P(n^{-1/2}) \text{ as } n \to \infty. \qquad (3.27)$$

Note now that, under condition (1.3), by (3.24)

$$\|E_{n,R_n}\|_{HS}\,\|\hat{H}_{n,R_n} - H_{R_n}\|_{HS} = o_P(n^{-1/2}). \qquad (3.28)$$

Since (1.3) implies

$$\sum_{i,j=1}^{R_n} \int_S \phi_i^2 \phi_j^2 dP = o(n),$$

we also have, by (3.20), that

$$\|E_{n,R_n}\|_{HS} = o_P(1) \text{ as } n \to \infty. \qquad (3.29)$$

Thus

$$\|E_{n,R_n}\|_{HS}^2\,\|\hat{H}_{n,R_n} - H_{R_n}\|_{HS} = o_P(n^{-1/2}). \qquad (3.30)$$

The bound (3.15), the relationships (3.26)–(3.30) and the fact that $\|\hat{H}_{n,R_n} - H_{R_n}\|_{HS} = o_P(1)$, show that

$$\|S_{n,R_n}\|_{HS} = o_P(n^{-1/2}) \text{ as } n \to \infty. \qquad (3.31)$$

By (3.23), we get under condition (1.3)

$$\mathbb{E}\|\hat{H}_{n,R_n} - H_{R_n}\|_{HS}^2 \leq n^{-1} \sum_{i,j=1}^{R_n} \int_S (\lambda_i^2 + \lambda_j^2)\phi_i^2\phi_j^2 dP$$

$$\leq 2\sup|\lambda_r|\left(n^{-1}\sum_{i,j=1}^{R_n}\int_S \phi_i^2\phi_j^2 dP\right)^{1/2}\left(n^{-1}\sum_{i,j=1}^{R_n}\int_S (\lambda_i^2+\lambda_j^2)\phi_i^2\phi_j^2 dP\right)^{1/2}$$

$$= o(n^{-1/2}),$$

which, taking into account (3.16), shows that

$$\|V_{n,R_n}\|_{HS} = o_P(n^{-1/2}) \text{ as } n \to \infty. \qquad (3.32)$$

Finally, (3.17) implies

$$\mathbb{E}\|\bar{V}_{n,R_n}\|_{HS} \leq \frac{3}{\delta_r^2}\mathbb{E}\|E_{n,R_n}\|_{HS}\big[\|E_{n,R_n}H_{R_n}\|_{HS} + \|H_{R_n}E_{n,R_n}\|_{HS}\big],$$

which, quite similarly, is of the order $o(n^{-1/2})$, so that

$$\|\bar{V}_{n,R_n}\|_{HS} = o_P(n^{-1/2}) \text{ as } n \to \infty. \tag{3.33}$$

Representation (3.19) can now be rewritten as

$$A_{n,R_n}^{-1} P_r(\hat{H}_{n,R_n}) A_{n,R_n}^{-1} = P_r(H_{R_n}) + L_{n,R_n}^{(r)} + M_{n,R_n}^{(r)},$$

where $M_{n,R_n}^{(r)} := S_{n,R_n} + \bar{S}_{n,R_n} + V_{n,R_n} + \bar{V}_{n,R_n}$, and, by (3.25) and (3.31)–(3.33),

$$\begin{aligned}
&\|M_{n,R_n}^{(r)}\|_{HS} \\
&\leq \|S_{n,R_n}\|_{HS} + \|\bar{S}_{n,R_n}\|_{HS} + \|V_{n,R_n}\|_{HS} + \|\bar{V}_{n,R_n}\|_{HS} \\
&= o_P(n^{-1/2}) \text{ as } n \to \infty.
\end{aligned} \tag{3.34}$$

Using the basic isometric representation (2.17), we get

$$\begin{aligned}
\langle P_r(\tilde{H}_{n,R_n})\tilde{f}, \tilde{g}\rangle_{L_2(P_n)} &= \langle A_{n,R_n}^{-1} P_r(\hat{H}_{n,R_n}) A_{n,R_n}^{-1} T_{n,R_n}f, T_{n,R_n}g\rangle_{L_2(P)} \\
&= \langle P_r(H_{R_n}) T_{n,R_n}f, T_{n,R_n}g\rangle_{L_2(P)} + \langle L_{n,R_n}^{(r)} T_{n,R_n}f, T_{n,R_n}g\rangle_{L_2(P)} \\
&\quad + \langle M_{n,R_n}^{(r)} T_{n,R_n}f, T_{n,R_n}g\rangle_{L_2(P)},
\end{aligned}$$

which implies

$$\begin{aligned}
&\sup_{f,g\in\mathcal{F}} \Big| \langle P_r(\tilde{H}_{n,R_n})\tilde{f}, \tilde{g}\rangle_{L_2(P_n)} - \langle P_r(H_{R_n}) T_{n,R_n}f, T_{n,R_n}g\rangle_{L_2(P)} \\
&\quad - \langle L_{n,R_n}^{(r)} T_{n,R_n}f, T_{n,R_n}g\rangle_{L_2(P)} \Big| \\
&= \sup_{f,g\in\mathcal{F}} \Big| \langle M_{n,R_n}^{(r)} T_{n,R_n}f, T_{n,R_n}g\rangle_{L_2(P)} \Big| \\
&\leq \|M_{n,R_n}^{(r)}\| \sup_{f\in\mathcal{F}} \|T_{n,R_n}f\|_{L_2(P)} \sup_{g\in\mathcal{F}} \|T_{n,R_n}g\|_{L_2(P)} \\
&\leq \|M_{n,R_n}^{(r)}\|_{HS} \sup_{f\in\mathcal{F}} \|T_{n,R_n}f\|_{L_2(P)}^2.
\end{aligned} \tag{3.35}$$

Note that, by condition (2.7),

$$\begin{aligned}
\sup_{f\in\mathcal{F}} \|T_{n,R_n}f\|_{L_2(P)}^2 &= \sup_{f\in\mathcal{F}} \sum_{j=1}^{R_n} \langle f, \phi_j\rangle_{L_2(P_n)}^2 \\
&\leq 2\sup_{f\in\mathcal{F}} \Big\{ \sum_{j=1}^{R_n} \big[\langle f, \phi_j\rangle_{L_2(P_n)} - \langle f, \phi_j\rangle_{L_2(P)}\big]^2 + \sum_{j=1}^{R_n} \langle f, \phi_j\rangle_{L_2(P)}^2 \Big\} \\
&\leq 2\sup_{f\in\mathcal{F}} \sum_{j=1}^{R_n} \Big[\int_S f\phi_j d(P_n - P)\Big]^2 + 2\sup_{f\in\mathcal{F}} \|f\|_{L_2(P)}^2 \\
&= 2n^{-1}\zeta_n^2(\mathcal{F}; R_n) + 2\sup_{f\in\mathcal{F}} \|f\|_{L_2(P)}^2 = 2\sup_{f\in\mathcal{F}} \|f\|_{L_2(P)}^2 + o_P(1) \text{ as } n \to \infty.
\end{aligned}$$

Thus, (3.34) and (3.35) imply

$$
\begin{aligned}
&\sup_{f,g\in\mathcal{F}}\Big|\big\langle P_r(\tilde{H}_{n,R_n})\tilde{f},\tilde{g}\big\rangle_{L_2(P_n)} - \big\langle P_r(H_{R_n})T_{n,R_n}f, T_{n,R_n}g\big\rangle_{L_2(P)} \\
&- \big\langle L^{(r)}_{n,R_n}T_{n,R_n}f, T_{n,R_n}g\big\rangle_{L_2(P)}\Big| \\
&= o_P(n^{-1/2}) \text{ as } n \to \infty.
\end{aligned}
\tag{3.36}
$$

STEP 4. APPROXIMATION BY KERNELS OF FINITE RANK. Since $P_r(H_{R_n}) = P_r(H)$ for all large n, now it is enough to compare

$$
\big\langle P_r(\tilde{H}_{n,R_n})\tilde{f},\tilde{g}\big\rangle_{L_2(P_n)} \quad\text{and}\quad \big\langle P_r(\tilde{H}_n)\tilde{f},\tilde{g}\big\rangle_{L_2(P_n)}.
$$

By isometry, $\lambda(\tilde{H}_{n,R_n}) = \lambda(\hat{H}_{n,R_n})$. Let $\delta < \delta_r - 2\varepsilon$. If

$$
\|\hat{H}_{n,R_n} - H_{R_n}\|_{HS} < \delta,
$$

then Lidskii's inequality implies

$$
\delta_2\big(\lambda(\tilde{H}_{n,R_n}), \lambda(H_{R_n})\big) < \delta.
$$

It follows that the δ_r-number of $\tilde{H}_{n,R_n}$ is larger than $\delta_r - \delta$. Applying Lemma A.1 iii), we get that under the conditions $\|\hat{H}_{n,R_n} - H_{R_n}\|_{HS} < \delta$ and $\|\tilde{H}_{n,R_n} - \tilde{H}_n\|_{HS} < \varepsilon$,

$$
\|P_r(\tilde{H}_{n,R_n}) - P_r(\tilde{H}_n)\|_{HS} \le \frac{2}{\delta_r - \delta}\|\tilde{H}_{n,R_n} - \tilde{H}_n\|_{HS}.
\tag{3.37}
$$

Clearly,

$$
\begin{aligned}
\mathbb{E}\|\tilde{H}_{n,R_n} - \tilde{H}_n\|^2_{HS} &= \mathbb{E}\sum_{i,j=1}^{n}\frac{(h - h_{R_n})^2(X_i, X_j)}{n^2} \\
&= n^{-2}\sum_{i\ne j}\mathbb{E}(h - h_{R_n})^2(X_i, X_j) + n^{-2}\sum_{i=1}^{n}\mathbb{E}(h - h_{R_n})^2(X_i, X_i) \\
&= \frac{n(n-1)}{n^2}\mathbb{E}(h - h_{R_n})^2(X, Y) + n^{-1}\mathbb{E}(h - h_{R_n})^2(X, X).
\end{aligned}
\tag{3.38}
$$

Under the conditions of the theorem, we have

$$
\mathbb{E}(h - h_{R_n})^2(X, Y) = \sum_{k=R_n+1}^{\infty}\lambda_k^2 = o(n^{-1}).
\tag{3.39}
$$

On the other hand,

$$
n^{-1}\mathbb{E}(h - h_{R_n})^2(X, X) = n^{-1}\int_S\Big(\sum_{k=R_n+1}^{\infty}\lambda_k\phi_k^2\Big)^2 dP = o(n^{-1}),
\tag{3.40}
$$

since $\sum_{k=1}^{\infty}\lambda_k\phi_k^2(x)$ converges to $h(x, x)$ in $L_2(P)$.

We also have $\|\hat{H}_{n,R_n} - H_{R_n}\|_{HS} = o_P(1)$. Thus (3.37)–(3.40) imply

$$\|P_r(\tilde{H}_{n,R_n}) - P_r(\tilde{H}_n)\|_{HS} = o_P(n^{-1/2}),\tag{3.41}$$

and we obtain

$$\sup_{f,g \in \mathcal{F}} \left| \langle P_r(\tilde{H}_{n,R_n})\tilde{f}, \tilde{g} \rangle_{L_2(P_n)} - \langle P_r(\tilde{H}_n)\tilde{f}, \tilde{g} \rangle_{L_2(P_n)} \right|$$

$$\leq \|P_r(\tilde{H}_{n,R_n}) - P_r(\tilde{H}_n)\|_{HS} \sup_{f \in \mathcal{F}} \|\tilde{f}\|_{L_2(P_n)} \sup_{g \in \mathcal{F}} \|\tilde{g}\|_{L_2(P_n)} \tag{3.42}$$

$$\leq \|P_r(\tilde{H}_{n,R_n}) - P_r(\tilde{H}_n)\|_{HS} \|\tilde{F}\|_{L_2(P_n)}^2 = o_P(n^{-1/2}),$$

by (3.41) and square integrability of the envelope F. Now (3.36) and (3.42) imply that

$$\sup_{f,g \in \mathcal{F}} \left| \langle P_r(\tilde{H}_n)\tilde{f}, \tilde{g} \rangle_{L_2(P_n)} - \langle P_r(H)T_{n,R_n}f, T_{n,R_n}g \rangle_{L_2(P)} \right.$$

$$\left. - \langle L_{n,R_n}^{(r)} T_{n,R_n}f, T_{n,R_n}g \rangle_{L_2(P)} \right| \tag{3.43}$$

$$= o_P(n^{-1/2}) \text{ as } n \to \infty.$$

STEP 5. CLT FOR HILBERT-SCHMIDT OPERATORS. By the Central Limit Theorem in one dimension, we have that for all $i \in \Delta_r, j \notin \Delta_r$

$$n^{1/2} \langle L_{n,R_n}^{(r)} \phi_i, \phi_j \rangle_{L_2(P)} = \frac{\lambda_j}{\mu_r - \lambda_j} n^{1/2} \int_S \phi_i \phi_j d(P_n - P)$$

converges weakly to $\frac{\lambda_j}{\mu_r - \lambda_j} G_P(\phi_i \phi_j)$. Similarly, for $i \notin \Delta_r, j \in \Delta_r$, the sequence

$$n^{1/2} \langle L_{n,R_n}^{(r)} \phi_i, \phi_j \rangle_{L_2(P)} = \frac{\lambda_i}{\mu_r - \lambda_i} n^{1/2} \int_S \phi_i \phi_j d(P_n - P)$$

converges weakly to $\frac{\lambda_i}{\mu_r - \lambda_i} G_P(\phi_i \phi_j)$. If $i \in \Delta_r, j \in \Delta_r$, or $i \notin \Delta_r, j \notin \Delta_r$, then

$$n^{1/2} \langle L_{n,R_n}^{(r)} \phi_i, \phi_j \rangle_{L_2(P)} = 0,$$

and so the limit is equal to 0. We also have (for large enough R, such that $\Delta_r \subset \{1, \dots, R\}$)

$$\mathbb{E} \sum_{i \vee j \geq R} \langle n^{1/2} L_{n,R_n}^{(r)} \phi_i, \phi_j \rangle_{L_2(P)}^2$$

$$= \sum_{i \in \Delta_r, j \notin \Delta_r, R \leq j \leq R_n} \left(\frac{\lambda_j}{\mu_r - \lambda_j} \right)^2 \mathbb{E} \left[n^{1/2} \int_S \phi_i \phi_j d(P_n - P) \right]^2$$

$$+ \sum_{j \in \Delta_r, i \notin \Delta_r, R \leq i \leq R_n} \left(\frac{\lambda_i}{\mu_r - \lambda_i} \right)^2 \mathbb{E} \left[n^{1/2} \int_S \phi_i \phi_j d(P_n - P) \right]^2$$

$$\leq \left(\frac{1}{\delta_r} \right)^2 \left[\sum_{i \in \Delta_r, j \notin \Delta_r, R \leq j \leq R_n} \lambda_j^2 \int_S \phi_i^2 \phi_j^2 dP + \sum_{j \in \Delta_r, i \notin \Delta_r, R \leq i \leq R_n} \lambda_i^2 \int_S \phi_i^2 \phi_j^2 dP \right]$$

$$= 2 \left(\frac{1}{\delta_r} \right)^2 \sum_{i \in \Delta_r, j \notin \Delta_r, R \leq j \leq R_n} \lambda_j^2 \int_S \phi_i^2 \phi_j^2 dP.$$

$$\tag{3.44}$$

By condition (2.6), this implies

$$\lim_{R\to\infty} \limsup_{n\to\infty} \mathbb{E} \sum_{i\vee j\geq R} \langle n^{1/2} L_{n,R_n}^{(r)} \phi_i, \phi_j \rangle_{L_2(P)}^2 = 0.$$

It shows that in fact $n^{1/2} L_{n,R_n}$ converges weakly in the Hilbert space $\mathcal{J}_2$ of all Hilbert-Schmidt operators on $L_2(P)$ to the Gaussian random Hilbert-Schmidt operator $\Gamma_P^{(r)}$. Since the map

$$\mathcal{J}_2 \ni A \mapsto \{\mathcal{F} \times \mathcal{F} \ni (f,g) \mapsto \langle Af, g \rangle_{L_2(P)}\} \in \ell^\infty(\mathcal{F} \times \mathcal{F})$$

is continuous, we also have the weak convergence of the sequence of random functions $n^{1/2} \langle L_{n,R_n}^{(r)} f, g \rangle_{L_2(P)}$, $f, g \in \mathcal{F}$ to the Gaussian random function $\langle \Gamma_P^{(r)} f, g \rangle_{L_2(P)}, f, g \in \mathcal{F}$ in the space in $\ell^\infty(\mathcal{F} \times \mathcal{F})$:

$$\begin{aligned}
&\left\{ n^{1/2} \langle L_{n,R_n}^{(r)} f, g \rangle_{L_2(P)}, \ f, g \in \mathcal{F} \right\} \\
&\longrightarrow_{\mathcal{L}} \left\{ \langle \Gamma_P^{(r)} f, g \rangle_{L_2(P)}, \ f, g \in \mathcal{F} \right\} \text{ in } \ell^\infty(\mathcal{F} \times \mathcal{F}).
\end{aligned} \tag{3.45}$$

STEP 6. CLT FOR SPECTRAL PROJECTIONS: THE CASE OF OPERATORS $\tilde{H}_n$.
Our goal is to use the relationship (3.43).
First we consider the term $\langle P_r(H) T_{n,R_n} f, T_{n,R_n} g \rangle_{L_2(P)}$. By simple calculations,

$$\begin{aligned}
&\langle P_r(H) T_{n,R_n} f, T_{n,R_n} g \rangle_{L_2(P)} \\
&= \sum_{i\in\Delta_r} \langle f, \phi_i \rangle_{L_2(P_n)} \langle g, \phi_i \rangle_{L_2(P_n)} \\
&= \sum_{i\in\Delta_r} \langle f, \phi_i \rangle_{L_2(P)} \langle g, \phi_i \rangle_{L_2(P)} + \sum_{i\in\Delta_r} [\langle f, \phi_i \rangle_{L_2(P_n)} - \langle f, \phi_i \rangle_{L_2(P)}] \langle g, \phi_i \rangle_{L_2(P)} \\
&\quad + \sum_{i\in\Delta_r} [\langle g, \phi_i \rangle_{L_2(P_n)} - \langle g, \phi_i \rangle_{L_2(P)}] \langle f, \phi_i \rangle_{L_2(P)} \\
&\quad + \sum_{i\in\Delta_r} [\langle f, \phi_i \rangle_{L_2(P_n)} - \langle f, \phi_i \rangle_{L_2(P)}][\langle g, \phi_i \rangle_{L_2(P_n)} - \langle g, \phi_i \rangle_{L_2(P)}] \\
&= \langle P_r(H) f, g \rangle_{L_2(P)} + \sum_{i\in\Delta_r} [\int_S f\phi_i d(P_n - P)] \langle g, \phi_i \rangle_{L_2(P)} \\
&\quad + \sum_{i\in\Delta_r} [\int_S g\phi_i d(P_n - P)] \langle f, \phi_i \rangle_{L_2(P)} \\
&\quad + \sum_{i\in\Delta_r} [\int_S f\phi_i d(P_n - P)][\int_S g\phi_i d(P_n - P)].
\end{aligned}$$

$$\tag{3.46}$$

Since $\mathcal{F}\phi_i \in CLT(P)$, we get, using the Cauchy inequality,

$$
\sup_{f,g\in\mathcal{F}} \Big| \sum_{i\in\Delta_r} \Big[\int_S f\phi_i d(P_n - P) \Big] \Big[\int_S g\phi_i d(P_n - P) \Big] \Big|
$$

$$
\leq \sup_{f\in\mathcal{F}} \Big(\sum_{i\in\Delta_r} \Big[\int_S f\phi_i d(P_n - P) \Big]^2 \Big)^{1/2} \sup_{g\in\mathcal{F}} \Big(\sum_{i\in\Delta_r} \Big[\int_S g\phi_i d(P_n - P) \Big]^2 \Big)^{1/2} \tag{3.47}
$$

$$
\leq \sum_{i\in\Delta_r} \sup_{f\in\mathcal{F}} \Big| \int_S f\phi_i d(P_n - P) \Big|^2 = O_P(n^{-1}).
$$

It follows from (3.46) and (3.47) that

$$
\langle P_r(H)T_{n,R_n}f, T_{n,R_n}g \rangle_{L_2(P)} - \langle P_r(H)f, g \rangle_{L_2(P)}
$$

$$
= \sum_{i\in\Delta_r} \Big[\int_S f\phi_i d(P_n - P) \Big] \langle g, \phi_i \rangle_{L_2(P)}
$$

$$
+ \sum_{i\in\Delta_r} \Big[\int_S g\phi_i d(P_n - P) \Big] \langle f, \phi_i \rangle_{L_2(P)} + O_P(n^{-1}) \tag{3.48}
$$

$$
= \int_S f P_r(H)g \, d(P_n - P) + \int_S g P_r(H)f \, d(P_n - P) + O_P(n^{-1})
$$

as $n \to \infty$ in $\ell^\infty(\mathcal{F} \times \mathcal{F})$.

Next we consider the term $\langle L^{(r)}_{n,R_n} T_{n,R_n}f, T_{n,R_n}g \rangle_{L_2(P)}$. The following representation is obvious:

$$
\langle L^{(r)}_{n,R_n} T_{n,R_n}f, T_{n,R_n}g \rangle_{L_2(P)} = \langle L^{(r)}_{n,R_n} T_{R_n}f, T_{R_n}g \rangle_{L_2(P)}
$$

$$
+ \langle L^{(r)}_{n,R_n}(T_{n,R_n}f - T_{R_n}f), T_{R_n}g \rangle_{L_2(P)} + \langle L^{(r)}_{n,R_n} T_{R_n}f, T_{n,R_n}g - T_{R_n}g \rangle_{L_2(P)}
$$

$$
+ \langle L^{(r)}_{n,R_n}(T_{n,R_n}f - T_{R_n}f), T_{n,R_n}g - T_{R_n}g \rangle_{L_2(P)}.
$$

Relationship (3.44) with $R = 1$ and condition (2.6) imply that

$$
\| L^{(r)}_{n,R_n} \|_{HS} = O_P(n^{-1/2}).
$$

Thus, by condition (2.7),

$$
\sup_{f,g\in\mathcal{F}} \Big| \langle L^{(r)}_{n,R_n}(T_{n,R_n}f - T_{R_n}f), T_{R_n}g \rangle_{L_2(P)} \Big|
$$

$$
\leq \| L^{(r)}_{n,R_n} \|_{HS} \sup_{f\in\mathcal{F}} \| T_{n,R_n}f - T_{R_n}f \|_{L_2(P)} \sup_{g\in\mathcal{F}} \| T_{R_n}g \|_{L_2(P)}
$$

$$
\leq \| L^{(r)}_{n,R_n} \|_{HS} \sup_{f\in\mathcal{F}} \Big(\sum_{j=1}^{R_n} \Big[\int_S f\phi_j d(P_n - P) \Big]^2 \Big)^{1/2} \sup_{g\in\mathcal{F}} \| g \|_{L_2(P)}
$$

$$
\leq \| L^{(r)}_{n,R_n} \|_{HS} \, n^{-1/2} \zeta_n(\mathcal{F}; R_n) \sup_{g\in\mathcal{F}} \| g \|_{L_2(P)} = o_P(n^{-1/2}) \text{ as } n \to \infty.
$$

Similarly,

$$\sup_{f,g\in\mathcal{F}}\left|\left\langle L_{n,R_n}^{(r)} T_{R_n} f, T_{n,R_n} g - T_{R_n} g\right\rangle_{L_2(P)}\right| = o_P(n^{-1/2})$$

and

$$\sup_{f,g\in\mathcal{F}}\left|\left\langle L_{n,R_n}^{(r)} (T_{n,R_n} f - T_{R_n} f), T_{n,R_n} g - T_{R_n} g\right\rangle_{L_2(P)}\right| = o_P(n^{-1/2}).$$

Hence

$$\left\langle L_{n,R_n}^{(r)} T_{n,R_n} f, T_{n,R_n} g\right\rangle_{L_2(P)}$$
$$= \left\langle L_{n,R_n}^{(r)} T_{R_n} f, T_{R_n} g\right\rangle_{L_2(P)} + o_P(n^{-1/2}) \text{ as } n \to \infty \text{ in } \ell^\infty(\mathcal{F}\times\mathcal{F}). \tag{3.49}$$

Now (3.43), (3.48) and (3.49) imply that

$$\left\langle P_r(\tilde{H}_n)\tilde{f}, \tilde{g}\right\rangle_{L_2(P_n)} = \left\langle P_r(H)f, g\right\rangle_{L_2(P)}$$
$$+ \int_S f P_r(H)g \, d(P_n - P) + \int_S g P_r(H)f \, d(P_n - P) \tag{3.50}$$
$$+ \left\langle L_{n,R_n}^{(r)} T_{R_n} f, T_{R_n} g\right\rangle_{L_2(P)} + o_P(n^{-1/2}) \text{ as } n \to \infty \text{ in } \ell^\infty(\mathcal{F}\times\mathcal{F}).$$

Finally, the conditions $\mathcal{F}\phi_i \in CLT(P), i \in \mathbb{Z}$ and the relationship (3.45) imply that the sequence of stochastic processes

$$n^{1/2}\left[\left\langle P_r(\tilde{H}_n)\tilde{f}, \tilde{g}\right\rangle_{L_2(P_n)} - \left\langle P_r(H)f, g\right\rangle_{L_2(P)}\right], \ f, g \in \mathcal{F}$$

converges weakly in $\ell^\infty(\mathcal{F}\times\mathcal{F})$ as $n \to \infty$ to the continuous Gaussian process

$$G_P\big(f P_r(H)g\big) + G_P\big(g P_r(H)f\big) + \left\langle \Gamma_P^{(r)} f, g\right\rangle_{L_2(P)}, \ f, g \in \mathcal{F}.$$

STEP 7. CLT FOR SPECTRAL PROJECTIONS: THE CASE OF OPERATORS H_n. To prove the result for the operators H_n, note that, by Lemma A.1. and quite similarly to (3.37), the conditions $\|\hat{H}_{n,R_n} - H_{R_n}\|_{HS} < \delta$ and $\|\tilde{H}_{n,R_n} - H_{n,R_n}\| < \varepsilon$ imply

$$\|P_r(\tilde{H}_{n,R_n}) - P_r(H_{n,R_n})\| \leq \frac{2}{\delta_r - \delta}\|\tilde{H}_{n,R_n} - H_{n,R_n}\|. \tag{3.51}$$

Since, by condition (1.4), $\sum_{k=1}^\infty |\lambda_k|\phi_k^2 \in L_2(P)$,

$$\|\tilde{H}_{n,R_n} - H_{n,R_n}\| = n^{-1} \max_{1\leq i\leq n} |h_{R_n}(X_i, X_i)|$$
$$\leq n^{-1} \max_{1\leq i\leq n} \sum_{k=1}^\infty |\lambda_k|\phi_k^2(X_i) = o_P(n^{-1/2}). \tag{3.52}$$

We have

$$\sup_{f,g\in\mathcal{F}}\left|\left\langle P_r(\tilde{H}_{n,R_n})\tilde{f},\tilde{g}\right\rangle_{L_2(P_n)} - \left\langle P_r(H_{n,R_n})\tilde{f},\tilde{g}\right\rangle_{L_2(P_n)}\right|$$

$$\leq \frac{2}{\delta_r - \delta}\|\tilde{H}_{n,R_n} - H_{n,R_n}\| \sup_{f\in\mathcal{F}}\|\tilde{f}\|_{L_2(P_n)} \sup_{g\in\mathcal{F}}\|\tilde{g}\|_{L_2(P_n)}$$

$$\leq \frac{2}{\delta_r - \delta}\|\tilde{H}_{n,R_n} - H_{n,R_n}\|\|\tilde{F}\|^2_{L_2(P_n)},$$

and it follows that

$$\sup_{f,g\in\mathcal{F}}\left|\left\langle P_r(\tilde{H}_{n,R_n})\tilde{f},\tilde{g}\right\rangle_{L_2(P_n)} - \left\langle P_r(H_{n,R_n})\tilde{f},\tilde{g}\right\rangle_{L_2(P_n)}\right| = o_P(n^{-1/2}) \qquad (3.53)$$

(we used (3.51), (3.52) and square integrability of F). Thus, in relationship (3.36), one can replace $\left\langle P_r(\tilde{H}_{n,R_n})\tilde{f},\tilde{g}\right\rangle_{L_2(P_n)}$ with $\left\langle P_r(H_{n,R_n})\tilde{f},\tilde{g}\right\rangle_{L_2(P_n)}$.

Instead of (3.38)–(3.40), we have

$$\mathbb{E}\|H_{n,R_n} - H_n\|^2_{HS} = \mathbb{E}\sum_{i\neq j}\frac{(h - h_{R_n})^2(X_i, X_j)}{n^2}$$

$$= \frac{n(n-1)}{n^2}\mathbb{E}(h - h_{R_n})^2(X, Y) = \sum_{k=R_n+1}^{\infty}\lambda_k^2 = o(n^{-1}),$$

which, similarly to (3.41), imply

$$\|P_r(H_{n,R_n}) - P_r(H_n)\|_{HS} = o_P(n^{-1/2}).$$

This bound allows one to complete the proof, using a relationship similar to (3.36) with $\tilde{H}_n$, $\tilde{H}_{n,R_n}$ replaced by H_n, H_{n,R_n}. $\qquad\square$

PROOF OF THEOREM 2.3. Again, we start with the case of a fixed number $R < +\infty$. The following identity is obvious:

$$A_{n,R}^{-1}P_r(\hat{H}_{n,R})A_{n,R}^{-1} - P_r(H_R)$$
$$= A_{n,R}^{-1}[P_r(\hat{H}_{n,R}) - P_r(H_R)]A_{n,R}^{-1} - [P_r(\hat{H}_{n,R}) - P_r(H_R)] \qquad (3.54)$$
$$+ [P_r(\hat{H}_{n,R}) - P_r(H_R)] + A_{n,R}^{-1}P_r(H_R)A_{n,R}^{-1} - P_r(H_R).$$

First we bound $\|P_r(\hat{H}_{n,R}) - P_r(H_R)\|$. Note that

$$\|P_r(\hat{H}_{n,R}) - P_r(H_R)\| \leq \|P_r(\hat{H}_{n,R})\| + \|P_r(H_R)\| \leq 2.$$

Using Lemma A1, iii) and the last bound, we get

$$\|P_r(\hat{H}_{n,R}) - P_r(H_R)\| \leq \frac{2}{\delta_r}\|\hat{H}_{n,R} - H_R\|_{HS}I_{\{\|\hat{H}_{n,R}-H_R\|_{HS}<\varepsilon\}}$$
$$+ 2I_{\{\|\hat{H}_{n,R}-H_R\|_{HS}\geq\varepsilon\}} \leq \frac{2}{\varepsilon}\|\hat{H}_{n,R} - H_R\|_{HS}. \qquad (3.55)$$

The bounds (3.23) and (3.55) yield

$$
\mathbb{E}\|P_r(\hat{H}_{n,R}) - P_r(H_R)\|^2 \le \frac{4}{\varepsilon^2}\mathbb{E}\|\hat{H}_{n,R} - H_R\|^2_{HS}
$$

$$
\le \frac{4}{\varepsilon^2}n^{-1}\sum_{i,j=1}^{R}(\lambda_i^2 + \lambda_j^2)\int_S \phi_i^2\phi_j^2 dP. \tag{3.56}
$$

Next we apply the method already used in the proof of bound (3.23) (see also the proof of Lemma 4.1 in Koltchinskii and Giné (1996)). Using this approach, it's easy to get, under the condition $\|E_{n,R}\| < 1/2$,

$$
\|A_{n,R}^{-1}[P_r(\hat{H}_{n,R}) - P_r(H_R)]A_{n,R}^{-1} - [P_r(\hat{H}_{n,R}) - P_r(H_R)]\|^2_{HS}
$$

$$
\le 4\Big(\|E_{n,R}[P_r(\hat{H}_{n,R}) - P_r(H_R)]\|^2_{HS} + \|[P_r(\hat{H}_{n,R}) - P_r(H_R)]E_{n,R}\|^2_{HS}\Big)
$$

$$
\le 8\|E_{n,R}\|^2_{HS}\|P_r(\hat{H}_{n,R}) - P_r(H_R)\|^2 \le 2\|P_r(\hat{H}_{n,R}) - P_r(H_R)\|^2.
$$

Therefore, the bound (3.56) yields

$$
\mathbb{E}\|A_{n,R}^{-1}[P_r(\hat{H}_{n,R}) - P_r(H_R)]A_{n,R}^{-1} - [P_r(\hat{H}_{n,R}) - P_r(H_R)]\|^2_{HS}I_{\{\|E_{n,R}\|<1/2\}}
$$

$$
\le 2\mathbb{E}\|P_r(\hat{H}_{n,R}) - P_r(H_R)\|^2 \le \frac{8}{\varepsilon^2}n^{-1}\sum_{i,j=1}^{R}(\lambda_i^2 + \lambda_j^2)\int_S \phi_i^2\phi_j^2 dP. \tag{3.57}
$$

As to the last term in (3.54), we can use the same approach to get under the condition $\|E_{n,R}\| < 1/2$

$$
\|A_{n,R}^{-1}P_r(H_R)A_{n,R}^{-1} - P_r(H_R)\|^2_{HS} \le 4\big[\|E_{n,R}P_r(H_R)\|^2_{HS} + \|P_r(H_R)E_{n,R}\|^2_{HS}\big].
$$

Now, (3.21) and (3.22) give

$$
\mathbb{E}\|A_{n,R}^{-1}P_r(H_R)A_{n,R}^{-1} - P_r(H_R)\|^2_{HS}I_{\{\|E_{n,R}\|<1/2\}}
$$

$$
\le 4\mathbb{E}\|E_{n,R}P_r(H_R)\|^2_{HS} + 4\mathbb{E}\|P_r(H_R)E_{n,R}\|^2_{HS}
$$

$$
\le 4n^{-1}\Big[\sum_{i\in\Delta_r}\sum_{j=1}^{R}\int_S \phi_i^2\phi_j^2 dP + \sum_{i=1}^{R}\sum_{j\in\Delta_r}\int_S \phi_i^2\phi_j^2 dP\Big]
$$

$$
\le \frac{4}{\mu_r^2}n^{-1}\Big[\sum_{i\in\Delta_r}\sum_{j=1}^{R}\lambda_i^2\int_S \phi_i^2\phi_j^2 dP + \sum_{i=1}^{R}\sum_{j\in\Delta_r}\lambda_j^2\int_S \phi_i^2\phi_j^2 dP\Big] \tag{3.58}
$$

$$
\le \frac{4}{\mu_r^2}n^{-1}\sum_{i,j=1}^{R}(\lambda_i^2 + \lambda_j^2)\int_S \phi_i^2\phi_j^2 dP.
$$

Note that condition (2.8) and the boundness of the λ-spectrum $\lambda(H)$ imply that

$$
\sum_{|i|\le R, |j|\le R}(\lambda_i^2 + \lambda_j^2)\int_S \phi_i^2\phi_j^2 dP = O(R^\alpha) \text{ as } R \to \infty.
$$

Assume in what follows that $R^\alpha \le n$.

The identity (3.54) and the bounds (3.56)–(3.58) imply, under condition (2.8), that with some constant $C > 0$

$$\mathbb{E}\|A_{n,R}^{-1}P_r(\hat{H}_{n,R})A_{n,R}^{-1} - P_r(H_R)\|^2 I_{\{\|E_{n,R}\|<1/2\}} \leq \frac{CR^\alpha}{n}, \quad R \geq 1. \qquad (3.59)$$

Next step is to use isometric representation (2.17). We have

$$\sup_{f,g\in\mathcal{F}} \left| \langle P_r(\tilde{H}_{n,R})\tilde{f}, \tilde{g} \rangle_{L_2(P_n)} - \langle P_r(H_R)f, g \rangle_{L_2(P)} \right|$$

$$= \sup_{f,g\in\mathcal{F}} \left| \langle A_{n,R}^{-1}P_r(\hat{H}_{n,R})A_{n,R}^{-1}T_{n,R}f, T_{n,R}g \rangle_{L_2(P)} - \langle P_r(H_R)f, g \rangle_{L_2(P)} \right|$$

$$\leq \sup_{f,g\in\mathcal{F}} \left| \langle A_{n,R}^{-1}P_r(\hat{H}_{n,R})A_{n,R}^{-1}T_{n,R}f, T_{n,R}g \rangle_{L_2(P)} - \langle P_r(H_R)T_{n,R}f, T_{n,R}g \rangle_{L_2(P)} \right|$$

$$+ \sup_{f,g\in\mathcal{F}} \left| \langle P_r(H_R)T_{n,R}f, T_{n,R}g \rangle_{L_2(P)} - \langle P_r(H_R)f, g \rangle_{L_2(P)} \right|.$$
$$(3.60)$$

Assuming that $\|E_{n,R}\| < 1/2$ and $\zeta_n^2(\mathcal{F}; R) < n$, we can bound the first term of the right hand side of (3.60) as follows:

$$\sup_{f,g\in\mathcal{F}} \left| \langle A_{n,R}^{-1}P_r(\hat{H}_{n,R})A_{n,R}^{-1}T_{n,R}f, T_{n,R}g \rangle_{L_2(P)} - \langle P_r(H_R)T_{n,R}f, T_{n,R}g \rangle_{L_2(P)} \right|$$

$$\leq \left\| A_{n,R}^{-1}P_r(\hat{H}_{n,R})A_{n,R}^{-1} - P_r(H_R) \right\| \sup_{f\in\mathcal{F}} \|T_{n,R}f\|_{L_2(P)}^2$$

$$= \left\| A_{n,R}^{-1}P_r(\hat{H}_{n,R})A_{n,R}^{-1} - P_r(H_R) \right\| \sup_{f\in\mathcal{F}} \sum_{j=1}^{R} \langle f, \phi_j \rangle_{L_2(P_n)}^2$$

$$\leq 2 \left\| A_{n,R}^{-1}P_r(\hat{H}_{n,R})A_{n,R}^{-1} - P_r(H_R) \right\|$$

$$\left\{ \sup_{f\in\mathcal{F}} \sum_{j=1}^{R} \left[\int_S f\phi_j d(P_n - P) \right]^2 + \sup_{f\in\mathcal{F}} \sum_{j=1}^{R} \langle f, \phi_j \rangle_{L_2(P)}^2 \right\}$$

$$\leq 2 \left[n^{-1}\zeta_n^2(\mathcal{F}; R) + \sup_{\mathcal{F}} \|f\|_{L_2(P)}^2 \right] \left\| A_{n,R}^{-1}P_r(\hat{H}_{n,R})A_{n,R}^{-1} - P_r(H_R) \right\|$$

$$\leq 2 \left[1 + \sup_{\mathcal{F}} \|f\|_{L_2(P)}^2 \right] \left\| A_{n,R}^{-1}P_r(\hat{H}_{n,R})A_{n,R}^{-1} - P_r(H_R) \right\|.$$
$$(3.61)$$

It follows from (3.59) and (3.61) that

$$\mathbb{E} \sup_{f,g\in\mathcal{F}} \left| \langle A_{n,R}^{-1}P_r(\hat{H}_{n,R})A_{n,R}^{-1}T_{n,R}f, T_{n,R}g \rangle_{L_2(P)} - \langle P_r(H_R)T_{n,R}f, T_{n,R}g \rangle_{L_2(P)} \right|^2$$

$$\times I_{\{\|E_{n,R}\|<1/2, \zeta_n^2(\mathcal{F};R)<n\}} \leq \frac{CR^\alpha}{n}, \quad R \geq 1, \ n \geq 1.$$
$$(3.62)$$

We also have, by a simple calculation,

$$
\langle P_r(H_R)T_{n,R}f, T_{n,R}g \rangle_{L_2(P)} - \langle P_r(H_R)f, g \rangle_{L_2(P)}
$$
$$
= \sum_{i \in \Delta_r} \left[\langle f, \phi_i \rangle_{L_2(P_n)} \langle g, \phi_i \rangle_{L_2(P_n)} - \langle f, \phi_i \rangle_{L_2(P)} \langle g, \phi_i \rangle_{L_2(P)} \right]
$$
$$
= \sum_{i \in \Delta_r} \int_S f \phi_i d(P_n - P) \langle g, \phi_i \rangle_{L_2(P)} + \sum_{i \in \Delta_r} \int_S g \phi_i d(P_n - P) \langle f, \phi_i \rangle_{L_2(P)}
$$
$$
+ \sum_{i \in \Delta_r} \int_S f \phi_i d(P_n - P) \int_S g \phi_i d(P_n - P).
$$

$$(3.63)$$

The Cauchy inequality gives the bound

$$
\left[\sum_{i \in \Delta_r} \int_S f \phi_i d(P_n - P) \langle g, \phi_i \rangle_{L_2(P)} \right]^2
$$
$$
\leq \sum_{i \in \Delta_r} \left[\int_S f \phi_i d(P_n - P) \right]^2 \sum_{i \in \Delta_r} \langle g, \phi_i \rangle^2_{L_2(P)} \leq \|g\|^2_{L_2(P)} \sum_{i \in \Delta_r} \left[\int_S f \phi_i d(P_n - P) \right]^2.
$$

Since $\mathcal{F}\phi_i \in CLT(P)$, we get

$$
\mathbb{E} \sup_{f,g \in \mathcal{F}} \left| \sum_{i \in \Delta_r} \int_S f \phi_i d(P_n - P) \langle g, \phi_i \rangle_{L_2(P)} \right|
$$
$$
\leq \sup_{g \in \mathcal{F}} \|g\|_{L_2(P)} \mathbb{E} \left(\sum_{i \in \Delta_r} \sup_{f \in \mathcal{F}} \left| \int_S f \phi_i d(P_n - P) \right|^2 \right)^{1/2} \tag{3.64}
$$
$$
\leq \sup_{g \in \mathcal{F}} \|g\|_{L_2(P)} \sum_{i \in \Delta_r} \mathbb{E}^{1/2} \sup_{f \in \mathcal{F}} \left| \int_S f \phi_i d(P_n - P) \right|^2 \leq \frac{C}{n^{1/2}}.
$$

Similarly,

$$
\mathbb{E} \sup_{f,g \in \mathcal{F}} \left| \sum_{i \in \Delta_r} \int_S g \phi_i d(P_n - P) \langle f, \phi_i \rangle_{L_2(P)} \right| \leq \frac{C}{n^{1/2}} \tag{3.65}
$$

and

$$
\mathbb{E} \sup_{f,g \in \mathcal{F}} \left| \sum_{i \in \Delta_r} \int_S f \phi_i d(P_n - P) \int_S g \phi_i d(P_n - P) \right| \leq \frac{C}{n}. \tag{3.66}
$$

In view of (3.64)–(3.66), the identity (3.63) implies that

$$
\mathbb{E} \sup_{f,g \in \mathcal{F}} \left| \langle P_r(H_R)T_{n,R}f, T_{n,R}g \rangle_{L_2(P)} - \langle P_r(H_R)f, g \rangle_{L_2(P)} \right| \leq \frac{C}{n^{1/2}}. \tag{3.67}
$$

Recalling (3.60) and using (3.62), (3.67), we obtain

$$
\mathbb{E} \sup_{f,g \in \mathcal{F}} \left| \langle P_r(\tilde{H}_{n,R})\tilde{f}, \tilde{g} \rangle_{L_2(P_n)} - \langle P_r(H_R)f, g \rangle_{L_2(P)} \right| I_{\{\|E_{n,R}\|<1/2, \varsigma_n^2(\mathcal{F};R) \leq n\}}
$$
$$
\leq \frac{CR^{\alpha/2}}{n^{1/2}}, \ R \geq 1, \ n \geq 1.
$$

$$(3.68)$$

On the other hand, the following bounds are trivial:

$$\sup_{f,g\in\mathcal{F}}\left|\left\langle P_r(\tilde{H}_{n,R})\tilde{f},\tilde{g}\right\rangle_{L_2(P_n)}\right| \leq \sup_{f\in\mathcal{F}}\|\tilde{f}\|^2_{L_2(P_n)} \leq \|\tilde{F}\|^2_{L_2(P_n)}$$

and

$$\sup_{f,g\in\mathcal{F}}\left|\left\langle P_r(H_R)f,g\right\rangle_{L_2(P)}\right| \leq \sup_{f\in\mathcal{F}}\|f\|^2_{L_2(P)} \leq \|F\|^2_{L_2(P)}.$$

Thus, using the condition that $F \in L_4(P)$ and applying the Cauchy and Markov inequalities, we get (for some constant $C > 0$)

$$\mathbb{E}\sup_{f,g\in\mathcal{F}}\left|\left\langle P_r(\tilde{H}_{n,R})\tilde{f},\tilde{g}\right\rangle_{L_2(P_n)}\right.$$
$$\left.-\left\langle P_r(H_R)f,g\right\rangle_{L_2(P)}\right|\left[I_{\{\|E_{n,R}\|\geq 1/2\}}+I_{\{\zeta^2_n(\mathcal{F};R)\geq n\}}\right]$$
$$\leq \mathbb{E}\left[\|\tilde{F}\|^2_{L_2(P_n)}+\|F\|^2_{L_2(P)}\right]\left[I_{\{\|E_{n,R}\|\geq 1/2\}}+I_{\{\zeta^2_n(\mathcal{F};R)\geq n\}}\right]$$
$$\leq \mathbb{E}^{1/2}\left[\|\tilde{F}\|^2_{L_2(P_n)}+\|F\|^2_{L_2(P)}\right]^2\left[\mathrm{Pr}^{1/2}\{\|E_{n,R}\|\geq 1/2\}+\mathrm{Pr}^{1/2}\{\zeta^2_n(\mathcal{F};R)\geq n\}\right]$$
$$\leq C\left[\mathbb{E}^{1/2}\|E_{n,R}\|^2_{HS}+\frac{\mathbb{E}^{1/2}\zeta^2_n(\mathcal{F};R)}{n^{1/2}}\right].$$

Using now (3.20) and conditions (2.8), (2.10), we obtain

$$\mathbb{E}\sup_{f,g\in\mathcal{F}}\left|\left\langle P_r(\tilde{H}_{n,R})\tilde{f},\tilde{g}\right\rangle_{L_2(P_n)}\right.$$
$$\left.-\left\langle P_r(H_R)f,g\right\rangle_{L_2(P)}\right|\left[I_{\{\|E_{n,R}\|\geq 1/2\}}+I_{\{\zeta^2_n(\mathcal{F};R)\geq n\}}\right] \qquad (3.69)$$
$$\leq \frac{CR^{\alpha/2}}{n^{1/2}},\ R\geq 1,\ n\geq 1.$$

Finally, combining (3.68) and (3.69) yields

$$\mathbb{E}\sup_{f,g\in\mathcal{F}}\left|\left\langle P_r(\tilde{H}_{n,R})\tilde{f},\tilde{g}\right\rangle_{L_2(P_n)}-\left\langle P_r(H_R)f,g\right\rangle_{L_2(P)}\right|\leq\frac{CR^{\alpha/2}}{n^{1/2}},\ R\geq 1,\ n\geq 1.$$
$$(3.70)$$

Similarly to (3.38)–(3.40),

$$\mathbb{E}\|\tilde{H}_{n,R}-\tilde{H}_n\|^2_{HS}=\frac{n(n-1)}{n^2}\mathbb{E}(h-h_R)^2(X,Y)+n^{-1}\mathbb{E}(h-h_R)^2(X,X)$$
$$\leq\sum_{k=R+1}^{\infty}\lambda^2_k+n^{-1}\int_S\left(\sum_{k=R+1}^{\infty}\lambda_k\phi^2_k\right)^2dP. \qquad (3.71)$$

Under conditions (2.9) and (1.4′), (3.71) implies that with some constant $C > 0$

$$\mathbb{E}\|\tilde{H}_{n,R}-\tilde{H}_n\|^2_{HS}\leq C(R^{-\beta}+n^{-1}). \qquad (3.72)$$

As in the proof of Theorem 2.2, we take $\delta := \delta_r/2-\varepsilon$. If $\|\hat{H}_{n,R}-H_R\|_{HS}<\delta$, then by Lidskii's inequality and by the isometry, $\delta_2\big(\lambda(\tilde{H}_{n,R}),\lambda(H_R)\big)\leq\delta$ with implication that the δ_r-number of $\tilde{H}_{n,R}$ is greater than or equal to $\delta_r - \delta > 2\varepsilon$. Using Lemma A.1 (in a way similar to getting (3.55)), we have

$$
\begin{aligned}
\|P_r(\tilde{H}_{n,R}) &- P_r(\tilde{H}_n)\| \\
&\leq \frac{2}{\delta_r - \delta}\|\tilde{H}_{n,R} - \tilde{H}_n\|_{HS} I_{\{\|\tilde{H}_{n,R}-\tilde{H}_n\|_{HS}<\varepsilon,\|\hat{H}_{n,R}-H_R\|_{HS}<\delta\}} \\
&\quad + 2I_{\{\|\tilde{H}_{n,R}-\tilde{H}_n\|_{HS}\geq\varepsilon\}} + 2I_{\{\|\hat{H}_{n,R}-H_R\|_{HS}\geq\delta\}} \\
&\leq \frac{2}{\varepsilon}\|\tilde{H}_{n,R} - \tilde{H}_n\| + \frac{2}{\delta}\|\hat{H}_{n,R} - H_R\|.
\end{aligned}
\tag{3.73}
$$

It follows from (3.23), (3.72), (3.73) and conditions of the theorem, that with some constant $C > 0$

$$
\mathbb{E}\|P_r(\tilde{H}_{n,R}) - P_r(\tilde{H}_n)\|^2 \leq C(n^{-1}R^\alpha + R^{-\beta} + n^{-1}).
\tag{3.74}
$$

Using (3.74) and the assumption that $F \in L_4(P)$, we get

$$
\begin{aligned}
\mathbb{E}\sup_{f,g\in\mathcal{F}}&\left|\left\langle P_r(\tilde{H}_{n,R})\tilde{f}, \tilde{g}\right\rangle_{L_2(P_n)} - \left\langle P_r(\tilde{H}_n)\tilde{f}, \tilde{g}\right\rangle_{L_2(P_n)}\right| \\
&\leq \mathbb{E}\|P_r(\tilde{H}_{n,R_n}) - P_r(\tilde{H}_n)\| \sup_{f\in\mathcal{F}}\|\tilde{f}\|_{L_2(P_n)}\sup_{g\in\mathcal{F}}\|\tilde{g}\|_{L_2(P_n)} \\
&\leq \mathbb{E}\|P_r(\tilde{H}_{n,R_n}) - P_r(\tilde{H}_n)\|\, \|\tilde{F}\|^2_{L_2(P_n)} \\
&\leq \mathbb{E}^{1/2}\|P_r(\tilde{H}_{n,R_n}) - P_r(\tilde{H}_n)\|^2\, \mathbb{E}^{1/2}\|\tilde{F}\|^4_{L_2(P_n)} \\
&\leq C\left(n^{-1/2}R^{\alpha/2} + R^{-\beta/2} + n^{-1/2}\right).
\end{aligned}
\tag{3.75}
$$

Since for large enough R $P_r(H_R) = P_r(H)$, we can conclude from (3.70) and (3.75) that for all $R \geq 1$

$$
\begin{aligned}
\mathbb{E}\sup_{f,g\in\mathcal{F}}&\left|\left\langle P_r(\tilde{H}_n)\tilde{f}, \tilde{g}\right\rangle_{L_2(P_n)} \right.\\
&\left. - \left\langle P_r(H)f, g\right\rangle_{L_2(P)}\right| \leq C\left[n^{-1/2}R^{\alpha/2} + R^{-\beta/2} + n^{-1/2}\right].
\end{aligned}
\tag{3.76}
$$

To get (2.12), it is enough now to set $R = n^{1/(\alpha+\beta)}$ in (3.76).

It remains to prove (2.11). First, note that under the condition (2.8),

$$
\begin{aligned}
\mathbb{E}\|\tilde{H}_{n,R} - H_{n,R}\|^2_{HS} &= n^{-2}\mathbb{E}\sum_{i=1}^{n}h_R^2(X_i, X_i) = n^{-1}\mathbb{E}h_R^2(X, X) \\
&= n^{-1}\sum_{i,j=1}^{R}\lambda_i\lambda_j\int_S \phi_i^2\phi_j^2 dP \leq n^{-1}\frac{1}{2}\sum_{i,j=1}^{R}(\lambda_i^2 + \lambda_j^2)\int_S \phi_i^2\phi_j^2 dP \leq \frac{CR^\alpha}{n}.
\end{aligned}
\tag{3.77}
$$

Again, similarly to (3.55) and (3.73), we show that

$$
\|P_r(\tilde{H}_{n,R}) - P_r(H_{n,R})\| \leq \frac{2}{\varepsilon}\|\tilde{H}_{n,R} - H_{n,R}\|_{HS} + \frac{2}{\delta}\|\hat{H}_{n,R} - H_R\|_{HS}.
\tag{3.78}
$$

By (3.77) and (3.78), the following bound holds with some constant $C > 0$

$$\mathbb{E}\|P_r(\tilde{H}_{n,R}) - P_r(H_{n,R})\|^2 \leq \frac{CR^\alpha}{n}.$$

It follows that

$$
\begin{aligned}
&\mathbb{E}\sup_{f,g\in\mathcal{F}}\left|\langle P_r(\tilde{H}_{n,R})\tilde{f}, \tilde{g}\rangle_{L_2(P_n)} - \langle P_r(H_{n,R})\tilde{f}, \tilde{g}\rangle_{L_2(P_n)}\right|\\
&\leq \mathbb{E}\|P_r(\tilde{H}_{n,R}) - P_r(H_{n,R})\|\sup_{f\in\mathcal{F}}\|\tilde{f}\|_{L_2(P_n)}\sup_{g\in\mathcal{F}}\|\tilde{g}\|_{L_2(P_n)}\\
&\leq \mathbb{E}\|P_r(\tilde{H}_{n,R}) - P_r(H_{n,R})\|\|\tilde{F}\|_{L_2(P_n)}^2\\
&\leq \mathbb{E}^{1/2}\|P_r(\tilde{H}_{n,R}) - P_r(H_{n,R})\|^2\mathbb{E}^{1/2}\|\tilde{F}\|_{L_2(P_n)}^4 \leq \frac{CR^{\alpha/2}}{n^{1/2}}.
\end{aligned}
\tag{3.79}
$$

Using (3.70) and (3.79), we obtain

$$\mathbb{E}\sup_{f,g\in\mathcal{F}}\left|\langle P_r(H_{n,R})\tilde{f}, \tilde{g}\rangle_{L_2(P_n)} - \langle P_r(H_R)f, g\rangle_{L_2(P)}\right| \leq \frac{CR^{\alpha/2}}{n^{1/2}}.\tag{3.80}$$

Now, instead of (3.71), we use the bound $\mathbb{E}\|H_{n,R} - H_n\|_{HS}^2 \leq \sum_{k=R+1}^{\infty}\lambda_k^2$, to get instead of (3.75)

$$\mathbb{E}\sup_{f,g\in\mathcal{F}}\left|\langle P_r(H_{n,R})\tilde{f}, \tilde{g}\rangle_{L_2(P_n)} - \langle P_r(H_n)\tilde{f}, \tilde{g}\rangle_{L_2(P_n)}\right| \leq C\left[n^{-1/2}R^{\alpha/2} + R^{-\beta/2}\right].\tag{3.81}$$

The bounds (3.80) and (3.81) allow one to complete the proof of (2.11). $\qquad\square$

Appendix. Some facts on Perturbation Theory.

We formulate here several lemmas of perturbation theory used in the proofs of the main results.

We start with a beautiful inequality due to Lidskii (see Kato (1982)). Recall that $\lambda^{\downarrow}(A)$ denotes the vector of eigenvalues (with multiplicities) of an operator (matrix) A arranged in the non-increasing order.

LIDSKII'S INEQUALITY. *Let A, B be symmetric $d\times d$ matrices Then, for any convex function φ on $\mathbb{R}$,*

$$\sum_{j=1}^{d}\varphi(\lambda_j^{\downarrow}(B) - \lambda_j^{\downarrow}(A)) \leq \sum_{j=1}^{d}\varphi(\lambda_j^{\downarrow}(B - A)).$$

In particular, for all $p \geq 1$

$$\left(\sum_{j=1}^{d}|\lambda_j^{\downarrow}(B) - \lambda_j^{\downarrow}(A)|^p\right)^{1/p} \leq \left(\sum_{j=1}^{d}|\lambda_j^{\downarrow}(B - A)|^p\right)^{1/p}.$$

For $p = +\infty$

$$\max_{1\leq j\leq d}|\lambda_j^{\downarrow}(B) - \lambda_j^{\downarrow}(A)| \leq \max_{1\leq j\leq d}|\lambda_j^{\downarrow}(B - A)|.$$

The case of Lidskii's inequality for $p = 2$ holds for normal matrices as well (but it is formulated differently since there is no natural ordering of complex eigenvalues). This extension is due to Hoffman and Wielandt.

A.1. LEMMA. *Let A be a symmetric $d \times d$ matrix with eigenvalues $\mu_1, \mu_2, \ldots, \mu_k$ of respective multiplicities $m_1, \ldots, m_k$ and eigenspaces $W_1, \ldots, W_k$. Let $P_j(A)$ be the orthogonal projection of $\mathbb{R}^d$ onto W_j, $1 \leq j \leq k$. For $r \leq k$, let*

$$\delta_r := \delta_r(A) :=$$

$$:= \frac{1}{2} \min\Big[\min\{|\mu_i - \mu_j| : 1 \leq i < j \leq r\}, \min\{|\mu_i - \mu_j| : 1 \leq i \leq r, r+1 \leq j \leq k\}\Big].$$

We refer to $\delta_r(A)$ as the δ_r-numbers of A. Let $0 < \varepsilon \leq \delta_r/2$ and let B be another symmetric linear operator with $\|B\|_{HS} < \varepsilon$. Then,

   i) the set of eigenvalues of $A + B$ partitions into $r + 1$ subsets $\Lambda_j(A + B)$, $j = 1, \ldots, r$, and R_r such that

$$\Lambda_j(A + B) \subset B(\mu_j, \varepsilon) \text{ and } dist\big(R_r, \{\mu_1, \ldots, \mu_r\}\big) > 2\delta_r - \varepsilon;$$

   ii) if $P_j(A + B)$ denotes the orthogonal projection onto the direct sum of the eigenspaces of $A + B$ with eigenvalues in the cluster $\Lambda_j(A + B)$, then

$$\text{Tr}\big(P_j(A + B)\big) = \text{Tr}\big(P_j(A)\big), \quad j = 1, \ldots, r;$$

  iii) For all $j = 1, \ldots, r$ $\|P_j(A + B) - P_j(A)\|_{HS} \leq 2\dfrac{\|B\|_{HS}}{\delta_r}$.

  iv) Let $e_1, \ldots, e_d$ be an orthonormal basis of $\mathbb{R}^d$ consisting of eigenvectors of A and let Δ_r denote the set of indices i such that $A e_i = \mu_r e_i$. Let $L_r(B)$ be the symmetric operator on $\mathbb{R}^d$ defined by the equations

$$\langle L_r(B) e_i, e_j \rangle := \begin{cases} \frac{1}{\mu_r - \lambda_j} \langle B e_i, e_j \rangle, & \text{if } i \in \Delta_r \text{ and } j \notin \Delta_r \\ \frac{1}{\mu_r - \lambda_i} \langle B e_i, e_j \rangle, & \text{if } i \notin \Delta_r \text{ and } j \in \Delta_r \\ 0, & \text{otherwise,} \end{cases}$$

where, for all $1 \leq j \leq d$, $\lambda_j = \mu_\ell$ if $A e_j = \mu_\ell e_j$. Then

$$P_r(A + B) = P_r(A) + L_r(B) + S_r,$$

where

$$\|S_r\|_{HS} \leq 2\frac{\|B\|_{HS}^2}{\delta_r^2}, \quad j = 1, \ldots, r.$$

In all bounds the Hilbert-Schmidt norm can be replaced by the operator norm.

The following two lemmas allow us to linearize expressions like $(I + E)^{1/2} K (I + E)^{1/2}$ and $(I + E)^{-1/2} K (I + E)^{-1/2}$.

A.2. LEMMA. *Let K, E be symmetric $d \times d$ matrices. Suppose that $I + E$ is nonnegative definite. Then*

$$\Big\|(I + E)^{1/2} K (I + E)^{1/2} - \big(K + \frac{EK + KE}{2}\big)\Big\|_{HS}$$

$$\leq 3\|E\|_{HS}\big[\|EK\|_{HS} + \|KE\|_{HS}\big].$$

A.3. LEMMA. *Let K, E be symmetric $d \times d$ matrices. Suppose that $\|E\| < 1/2$. Then*

$$\left\| (I + E)^{-1/2} K (I + E)^{-1/2} - \left(K - \frac{EK + KE}{2} \right) \right\|_{HS}$$
$$\leq 6 \|E\|_{HS} \left[\|EK\|_{HS} + \|KE\|_{HS} \right].$$

The proofs of the above (or very similar) statements can be found in Koltchinskii and Giné (1996). They are based on well known facts and methods of perturbation theory for linear operators in finite dimensional Euclidean spaces (see Kato (1982)).

References

[1]   Davies, E.B. *Heat kernels and spectral theory.* Cambridge University Press (1989), Cambridge.

[2]   Dauxois, J., Pousse, A. and Romain, Y. *Asymptotic theory for the principal component analysis of a vector random function: some applications to statistical inference.* J. Multivariate Analysis, 12 (1982), 136-154.

[3]   Dudley, R.M. *A course on empirical processes. Lecture Notes in Matthematics.* 1097 (1984), 1-142. Springer, New York.

[4]   Koltchinskii, V. and Giné, E. *Random matrix approximation of spectra of integral operators.* Preprint (1996).

[5]   Giné, E. and Zhang, C.-H. *On integrability in the LIL for degenerate U-statistics.* J. Theoretical Probab. 9 (1996), 385-412.

[6]   Girko, V.L. *Theory of Random Determinants.* Kluwer Academic Publishers (1990), Dordrecht.

[7]   Kato, T. *A short introduction to perturbation theory for linear operators.* Springer (1982), New York.

[8]   Mehta, M.L. *Random Matrices.* Academic Press (1991). New York.

[9]   Simon, B. *Functional integration and quantum physics.* Academic Press (1979), New York.

[10] van der Vaart, A.W. and Wellner, J. *Weak Convergence and Empirical Processes with applications to Statistics.* Springer (1996), New York.

Vladimir I. Koltchinskii
Dept. of Mathematics and Statistics
The University of New Mexico
Albuquerque, NM 87131-1141
vlad@math.unm.edu

Progress in Probability, Vol. 43
© 1998 Birkhäuser Verlag Basel/Switzerland

A Short Proof of the
Gaussian Isoperimetric Inequality

Michel Ledoux

ABSTRACT. We review, in a self-contained way, a short and simple proof of the Gaussian isoperimetric inequality drawn from the recent works by S. Bobkov [B2] and D. Bakry and the author [B-L].

Let γ denote the canonical Gaussian measure on $\mathbb{R}^n$ with density with respect to Lebesgue measure $(2\pi)^{-n/2} \exp(-|x|^2/2)$. The Gaussian isoperimetric inequality states that among all Borel sets in $\mathbb{R}^n$ with fixed Gaussian measure, half-spaces achieves the minimal (Gaussian) surface measure. In other words, define the Gaussian surface measure of a Borel set A in $\mathbb{R}^n$ as

$$\gamma_s(\partial A) = \liminf_{r \to 0} \frac{1}{r} \big[\gamma(A_r) - \gamma(A)\big] \qquad (1)$$

where A_r is the r-Euclidean open neigborhood of A. Then, if H is a half-space in $\mathbb{R}^n$, that is $H = \{x \in \mathbb{R}^n; \langle x, u \rangle < a\}$, where $|u| = 1$ and $a \in [-\infty, +\infty]$, and if $\gamma(A) = \gamma(H)$, then

$$\gamma_s(\partial A) \geq \gamma_s(\partial H).$$

Let $\Phi(t) = (2\pi)^{-1/2} \int_{-\infty}^{t} e^{-x^2/2} dx$, $t \in [-\infty, +\infty]$, be the distribution function of the canonical Gaussian measure in dimension one and let $\varphi = \Phi'$. Then $\gamma(H) = \Phi(a)$ and $\gamma_s(\partial H) = \varphi(a)$ so that,

$$\gamma_s(\partial A) \geq \varphi(a) = \varphi \circ \Phi^{-1}\big(\gamma(A)\big). \qquad (2)$$

Moreover, half-spaces are the extremal sets in this inequality. In applications, the Gaussian isoperimetric inequality is often used in its integrated version, namely, if $\gamma(A) = \gamma(H) = \Phi(a)$ (or only $\gamma(A) \geq \Phi(a)$), for every $r \geq 0$,

$$\gamma(A_r) \geq \gamma(H_r) = \Phi(a + r). \qquad (3)$$

To see that (2) implies (3), we may assume, by a simple approximation, that A is given by a finite union of open balls. The family of such sets A is closed under the operation $A \to A_r$, $r \geq 0$. Then, the lim inf in (1) is a true limit. Actually, the boundary ∂A of A is a finite union of piecewise smooth $(n-1)$-dimensional surfaces in $\mathbb{R}^n$ and $\gamma_s(\partial A)$ is given by the integral of the Gaussian density along ∂A with respect to Lebesgue's measure on ∂A. Now, by (2), the function $v(r) = \Phi^{-1} \circ \gamma(A_r)$, $r \geq 0$, satisfies

$$v'(r) = \frac{\gamma_s(\partial A_r)}{\varphi \circ \Phi^{-1}(\gamma(A_r))} \geq 1$$

so that $v(r) = \int_0^r v'(s)ds \geq r$, which is (3). (Alternatively, see [B1].)

The Gaussian isoperimetric inequality has been established in 1974 independently by C. Borell [Bo] and V. N. Sudakov and B. S. Tsirel'son [S-T] on the basis of the isoperimetric inequality on the sphere and a limiting procedure known as Poincaré's lemma. A proof using Gaussian symmetrizations was developed by A. Ehrhard in 1983 [E]. The aim of this note is to present a short self-contained proof of this inequality. Our approach will be functional. Denote by $\mathcal{U} = \varphi \circ \Phi^{-1}$ the Gaussian isoperimetric function. In a recent striking paper, S. Bobkov [B2] showed that for every smooth enough function f with values in the unit interval $[0, 1]$,

$$\mathcal{U}\left(\int f d\gamma\right) \leq \int \sqrt{\mathcal{U}^2(f) + |\nabla f|^2}\, d\gamma \tag{4}$$

where $|\nabla f|$ denotes the Euclidean length of the gradient ∇f of f. It is easily seen that inequality (4) is a functional version of the Gaussian isoperimetric inequality (2). Namely, if (4) holds for all smooth functions, it holds for all Lipschitz functions with values in $[0, 1]$. Assume again that the set A in (2) is a finite union of nonempty open balls. In particular, $\gamma(\partial A) = 0$. Apply then (4) to $f_r(x) = (1 - \frac{1}{r}d(x, A))^+$ (where d is the Euclidean distance function). Then $f_r \to I_A$ and $\mathcal{U}(f_r) \to 0$ almost everywhere since $\gamma(\partial A) = 0$ and $\mathcal{U}(0) = \mathcal{U}(1) = 0$. Moreover, $|\nabla f_r| = 0$ on A and on the complement of the closure of A_r, and $|\nabla f_r| \leq \frac{1}{r}$ everywhere. Note that the sets $\partial(A_r)$ are of measure zero for every $r \geq 0$. Therefore

$$\mathcal{U}\big(\gamma(A)\big) \leq \liminf_{r \to 0} \int |\nabla f_r| d\gamma \leq \liminf_{r \to 0} \frac{1}{r}\big[\gamma(A_r) - \gamma(A)\big] = \gamma_s(\partial A).$$

To prove (4), S. Bobkov first establishes the analogous inequality on the two-point space and then uses the central limit theorem, very much as L. Gross in his proof of the Gaussian logarithmic Sobolev inequality [G]. A direct proof of (4) (or, rather, of a slight weakening of (4)) already appeared in the introduction of the joint work [B-L] with D. Bakry. A somewhat different argument is further developed in the second part of [B-L] in the abstract framework of Markov diffusion generators and semigroups. In this note, we simply would like to emphasize this argument in the preceding concrete Gaussian setting, with only elementary calculus and no previous knowledge of whatsover abstract diffusion semigroups. We refer to [B-L] for more on this proof and for extensions and applications. A stochastic calculus version of this proof may be found in the recent work [C-H-L] (in a much broader setting).

Our main tool will be the so-called Ornstein-Uhlenbeck or Hermite semigroup with invariant measure the Gaussian measure γ. For every f, say in $L^1(\gamma)$, set

$$P_t f(x) = \int_{\mathbb{R}^n} f\big(e^{-t}x + (1 - e^{-2t})^{1/2}y\big) d\gamma(y), \quad x \in \mathbb{R}^n, \quad t \geq 0.$$

The operators P_t are contractions on all $L^p(\gamma)$-spaces, and are symmetric and invariant with respect to γ. This means that for every sufficiently integrable functions f and g, and every $t \geq 0$, $\int f P_t g\, d\gamma = \int g P_t f\, d\gamma$. The family $(P_t)_{t \geq 0}$ is a semigroup $(P_s \circ P_t = P_{s+t})$. P_0 is the identity operator whereas, for any f in

$L^1(\gamma)$, $P_t(f)$ converges towards $\int f d\gamma$ as t tends to infinity. All these properties are immediately checked on the preceding integral representation of P_t together with the elementary properties of Gaussian measures. The infinitesimal generator of the semigroup $(P_t)_{t\geq 0}$, that is the operator L such that

$$\frac{d}{dt} P_t f = P_t L f = L P_t f,$$

is acting on all smooth functions f on $\mathbb{R}^n$ by

$$Lf(x) = \Delta f(x) - \langle x, \nabla f(x) \rangle.$$

Moreover, the integration by parts formula for L indicates that, for f and g smooth enough on $\mathbb{R}^n$,

$$\int f(-Lg)d\gamma = \int \langle \nabla f, \nabla g \rangle d\gamma. \tag{5}$$

Let now f be a fixed smooth function on $\mathbb{R}^n$ with values in $[0,1]$. It might actually be convenient to assume throughout the argument that $0 < \varepsilon \leq f \leq 1 - \varepsilon$ and let then ε tend to zero. Recall $\mathcal{U} = \varphi \circ \Phi^{-1}$. To prove (4) it will be enough to show that the function

$$F(t) = \int \sqrt{\mathcal{U}^2(P_t f) + |\nabla P_t f|^2}\, d\gamma$$

is non-increasing in $t \geq 0$. Indeed, if this is the case, $F(\infty) \leq F(0)$, which, together with the elementary properties of P_t recalled above, amounts to (4). To establish this result, we take the derivative of F, and simply check, by elementary calculus, that it is non-positive!

We could stop here since our program is actually fulfilled in this way. We however indicate, in the next few lines, some details on the so-called "elementary calculus". In particular, we strongly emphasize the basic property of the Gaussian isoperimetric function $\mathcal{U}$ that will be used in the argument, namely that $\mathcal{U}$ satisfies the fundamental differential equality $\mathcal{U}\mathcal{U}'' = -1$ (exercise). We now have

$$\frac{dF}{dt} = \int \frac{1}{\sqrt{\mathcal{U}^2(P_t f) + |\nabla P_t f|^2}} \left[\mathcal{U}\mathcal{U}'(P_t f) L P_t f + \langle \nabla(P_t f), \nabla(L P_t f)\rangle \right] d\gamma.$$

To ease the notation, write f for $P_t f$. We also set $K(f) = \mathcal{U}^2(f) + |\nabla f|^2$. Therefore,

$$\frac{dF}{dt} = \int \frac{1}{\sqrt{K(f)}} \left[\mathcal{U}\mathcal{U}'(f) L f + \langle \nabla f, \nabla(L f)\rangle \right] d\gamma. \tag{6}$$

For simplicity in the exposition, let us assume that the dimension n is one, the general case being entirely similar, though notationaly a little bit heavier. By the integration by parts formula (5),

$$\int \frac{1}{\sqrt{K(f)}} \mathcal{U}\mathcal{U}'(f) L f d\gamma = -\int \left(\frac{\mathcal{U}\mathcal{U}'(f)}{\sqrt{K(f)}}\right)' f' d\gamma$$

$$= -\int \frac{1}{\sqrt{K(f)}} \left[\mathcal{U}'^2(f) - 1\right] f'^2 d\gamma$$

$$+ \int \frac{\mathcal{U}\mathcal{U}'(f)f'}{K(f)^{3/2}} \left[\mathcal{U}\mathcal{U}'(f)f' + f'f''\right] d\gamma$$

where we used that $\mathcal{U}\mathcal{U}'' = -1$ and that

$$K(f)' = 2\mathcal{U}\mathcal{U}'(f)f' + \left(f'^2\right)' = 2\mathcal{U}\mathcal{U}'(f)f' + 2f'f''. \tag{7}$$

In order to handle the second term on the right-hand-side of (6), let us note that

$$\langle \nabla f, \nabla(Lf) \rangle = f'\left(f'' - xf'\right)' = -f'^2 + f'Lf'.$$

Hence, again by the integration by parts formula (5), and by (7),

$$\int \frac{1}{\sqrt{K(f)}} \langle \nabla f, \nabla(Lf) \rangle d\gamma = -\int \frac{f'^2}{\sqrt{K(f)}} d\gamma + \int \frac{f'}{\sqrt{K(f)}} Lf' d\gamma$$

$$= -\int \frac{f'^2}{\sqrt{K(f)}} d\gamma - \int \frac{f''^2}{\sqrt{K(f)}} d\gamma$$

$$+ \int \frac{f'f''}{K(f)^{3/2}} \left[\mathcal{U}\mathcal{U}'(f)f' + f'f''\right] d\gamma.$$

Putting these equations together, we get, after some algebra,

$$\frac{dF}{dt} = -\int \frac{1}{K(f)^{3/2}} \left[\mathcal{U}'^2(f)f'^4 - 2\mathcal{U}\mathcal{U}'(f)f'^2 f'' + \mathcal{U}^2(f)f''^2\right] d\gamma$$

and the result follows since

$$\mathcal{U}'^2(f)f'^4 - 2\mathcal{U}\mathcal{U}'(f)f'^2 f'' + \mathcal{U}^2(f)f''^2 = \left(\mathcal{U}'(f)f'^2 - \mathcal{U}(f)f''\right)^2 \geq 0.$$

References

[B-L] D. Bakry, M. Ledoux. Lévy-Gromov's isoperimetric inequality for an infinite dimensional diffusion generator. Invent. math. 123, 259–281 (1996)

[B1] S. Bobkov. A functional form of the isoperimetric inequality for the Gaussian measure. J. Funct. Anal. 135, 39–49 (1996)

[B2] S. Bobkov. An isoperimetric inequality on the discrete cube and an elementary proof of the isoperimetric inequality in Gauss space (1995). Ann. Probability, to appear

[Bo] C. Borell. The Brunn-Minkowski inequality in Gauss space. Invent. math. 30, 207–216 (1975)

[C-H-L] M. Capitaine, E. P. Hsu, M. Ledoux. Martingale representation and a simple proof of logarithmic Sobolev inequalities on path spaces. Preprint (1997)

[E] A. Ehrhard. Symétrisation dans l'espace de Gauss. Math. Scand. 53, 281–301 (1983)

[G] L. Gross. Logarithmic Sobolev inequalities. Amer. J. Math. 97, 1061–1083 (1975)

[S-T] V. N. Sudakov, B. S. Tsirel'son. Extremal properties of half-spaces for spherically invariant measures. J. Soviet. Math. 9, 9–18 (1978); translated from Zap. Nauch. Sem. L.O.M.I. 41, 14–24 (1974)

Département de Mathématiques,
Laboratoire de Statistique et
Probabilités associé au C.N.R.S.,
Université Paul-Sabatier,
31062 Toulouse, France
ledoux@cict.fr

Progress in Probability, Vol. 43
© 1998 Birkhäuser Verlag Basel/Switzerland

Some Shift Inequalities for Gaussian Measures

WENBO V. LI* AND JAMES KUELBS*

ABSTRACT. Let μ be a centered Gaussian measure on a Banach space B and suppose $h \in H_\mu$, the generating Hilbert space of μ. If E is a Borel subset of B, we establish some inequalities between $\mu(E)$ and $\mu(E+h)$ which are similar in spirit to the isoperimetric inequality for Gaussian measures. We also include some applications to precise large deviation probabilities for μ.

1. Introduction

The Cameron-Martin formula provides a precise equality for shifts of a Gaussian measure, and here we present a related inequality. If γ_n is the canonical Gaussian measure on $\mathbb{R}^n$ the shift inequality takes the following form. Throughout $\Phi(\cdot)$ denotes the standard normal distribution on $\mathbb{R}^1$ and $\|\cdot\|_2$ the usual Euclidean norm on $\mathbb{R}^n$.

Theorem 1. Let γ_n be the canonical Gaussian measure on $\mathbb{R}^n$ and assume E is a Borel subset of $\mathbb{R}^n$. If $\theta \in (-\infty, \infty)$ is such that $\mu(E) = \Phi(\theta)$, then for every $\lambda \in \mathbb{R}^1$ and $h \in \mathbb{R}^n$ with $\|h\|_2 = 1$ we have

$$\Phi(\theta - |\lambda|) \le \mu(E + \lambda h) \le \Phi(\theta + |\lambda|). \qquad (1.1)$$

Remark. If $\gamma_n(E) = 0$ (or $\gamma_n(E) = 1$), then by taking $\Phi(-\infty) = 0$ and $\Phi(+\infty) = 1$ we see (1.1) is valid for all $\lambda \in \mathbb{R}^1$. Hence we assume $0 < \mu(E) < 1$ throughout. It is also easy to see from the proof that both inequalities in (1.1) can only be achieved by a half-space perpendicular to h, one for the upper bound and one for the lower bound, but as $|\lambda| \to \infty$ at least one side of the inequality always becomes trivial. Of course, the parameter λ in (1.1) can always be absorbed into the vector h without loss of generality, see (1.3) below.

By a monotone class argument, and some well-known properties of Gaussian measures, one can easily extend Theorem 1 to a Banach space. Hence we will not include details of the proof of this extension, but restrict ourself to a precise statement. Here B denotes a real separable Banach space with norm $\|\cdot\|$ and topological dual B^*, and X is a centered B-valued Gaussian random vector with $\mu = \mathcal{L}(X)$. Hence μ is generated by a Hilbert space H_μ which is the closure of $S(B) = \{\int_B x f(x) d\mu(x) : f \in B^*\}$ in the inner product norm given on $S(B^*)$ by

$$\langle Sf, Sg \rangle_\mu = \int_B f(x)g(x)d\mu(x). \qquad (1.2)$$

*)   Supported in part by NSF grant DMS-9400024, DMS-9503458 & DMS-9627494.

We use $\| \cdot \|_\mu$ to denote the inner product norm induced on H_μ, and for well known properties, and various relationships between μ, H_μ, and B, consider Lemma 2.1 in [K]. These properties are used freely throughout the paper, as well as the fact that the support of μ is $\bar{H}_\mu$; the B-closure of H_μ. Then Theorem 1 implies:

Theorem 1′ Let μ be a centered Gaussian measure on B and assume E is a Borel subset of B. If $\theta \in (-\infty, \infty)$ is such that $\mu(E) = \Phi(\theta)$, then for every $h \in H_\mu$ we have

$$\Phi(\theta - \|h\|_\mu) \le \mu(E + h) \le \Phi(\theta + \|h\|_\mu). \tag{1.3}$$

Although we have never seen Theorem 1 or 1' in print, they are perhaps known by some experts. We learned this after circulating our Gaussian symmetrization proof of the result, and eventually several much simpler proofs emerged in discussions with Michel Ledoux. We thank him for his interest and contributions to these results. We present the simplest of these proofs below, and also some applications of the shift inequality in hopes that it will become more widely known. Our first application deals with the relationship of large deviation results for a Gaussian measure and the shift inequality. This gives another perspective to [KL]. The other applications are intuitive and easily believed, but we do not know how to prove them without the shift inequality. Also, our Theorem 2 in Section 3 provides a sharper result than Theorem 1 when the set E is convex or bounded. All these results are of isoperimetric type over different classes of sets.

2. Proof of Theorem 1

As mentioned earlier, our first proof used the Gaussian symmetrization of sets, but one can also prove the result using Ehrhard's symmetrization of functions developed in [E]. The proof we give here is based on the Cameron-Martin translation theorem as used in the proof of Theorem 3 by [KS].

Let $\langle x, y \rangle$ denote the canonical inner product on $\mathbb{R}^n$ and take $F = \{x \in \mathbb{R}^n : \langle x, h \rangle \le \theta\}$ where $\gamma_n(F) = \Phi(\theta) = \gamma_n(E)$. Then by the Cameron-Martin theorem

$$\begin{aligned}
\gamma_n(E + \lambda h) e^{\lambda^2 \|h\|_2^2 / 2} &= \int_E e^{-\lambda \langle x, h \rangle} d\gamma_n(x) \\
&\le \int_{E \cap F} e^{-\lambda \langle x, h \rangle} d\gamma_n(x) + \int_{E \cap F^c} e^{-|\lambda|\theta} d\gamma_n(x),
\end{aligned} \tag{2.1}$$

and, similarly,

$$\gamma_n(F + \lambda h) e^{\lambda^2 \|h\|_2^2 / 2} \ge \int_{E \cap F} e^{-\lambda \langle x, h \rangle} d\gamma_n(x) + \int_{E^c \cap F} e^{-|\lambda|\theta} d\gamma_n(x) \tag{2.2}$$

Since $\gamma_n(E) = \gamma_n(F)$, $\gamma_n(E \cap F^c) = \gamma_n(E) - \gamma_n(E \cap F)$, and $\gamma_n(E^c \cap F) = \gamma_n(F) - \gamma_n(E \cap F)$, we have

$$\gamma_n(E \cap F^c) = \gamma_n(E^c \cap F). \tag{2.3}$$

Combining (2.1), (2.2) and (2.3) we see

$$\gamma_n(E + \lambda h) \leq \gamma_n(F + \lambda h) = \Phi(\theta + |\lambda|). \tag{2.4}$$

Thus the upper bound in (1.1) holds, and (1.1) follows from the following lemma.

Lemma. If the upper bound in (1.1) holds for all Borel subsets E and all $\lambda > 0$, then (1.1) holds as stated.

Proof. As mentioned previously, if $\mu(E) = 0$ (or $\mu(E) = 1$) the result holds for all $\lambda \in \mathbb{R}$ by setting $\theta = -\infty$ (or $\theta = +\infty$), so we assume $0 < \mu(E) < 1$. For any Borel set A with $0 < \mu(A) < 1$, we let $\theta(A) \in \mathbb{R}^1$ denote the unique number satisfying $\Phi(\theta(A)) = \mu(A)$.

Now $\mu(E + \lambda h) = 1 - \mu(E^c + \lambda h)$, and if $\lambda > 0$ the upper bound in (1.1) implies

$$\mu(E^c + \lambda h) \leq \Phi(\theta(E^c) + \lambda).$$

Now by symmetry we also have $\Phi(\theta(E^c) + \lambda) = 1 - \Phi(\theta(E) - |\lambda|)$ for $\lambda > 0$, so by combining the above when $\lambda > 0$ we have

$$\mu(E + \lambda h) \geq \Phi(\theta(E) - |\lambda|).$$

Thus the lower bound also holds for all $\lambda > 0$.

Now if both the upper bound and lower bound in (1.1) hold for all Borel sets E and all $\lambda > 0$, then it also holds for $\lambda < 0$. This follows since $\lambda < 0$ and μ symmetric implies

$$\mu(E + \lambda h) = \mu(E - |\lambda| h) = \mu(-E + |\lambda| h).$$

Symmetry also implies

$$\mu(-E) = \mu(E) = \Phi(\theta).$$

Hence (1.1) for $\lambda > 0$ implies

$$\Phi(\theta - |\lambda|) \leq \mu(-E + |\lambda| h) \leq \Phi(\theta + |\lambda|),$$

and combining the above we thus have for $\lambda < 0$ that

$$\Phi(\theta - |\lambda|) \leq \mu(E + \lambda h) \leq \Phi(\theta + |\lambda|).$$

Thus (1.1) holds for $\lambda < 0$ as well, and the lemma has been proved. Hence Theorem 1 is proven. $\square$

3. Additional Comments and Comparisons

Let A denote a symmetric (not necessarily convex) subset of $\mathbb{R}^n, h \in R^n$, and $S = \{x : |\langle x, h \rangle| \leq a\}$. If γ_n is the canonical Gaussian measure on $\mathbb{R}^n$ and $a \geq 0$ is such that $\gamma_n(A) = \gamma_n(S)$, then Theorem 3 of [KS] implies

$$\gamma_n(S + h) \leq \gamma_n(A + h). \tag{3.1}$$

Of course, this result also extends to their Theorem 3 using standard arguments just as Theorem 1 implies Theorem 1'. Combining (3.1) and the ideas in the proof of the previous lemma, we see

$$\gamma_n(S + h) \leq \gamma_n(A + h) \leq \gamma_n(T + h) \tag{3.2}$$

where $T = \{x : |\langle x, h \rangle| \geq b\}$ and $b \geq 0$ is such that $\gamma_n(A) = \gamma_n(T)$. Our Theorem 1 is equivalent to the following:

Theorem 1''. If E is a Borel subset of $\mathbb{R}^n$, $h \in \mathbb{R}^n$, $H_- = \{x : \langle x, h \rangle \leq a\}$, and $H_+ = \{x : \langle x, h \rangle \geq b\}$ where a and b are such that $\gamma_n(E) = \gamma_n(H_-) = \gamma_n(H_+)$, then

$$\gamma_n(H_+ + h) \leq \gamma_n(E + h) \leq \gamma_n(H_- + h). \tag{3.3}$$

Our next result is a more general form of Theorem 1''. In particular, it provides sharper estimates than Theorem 1'' when the set E is convex or bounded.

Theorem 2. Let E be a Borel subset of $\mathbb{R}^n$, $h \in \mathbb{R}^n$, and suppose

$$E \subseteq \{x : a \leq \langle x, h \rangle \leq d\}. \tag{3.4}$$

If $S_- = \{x : a \leq \langle x, h \rangle \leq b\}, S_+ = \{x : c \leq \langle x, h \rangle \leq d\}$, are such that

$$\gamma_n(S_-) = \gamma_n(S_+) = \gamma_n(E),$$

then

$$\gamma_n(S_+ + h) \leq \gamma_n(E + h) \leq \gamma_n(S_- + h). \tag{3.5}$$

Proof. By the Cameron-Martin formula we have

$$\gamma_n(E + h) = e^{-\|h\|_2^2/2} \int_E e^{-\langle x, h \rangle} d\gamma_n(x) \tag{3.6}$$

and

$$\gamma_n(S_- + h) = e^{-\|h\|_2^2/2} \int_{S_-} e^{-\langle x, h \rangle} d\gamma_n(x). \tag{3.7}$$

Furthermore,

$$\int_E e^{-\langle x, h \rangle} d\gamma_n(x) = \int_{E \cap S_-} e^{-\langle x, h \rangle} d\gamma_n(x) + \int_{E \cap S_-^c} e^{-\langle x, h \rangle} d\gamma_n(x), \tag{3.8}$$

$$\int_{S_-} e^{-\langle x, h \rangle} d\gamma_n(x) = \int_{E \cap S_-} e^{-\langle x, h \rangle} d\gamma_n(x) + \int_{E^c \cap S_-} e^{-\langle x, h \rangle} d\gamma_n(x), \tag{3.9}$$

and for all $x \in E \cap (S_-)^c, y \in S_- \cap E^c$ we have $\langle x, h \rangle \geq b \geq \langle y, h \rangle \geq a$. Hence

$$
\int_{E \cap S_-^c} e^{-\langle x, h \rangle} d\gamma_n(x) \leq e^{-b} \gamma_n(E \cap S_-^c)
$$

$$
= e^{-b} \gamma_n(S_- \cap E^c) \tag{3.10}
$$

$$
\leq \int_{S_- \cap E^c} e^{-\langle y, h \rangle} d\gamma_n(y),
$$

since $\gamma_n(E \cap S_-^c) = \gamma_n(S_- \cap E^c)$. Combining (3.6)–(3.10) we thus have

$$
\gamma_n(E + h) \leq \gamma_n(S_- + h),
$$

which is the upper bound in (3.5). The proof of the lower bound is similar, and hence Theorem 2 is proven. $\qquad \square$

As a consequence of Theorem 2, we give the following result which provides a sharper estimate than Anderson's inequality for the upper bound, and the well known result for the lower bound regarding the shift of symmetric convex set. Of course, Theorem 2 applies to arbitrary Borel (not necessarily symmetric) sets as well.

Corollary. Let C be a symmetric convex subset of $\mathbb{R}^n$ and $h \in \mathbb{R}^n$. Then

$$
\max(\gamma_n(S_+ + h), \exp(-\|h\|_2^2/2)\gamma_n(C)) \leq \gamma_n(C + h) \leq \min(\gamma_n(S_- + h), \gamma_n(C)) \tag{3.11}
$$

where

$$
S_- = \left\{ x : -\max_{y \in C} \langle y, h \rangle \leq \langle x, h \rangle \leq b \right\}, S_+ = \left\{ x : c \leq \langle x, h \rangle \leq \max_{y \in C} \langle y, h \rangle \right\},
$$

are such that $\gamma_n(S_-) = \gamma_n(S_+) = \gamma_n(C)$.

Proof. Since C is symmetric convex, we have

$$
C \subseteq \left\{ x : -\max_{y \in C} \langle y, h \rangle \leq \langle x, h \rangle \leq \max_{y \in C} \langle y, h \rangle \right\}.
$$

Thus (3.11) follows from Theorem 2 and the following well known facts about the shift of symmetric convex set (see, for example, [DHS]):

$$
\exp(-\|h\|_2^2/2)\gamma_n(C) \leq \gamma_n(C + h) \leq \gamma_n(C). \tag{3.12}
$$

Note that if the symmetric convex set C is unbounded in the direction of h, then we can simply take $d = \infty$ in (3.4) and the set S_+ is a half space in this case.

It is easy to see that in a variety of cases our new bounds are better than the simple but very useful facts given in (3.12), in particular when $\|h\|_2$ is large. This is obvious in terms of the upper bound, but the lower bound is also better when $c < 0$ and $\|h\|_2$ is large. This is interesting since one also knows that

$$
\gamma_n(C + h) \sim \exp(-\|h\|_2^2/2)\gamma_n(C) \text{ as } \|h\|_2 \to \infty.
$$

As can be seen from Theorem 2, the bounds provided here take into account the relative size of C in the direction of the shift, as well as the magnitude of $\gamma_n(C)$ and $\|h\|_2$.

Now we examine how these results relate to "isoperimetric inequalities" over different classes of sets, and mention some related open problems from this point of view. Let $\mathcal{B}(\mathbb{R}^n)$, $\mathcal{S}(\mathbb{R}^n)$, and $\mathcal{C}(\mathbb{R}^n)$ denote the class of Borel sets, symmetric Borel sets and convex sets in $\mathbb{R}^n$, respectively. Fix $0 \le \alpha \le 1$. Theorem 1 tells us

$$\sup\{\gamma_n(E + h) \,:\, \gamma_n(E) = \alpha, E \in \mathcal{B}(\mathbb{R}^n)\} = \gamma_n(H_- + h) \qquad (3.13)$$

for any $h \in \mathbb{R}^n$ where $H_- = \{x : \langle x, h \rangle \le a\}$ is the half space such that $\gamma_n(H_-) = \alpha$, and

$$\inf\{\gamma_n(E + h) \,:\, \gamma_n(E) = \alpha, E \in \mathcal{B}(\mathbb{R}^n)\} = \gamma_n(H_+ + h) \qquad (3.14)$$

for any $h \in \mathbb{R}^n$ where $H_+ = \{x : \langle x, h \rangle \ge b\}$ is the half space such that $\gamma_n(H_+) = \alpha$. The extremal set in (3.13) is given by H_- and by H_+ in (3.14). They are unique up to sets of measure zero.

Our Theorem 2 implies in particular that for any $h \in \mathbb{R}^n$

$$\sup\{\gamma_n(C + h) \,:\, \gamma_n(C) = \alpha, C \in \mathcal{C}(\mathbb{R}^n) \text{ and } C \subseteq \{x : a \le \langle x, h \rangle \le d\}\}$$
$$= \gamma_n(S_- + h) \qquad (3.15)$$

where $S_- = \{x : a \le \langle x, h \rangle \le b\}$ is the slab such that $\gamma_n(S_-) = \alpha$, and for any $h \in \mathbb{R}^n$

$$\inf\{\gamma_n(C + h) \,:\, \gamma_n(C) = \alpha, C \in \mathcal{C}(\mathbb{R}^n) \text{ and } C \subseteq \{x : a \le \langle x, h \rangle \le d\}\}$$
$$= \gamma_n(S_+ + h) \qquad (3.16)$$

where $S_+ = \{x : c \le \langle x, h \rangle \ge d\}$ is the slab such that $\gamma_n(S_+) = \alpha$. The extremal sets here are given by S_- in (3.15) and by S_+ in (3.16), and again are unique in the sense indicated above.

Equation (3.2) implies in particular that

$$\sup\{\gamma_n(A + h) \,:\, \gamma_n(A) = \alpha, A \in \mathcal{S}(\mathbb{R}^n)\} = \gamma_n(P_- + h) \qquad (3.17)$$

for any $h \in \mathbb{R}^n$ where $P_- = \{x : |\langle x, h \rangle| \le a\}$ and a is such that $\gamma_n(P_-) = \alpha$, and

$$\inf\{\gamma_n(A + h) \,:\, \gamma_n(A) = \alpha, A \in \mathcal{S}(\mathbb{R}^n)\} = \gamma_n(P_+ + h) \qquad (3.18)$$

for any $h \in \mathbb{R}^n$ where $P_+ = \{x : |\langle x, h \rangle| \ge b\}$ and b is such that $\gamma_n(P_+) = \alpha$. The extremal sets are given by P_- in (3.17) and by P_+ in (3.18), and are unique as before.

Thus for the simple shift operation there are a variety of isoperimetric results over different classes of sets. On the other hand, there are other useful operations on sets where much less is known at present. Here we only mention two well known ones, namly, addition and dilation of sets.

The isoperimetric property for Gaussian measures states that

$$\inf\{\gamma_{n*}(E + \lambda K) \, : \, \gamma_n(E) = \alpha, E \in \mathcal{B}(\mathbb{R}^n)\} = \gamma_n(H + \lambda K) \tag{3.19}$$

for any $\lambda \geq 0$ where $K = \{x \in \mathbb{R}^n \, : \, \|x\|_2 \leq 1\}$ and H is a half space such that $\gamma_n(H) = \alpha$. Here $A + \lambda K = \{a + \lambda k : a \in A, k \in K\}$, and $\gamma_{n*}(\cdot)$ is the inner measure obtained from γ_n. The relation (3.19) is very powerful, and provides the best results in a variety of settings. It is due independently to Borell[Bo] and Sudakov-Tsierlson[ST]. A beautiful extension for convex Borel sets using Gaussian symmetrizations was given by Ehrhard in [E]. If the inf in (3.19) is replaced by sup, then the righthand term easily is seen to be one. On the other hand, it seems to be a very hard problem (also for sup instead of inf) if we replace $E \in \mathcal{B}(\mathbb{R}^n)$ in (3.19) by $E \in \mathcal{S}(\mathbb{R}^n) \cap \mathcal{C}(\mathbb{R}^n)$. In particular, the extremal sets depend on the parameter α and the number $\lambda > 0$ as can be easily seen in $\mathbb{R}^2$. For example, let $K_b = \{(x, y) : x^2 + y^2 \leq b^2\}$ and $S_a = \{(x, y) : |x| \leq a\}$ where a and b are such that $f(b) = \gamma_2(K_b) = \gamma_2(S_a) = g(a)$. Note that

$$f(b) = \frac{1}{2\pi} \int \int_{x^2 + y^2 \leq b^2} e^{-(x^2 + y^2)/2} dx dy = 1 - e^{-b^2/2},$$

$$g(a) = 1 - \frac{2}{\sqrt{2\pi}} \int_0^a e^{-x^2/2} dx$$

and

$$\gamma_2(K_b + \lambda K) = \gamma_2(K_{b+\lambda}) = f(b + \lambda), \quad \gamma_2(S_a + \lambda K) = \gamma_2(S_{a+\lambda}) = g(a + \lambda).$$

It is easy to see that with fixed $b > a > 0$ from $f(b) = g(a)$, $f(b + \lambda) > g(a + \lambda)$ for λ sufficiently large. On the other hand, we have

$$\frac{\partial}{\partial \lambda} f(b + \lambda)\Big|_{\lambda=0} = be^{-b^2/2} = b(1 - f(b)) = b(1 - g(a)) = \frac{2b}{\sqrt{2\pi}} \int_a^\infty e^{-x^2/2} dx$$

$$< \frac{2}{\sqrt{2\pi}} e^{-a^2/2} = \frac{\partial}{\partial \lambda} g(a + \lambda)\Big|_{\lambda=o}$$

for $a > 0$ sufficiently small since $b > a$ is also very small. It should also be noted that the first inequality conjectured in Problem 3 of the book [Lif], page 277, is false. The extremal set is neither a slab nor a ball depending on different values of α and λ as seen in the above example.

For the dilation operation it is known that any fixed $0 \leq \alpha \leq 1$

$$\inf\{\gamma_{n*}(\lambda E) \, : \, \gamma_n(E) = \alpha, E \in \mathcal{B}(\mathbb{R}^n)\} = \gamma_n(\lambda H) \tag{3.20}$$

for any $\lambda \geq 1$ where H is a half space such that $\gamma_n(H) = \alpha$. The relation was first given in [LS] for $\gamma_n(E) \geq 1/2$ in connection to the exponential integrability of seminorms of Gaussian random vectors. On the other hand, it is still a conjecture if we replace $E \in \mathcal{B}(\mathbb{R}^n)$ in (3.20) by $E \in \mathcal{S}(\mathbb{R}^n) \cap \mathcal{C}(\mathbb{R}^n)$. It was shown in [KS] that if the set E is totally symmetric, that is, symmetric with respect to each coordinate, then the conjecture holds.

4. Some Applications of the Shift Theorem

Our first application involves the connection between large deviation probabilities and the shift inequality for Gaussian measures. If a_t and b_t are non-negative, we write $a_t << b_t$ if $\overline{\lim}_{t \to \infty} a_t/b_t < \infty$, and $a_t \approx b_t$ if both $a_t << b_t$ and $b_t << a_t$. Here we assume D to be an open convex subset of B and μ is a centered Gaussian measure on B. The parameter t is strictly positive. If $D \cap \bar{H}_\mu \neq$ and $0 \notin \bar{D}$, then Proposition 1 of [KL] showed there exists a unique point $h \in \partial D$ and $f \in B^*$ such that $h = Sf$, $D \subseteq \{x : f(x) > f(h)\}$, and $\inf_{x \in D} \|x\|_\mu^2 = \inf_{x \in \bar{D}} \|x\|_\mu^2 = \|h\|_\mu^2$. The point h is called a dominating point of D. Hence by the Cameron-Martin theorem

$$\mu(tD) = \mu(t(D-h)+th) = \exp\left\{-t^2\|h\|_\mu^2/2\right\} \int_{t(D-h)} e^{-tf(x)} d\mu(x) \qquad (4.1)$$

where $h = Sf$. Since $f(x) > f(h)$ for all $x \in D$ and f is a centered Gaussian variable it is easy to see that

$$\mu(tD) << t^{-1} \exp\left\{-t^2\|h\|_\mu^2/2\right\}, \qquad (4.2)$$

which provides an upper bound on $\mu(tD)$.

For a lower bound we consider the lower bound in the shift inequality applied to the middle term in (4.1). This implies

$$\mu(tD) \geq \Phi(\theta_t - t\|h\|_\mu) \qquad (4.3)$$

where θ_t satisfies

$$\mu(t(D-h)) = \Phi(\theta_t). \qquad (4.4)$$

Since $t(D-h)$ is a subset of $\{x : f(x) > 0\}$ for $t > 0$ we have $\mu(t(D-h)) \leq 1/2$ and thus $\theta_t \leq 0$. Hence if D is also an open ball in a 2-smooth Banach space, then Corollary 1 of [KL] implies

$$\overline{\lim_{t \to \infty}} \, t(\frac{1}{2} - \mu(t(D-h))) < \infty. \qquad (4.5)$$

Now (4.4) and (4.5) with $\theta_t \leq 0$ combine to imply $0 \leq -\theta_t \leq t^{-1}$ as $t \to \infty$. Hence (4.2) and (4.3) imply as $t \to \infty$

$$\mu(tD) \approx t^{-1} \exp\{(\theta_t - t\|h\|_\mu)^2/2\} \approx t^{-1} \exp\{-t^2\|h\|_\mu^2/2\}.$$

Of course, the crucial things to prove in the above are that the dominating point exists, and (4.5) holds. This is done in [KL], but once these things are known, the shift inequality applies nicely for the lower bound. This is the more delicate part of the argument, and what was done in [KL] was to show the integral in (4.1) to be of size t^{-1} as $t \to \infty$. These approaches are essentially equivalent, but the shift inequality is definitely more direct.

If U is the open unit ball of B in the norm $\|\cdot\|$, then easy examples in $\mathbb{R}^n$ show that for various norms $\|\cdot\|$ we do not have

$$\lim_{t \to \infty} P(X \in t(U-p)) = 1/2 \qquad (4.6)$$

for every $p \in \mathbb{R}^n$, $\|p\| = 1$. For example, if $\mathcal{L}(X) = \mu$, where μ is the standard normal distribution on $\mathbb{R}^n$, and $\|\cdot\|$ is an ℓ^∞ or ℓ^1 norm, then the limit in (4.6) is 2^{-d} when p is one of the corners of the closed unit ball $\overline{U}$. On the other hand, our next result will show that there always exists p, $\|p\| = 1$, such that (4.6) holds. Hence at such p the boundary of U is rather flat. This is obvious in many cases, and intuitively clear, but the result below is completely general. Also, we do not know how to prove this result without the shift inequality.

Proposition. Let U be the open unit ball of B. Then there exists $h \in \partial U$ such that

$$\lim_{t \to \infty} \mu(t(U - h)) = 1/2. \tag{4.7}$$

Proof. Let K be the unit ball of H_μ, $D = U^c$, and recall μ is a centered non-degenerate Gaussian measure. Hence $K \neq \{0\}$ and if

$$\lambda_0 = \sup\{\lambda > 0 : \lambda K \subseteq \overline{U}\},$$

we have $\lambda_0 K \cap D \neq \phi$. Hence take $h \in \lambda_0 K \cap D$. Then $\|h\|_\mu = \lambda_0 > 0$, $\|h\| = 1$, and by Lemma 2.1 in [K] we have

$$\sigma^2 = \sup_{\|f\|_{B^*} \leq 1} \int_B f^2(x) d\mu(x) = \sup_{x \in K} \|x\|^2.$$

Thus $\sigma^2 = \lambda_0^{-2}$, and [LT, p. 87] implies that, as $t \to \infty$,

$$\mu(tD) = \exp\{-t^2\|h\|_\mu^2/2 + \epsilon(t)t\} \tag{4.8}$$

where $\lim_{t \to \infty} \epsilon(t) = 0$. Hence if $\mu(tD) = \Phi(\theta_t)$, then $\lim_{t \to \infty} \theta_t = -\infty$ and we also have

$$\mu(tD) = \exp\{-\theta_t^2/2 - \log\sqrt{2\pi} - \log|\theta_t| + \delta(t)\} \tag{4.9}$$

where $\lim_{t \to \infty} \delta(t) = 0$. Combining (4.8) and (4.9) it is easy to check that

$$\theta_t = -t\|h\|_\mu + \gamma(t) \tag{4.10}$$

where $\lim_{t \to \infty} \gamma(t) = 0$. Thus

$$\epsilon(t)t = t\|h\|_\mu\gamma(t) - \gamma^2(t)/2 - \log\sqrt{2\pi} - \log|-t\|h\|_\mu + \gamma(t)| + \delta(t),$$

and hence

$$\gamma(t) = \frac{\epsilon(t)}{\|h\|_\mu} + t^{-1}\log(\sqrt{2\pi}\|h\|_\mu t) + o(t^{-1}) \tag{4.11}$$

as $t \to \infty$. Now $\mu(t(U - h)) = 1 - \mu(t(D - h))$, and by the shift inequality and (4.10) we have

$$1/2 \leq \mu(t(D - h)) \leq \Phi(\theta_t + |t|\|h\|_\mu) = \Phi(\gamma(t)) \leq 1/2 + \gamma(t). \tag{4.12}$$

Furthermore, (4.12) implies $\gamma(t)$ as given in (4.1) is non-negative as $t \to \infty$. Taking complements, as $t \to \infty$

$$1/2 - \gamma(t) \le \mu(t(U - h)) \le 1/2$$

where $\lim_{t \to \infty} \gamma(t) = 0$. Thus (4.7) holds. $\qquad\qquad\qquad\qquad\qquad\qquad\qquad\qquad \square$

Finally we present two miscellaneous inequalities which are intuitively obvious, but seemingly not so easy to prove without the shift inequality. They are as follows.

I. If E is *any* norm bounded Borel subset of B with $\mu(E) \ge 1/2$ and $h \in H_\mu$, $\|h\|_\mu = 1$, then $\mu(E^c + \lambda h)$ converges to one as $\lambda \to \infty$ at a rate slower than the μ-measure of the half space $\{x \in B : \langle x, h \rangle^\sim \le \lambda\}$. This follows immediately from Theorem 1 since for $\lambda > 0$

$$\mu(E^c + \lambda h) \le \Phi(\theta + \lambda)$$

where $\Phi(\theta) = \mu(E^c) \le 1/2$. Thus $\theta \le 0$ and

$$\mu(E^c + \lambda h) \le \Phi(\lambda) = \mu(x : \langle x, h \rangle^\sim \le \lambda)$$

since $\langle x, h \rangle^\sim$ is $N(0,1)$ when $\|h\|_\mu = 1$.

II. Let C be an open cone strictly smaller than a half space with vertex at the zero vector in B and assume $\mu(C) > 0$. If $h \in H_\mu \cap C$, $\|h\|_\mu = 1$, it is easy to see that $\lim_{\lambda \to \infty} \mu(C + \lambda h) = 0$ and $\lim_{\lambda \to \infty} \mu(C + \lambda(-h)) = 1$. Since the map $\lambda \to \mu(C + \lambda h)$ is easily seen to be continuous by the Cameron-Martin formula and the dominated convergence theorem, we take λ_0 such that $\mu(C + \lambda_0 h) = 1/2$. Then, as $\lambda \to \infty$,

$$\begin{aligned}
\mu((C + \lambda(-h))^c) &= 1 - \mu(C + \lambda(-h)) \\
&= 1 - \mu(C + \lambda_0 h + (\lambda + \lambda_0)(-h)) \\
&\ge 1 - \Phi(\lambda + \lambda_0)
\end{aligned}$$

by the upper bound in the shift inequality for $E = C + \lambda_0 h$. Hence as $\lambda \to \infty$, $\mu((C + \lambda(-h))^c)$ goes to zero faster than the μ-measure of the half space $\{x : \langle x, h \rangle^\sim \le -\lambda_0 - \lambda\}$.

References

[B]   Borell, C. (1975). The Brunn-Minkowski inequality in Gauss space. *Invent. Math.* **30** 207–216.

[E]   Ehrhard, A. (1983). Symétrisation dans l'espace de Gauss. *Math. Scand.* **53** 281–301.

[DHS] Hoffmann-Jørgensen, J., Shepp, L. A. and Dudley, R. M. (1979). On the lower tail of Gaussian seminorms. *Ann. Probab.* **9** 319–342.

[K]   Kuelbs, J. (1976). A strong convergence theorem for Banach space valued random variables. *Ann. Probab.* **4** 744–771.

[KL]   Kuelbs, J. and Li, W.V. (1994). Some large deviation results for Gaussian measures, *Progress in Probability*, Vol. **35**, 251–270, Birkhäuser, Boston.

[KLL]   Kuelbs, J., Li, W.V. and Linde, W. (1994). The Gaussian measure of shifted balls. *Probab. Theory Rel. Fields* **98** 143–162.

[KS]   Kwapień, S. and Sawa, J. (1993), On some conjecture concerning Gaussian measures of dilatations of convex symmetric sets, *Studia Mathematica*, **105**, 173–187.

[LS]   Landau, H.J. and Shepp, L.A. (1970). On the supremum of a Gaussian process, *Sankhya* **32**, 369–378.

[LT]   Ledoux, M. and Talagrand M. (1991). Probability in Banach Spaces, Ergebnisse der Mathematik und ihrer Grenzgebiete, 3. Folge Band 23, Springer-Verlag, Berlin.

[Lif]   Lifshits, M.A. (1995). Gaussian Random Functions, Kluwer Academic Publishers, Boston.

[ST]   Sudakov, V.N. and Tsirel'son, B.S. (1978). Extremal properties of half-spaces for spherically invariant measures, *J. Soviet Math.* **9**, 9–18 (translated from Zap. Nauch. Sem. L.O.M.I. 41, 14–24 (1974)).

Wenbo V. Li
Department of Mathematical Sciences
University of Delaware
Newark, DE 19716, USA
wli@math.udel.edu

James Kuelbs
Department of Mathematics
University of Wisconsin-Madison
Madison, WI 53706, USA
kuelbs@math.wisc.edu

Progress in Probability, Vol. 43
© 1998 Birkhäuser Verlag Basel/Switzerland

A Central Limit Theorem for the Sock-Sorting Problem

WENBO V. LI AND GEOFFREY PRITCHARD

ABSTRACT. The problem of arranging $2n$ objects into n pairs in a prescribed way, when the objects are presented one at a time in random order, is considered. Using tools from the theory of empirical processes, we derive a functional central limit theorem, with a limiting Gaussian process closely related to the Brownian sheet.

1. Introduction and statement of results

This paper considers the following problem, which seems to have been first proposed in [5]; see also [8] and [10]. (For a related problem, see [6].) A collection of n different pairs of socks are scrambled in a laundry bag. Socks are drawn one at a time from the bag in random order, and laid on a table. When the mate of a sock on the table is drawn, the two are paired and put in a drawer. How much table space is required?

Our main result is a functional central limit theorem for the table-space usage as a function of the number of socks so far drawn. The Gaussian process occurring in the limit is an interesting one which can be described as the diagonal of a constrained Brownian sheet, among other representations. As a corollary, we obtain an asymptotic distribution for the maximum space required, as n becomes large. Simulations have shown that this asymptotic distribution closely approximates the true one as soon as n is greater than about 20.

We now state our main result. Let L_k be the number of socks on the table after k socks have been drawn.

THEOREM 1.1. Let $X_n \in C[0,1]$ be the usual piecewise linear interpolation of the sequence $\{L_k\}_{k=0}^{2n}$, with $X_n(k/2n) = L_k$. Let $q(t) = 2t(1-t)$. Then

$$\frac{X_n(t) - nq(t)}{2\sqrt{n}} \implies \Delta(t).$$

Here $\implies$ denotes weak convergence with respect to the uniform norm topology on $C[0,1]$, and $\Delta(t) = \sigma(t,t)$ with σ the Brownian sheet on the unit square constrained to be 0 on the whole boundary of the square. In other words,

$$\sigma(s,t) = V(s,t) - sV(1,t) - tV(s,1) + stV(1,1)$$

where V is the usual Brownian sheet with $\mathrm{Cov}\,(V(s,t), V(s',t')) = (s \wedge s')(t \wedge t')$.

The proof we will present of Theorem 1.1 will proceed by embedding the process in the uniform distribution on $[0,1]^{2n}$, and using empirical-process theory. This is the "slickest" proof known to us, although it is not the only one. One alternative (our original proof of this result) proceeds by finding an integral operator T such that when T is applied to (a close approximation of) $X_n - nq$ the result is a martingale, making use of a central limit theorem for martingale triangular arrays, and finally transforming back to the original problem by applying the inverse of T. This proof still uses the embedding in the uniform distribution. After we had completed this first proof, it was pointed out to us by D. Mason and Z. Shi that the embedding idea is powerful enough to give a direct empirical-process proof. A third possible approach to this problem (suggested by T. Kurtz) is to apply a diffusion limit theorem for Markov chains (see, for example, Chapter 11 of [7]).

The process σ also appears in empirical-process theory as the limit of the Hoeffding, Blum, Kiefer, Rosenblatt empirical process; see [2].

In §2 we also consider other representations of the limit process Δ, and return to the original question of the maximum table-space usage.

2. Proof and remarks

Proof of Theorem 1.1. Let $\{X_i\}_{i=1}^n, \{Y_i\}_{i=1}^n$ be independent i.i.d. random samples from the uniform distribution on $[0, 1]$. (X_i, Y_i may be thought of as the "times" at which the socks of the ith pair are drawn.) Define empirical distribution functions F_n, G_n by

$$G_n(s, t) = \frac{1}{n} \sum_{i=1}^n 1_{\{X_i \le s, Y_i \le t\}} \quad \text{and} \quad F_n(t) = \frac{1}{2n} \sum_{i=1}^n \left(1_{\{X_i \le t\}} + 1_{\{Y_i \le t\}}\right),$$

for $s, t \in [0, 1]$. Let Q_n denote the quantile function corresponding to F_n, i.e. $Q_n(u) = \inf\{t : F_n(t) \ge u\}$. Then $nG_n(Q_n(k/2n), Q_n(k/2n))$ is the number of completed pairs after k socks have been drawn, i.e. is $(k - L_k)/2$. We thus see that $(X_n(u) - nq(u))/2\sqrt{n}$ is within $O(n^{-1})$ of

$$\frac{2nu - 2nG_n(Q_n(u), Q_n(u)) - nq(u)}{2\sqrt{n}} \tag{1}$$

uniformly in u. So it will suffice to show that quantity (1) converges weakly to $\Delta(u)$, using the Skorokhod topology on the space $D[0, 1]$. To this end, write (1) as

$$\sqrt{n}\left(Q_n(u)^2 - G_n(Q_n(u), Q_n(u))\right) - \sqrt{n}\left(Q_n(u) - u\right)\left(Q_n(u) + u\right).$$

For the purpose of establishing the weak convergence, this may be replaced by

$$\sqrt{n}\left(u^2 - G_n(u, u)\right) - \sqrt{n}\left(u - F_n(u)\right)2u.$$

The replacement of $Q_n(u)$ by u in the first term is a "random time change" (see [1], p.144), justified by the Glivenko-Cantelli theorem ($Q_n(u) \to u$ a.s.; see [4], p.56). The replacement of $Q_n(u) - u$ by $u - F_n(u)$ is justified by the Bahadur-Kiefer theorem (see e.g. [3]). The expression we now have can be written

$$\sqrt{n}\left(u^2 - G_n(u, u)\right) - u\sqrt{n}(u - G_n(u, 1)) - u\sqrt{n}(u - G_n(1, u)).$$

The following functional central limit theorem is well-known (see e.g. [9]):

$$\sqrt{n}(G_n(s,t) - st) \implies B(s,t)$$

where $B(s,t) = V(s,t) - stV(1,1)$, with V the usual Brownian sheet. Since we have reduced our process to a linear transformation of this one, we conclude that it converges weakly to

$$-B(u,u) + uB(1,u) + uB(u,1),$$

which is $-\Delta(u)$. This has the same law as $\Delta(u)$. □

The limiting process Δ can be described in several ways, one of which was given in the statement of Theorem 1.1. Another representation (discovered as a consequence of the alternative proof of Theorem 1.1 mentioned in the introduction) is the stochastic integral:

$$\Delta(t) = (1-t)^2 \int_0^t (2s)^{1/2}(1-s)^{-3/2}dW(s),$$

where W is the usual Wiener process. To check this, it suffices to note that both processes are Gaussian, and observe that their covariance structures are the same, namely

$$\mathrm{Cov}\left(\Delta(s), \Delta(t)\right) = (s \wedge t - st)^2.$$

Another way to view Δ is as a diffusion process, described by the associated stochastic differential equation. This may be easily found from the stochastic integral representation to be

$$d\Delta(t) = -2\left(\frac{\Delta(t)}{1-t}\right) dt + \sqrt{2t(1-t)}dW(t), \qquad \Delta(0) = 0.$$

Compare this with the equation for the more common Brownian bridge process B:

$$dB(t) = -\left(\frac{B(t)}{1-t}\right) dt + dW(t), \qquad B(0) = 0.$$

Though Δ is similar to B in vanishing at the endpoints 0 and 1, its equation has a stronger "drift" coefficient and a variable "speed" coefficient.

Finally, we return to the original problem – that of the maximum table-space usage. An asymptotic law for this quantity is now an easy corollary of our main result.

PROPOSITION 2.1. Let $K_n = \max_{k=0}^{2n} L_k$. Then

$$\frac{K_n - n/2}{\sqrt{n/4}} \implies Y,$$

where Y has a standard normal distribution.

Remark. This says that K_n has the same asymptotic behaviour as L_n; essentially because for large n the maximum L_k will occur for $k \approx n$.

Proof. Let $\Phi(\alpha) = P\left(Y \le \alpha\right)$. We need to show $P\left((K_n - \frac{1}{2}n)/\frac{1}{2}\sqrt{n} > \alpha\right) \to \Phi(\alpha)$ for all $\alpha \in \mathbb{R}$. Noting $K_n \ge L_n$ gives $\underline{\lim}_n P\left((K_n - \frac{1}{2}n)/\frac{1}{2}\sqrt{n} > \alpha\right) \ge \Phi(\alpha)$. Also, with X_n as in Theorem 1.1 and any $c > 0$,

$$\overline{\lim_n} P\left(K_n > \frac{1}{2}n + \frac{1}{2}\sqrt{n}\alpha \right)$$

$$\leq \overline{\lim_n} P\left(\sup_{|t-\frac{1}{2}|\leq c} X_n(t) > \frac{1}{2}n + \frac{1}{2}\sqrt{n}\alpha \right) + \overline{\lim_n} P\left(\sup_{|t-\frac{1}{2}|\geq c} X_n(t) > \frac{1}{2}n + \frac{1}{2}\sqrt{n}\alpha \right)$$

$$\leq \overline{\lim_n} P\left(\sup_{|t-\frac{1}{2}|\leq c} \frac{X_n(t) - nq(t)}{2\sqrt{n}} > \frac{\alpha}{4} \right)$$

$$+ \overline{\lim_n} P\left(\sup_{|t-\frac{1}{2}|\geq c} \frac{X_n(t) - nq(t)}{2\sqrt{n}} > \frac{\frac{1}{2}n + \frac{1}{2}\sqrt{n}\alpha - nq(\frac{1}{2}+c)}{2\sqrt{n}} \right)$$

$$\leq P\left(\sup_{|t-\frac{1}{2}|\leq c} \Delta(t) > \frac{\alpha}{4} \right) + \overline{\lim_n} P\left(\sup_{0\leq t\leq 1} \Delta(t) > c^2\sqrt{n} + \frac{\alpha}{4} \right).$$

The second term is 0 for any $c > 0$, since $\sup_{0\leq t\leq 1}\Delta(t) < \infty$ a.s. The first term goes to $\Phi(\alpha)$ as $c \to 0$ by the Dominated Convergence Theorem, noting that $\lim_{c\to 0}\sup_{|t-\frac{1}{2}|\leq c}\Delta(t) = \Delta(1/2)$ a.s. $\square$

Acknowledgement. The authors wish to thank D. Mason, Z. Shi, and T. Kurtz for useful discussions.

References

[1] P. Billingsley, *Convergence of probability measures* Wiley, New York, 1968.

[2] Csörgo, *Strong approximations of the Hoeffding, Blum, Kiefer, Rosenblatt multivariate empirical process*, J. Multivariate Analysis **9** (1979), 84–100.

[3] P. Deheuvels and D. Mason, *Bahadur-Kiefer-type processes*, Ann. Prob. **18** (1990), 669–697.

[4] R. Durrett, *Probability: Theory and examples* 1st ed., Wadsworth, Pacific Grove, California, 1991.

[5] M. P. Eisner, Problem 216, College Mathematics Journal **13** (1982), 206.

[6] D. M. Friedlen, Problem E3265, Amer. Math. Monthly **97** (1990), 242–244.

[7] S. Ethier and T. Kurtz, *Markov processes: characterization and convergence* Wiley, New York, 1986.

[8] R. Luttmann, Problem E3148, Amer. Math. Monthly **95** (1988), 357–358.

[9] E. Giné and J. Zinn, *Some limit theorems for empirical processes*, Ann. Prob. **12** (1984), 929–989.

[10] S. Rabinowitz *Index to Mathematical Problems 1980–1984*, MathPro Press, Westford, MA, 1992.

Wenbo V. Li
Department of Mathematical Sciences
University of Delaware
Newark, DE 19716, USA
wli@math.udel.edu

Geoffrey Pritchard
Department of Mathematics
Texas A&M University,
College Station, TX 77843, USA
Geoffrey.Pritchard@math.tamu.edu

Progress in Probability, Vol. 43
© 1998 Birkhäuser Verlag Basel/Switzerland

Oscillations of Gaussian Stein's Elements

MIKHAIL LIFSHITS* AND MICHEL WEBER

ABSTRACT. In this paper we investigate the properties of a remarkable class of Gaussian sequences arising from E.M. Stein's probabilistic approach to continuity principle in ergodic theory followed by spectacular applications in real analysis due to J. Bourgain and further investigations of the second named author. For such sequences, we obtain a nearly complete picture of the properties of the oscillations and describe the weak and strong convergence properties of associated sojourn times.

1. Introduction and Results

Let $(X, \mathcal{A}, \mu)$ be a measure space with $\mu(X) = 1$, endowed with an ergodic measure-preserving transformation $T : X \rightarrow X$. Let also $\{g_n, n \in \mathbf{N}\}$ be an isonormal sequence defined on another probability space $(\Omega, \mathcal{B}, P)$. To each element $f \in L^p(\mu)$ we associate the following $L^p(\mu)$-valued random sequence:

$$\forall J \geq 1, \ \forall(\omega, x) \in \Omega \times X, \ F_J(\omega, x) = F_{J,f}(\omega, x) = \frac{1}{\sqrt{J}} \sum_{j \leq J} g_j(\omega) f \circ T^j(x). \quad (1.1)$$

The similar elements were introduced by E.M. Stein [S] in the study of the continuity principle for ergodic transformations where they played a key role in the probabilistic proof of his main result. We refer the reader to the Theorem 1 of [S] for details concerning the randomization technique involving these random elements. Recently, by combining this randomization technique with the theory of Gaussian processes, J. Bourgain ([B], Propositions 1 and 2) discovered a remarquable and useful entropy criterion for the study of the almost everywhere convergence in ergodic theory, completing in some sense Stein's continuity principle. The relative compacity properties of that sequence finally led to obtain various refinements of Bourgain's entropy criterion in [W2, W3].

Although the role of these Gaussian elements in ergodic theory is now self-evident, they constitute a remarkable class of Gaussian sequences independently of this kind of applications. In our opinion, the study of their behavior may successfully contribute to the general theory of Gaussian random functions. In particular, one can adopt a "J-trajectory approach" (with fixed $x \in X$ and varying integer parameter J). From this point of view, the observed structure is similar to that of the averages related to the central limit theorem. We are going to develop this approach in the present publication.

*) Supported by International Science Foundation and Russian Foundation for Basic Research

In order to motivate the work, some elementary considerations are necessary. Since

$$||F_J(.,x)||_{2,P}^2 = \frac{1}{J}\sum_{j=1}^{J}(f \circ T^j(x))^2,$$

by Birkhoff's ergodic theorem

$$\mu\{\lim_{J\to\infty}||F_J(.,x)||_{2,P} = ||f||_{2,\mu}\} = 1. \tag{1.2}$$

Thus, for any real M

$$P\{\limsup_{J\to\infty}|F_J(.,x)| > M\} \geq P\{|\mathcal{N}(0,||f||_{2,\mu})| > M\} > 0,$$

μ a.s. x. And since the law of $\limsup_{J\to\infty}|F_J(.,x)|$ does not depend on the first g_n's, this implies by the 0–1 law that

$$P\{\limsup_{J\to\infty}|F_J(.,x)| = \infty\} = 1, \tag{1.3}$$

μ a.s. x. The regularity of the F_J's will be thus reflected by the magnitude of their oscillations. The study of these oscillations is the main purpose of the present work. Introduce some convenient notations

$$\hat{J} = \hat{J}_f(x) = \sum_{j=1}^{J}(f \circ T^j(x))^2;$$

$$A_J = A_{J,f}(x) = \frac{\hat{J}}{J};$$

$$A_f = A_f(x) = \sup_J A_J;$$

$$\Delta_f = \Delta_f(x) = \left(\sum_{J=1}^{\infty}(A_{J+1} - A_J)^2\right)^{\frac{1}{2}}.$$

We will prove

THEOREM 1.1. *(Boundedness of the oscillations) Let (J_k) be an increasing sequence of positive integers such that the series*

$$Q_1 = \sum_{k=1}^{\infty}\exp\{-MJ_k/(J_{k+1} - J_k)\}$$

converges for some $M > 0$. Then $Q_2 = \sup_k \frac{J_{k+1}-J_k}{J_k} < \infty$ and for each $f \in L^2(X,\mu)$, we have

$$\int_X E \sup_k \sup_{\theta_1,\theta_2\in[J_k,J_{k+1}]}|F_{\theta_1,f} - F_{\theta_2,f}|\,d\mu \leq K, \tag{1.4}$$

where the finite constant K does depend on $M, Q_1, Q_2, \int A_f^{\frac{3}{4}}d\mu$, and $\int \Delta_f^{\frac{3}{4}}d\mu$ only. In particular, we have

$$\sup_k \sup_{\theta_1,\theta_2\in[J_k,J_{k+1}]}|F_{\theta_1,f} - F_{\theta_2,f}| < \infty \quad (\mu \times P) - almost\ surely. \tag{1.5}$$

The size of blocks in this statement is nearly the best possible. Indeed, we will also prove

THEOREM 1.2. *Let (J_k) be an increasing sequence of positive integers satisfying the two following assumptions*

$$\text{The mapping } k \;\rightarrow\; J_{k+1} - J_k \quad \text{is nondecreasing,} \tag{$\mathcal{H}_1$}$$

$$\text{The mapping } k \;\rightarrow\; \frac{J_{k+1} - J_k}{J_k} \quad \text{is nonincreasing.} \tag{$\mathcal{H}_2$}$$

Assume that there exists some ergodic dynamical system $(X, \mathcal{A}, \mu, T)$ and a function $f \in L^2(X, \mu)$, $f \neq 0$ such that (1.5) holds. Then, there exists an $M \in (0, \infty)$ such that

$$Q_1 = \sum_{k=1}^{\infty} \exp\{-M J_k / (J_{k+1} - J_k)\} < \infty. \tag{1.6}$$

These results on oscillations will be complemented by a study of the sojourn time of the sequence F_J in a given measurable subset $\Delta \subset \mathbf{R}^1$. We investigate the "density"

$$d_l(\Delta, x, \omega) = l^{-1} \sum_{J=1}^{l} \mathbf{1}_{F_J(x,\omega) \in \Delta}, \quad l \to \infty. \tag{1.7}$$

Let λ denote Lebesgue measure, $\partial\Delta$ denote the boundary of the set Δ, and $\xrightarrow{\mathcal{D}}$ stand for weak convergence of random variables. We prove

PROPOSITION 1.3 *(Invariance principle) Let $f \in L^2(X, \mu)$ with $||f||_2 = 1$. Let $\Delta \subset \mathbf{R}^1$ be such that $\lambda(\partial\Delta) = 0$. Then, for μ-almost all $x \in X$*

$$d_l(\Delta, x, .) \xrightarrow{\mathcal{D}} I = \int_0^1 \mathbf{1}_\Delta(t^{-\frac{1}{2}} W(t)) \, dt, \quad l \to \infty, \tag{1.8}$$

where (W_t) is the Wiener process.

As a corollary we get

COROLLARY 1.4. *For any $\underline{interval}$ Δ, we have $\mu \times P$-almost surely*

$$\liminf_{l \to \infty} d_l(\Delta, x, \omega) = 0, \qquad \limsup_{l \to \infty} d_l(\Delta, x, \omega) = 1. \tag{1.9}$$

Latter corollary suggests a refinement by substituting to the density (1.7) the logarithmic density:

$$\Theta_l(\Delta, x, \omega) = \left(\log l\right)^{-1} \sum_{J=1}^{l} J^{-1} \mathbf{1}_{F_J(x,\omega) \in \Delta}. \tag{1.10}$$

We indeed have the following

THEOREM 1.5. *(LAW OF THE ITERATED LOGARITHM) Let $f \in L^2(X, \mu)$ with $||f||_2 = 1$. Then, for any Borel set $\Delta \subset \mathbf{R}^1$, we have*

$$\forall \delta > 0, \qquad \lim_{k \to \infty} \frac{\left| \sum_{J=1}^{2^k} \frac{1}{J} \left[\mathbf{1}_{\{F_J(x,\omega) \in \Delta\}} - P\{F_J(x, \omega) \in \Delta\} \right] \right|}{(\log 2^k)^{\frac{1}{2}} (\log \log 2^k)^{2+\delta}} = 0,$$

for μ-almost all $x \in X$ and P-almost all $\omega \in \Omega$.

And as an immediate corollary of the above theorem, we get

COROLLARY 1.6. (STRONG LAW OF LARGE NUMBERS) *Under the above assumptions*

$$\lim_{N \to \infty} \frac{1}{\log N} \sum_{J=1}^{N} \frac{1}{J} 1_{\{F_J(x,\omega) \in \Delta\}} = P\{\zeta \in \Delta\},$$

for μ-almost all $x \in X$ and P-almost all $\omega \in \Omega$, where ζ is a $\mathcal{N}(0,1)$ random variable.

2. Oscillations

2.1 Sufficient conditions. In this section, we prove Theorem 1.1. By Riesz's maximal lemma, the maximal operator A is weak-(2,1):

$$\mu\{A_f > B\} \le B^{-1} \|f\|_2^2, \quad (B > 0). \tag{2.1}$$

According to a recent result due to Jones, Ostrovskii and Rosenblatt [JOR], (see also [LW]) we also know that the second operator Δ is strong-(2,2):

$$\|\Delta_f\|_2 \le C\|f\|_2, \tag{2.2}$$

where $C = \pi\sqrt{6}$. This clearly shows that the constant K occuring in (1.4) depends on *finite* parameters; and in turn depends on M, Q_1, Q_2, and $\|f\|_2$ only. We can now pass to the

Proof of Theorem 1.1: Fix some $x \in X$; and let $W(.) = W^x(.)$ be a Wiener process such that for any J

$$W(\hat{J}) = \sum_{j=1}^{J} f(T^j x) g_j = J^{\frac{1}{2}} F_J. \tag{2.3}$$

Then, for any integer k and $\theta_1, \theta_2 \in [J_k, J_{k+1}]$ we have

$$F_{\theta_1} - F_{\theta_2} = \theta_1^{-\frac{1}{2}} W(\hat{\theta}_1) - \theta_2^{-\frac{1}{2}} W(\hat{\theta}_2) + (\theta_2^{-\frac{1}{2}} - \theta_2^{-\frac{1}{2}}) W(\hat{\theta}_1)$$

$$= (\theta_1^{-\frac{1}{2}} - \theta_2^{-\frac{1}{2}}) W(\hat{\theta}_1) + \theta_2^{-\frac{1}{2}} (W(\hat{\theta}_1) - W(\hat{\theta}_2)))$$

$$\le 2^{-1} J_k^{-\frac{3}{2}} (J_{k+1} - J_k) \sup_{[0,\hat{j}_{k+1}]} |W| + 2 J_k^{-\frac{1}{2}} \sup_{u \in [\hat{j}_k, \hat{j}_{k+1}]} |W(u) - W(\hat{J}_k)|.$$

Concerning the first half of the last expression, we have

$$2^{-1} J_k^{-\frac{3}{2}} (J_{k+1} - J_k) \sup_{[0,\hat{j}_{k+1}]} |W| = 2^{-1} \frac{J_{k+1} - J_k}{J_k} \frac{J_{k+1}^{\frac{1}{2}}}{J_k^{\frac{1}{2}}} \frac{\hat{J}_{k+1}^{\frac{1}{2}}}{J_{k+1}^{\frac{1}{2}}} \hat{J}_{k+1}^{-\frac{1}{2}} \sup_{[0,\hat{j}_{k+1}]} |W|$$

$$\le 2^{-1} \left(\frac{J_{k+1} - J_k}{J_k} \right)^{\frac{1}{2}} Q_2^{\frac{1}{2}} (Q_2 + 1)^{\frac{1}{2}} A^{\frac{1}{2}} \hat{J}_{k+1}^{-\frac{1}{2}} \sup_{[0,\hat{j}_{k+1}]} |W|$$

$$= K \left(\frac{J_{k+1} - J_k}{J_k} \right)^{\frac{1}{2}} A^{\frac{1}{2}} \sup_{[0,1]} |W_{1,k}|,$$

where $W_{1,k}$ is a Wiener process. Concerning the second half, we observe

$$2J_k^{-\frac{1}{2}} \sup_{u \in [\hat{J}_k, \hat{J}_{k+1}]} |W(u) - W(\hat{J}_k)|$$

$$= \left(\frac{\hat{J}_{k+1} - \hat{J}_k}{J_k}\right)^{\frac{1}{2}} \sup_{[0,1]} |W_{2,k}|,$$

where $W_{2,k}$ is another Wiener process. Moreover,

$$\frac{\hat{J}_{k+1} - \hat{J}_k}{J_k} + \frac{\hat{J}_{k+1}}{J_{k+1}} - \frac{\hat{J}_{k+1}}{J_{k+1}} = A_{k+1} - A_k + \frac{J_{k+1} - J_k}{J_k} \frac{\hat{J}_{k+1}}{J_{k+1}}$$

$$\leq |A_{k+1} - A_k| + \frac{J_{k+1} - J_k}{J_k} A.$$

Putting now all our estimations together, leads us to

$$\mathrm{E} \sup_k \sup_{\theta_1, \theta_2 \in [J_k, J_{k+1}]} |F_{\theta_1} - F_{\theta_2}|$$

$$\leq K A^{\frac{1}{2}} \mathrm{E} \sup_k \left(\frac{J_{k+1} - J_k}{J_k}\right)^{\frac{1}{2}} \sup_{0 \leq u \leq 1} |W_{1,k}(u)| + 2\mathrm{E} \sup_k |A_{k+1} - A_k|^{\frac{1}{2}} \sup_{[0,1]} |W_{2,k}|,$$
$$(2.4)$$

where $W_{i,k}$ are Wiener processes (there is no assumption concerning their mutual independence). Now, we are ready to apply the following lemma which goes back to more general results on Gaussian processes.

LEMMA 2.1. *Let (ζ_k) be a sequence of standard $\mathcal{N}(0,1)$ random variables and $\eta_k = |\zeta_k|$. Let $\sigma_k > 0$ and $\sigma = \sup_k \sigma_k$. Then, for each $m > 0$, we have*

$$\mathrm{E} \sup_k \sigma_k \eta_k \leq \left(2 \log_+ \sum_{k=1}^\infty \exp\{-m\sigma_k^{-2}\}\right)^{\frac{1}{2}} \sigma + 3m^{\frac{1}{2}} + 2\sigma. \qquad (2.5)$$

Proof of the lemma: Let $\mathcal{K}_n = \{k : 2^{-n-1}\sigma < \sigma_k \leq 2^{-n}\sigma\}$. Let $N_n = Card(\mathcal{K}_n)$. Set $L_n = 2^{-n}\sigma(2\log N_n)^{\frac{1}{2}}$ and $S_n = \sup_{k \in \mathcal{K}_n} \sigma_k \eta_k$. Then,

$$\mathrm{E}\, S_n \mathbf{1}_{\{S_n > L_n\}} \leq \sum_{k \in \mathcal{K}_n} \mathrm{E} \sigma_k \eta_k \mathbf{1}_{\{\sigma_k \eta_k > L_n\}}$$

$$\leq 2^{-n}\sigma N_n \mathrm{E}\, \eta_1 \mathbf{1}_{\{\eta_1 > (2\log N_n)^{\frac{1}{2}}\}} = 2^{-n}\sigma N_n (2/\pi)^{\frac{1}{2}} N_n^{-1} \leq 2^{-n}\sigma.$$

Further,

$$\mathrm{E} \sup_k \sigma_k \eta_k = \mathrm{E} \sup_{n: N_n > 0} S_n \leq \sup_{n: N_n > 0} L_n + \sum_{n: N_n > 0} \mathrm{E} S_n \mathbf{1}_{\{S_n > L_n\}}$$

$$\leq \sup_{n: N_n > 0} L_n + \sum_{n=0}^\infty 2^{-n}\sigma \leq \sup_{n: N_n > 0} L_n + 2\sigma.$$

Moreover, for each n such that $N_n > 0$, we have

$$\left(2\log_+ \sum_{k=1}^\infty \exp\{-m\sigma_k^{-2}\}\right)^{\frac{1}{2}}\sigma \geq \left(2\log_+(N_n \exp\{-2^{2n+2}m\sigma^{-2}\})\right)^{\frac{1}{2}} 2^{-n}\sigma$$

$$\geq \left(2\log N_n - 2^{2n+3}m\sigma^{-2}\right)_+^{\frac{1}{2}} 2^{-n}\sigma \geq L_n - 2^{\frac{3}{2}}m^{\frac{1}{2}},$$

which proves the lemma. $\qquad \square$

Applying then (2.5) to the first part of (2.4), with the choices $m = M$, $\sigma_k = (\frac{J_{k+1} - J_k}{J_k})^{\frac{1}{2}}$, produces a bound equal to $KA^{\frac{1}{2}}$. Applying next (2.5) to the second half of (2.4) with the choices $m = 1$, $\sigma_k = |A_{k+1} - A_k|^{\frac{1}{2}}$, $\sigma \leq A^{\frac{1}{2}}$, also leads to the bound

$$KA^{\frac{1}{2}} \left(2\log_+ \sum_{k=1}^{\infty} \exp\{-|A_{k+1} - A_k|^{-1}\} \right)^{\frac{1}{2}} \sigma + KA^{\frac{1}{2}} + K.$$

Now, we apply the obvious inequality $\exp\{-u\} \leq u^{-2}$ with $u = |A_{k+1} - A_k|^{-1}$ and thus replace the sum in the last expression by $\Delta^2 = \sum_{k=1}^{\infty} |A_{k+1} - A_k|^2$. It remains then to study the integral

$$\int_X A(x)^{\frac{1}{2}} [\log_+ \Delta(x) + 1] \mu(dx).$$

We use the inequality $\log_+ \Delta \leq 2\Delta^{\frac{1}{4}}$, and apply next Hölder's inequality, which provides

$$\int_X A(x)^{\frac{1}{2}} \Delta(x)^{\frac{1}{4}} \mu(dx) \leq \left(\int_X A(x)^{\frac{3}{4}} \mu(dx) \right)^{\frac{2}{3}} \left(\int_X \Delta(x)^{\frac{3}{4}} \mu(dx) \right)^{\frac{1}{3}} \leq K.$$

Theorem 1.1 is thus proved. □

2.2 Necessary Conditions.
In this section, we prove Theorem 1.2. We split the proof in four steps.

Step 1: *(Exponential consolidation of the sequence (J_k))*
Under assumption $(\mathcal{H}_2)$, $b = \sup_k \frac{J_{k+1} - J_k}{J_k} < \infty$. Put $B = (b+1)^2$ and for each integer l

$$\mathcal{J}_l = \{k : J_k \in [B^l, B^{l+1})\}, \qquad N_l = Card(\mathcal{J}_l).$$

Let $k, k' \in \mathcal{J}_l$ with $k \leq k'$. Then, by $(\mathcal{H}_1)$, $(\mathcal{H}_2)$, we have

$$J_{k+1} - J_k \leq J_{k'+1} - J_{k'} \leq \frac{J_{k'+1} - J_{k'}}{J_{k'}} \cdot J_{k'} \cdot \frac{J_k}{B^l}$$

$$\leq \frac{J_{k+1} - J_k}{J_k} \cdot J_k \cdot \frac{J_{k'}}{B^l} \leq B(J_{k+1} - J_k).$$

Thus,

$$\max_{k \in \mathcal{J}_l}(J_{k+1} - J_k) \leq B \min_{k \in \mathcal{J}_l}(J_{k+1} - J_k). \qquad (2.6)$$

By the definition of B we also have for each $k \in \mathcal{J}_l$

$$J_{k+1} = J_k + (J_{k+1} - J_k) \leq J_k(1 + b) \leq J_k B^{1/2} \leq B^{l+3/2},$$

$$J_k = J_{k-1} + (J_k - J_{k-1}) \leq J_{k-1}(1 + b) \leq J_{k-1} B^{1/2}.$$

It follows that $\sup_{k \in \mathcal{J}_l} J_{k+1} \le B^{l+3/2}$ and $\inf_{k \in \mathcal{J}_l} J_k \le B^{l+1/2}$. Hence the following chain of inequalities is true

$$\frac{N_l}{\max_{k \in \mathcal{J}_l} \frac{J_k}{(J_{k+1} - J_k)}} \ge \frac{N_l \cdot \min_{k \in \mathcal{J}_l}(J_{k+1} - J_k)}{B^{l+1}} \ge \frac{N_l \cdot \max_{k \in \mathcal{J}_l}(J_{k+1} - J_k)}{B^{l+2}}$$

$$\ge \frac{\sum_{k \in \mathcal{J}_l}(J_{k+1} - J_k)}{B^{l+2}} \ge \frac{B^{l+1} - B^{l+\frac{1}{2}}}{B^{l+2}} = \frac{B - \sqrt{B}}{B^2}. \qquad (2.7)$$

Similarly, we also have

$$\frac{N_l}{\min_{k \in \mathcal{J}_l} \frac{J_k}{(J_{k+1} - J_k)}} \le \frac{N_l \cdot \max_{k \in \mathcal{J}_l}(J_{k+1} - J_k)}{B^l} \le \frac{B \cdot N_l \cdot \min_{k \in \mathcal{J}_l}(J_{k+1} - J_k)}{B^l}$$

$$\le \frac{B \cdot \sum_{k \in \mathcal{J}_l}(J_{k+1} - J_k)}{B^l} \le \frac{B(B^{l+\frac{3}{2}} - B^l)}{B^l} \le B(B^{\frac{3}{2}} - 1). \qquad (2.8)$$

Consequently, condition (1.6) can be rewritten in the following more convenient form: for some $M \in (0, \infty)$

$$\sum_{l \ge 1} N_l \exp\{-M N_l\} < \infty. \qquad (1.6^*)$$

Indeed,

$$(1.6) \quad \Leftrightarrow \sum_{k=1}^{\infty} \exp\{-M J_k/(J_{k+1} - J_k)\} < \infty$$

$$\Leftrightarrow \sum_l \sum_{k \in \mathcal{J}_l} \exp\{-M J_k/(J_{k+1} - J_k)\} < \infty;$$

and thus (1.6) implies

$$\sum_{l \ge 1} N_l \exp\{-M \frac{B^2}{B - \sqrt{B}} N_l\} < \infty.$$

In the opposite direction, we also have

$$(1.6^*) \quad \Leftrightarrow \sum_{l \ge 1} N_l \exp\{-M B(B^{\frac{3}{2}} - 1) \min_{k \in \mathcal{J}_l} \frac{J_k}{J_{k+1} - J_k}\} < \infty$$

$$\Rightarrow \sum_{l \ge 1} \sum_{k \in \mathcal{J}_l} \exp\{-M B(B^{\frac{3}{2}} - 1) \frac{J_k}{J_{k+1} - J_k}\} < \infty$$

$$\Leftrightarrow \sum_k \exp\{-M B(B^{\frac{3}{2}} - 1) \frac{J_k}{J_{k+1} - J_k}\} < \infty.$$

Step 2: (Reduction to bounded functions)
Without loss of generality, we can indeed assume that our function f belongs to

$L^\infty(X, \mu)$. This follows from the general Fernique-Sudakov comparison principle (*cf* Chapter 14 in [L]), by means of which (1.5) implies that the same property holds for the sequence (F_{J,f_A}) where $f_A = f\mathbf{1}_{\{|f| \le A\}}$. Consequently, we will and do assume in what follows that $||f||_\infty < \infty$. In this case, we also have $||f||_2 \in (0, \infty)$.

Step 3: (Convergence of some random series)
By our assumption (1.5), there is a measurable set X' of x's of unit μ-measure, such that for any $x \in X'$

$$P\{\sup_k |F_{J_{k+1}} - F_{J_k}| < \infty\} = 1.$$

We fix the variable x in X'. Then, writing $F_J(x) = F_J$, $J \ge 1$, in what follows, we have for each A large enough,

$$0 < P\{\sup_k |F_{J_{k+1}} - F_{J_k}| \le A\}.$$

Let us consider the conditional variance

$$\sigma_k^2 = \mathrm{Var}\Big(F_{J_{k+1}} - F_{J_k} \mid \{F_{J_{k'+1}} - F_{J_{k'}}\}, 1 \le k' < k\Big).$$

Then

$$\sigma_k^2 \ge \mathrm{Var}\Big(F_{J_{k+1}} - F_{J_k} \mid \{g_j, 1 \le j \le J_k\}\Big)$$

$$= \frac{1}{J_{k+1}} \sum_{j=J_k+1}^{J_{k+1}} f(T^j x)^2 = \frac{\hat{J}_{k+1} - \hat{J}_k}{J_{k+1}}.$$

By Proposition 3.3 of [L], it follows that

$$P\{\sup_k |F_{J_{k+1}} - F_{J_k}| \le A\} \le \Pi_{k=2}^\infty P\{\sigma_k|\zeta| \le A\},$$

where we note by ζ a standard $\mathcal{N}(0, 1)$ random variable. And therefore,

$$\sum_k \exp\{-\frac{A^2}{2\sigma_k^2}\} < \infty,$$

which implies

$$\sum_k \exp\{-\frac{A^2 J_{k+1}}{2(\hat{J}_{k+1} - \hat{J}_k)}\} < \infty. \qquad (2.9)$$

To conclude, we only have to replace $\hat{J}_k$ by J_k in the above expression.

Step 4: (Conclusion)
We will deduce (1.6*) from (2.9). We fix l and put $\Delta_k = \hat{J}_{k+1} - \hat{J}_k$, $k \in \mathcal{J}_l$. Let $\{\Delta_k^*, 1 \le k \le N_l\}$ be the nondecreasing rearrangement of the sequence $\{\Delta_k, 1 \le$

$k \leq N_l\}$. Fix also some $\varepsilon > 0$. According to Birkhoff's theorem, the following inequality

$$\mid \frac{\hat{J}_k}{J_k} - ||f||_2^2 \mid \leq \varepsilon,$$

holds for μ-almost all x in X provided that k is large enough ($k \geq k_0(x)$). Moreover,

$$\Delta_k = \sum_{j=J_k+1}^{J_{k+1}} f(T^j x)^2 \leq (J_{k+1} - J_k)||f||_\infty^2.$$

Then,

$$\sum_{k=1}^{N_l} \Delta_k^* = \sum_{k \in \mathcal{J}_l} \Delta_k = \sum_{k \in \mathcal{J}_l} (\hat{J}_{k+1} - \hat{J}_k) = \sup_{k \in \mathcal{J}_l} \hat{J}_{k+1} - \inf_{k \in \mathcal{J}_l} \hat{J}_k$$

$$\geq (||f||^2 - \varepsilon) \sup_{k \in \mathcal{J}_l} J_{k+1} - (||f||^2 + \varepsilon) \inf_{k \in \mathcal{J}_l} J_k \geq (||f||^2 - \varepsilon)B^{l+1} - (||f||^2 + \varepsilon)B^{l+\frac{1}{2}}$$

$$\geq [||f||^2(B - \sqrt{B}) - \varepsilon(B + \sqrt{B})]B^l = CB^l,$$

with $C > 0$, provided that ε is chosen small enough, that we do.

Let now $0 < \alpha < 1$ be fixed. Then,

$$\sum_{k=1}^{N_l} \Delta_k^* = \sum_{k=1}^{(1-\alpha)N_l} \Delta_k^* + \sum_{k=(1-\alpha)N_l+1}^{N_l} \Delta_k^* \leq N_l \Delta_{(1-\alpha)N_l}^* + \alpha N_l \cdot \sup_{k \in \mathcal{J}_l} \Delta_k$$

$$\leq N_l \Delta_{(1-\alpha)N_l}^* + \alpha N_l ||f||_\infty^2 \sup_{k \in \mathcal{J}_l} (J_{k+1} - J_k).$$

Recall, according to (2.6), that

$$N_l \cdot \sup_{k \in \mathcal{J}_l} (J_{k+1} - J_k) \leq BN_l \inf_{k \in \mathcal{J}_l} (J_{k+1} - J_k) \leq B \sum_{k \in \mathcal{J}_l} (J_{k+1} - J_k)$$

$$\leq B(B^{l+\frac{3}{2}} - B^l) = (B^{\frac{5}{2}} - B)B^l.$$

We thus have,

$$N_l \Delta_{(1-\alpha)N_l}^* \geq \sum_{k=1}^{N_l} \Delta_k^* - \alpha ||f||_\infty^2 (B^{\frac{5}{2}} - B)B^l$$

$$\geq [C - \alpha ||f||_\infty^2 (B^{\frac{5}{2}} - B)]B^l = C_1 B^l,$$

where $C_1 > 0$ provided that α is chosen sufficiently small, that we do assume. The implication $(2.9) \Leftrightarrow (1.6^*)$ will finally result from the following estimates:

$$\sum_{k \in \mathcal{J}_l} \exp\{-\frac{A^2 J_{k+1}}{2(\hat{J}_{k+1} - \hat{J}_k)}\} \geq \sum_{k \in \mathcal{J}_l} \exp\{-\frac{A^2 B^{l+\frac{3}{2}}}{2\Delta_k}\}$$

$$\geq \sum_{(1-\alpha)N_l \leq k \leq N_l} \exp\{-\frac{A^2 B^{l+\frac{3}{2}}}{2\Delta_k^*}\} \geq \alpha N_l \exp\{-\frac{A^2 B^{l+\frac{3}{2}}}{2\Delta_{(1-\alpha)N_l}^*}\}$$

$$\geq \alpha N_l \exp\{\frac{-A^2 B^{\frac{3}{2}} B^l N_l}{2C_1 B^l}\}\} = \alpha N_l \exp\{\frac{-A^2 B^{\frac{3}{2}}}{2C_1} N_l\}. \qquad \square$$

3. Densities

3.1 Mean Density. In this section, we give the proofs of Proposition 1.3 and its Corollary 1.4. We start with the

Proof of Proposition 1.3. We still use $\hat{J}(x) = \hat{J}_f(x)$, defined in Section 1 for any positive integer J. Recall that, by virtue of Birkhoff ergodic theorem:

$$\frac{\hat{J}(x)}{J} \longrightarrow 1, \tag{3.1}$$

as J tends to infinity, μ-almost surely. Fix an x satisfying the above property.

We will use the natural embedding of F_J into the Wiener process. More precisely, if $\tilde{W}$ is a Wiener process, then we have the equalities of the laws

$$\{F_J(x, .)\}_{J \geq 1} \overset{\mathcal{D}}{=} \Big\{ J^{-\frac{1}{2}} \tilde{W}\Big(\sum_{j=1}^{J} f(T^j x)^2\Big)\Big\}_{J \geq 1}$$

$$= \big\{ J^{-\frac{1}{2}} \tilde{W}(\hat{J}(x)) \big\}_{J \geq 1} \overset{\mathcal{D}}{=} \big\{ (J/l)^{-\frac{1}{2}} W(\hat{J}(x)/l) \big\}_{J \geq 1},$$

where $W(u) = \tilde{W}(lu)l^{-\frac{1}{2}}$ also is a Wiener process. Thus,

$$d_l(\Delta, x, .) \overset{\mathcal{D}}{=} d_l^W = l^{-1} \sum_{J=1}^{l} \mathbf{1}_{\{(J/l)^{-\frac{1}{2}} W(\hat{J}(x)/l) \in \Delta\}}.$$

It will be more convenient to work with the following object

$$\hat{d}_l^W = l^{-1} \sum_{J=1}^{l} \mathbf{1}_{\{(\hat{J}/l)^{-\frac{1}{2}} W(\hat{J}(x)/l) \in \Delta\}}.$$

This one can be viewed as

$$\hat{d}_l^W = \lambda_l(V),$$

where

$$\lambda_l = \lambda_l(x) = l^{-1} \sum_{J=1}^{l} \delta_{\theta_j/l}$$

is a deterministic nonnegative measure on $\mathbf{R}^+$, in the definition of which δ_a stands for the Dirac measure at the point a and

$$V = V(\omega) = \{t \in [0, 1] \ : \ t^{-\frac{1}{2}} W(t) \in \Delta\}.$$

Then, as a direct consequence of (3.1), we have that λ_l converges weakly to the restricted Lebesgue measure $\lambda^{(1)}(dt) = \mathbf{1}_{[0,1]}(t)dt$, as l tends to infinity. Since $P\{\lambda^{(1)}(\partial A) = 0\} = 1$, weak convergence implies

$$\hat{d}_l^W = \lambda_l(V) \overset{\mathcal{D}}{\longrightarrow} \lambda^{(1)}(V) = I, \tag{3.2}$$

almost surely. Thus, we deduce that $\hat{d}_l^W \to I$, almost surely.

Moreover, the property (3.1) together with the condition $\lambda(\partial \Delta) = 0$ easily imply

$$\lim_{l \to \infty} \mathbf{E} \, |d_l^W - \hat{d}_l^W| = 0.$$

It follows now from (3.2) that $d_l^W \overset{\mathcal{D}}{\longrightarrow} I$ as l tends to infinity. Therefore $d_l(\Delta, x, .)$ $\overset{\mathcal{D}}{\longrightarrow} I$ as l tends to infinity. $\qquad \square$

Proof of Corollary 1.4. Let $0 < \varepsilon < 1$ be fixed. It follows from the Proposition 1.3, that

$$P\{\omega \ : \ \limsup_{l \to \infty} d_l(\Delta, x, \omega) \geq 1 - \varepsilon\} \geq \limsup_{l \to \infty} P\{\omega \ : \ d_l(\Delta, x, \omega) \geq 1 - \varepsilon\}$$

$$= P\{I \geq 1 - \varepsilon\} > 0.$$

And this, by applying the $0 - 1$ law, shows that

$$P\{\omega \ : \ \limsup_{l \to \infty} d_l(\Delta, x, \omega) = 1\} = 1.$$

The proof is thus achieved. $\qquad\square$

3.2 Logarithmic Density.

In this section, we prove Theorem 1.5 and its Corollary 1.6. We start with the

Proof of the Theorem 1.4. The following inequality of I.S. Gaal and J.F. Koksma will be our basic tool.

LEMMA 3.1. ([PS], Theorem A1, p.134) *Let* (B_i) *be a sequence of centered random variables such that* $\mathrm{E}\, B_i^2 < \infty$ *for all* $i \geq 1$. *Assume that there exist two constants* $\sigma > 0$ *and* $C > 0$ *such that*

$$\forall m \geq 0, \ \forall n > 0, \qquad \mathrm{E}\, \Big(\sum_{i=m+1}^{m+n} B_i \Big)^2 \leq C\big((m+n)^{\sigma} - n^{\sigma}\big). \qquad (3.3)$$

Then, for any $\delta > 0$,

$$\sum_{i=1}^{k} B_i = O\Big(k^{\frac{\sigma}{2}} \big(\log k\big)^{2+\delta}\Big), \qquad k \to \infty, \qquad (3.4)$$

almost surely.

Fix an x such that the relation (3.1) holds. Thus $A_f(x) < \infty$.
We will apply the above lemma to the sequence of random variables

$$B_i = \sum_{2^i < J \leq 2^{i+1}} \frac{1}{J} e_J \qquad (i \geq 0),$$

where

$$e_J = \mathbf{1}_{\{F_J(x,.) \in \Delta\}} - P\{F_J(x,.) \in \Delta\} \qquad (J \geq 1) \qquad .$$

A control of $\mathrm{Cov}(e_I, e_J)$ is firstly needed. It will be derived from the elementary and classical lemma,

LEMMA 3.2. (see, for instance, [W1], Lemma 1) *Let* (g_1, g_2) *be jointly Gaussian centered random variables with unit variance and let* $\rho = \mathrm{Cov}(g_1, g_2)$.
Then, for each pair of intervals Δ_1, Δ_2

$$\Big\| \mathrm{Cov}\Big(\mathbf{1}_{\{g_1 \in \Delta_1\}} - P\{g_1 \in \Delta_1\}, \mathbf{1}_{\{g_2 \in \Delta_2\}} - P\{g_2 \in \Delta_2\}\Big) \Big\|$$

$$= \big|P\{g_1 \in \Delta_1, g_2 \in \Delta_2\} - P\{g_1 \in \Delta_1\}P\{g_2 \in \Delta_2\}\big| \leq C_1 |\rho|, \qquad (3.5)$$

where $0 < C_1 < \infty$ *is an absolute constant.*

Since for any $I \leq J$,

$$\mathrm{Cov}(F_I, F_J) = \frac{1}{(IJ)^{\frac{1}{2}}} \sum_{i \leq I} f(T^i x)^2 \leq (I/J)^{\frac{1}{2}} A_f(x),$$

we get, as a direct consequence of Lemma 3.2,

$$\mathrm{Cov}(e_I, e_J) \leq C_1 (I/J)^{\frac{1}{2}} A_f(x).$$

Hence, for any $i < l$

$$\mathrm{Cov}(B_i, B_l) \leq C_1 A_f(x) \leq 2^{-(l-i)/2}. \tag{3.6}$$

Similarly,

$$\mathrm{E}\, B_i^2 \leq 4 C_1 A_f(x) + 1 = C_2.$$

Consequently, by (3.6),

$$\mathrm{E}\left(\sum_{i=m+1}^{m+n} B_i \right)^2 \leq n C_2 + 2 C_1 A_f(x) \sum_{m+1 \leq i < l \leq m+n} 2^{-(l-i)/2}$$

$$\leq n C_2 + 4 C_1 A_f(x) n = C_3 n.$$

Thus, condition (3.4) holds with $\sigma = 1$ and $C = C_3$. This implies, for any $\delta > 0$,

$$\left| \sum_{i=0}^{k} B_i \right| = \left| \sum_{I \leq 2^{k+1}} (e_I/I) \right| = O\left(\left(k^{\frac{1}{2}} (\log k)^{2+\delta} \right) \right),$$

as N tends to infinity, almost surely. This is exactly what we wanted to prove.

$\square$

Proof of the Corollary 1.5. Since $\mathrm{Var}\, F_J \to 1$, we infer from Theorem 1.5 that

$$\sum_{J=1}^{2^k} \frac{1}{J} \mathbf{1}_{\{F_J(x,\omega)\in\Delta\}} = P\{\zeta \in \Delta\} k \log 2 (1 + o(1)), \quad k \to \infty. \tag{3.7}$$

For each $N \in [2^k, 2^{k+1})$ we have

$$\frac{1}{(k+1)\log 2} \sum_{J=1}^{2^k} \frac{1}{J} \mathbf{1}_{\{F_J(x,\omega)\in\Delta\}} \leq \frac{1}{\log N} \sum_{J=1}^{N} \frac{1}{J} \mathbf{1}_{\{F_J(x,\omega)\in\Delta\}}$$

$$\leq \frac{1}{k \log 2} \sum_{J=1}^{2^{k+1}} \frac{1}{J} \mathbf{1}_{\{F_J(x,\omega)\in\Delta\}},$$

and our corollary follows from (3.7).

References

[B]   Bourgain, J., Almost sure convergence and bounded entropy, *Israel J. of Math.*, **63**, 1988, p. 79–95.

[JOR] Jones, R., Ostrovskii, I., Rosenblatt, J., Square functions in ergodic theory, *preprint*, 1996.

[L]   Lifshits, M., Gaussian random functions. *Dordrecht, Kluwer*, 1995.

[LW] Lifshits, M., Weber, M. Régularisation spectrale en théorie ergodique et en théorie des probabilités, submitted to *Comptes Rendus de l'Académie des Sciences, Paris*, 1996.

[PS]   Philipp, W., Stout, W., *Invariance principles for sums of weakly dependent random variables*, Mem. Amer. Math. Soc., **161**, 1975.

[S]   Stein, E.M., On limits of sequences of operators *Ann. Math.*, **74**, 1961, p. 140–170.

[W1] Weber, M., *Sur un théorème de Maruyama*, Séminaire de Probabilités XIV, Lectures Notes in Math. **784**, 1980, p. 475–488.

[W2] Weber, M., The Stein randomization procedure, *Rendi conti di Matematica (Roma)*, **16**, 1996, p.569–605.

[W3] Weber, M., GB and GC sets in ergodic theory, *Proc. of the IXth Conference on Prob. in Banach Spaces, Sandberg, August 1993, Denmark, Progress in Prob., Basel, Birkhäuser*, **35**, 1994, p.129–151.

[W4] Weber, M., GC sets, Stein's elements and matrix summation methods. *Prepublication IRMA 1993/027.*

Mikhail Lifshits
Komendantskii prospect, 22-2-49,
197372, St-Petersbourg,
Russia
mikhail@lifshits.spb.su

Michel Weber
Mathématique, Université Louis-Pasteur,
7 rue René Descartes,
67084 Strasbourg Cedex, France
weber@math.u-strasbg.fr

Progress in Probability, Vol. 43
© 1998 Birkhäuser Verlag Basel/Switzerland

A Sufficient Condition for the Continuity of High Order Gaussian Chaos Processes

MICHAEL B. MARCUS*

1. Introduction

We obtain sufficient conditions for continuity and boundedness of m-th order Gaussian chaos processes, $m \geq 2$. The results obtained are stated in Theorem 11.22, [2] in the case $m = 2$. However the proof given in [2] is not correct and it does not seem as though the approach taken can be modified to give a correct proof. We discuss in Remark 3.1, at the end of this paper, how the error in Theorem 11.22 was discovered and how this paper came about.

Our main result is Theorem 2.1, which is not stated for Gaussian chaos processes explicitly, but for processes with an increments condition that depends on several metrics. We show in Section 3. that this condition is satisfied by m-th order Gaussian chaos processes. This paper depends on ideas developed by M. Talagrand in [6] and [7], after [2] was written. The extension to Gaussian chaos processes of order greater than 2 uses many results in [1], by M. Arcones and E. Giné.

2. Multiple metrics

In this section we develop some Lemmas which will enable us to obtain sufficient conditions for boundedness and continuity of a stochastic process $\{X_t, t \in T\}$ when we have conditions on the distribution of $X_s - X_t$, $s, t \in T$, that depend on several metrics on T. Let T be a finite set and let μ be a probability measure on T. Let $\{d_p\}_{p=1}^{m}$, $m \geq 2$, be a family of metrics on T. We denote the diameter of $A \subset T$, with respect to d_p, by $\Delta_p(A)$. A ball, centered at $t \in T$, of radius a with respect to the metric d_p, is denoted by $B_p(t, a)$.

LEMMA 2.1 *Fix m and $1 \leq p \leq m - 1$. Let $r = 2^{2(m-1)!}$ and let i be the largest integer such that $r^{-i} \geq \Delta_p(T)$. Define*

$$n(t, j) \overset{def}{=} \inf\{n \geq 0 : \mu(B_p(t, r^{-j}2^{(m-p)n/m})) \geq \exp(-2^{2n/m})\} \qquad (2.1)$$

and

$$M(T) \overset{def}{=} \sup_{t \in T} \sum_{j \geq i} r^{-j} 2^{n(t,j)}. \qquad (2.2)$$

*)   This research was supported, in part, by grants from the National Science Foundation and PSC-CUNY.

Then there exists a family of partitions $\{\mathcal{A}_j\}_{j\geq i}$ of T, with $\mathcal{A}_i = T$, and a probability measure $\widetilde{m}$ on T such that, for all $t \in T$

$$\sum_{j\geq i}\left(r^{pj/(m-p)}\Delta_p^{m/(m-p)}(A_j(t)) + r^{-j}\left(\log\frac{1}{\widetilde{m}(A_j(t))}\right)^{m/2}\right) \tag{2.3}$$

$$\leq KM(T)$$

where $A_j(t)$ is the element of $\mathcal{A}_j$ that contains t. Furthermore

$$M(T) \leq K\left(\Delta_p(T) + \sup_{t\in T}\int_0^{\Delta_p(T)}\left(\log\frac{1}{\mu(B_p(t,\epsilon))}\right)^{p/2}d\epsilon\right). \tag{2.4}$$

Clearly the partitions, the measures μ and $\widetilde{m}$, and the important quantity $n(t,j)$, also depend on p. We have suppressed this reference, at this stage, to simplify the notation.

Proof. We construct $\{\mathcal{A}_j\}_{j\geq i}$ and show that (2.3) is satisfied. Let $\mathcal{A}_i = T$ and assume that we have constructed $\mathcal{A}_{j-1}$. We get $\mathcal{A}_j$ by partitioning each element of $\mathcal{A}_{j-1}$ as follows: Consider the set $T_n \overset{def}{=} T_{n,j} = \{t \in T : n(t,j) = n\}$. If $t \in T_n$, then $\mu(B_p(t, r^{-j}2^{(m-p)n/m})) \geq \exp(-2^{2n/m})$. Let G_n be a subset of T_n that has maximal cardinality subject to the condition that $d_p(u,v) \geq 2r^{-j}2^{(m-p)n/m}$ for $u,v \in G_n$. Note that

$$\sum_{u\in G_n}\mu(B_p(t, r^{-j}2^{(m-p)n/m}) \leq \mu(T_n) \tag{2.5}$$

since these balls are disjoint subsets of T_n. Therefore, since $\mu(T_n) \leq 1$ and $\mu(B_p(t, r^{-j}2^{(m-p)n/m})) \geq \exp(-2^{2n/m})$, we see that $\text{card}(G_n) \leq \exp(2^{2n/m})$. It is also easy to see that T_n is covered by balls centered in G_n with radius $4r^{-j}2^{(m-p)n/m}$ and consequently diameter $8r^{-j}2^{(m-p)n/m}$, since otherwise G_n would not have maximal cardinality. We can continue this procedure for all j even when the partitions of T each consists of only a single point.

Let $C \in \mathcal{A}_{j-1}$. By the above we see that $T_n \cap C$ is covered by at most $\exp(2^{2n/m})$ balls of diameter $8r^{-j}2^{(m-p)n/m}$. Using these balls we can construct a partition of $T_n \cap C$ by at most $\exp(2^{2n/m})$ sets of diameter less than or equal to $8r^{-j}2^{(m-p)n/m}$. We do this for each $n \geq 0$, there are only a finite number of them since, as n increases, T_n is eventually empty. This gives us a partition of C. We repeat this procedure for each element in $\mathcal{A}_{j-1}$ and thus obtain $\mathcal{A}_j$.

We now construct the measure $\widetilde{m}$. To do this we first construct a set of weights for the elements of each partition $\mathcal{A}_j$. As defined on page 313, [2], a set of weights for a collection of sets $\{C_j\}$ is a set of positive numbers $\{w(C_j)\}$ such that $\sum_j w(C_j) \leq 1$. Suppose that we have a set of weights for $\mathcal{A}_{j-1}$. Let $C \in \mathcal{A}_{j-1}$. Denote the weight associated with it by $w_{j-1}(C)$. To each set C' in $T_n \cap C$ we assign the weight

$$w_j(C') = \left(\exp(-2^{2n/m})\right)^2 w_{j-1}(C). \tag{2.6}$$

In this way we get a set of weights for $\mathcal{A}_j$. (To begin this process, since $\mathcal{A}_i = T$, we take $w_i(\mathcal{A}_i) = 1$.) Let us verify that we do indeed have weights, as defined above. To do this we need to check that $\sum_{C' \in \mathcal{A}_j} w_j(C') \le 1$. To see this note that since the number of sets in $T_n \cap C$ is at most $\exp(2^{2n/m})$

$$\sum_{C' \in \mathcal{A}_j} w_j(C') = \sum_{C \in \mathcal{A}_{j-1}} \sum_n \sum_{C' \in T_n \cap C} w_j(C') \tag{2.7}$$

$$\le \sum_{C \in \mathcal{A}_{j-1}} \sum_n \exp(-2^{2n/m}) w_{j-1}(C)$$

$$\le \sum_{C \in \mathcal{A}_{j-1}} w_{j-1}(C)$$

which, continued recursively, gives us what we want.

For the convenience of the reader we repeat the argument on page 313, [2] which shows how to construct $\widetilde{m}$ from the family of weights $\{w_j\}$. Let $\{C_{j,k}\}$ denote the sets in $\mathcal{A}_j$. Fix a point $t_{j,k} \in C_{j,k}$. Consider the sub-probability measure $\widetilde{m}'$ given by

$$\widetilde{m}' = \sum_{j=i}^{\infty} 4^{-j+i-2} \sum_k w_j(C_{j,k}) \delta_{t_{j,k}} \tag{2.8}$$

where δ_t is the Dirac measure at t and j_1 is the smallest integer such that every set in the partition $\mathcal{A}_{j_1}$ is a point. Multiplying $\widetilde{m}'$ by a constant we get a probability measure $\widetilde{m}$ on T which satisfies

$$\widetilde{m}(C_{j,k}) \ge 4^{-j+i-2} w_j(C_{j,k}). \tag{2.9}$$

We now verify (2.3). The diameter of $A_j(t)$, $\Delta_p(A_j(t))$, is constructed to be less than or equal to $8r^{-j}2^{(m-p)n(t,j)/m}$. Consequently

$$r^{pj/(m-p)} \Delta_p^{m/(m-p)}(A_j(t)) \le 8^m r^{-j} 2^{n(t,j)}. \tag{2.10}$$

To deal with the next term in (2.3) we note that by (2.9)

$$\left(\log \frac{1}{\widetilde{m}(A_j(t))} \right)^{m/2} \tag{2.11}$$

$$\le 2^{m/2+1} \left((j-i+2)^{m/2} (\log 4)^{m/2} + \left(\log \frac{1}{w_j(A_j(t))} \right)^{m/2} \right).$$

Clearly

$$\sum_{j \ge i} r^{-j} (j-i+2)^{m/2} \le Kr^{-i} \tag{2.12}$$

$$\le KM(T)$$

for some constant K depending only on m. Considering (2.10) (2.11) and (2.12) we see that in order to verify (2.3) we need only show that

$$\sum_{j \ge i} r^{-j} \left(\log \frac{1}{w_j(A_j(t))} \right)^{m/2} \le KM(T) \tag{2.13}$$

for some constant K depending only on m. To do this we first note that by (2.6)

$$\left(\log\frac{1}{w_j(A_j(t))}\right)^{m/2} \leq 2^m 2^{n(t,j)} + 2^{m/2}\left(\log\frac{1}{w_{j-1}(A_{j-1}(t))}\right)^{m/2}. \tag{2.14}$$

Iterating this procedure, and taking into account the fact that $w_i(A_i) = 1$, we see that

$$\left(\log\frac{1}{w_j(A_j(t))}\right)^{m/2} \leq 2^m\left(\sum_{k=0}^{j-i}(2^{m/2})^k 2^{n(t,j-k)}\right) \tag{2.15}$$

and consequently

$$r^{-j}\left(\log\frac{1}{w_j(A_j(t))}\right)^{m/2} \leq 2^m\left(\sum_{k=0}^{j-i}\left(\frac{2^{m/2}}{r}\right)^k r^{-(j-k)}2^{n(t,j-k)}\right). \tag{2.16}$$

Thus

$$\sum_{j\geq i}r^{-j}\left(\log\frac{1}{w_j(A_j(t))}\right)^{m/2} \leq 2^m\left(\sum_{k=0}^{\infty}\left(\frac{2^{m/2}}{r}\right)^k\right)\left(\sum_{j\geq i}r^{-j}2^{n(t,j)}\right). \tag{2.17}$$

Since $r = 2^{2(m-1)!}$ and obviously

$$\sum_{j\geq i}r^{-j} \leq \sum_{j\geq i}r^{-j}2^{n(t,j)} \tag{2.18}$$

get (2.13). Thus we verify (2.3).

We complete the proof by verifying (2.4). Since $r = 2^{2(m-1)!}$ it is clear from the definition of $n(t,j)$ that

$$r^{-j}2^{(m-p)n(t,j)/m} \geq r^{-(j+1)}2^{(m-p)n(t,j+1)/m}. \tag{2.19}$$

Let $k_0 = 2(m-1)!i$ so that $r^{-i} = 2^{-k_0}$, then $r^{-j}2^{(m-p)n(t,j)/m} \leq 2^{-k_0}$, for all $j \geq i$. Also, by the definition of $n(t,j)$, when $n(t,j) \geq 1$

$$\mu\left(B_p(t, r^{-j}2^{(m-p)(n(t,j)-1)/m})\right) \leq \exp(-2^{2(n(t,j)-1)/m}) \tag{2.20}$$

and consequently

$$r^{-j}2^{(m-p)(n(t,j)-1)/m}\left(\log\frac{1}{\mu\left(B_p(t, r^{-j}2^{(m-p)(n(t,j)-1)/m})\right)}\right)^{p/2} \tag{2.21}$$

$$\geq (1/2)r^{-j}2^{n(t,j)}.$$

Let

$$U_k \stackrel{def}{=} \{j : r^{-j}2^{(m-p)(n(t,j)-1)/m} \in [2^{-k}, 2^{-(k+1)})\}. \tag{2.22}$$

and set $u_k \overset{def}{=} \max\{j : j \in U_k\}$. Since for each $t \in T$ there exists a $j_0(t)$ such that $n(t,j) = 0$ for all $j \geq j_0$, $u_k < \infty$. For each $j \in U_k$

$$r^{-j}2^{n(t,j)} \leq 2^{-(k-1)}\left(\log \frac{1}{\mu\left(B_p(t, 2^{-(k+1)})\right)}\right)^{p/2} \tag{2.23}$$

Also, by (2.22) for each $j \in U_k$

$$r^{-j}2^{n(t,j)} \leq 2r^{pj/(m-p)}2^{-mk/(m-p)}. \tag{2.24}$$

Therefore

$$\sum_{\{j\in U_k \,:\, n(t,j)\geq 1\}} r^{-j}2^{n(t,j)} \leq 4r^{pu_k/(m-p)}2^{-mk/(m-p)}. \tag{2.25}$$

Using the upper bound for $2^{-(k+1)}$ given by (2.22), which is valid for $j = u_k$, since $u_k \in U_k$ we see that

$$\sum_{\{j\in U_k \,:\, n(t,j)\geq 1\}} r^{-j}2^{n(t,j)} \leq 4r^{-u_k}2^{n(t,u_k)} \tag{2.26}$$

$$\leq 2^{-(k-3)}\left(\log \frac{1}{\mu\left(B_p(t, 2^{-(k+1)})\right)}\right)^{p/2}$$

where, at the last step we use (2.23) which is valid for $u_k \in U_k$. Therefore

$$\sum_{\{j\geq i \,:\, n(t,j)\geq 1\}} r^{-j}2^{n(t,j)} = \sum_{k\geq k_0}\sum_{\{j\in U_k \,:\, n(t,j)\geq 1\}} r^{-j}2^{n(t,j)} \tag{2.27}$$

$$\leq 16\sum_{k\geq k_0} 2^{-k}\left(\log \frac{1}{\mu\left(B_p(t, 2^{-k})\right)}\right)^{p/2}$$

$$\leq 32\int_0^{\Delta_p(T)}\left(\log \frac{1}{\mu(B_p(t, \epsilon))}\right)^{p/2} d\epsilon.$$

Obviously,

$$\sum_{\{j\geq i \,:\, n(t,j)=0\}} r^{-j}2^{n(t,j)} \leq 2\Delta_p(T). \tag{2.28}$$

Thus we get (2.4). This completes the proof.

Lemma 2.1 deals with cases $p = 1, \ldots, m-1$ of a scheme developed below. The case $p = m$ is given by the following lemma which is simply a restatement of Proposition 11.10, [2].

LEMMA 2.2 *Let* $r = 2^{2(m-1)!}$ *and let* i *be the largest integer such that* $r^{-i} \geq \Delta_m(T)$. *Assume that*

$$\int_0^{\Delta_m(T)}\left(\log \frac{1}{\mu(B_m(t, \epsilon))}\right)^{m/2} d\epsilon < \infty. \tag{2.29}$$

Then there exists a family of partitions $\{\mathcal{C}_j\}_{j\geq i}$, with $\mathcal{C}_i = T$, and a probability measure $\widetilde{m}_m$ on T such that, for all $t \in T$, $\Delta_m(\mathcal{C}_j(t)) \leq r^{-j}$ and

$$\sum_{j\geq i} r^{-j} \log\left(\frac{1}{\widetilde{m}_m(\mathcal{C}_j(t))}\right)^{m/2} \tag{2.30}$$

$$\leq K\left(\Delta_m(T) + \int_0^{\Delta_m(T)} \left(\log\frac{1}{\mu(B_m(t,\epsilon))}\right)^{m/2} d\epsilon\right).$$

Proof. This is Proposition 11.10, [2]. However, note that m and μ in the notation of Proposition 11.10 correspond to μ and $\widetilde{m}_m$ in our notation. Also instead of working with the sequence 4^{-l} we use r^{-l}, (except that we use the variable j instead of l). For the partitions $\{\mathcal{C}_j\}_{j\geq i}$ we take the sets $\{T_{l,i}\}_{l\geq l_0}$ in Proposition 11.10. Our $\widetilde{m}_m$ is the measure μ constructed in Proposition 11.10 and (2.30) is the final inequality in Proposition 11.10.

In the following theorem we continue the notation introduced at the beginning of this section.

THEOREM 2.1 *Let $\{X_t, t \in T\}$ be a real valued stochastic process that satisfies*

$$P(|X_s - X_t| \geq x) \leq K \exp\left(-\min_{1\leq p\leq m}\left(\frac{x}{d_p(s,t)}\right)^{2/p}\right) \tag{2.31}$$

for some constant K. Furthermore assume that there exist probability measures μ_p on T, $1 \leq p \leq m$, such that

$$\sup_{t\in T} \int_0^{\Delta_p(T)} \left(\log\frac{1}{\mu_p(B_p(t,\epsilon))}\right)^{p/2} d\epsilon < \infty. \tag{2.32}$$

Then

$$E\sup_{s,t\in T} |X_s - X_t| \tag{2.33}$$

$$\leq K\left(\sup_{1\leq p\leq m} \Delta_p(T) + \sum_{p=1}^m \sup_{t\in T}\int_0^{\Delta_p(T)}\left(\log\frac{1}{\mu_p(B_p(t,\epsilon))}\right)^{p/2} d\epsilon\right).$$

where K is a constant depending only on m.

Proof. We use the families of partitions and measures that appear in Lemma 2.1 and the important quantity $n(t,j)$. For each $1 \leq p \leq m-1$, we now include the index p to distinguish them. We also use the partitions $\{\mathcal{C}_j\}_{j\geq i}$ constructed in Lemma 2.2 but relabel them $\{\mathcal{A}_{m,j}\}_{j\geq i}$ and relabel the measures $\widetilde{m}$ and μ in Lemma 2.2, $\widetilde{m}_m$ and μ_m.

In Lemmas 2.1 and 2.2 we begin the partitions at index i where $r^{-i} \geq \Delta_p(T)$. We now relabel these initial points i_p. Let $i_0 \overset{def}{=} \inf\{i_p : 1 \leq p \leq m\}$. Add to each family of partitions $\{\mathcal{A}_{p,j}\}_{j\geq i_p}$, the trivial partitions $\{\mathcal{A}_{p,j}\}_{j=i_0}^{j=i_p-1}$, if $i_p > i_0$, where each $\mathcal{A}_{p,j} = T$, $i_0 \leq j \leq i_p - 1$. We now use the m families of partitions $\{\mathcal{A}_{p,j}\}_{j \geq i_0}$, $1 \leq p \leq m$, to construct another family of partitions $\{\mathcal{D}_j\}_{j \geq i_0}$ by, for each j, portioning T into disjoint subsets of the form

$$D_{i_1,\ldots,i_m;j} = \cap_{p=1}^m A_{i_p,j} \tag{2.34}$$

where $A_{i_p,j} \in \mathcal{A}_{p,j}$. We define another measure $\tilde{m}$ on T by taking

$$\tilde{m}(D_{i_1,\ldots,i_m;j}) = \prod_{p=1}^m \tilde{m}_p(A_{i_p,j}). \tag{2.35}$$

As above $D_j(t)$ denotes the element of $\mathcal{D}_j$ that contains t. If $u, v \in D_j(t)$ then $d_p(u,v) \leq \Delta_p(D_j(t))$. Thus it follows from (2.31) that for $u, v \in D_j(t)$

$$P(|X_u - X_v| \geq x) \leq \exp\left(-\min_{1 \leq p \leq m}\left(\frac{x}{\Delta_p(D_j(t))}\right)^{2/p}\right). \tag{2.36}$$

For $j \geq i_0$ set

$$u_j = \sum_{p=1}^{m-1} r^{pj/(m-p)}\Delta_p^{m/(m-p)}(D_j(t)) + r^{-j}\left(\log\frac{2^{j-i_0+2}}{\tilde{m}(D_{j+1}(t))}\right)^{m/2}. \tag{2.37}$$

For $a_i, i = 1,\ldots,m$ and b positive, $(a_1 + \cdots + a_m + b)^{2/p} \geq a^{2(m-p)/mp}b^{2/m}$. Considering the p-th term in the sum in (2.37) as a_p and the last term in (2.37) as b, we see that for $1 \leq p \leq m-1$

$$\left(\frac{u_j}{\Delta_p(D_j(t))}\right)^{2/p} \geq \log\left(\frac{2^{j-i_0+2}}{\tilde{m}(D_{j+1}(t))}\right). \tag{2.38}$$

When $p = m$

$$u_j \geq r^{-j}\left(\log\frac{2^{j-i_0+2}}{\tilde{m}(D_{j+1}(t))}\right) \tag{2.39}$$

since the log term in (2.37)is greater than or equal to one. By (2.34), $D_j(t) \subset A_{m,j}(t)$, the set in $\mathcal{A}_{m,j}$ that contains t. (Recall that $\{A_{m,j}\}$ is a relabeling of $\{C_j\}$ defined in Lemma 2.2 and that $\Delta_m(A_{m,j}(t)) \leq r^{-j}$). Thus $\Delta_m(D_j(t)) \leq r^{-j}$ and consequently, just as in (2.38)

$$\left(\frac{u_j}{\Delta_m(D_j(t))}\right)^{2/m} \geq \log\left(\frac{2^{j-i_0+2}}{\tilde{m}(D_{j+1}(t))}\right). \tag{2.40}$$

The inequalities in the preceding paragraph show us that for $u, v \in D_j(t)$ and $w \geq 1$

$$P\left(\frac{|X_u - X_v|}{u_j} \geq w\right) \leq K\tilde{m}(D_{j+1}(t))2^{-(j-i_0+2)w^{2/m}}. \tag{2.41}$$

Let D_{j,k_j} be an enumeration of the sets that make up the partition $\mathcal{D}_j$. Select a point x_{j,k_j} in each D_{j,k_j}. Each point $t \in T$ is in some D_{j,k_j}, for each $j \geq i_0$. Let

$x_{j,k_j}(t)$ denote the selected point of the set D_{j,k_j} that contains t. Recall that $\mathcal{D}_{i_0}$, the initial partition consists of a single set. Consequently there is only one point $x_{i_0}(t)$ associated with all the $t \in T$. Let $X(t) \stackrel{def}{=} X_t$. We can write

$$X(t) - X(x_{i_0}(t)) = \sum_{j \geq i_0} (X(x_{j,k_j}(t)) - X(x_{j+1,k_{j+1}}(t))). \tag{2.42}$$

This is actually only a finite sum since, for each t, D_{j,k_j} is ultimately reduced to a single point $x_{i_0,1} \stackrel{def}{=} x_{i_0}$.

Using (2.42) we see that

$$P\left(\sup_{t \in T} |X(t) - X(x_{i_0}(t))| \geq w \sum_{j \geq i_0} u_j\right) \tag{2.43}$$

$$\leq P\left(\sum_{j \geq i_0} \sup_{t \in T} |X(x_{j,k_j}(t)) - X(x_{j+1,k_{j+1}}(t))| \geq \sum_{j \geq i_0} wu_j\right)$$

$$\leq P\left(\sup_{j \geq i_0} \sup_{t \in T} |X(x_{j,k_j}(t)) - X(x_{j+1,k_{j+1}}(t))| \geq wu_j\right)$$

$$\leq \sum_{j \geq i_0} P\left(\sup_{t \in T} |X(x_{j,k_j}(t)) - X(x_{j+1,k_{j+1}}(t))| \geq wu_j\right).$$

The $\sup_{t \in T} |X(x_{j,k_j}(t)) - X(x_{j+1,k_{j+1}}(t))|$ is actually only taken over the selected points $x_{j+1,k_{j+1}}(t)$, one from each of the sets $D_{j+1,k_{j+1}}$. Therefore, using (2.41) and the fact that $\widetilde{m}$ is a probability measure we see that

$$P\left(\sup_{t \in T} |X(x_{j,k_j}(t)) - X(x_{j+1,k_{j+1}}(t))| \geq wu_j\right)$$

$$\leq K \sum_{k_{j+1}} \widetilde{m}(D_{j+1}(t)) 2^{-(j-i_0+2)w^2/m} \tag{2.44}$$

$$\leq K 2^{-(j-i_0+2)w^2/m}.$$

Thus by (2.43) and (2.44)

$$\int_1^\infty P\left(\sup_{t \in T} |X(t) - X(x_{i_0}(t))| \geq w \sum_{j \geq i_0} u_j\right) dw \leq K. \tag{2.45}$$

and consequently

$$E\left(\sup_{t \in T} |X(t) - X(x_{i_0}(t))|\right) \leq K \sum_{j \geq i_0} u_j. \tag{2.46}$$

Using (2.35) and the fact that for all $1 \leq p \leq m-1$, $\Delta_p(D_j(t)) \leq \Delta_p(A_{p,j}(t))$ we see that

$$\sum_{j \geq i_0} u_j \leq \sum_{p=1}^{m-1} \left(\sum_{j \geq i_0} \left(r^{pj/(m-p)} \Delta_p^{m/(m-p)}(A_{p,j}(t)) \right.\right. \tag{2.47}$$

$$+r^{-j}\left(\log\frac{1}{\widetilde{m}_p(A_{p,j}(t))}\right)^{m/2}\right)\right)+\sum_{j\geq i_0}(j-i_0+2)r^{-j}$$

$$+\sum_{j\geq i_0}r^{-j}\left(\log\frac{1}{\widetilde{m}_m(A_{m,j}(t))}\right)^{m/2}.$$

Since $A_{p,j}(t)=\mathcal{A}_{p,j}=T$, $i_0\leq j\leq i_p-1$, $\Delta_p(T)\leq r^{-i_p}\leq r\Delta_p(T)$ and $\widetilde{m}_p(A_{p,j}(t))=1$, $i_0\leq j\leq i_p-1$, it follows from Lemma 2.1 that

$$\sum_{j\geq i_0}\left(r^{pj/(m-p)}\Delta_p^{m/(m-p)}(A_{p,j}(t))+r^{-j}\left(\log\frac{1}{\widetilde{m}_p(A_{p,j}(t))}\right)^{m/2}\right)$$

$$\leq K\left(\Delta_p(T)+\sum_{j\geq i_p}\left(r^{pj/(m-p)}\Delta_p^{m/(m-p)}(A_{p,j}(t))\right.\right. \tag{2.48}$$

$$\left.\left.+r^{-j}\left(\log\frac{1}{\widetilde{m}_p(A_{p,j}(t))}\right)^{m/2}\right)\right).$$

So that

$$\sum_{p=1}^{m-1}\left(\sum_{j\geq i_0}\left(r^{pj/(m-p)}\Delta_p^{m/(m-p)}(A_{p,j}(t))+r^{-j}\left(\log\frac{1}{\widetilde{m}_p(A_{p,j}(t))}\right)^{m/2}\right)\right)$$

$$\leq K\left(\sup_{1\leq p\leq m-1}\Delta_p(T)+\sum_{p=1}^{m-1}\int_0^{\Delta_p(T)}\left(\log\frac{1}{\mu_p(B_p(t,\epsilon))}\right)^{p/2}d\epsilon\right).$$
$$\tag{2.49}$$

Also

$$\sum_{j\geq i_0}(j-i_0+2)r^{-j}\quad\leq\quad Kr^{-i_0}$$

$$\leq\quad K\sup_{1\leq p\leq m}\Delta_p(T). \tag{2.50}$$

Finally (2.30) gives us the desired bound for the last term in (2.47). This completes the proof of Theorem 2.1.

3. Gaussian chaos

Consider a real valued m–th order Gaussian chaos process

$$X(t)=\sum_{i_1,\ldots,i_m}z_{i_1,\ldots,i_m}(t)\prod_{j\geq1}H_{m_j(i_1,\ldots,i_m)}(g_j)\qquad t\in T \tag{3.1}$$

defined on $L_2(\Omega,\mathcal{A},\mathcal{P})$. Here H_m is the Hermite polynomial of degree m normalized so that it has leading coefficient 1 and $m_j(i_1,\ldots,i_m)=\sum_{r=1}^m I(i_r=j)$. (See e.g.

[1] for details). We assume that $z_{i_1,\ldots,i_m}$ is unchanged under a permutation of its indices. Setting $\xi_{i_1,\ldots,i_m}(s,t) = z_{i_1,\ldots,i_m}(s) - z_{i_1,\ldots,i_m}(t)$ we write

$$X(s) - X(t) = \sum_{i_1,\ldots,i_m} \xi_{i_1,\ldots,i_m}(s,t) \prod_{j\geq 1} H_{m_j(i_1,\ldots,i_m)}(g_j). \qquad (3.2)$$

Let

$$
\begin{aligned}
d_0(s,t) &= \left(E(X(s) - X(t))^2\right)^{1/2} \\
&= \left(E\left(\sum_{i_1,\ldots,i_m} \xi_{i_1,\ldots,i_m}(s,t) \prod_{j\geq 1} H_{m_j(i_1,\ldots,i_m)}(g_j)\right)^2\right)^{1/2}. \qquad (3.3)
\end{aligned}
$$

By a decoupling theorem, Theorem 2.2, [1] there exists a constant $0 < C_m < \infty$, depending only on m, such that

$$(3.4)$$

$$C_m^{-1} d_0(s,t) \leq \left(E\left(\sum_{i_1,\ldots,i_m} \xi_{i_1,\ldots,i_m}(s,t) g_{i_1}^{(1)} \cdots g_{i_m}^{(m)}\right)^2\right)^{1/2} \leq C_m d_0(s,t)$$

where $\{g_i^{(j)}\}_i$ are independent copies of $\{g_i\}_i$. Calculating the expectation in (3.4) we see that

$$C_m^{-1} d_0(s,t) \leq \left(\sum_{i_1,\ldots,i_m} \xi_{i_1,\ldots,i_m}^2(s,t)\right)^{1/2} \leq C_m d_0(s,t). \qquad (3.5)$$

Let $\{h_i\}$ be a sequence of real numbers with $|h| \overset{def}{=} (\sum h_i^2)^{1/2} \leq 1$. Set, for $0 \leq r \leq m-1$

$$d_{m-r}(s,t) \qquad (3.6)$$

$$= E \sup_{|h|\leq 1} \left| \sum_{i_1,\ldots,i_m} \xi_{i_1,\ldots,i_m}(s,t) \prod_{j\geq 1} H_{r_j(i_1,\ldots,i_r)}(g_j) h_{i_{r+1}} \cdots h_{i_m} \right|$$

$$= E \sup_{|h|\leq 1} \left| \sum_{i_1,\ldots,i_r} \left(\sum_{i_{r+1},\ldots,i_m} \xi_{i_1,\ldots,i_m}(s,t) h_{i_{r+1}} \cdots h_{i_m} \right) \right.$$

$$\left. \prod_{j\geq 1} H_{r_j(i_1,\ldots,i_r)}(g_j) \right|.$$

Viewed this way we see that $d_{m-r}(s,t)$ is a norm on an r-th order Gaussian chaos process and hence by the same decoupling theorem we see that there exists a constant K such that for all $0 \leq r \leq m-1$

$$d_{m-r}(s,t) \leq K d_0(s,t). \qquad (3.7)$$

We now show that the metrics $d_0(s,t)$ and $d_1(s,t)$ are equivalent. Since $d_1(s,t)$ is a norm on an $m-1$-st order Gaussian chaos, it follows from Lemma 3.2, [1] that it is equivalent to $\hat{d}_1(s,t)$ where

$$\hat{d}_1^2(s,t) \tag{3.8}$$

$$= E \sup_{|h|\leq 1} \left(\sum_{i_n} \left(\sum_{i_1,\ldots,i_{n-1}} \xi_{i_1,\ldots,i_m}(s,t) \right. \right.$$

$$\left. \left. \prod_{j\geq 1} H_{(n-1)_j(i_1,\ldots,i_{n-1})}(g_j) \right) h_{i_m} \right)^2$$

$$= E \sum_{i_n} \left(\sum_{i_1,\ldots,i_{n-1}} \xi_{i_1,\ldots,i_m}(s,t) \prod_{j\geq 1} H_{(n-1)_j(i_1,\ldots,i_{n-1})}(g_j) \right)^2$$

$$\geq C \sum_{i_1,\ldots,i_m} \xi_{i_1,\ldots,i_m}^2(s,t)$$

where, for the last step, we use the same calculations, but for an $m-1$-st order Gaussian chaos, that we used to obtain (3.5). Combining (3.7) and (3.8) we see that $d_0(s,t)$ and $d_1(s,t)$ are equivalent. Using this and Theorem 4.3, [1], we obtain

$$P\left(|X(s) - X(t)| \geq \sum_{p=1}^{m} C_p u^p d_p(s,t)\right) \leq K \exp\left(-\frac{u^2}{K}\right) \tag{3.9}$$

for all $u \geq 0$, where K is a constant depending only on m. (The passage from the quantiles defined in (4.7), [1] to the metrics in (3.9) follows from Chebychev's inequality.)

Let $x = \sum_{p=1}^{m} C_p u^p d_p(s,t)$ then $C_p u^p d_p(s,t) \geq x/m$ for at least one $1 \leq p \leq m$. This implies that there exists a constant K such that

$$u^2 \geq K \left(\frac{u}{d_p(s,t)}\right)^{2/p} \tag{3.10}$$

for a least one $1 \leq p \leq m$. Thus it follows from (3.9) that

$$P(|X(s) - X(t)| \geq x) \leq K \exp\left(-\min_{1\leq p\leq m}\left(\frac{u}{K d_p(s,t)}\right)^{2/p}\right). \tag{3.11}$$

for some constant K which depends only on m.

We see from (3.11) that Theorem 2.1 can be applied to an m-th order Gaussian chaos. We now give a stronger version of this Theorem which can also be used to obtain moduli of continuity.

THEOREM 3.1 *Let $\{X_t, t \in T\}$ be an m-th order Gaussian chaos as defined in (3.1), with associated metrics d_p, $1 \leq p \leq m$, as defined in (3.6). Assume that* *there exist probability measures μ_p on T, $1 \leq p \leq m$, such that (2.32) is satisfied. Then*

$$E \sup_{\substack{s,t \in T \\ d_0(s,t) \leq \eta}} |X_s - X_t| \leq K \left(\eta + \sum_{p=1}^{m} \sup_{t \in T} \int_0^{K'\eta} \left(\log \frac{1}{\mu_p(B_p(t,\epsilon))} \right)^{p/2} d\epsilon \right)$$

(3.12)

where K and K' are constants depending only on m.

Proof. Define $\widetilde{T} = \{(s,t) \in T, d_0(s,t) \leq \eta\}$ and $Z(s,t) = X_s - X_t$. The stochastic process $Z = \{Z(s,t), (s,t) \in \widetilde{T}\}$ is also an m-th order Gaussian chaos. Let $\tilde{d}_p$, $1 \leq p \leq m$, be the metrics associated with Z as defined in (3.6). Note that both d_p and $\tilde{d}_p$ are norms. It follows from (3.5) and the triangle inequality that

$$\begin{aligned} \tilde{d}_p((s_1,t_1),(s_2,t_2)) &\leq (d_p(s_1,s_2) + d_p(t_1,t_2)) \wedge (d_p(s_1,t_1) + d_p(s_2,t_2)) \\ &\leq (d_p(s_1,s_2) + d_p(t_1,t_2)) \wedge K\eta. \end{aligned}$$

(3.13)

Since Z is also an m-th order Gaussian chaos, (3.11) holds for Z with respect to the metrics $\tilde{d}_p$ and hence so does (2.31). (It may be necessary to divide Z by a constant but that will not affect the final result (3.12)).

We now show that (2.32) is satisfied by Z and $\tilde{d}_p$. That is, that there exist probability measures $\tilde{\mu}_p$ on $\widetilde{T}$, $1 \leq p \leq m$, such that

$$\sup_{(s,t) \in \widetilde{T}} \int_0^{\widetilde{\Delta}_p(\widetilde{T})} \left(\log \frac{1}{\tilde{\mu}_p(\widetilde{B}_p((s,t),\epsilon))} \right)^{p/2} d\epsilon < \infty.$$

(3.14)

where now $\widetilde{B}_p((s,t),\epsilon)$ and $\widetilde{\Delta}_p(\widetilde{T})$ are taken with respect to $\tilde{d}_p$. By (3.7) $\widetilde{\Delta}_p(\widetilde{T}) \leq K\eta$. For $\epsilon < K\eta$ set

(3.15)

$$\tilde{\nu}_p(\{(u,v) \in \widetilde{T} : u \in B_p(s,\epsilon), v \in B_p(t,\epsilon)\}) = \mu_p(B_p(s,\epsilon))\mu_p(B_p(t,\epsilon)).$$

Dividing $\tilde{\nu}_p$ by its total mass on $\widetilde{T}$ we get a probability measure $\tilde{\mu}_p$ on $\widetilde{T}$. Since $\{(u,v) \in \widetilde{T} : u \in B_p(s,\epsilon/2), v \in B_p(t,\epsilon/2)\} \subset \widetilde{B}_p((s,t),\epsilon)$ we see that

$$\int_0^{K'\eta} \left(\log \frac{1}{\tilde{\mu}_p(\widetilde{B}_p((s,t),\epsilon))} \right)^{p/2} d\epsilon \leq K \left(\eta \right.$$

(3.16)

$$\left. + \int_0^{K'\eta} \left(\log \frac{1}{\mu_p(B_p(s,\epsilon))} \right)^{p/2} d\epsilon + \int_0^{K'\eta} \left(\log \frac{1}{\mu_p(B_p(t,\epsilon))} \right)^{p/2} d\epsilon \right)$$

for constants K and K'.

Since (2.32) is satisfied for μ_p with respect to d_p, we see that (3.14) holds. Therefore, by Theorem 2.1

$$E \left(\sup_{(s_1,t_1),(s_2,t_2) \in \widetilde{T}} |X_{(s_1,t_1)} - X_{(s_2,t_2)}| \right)$$

(3.17)

$$\leq K \left(\eta + \sum_{p=1}^{m} \sup_{(s,t)\in\widetilde{T}} \int_0^{K'\eta} \left(\log \frac{1}{\widetilde{\mu}_p(B_p((s,t),\epsilon))} \right)^{p/2} d\epsilon \right).$$

Clearly

$$E \left(\sup_{(s_1,t_1)\in\widetilde{T}} |X_{(s_1,t_1)} - X_{(s_2,t_2)}| \right) \leq E \left(\sup_{(s_1,t_1),(s_2,t_2)\in\widetilde{T}} |X_{(s_1,t_1)} - X_{(s_2,t_2)}| \right).$$

(3.18)

Using this, (3.16) and (3.17) we get (3.12).

So far we have taken T to be finite. However, since none of the constants depend on the dimension of T the conclusions of Theorems 2.1 and 3.1 remain valid if T is countable. By strengthening (2.32) we can obtain a sufficient condition for the continuity of an m-th order Gaussian chaos.

THEOREM 3.2 *Let* $\{X_t, t \in T\}$ *be an* m-*th order Gaussian chaos as defined in (3.1), with associated metrics* d_p, $1 \leq p \leq m$, *as defined in (3.6). Assume that there exist probability measures* μ_p *on* T, $1 \leq p \leq m$, *such that*

$$\lim_{\eta\to 0} \sup_{t\in T} \int_0^\eta \left(\log \frac{1}{\mu_p(B_p(t,\epsilon))} \right)^{p/2} d\epsilon = 0.$$

(3.19)

Then X_t *has a version with continuous paths on* (T, d_0).

Proof. Using (3.12) we can find a convergent sequence $\{u_k\}$ and a sequence $\{\eta_k\}$ for which $\lim_{k\to\infty} \eta_k = 0$ such that

$$E \sup_{\substack{s,t\in T \\ d_0(s,t)\leq\eta_k}} |X_s - X_t| \leq u_k.$$

(3.20)

The proof now follows from Chebychev's inequality and the Borel-Cantelli lemma.

This same approach can also be used to obtain moduli of continuity. Thus, when these results apply, they can be used to obtain uniform continuity on countably dense sets. In many cases that interest us they can then be extended to show that $\{X_t, t \in T\}$ has a continuous version on (T, d_0) even when T is not countable. Then $\{X_t, t \in T\}$ also has a continuous version on (T, τ) if d_0 is continuous on (T, τ).

These results appear to be difficult to apply because it is difficult to get good estimates of d_p, for $p > 1$, other than (3.7). However, roughly speaking, the metrics d_p must get smaller as p increases because the power of the logarithm term, such as in (3.19), gets larger. In [5], which considers sufficient conditions for continuity of a certain class of second order Gaussain chaos processes, good estimates are given for the d_2 metric, which then enables us to give a square root metric entropy condition, with respect to the L^2 metric d_0, for the continuity of these processes. These results are used in [3] to obtain necessary and sufficient conditions for continuity of some continuous additive functionals of certain Lévy processes. In [4] it is shown that certain families of renormalized self-intersection local times are continuous almost surely if corresponding m-th order Gaussain chaos processes are continuous. Now we have sufficient conditions for continuity of these processes which we can apply if we can figure out how to estimate the metrics d_p.

REMARK 3.1 The first draft of this paper added to the proof of Theorem 11.22, [2] in an attempt to extend the Theorem to m-th order chaos processes. I am very grateful to the referee who, in a very careful reading of the draft, found that there was a gap in the proof of Theorem 11.22 itself. I tried, unsuccessfully, to close the gap. In later conversations with Michel Talagrand, he told me that he didn't think that the proof could be fixed, but that he had developed stronger methods for dealing with multiple metric conditions in [6] and [7]. He also outlined a proof of Theorem 2.1 in this paper for the case $m = 2$. This is the critical part of the proof for general m. The extension is relatively straight forward. Michel Talagrand told me that all the ingredients needed to prove Theorem 2.1 in this paper in the case $m = 2$ are already contained in [6] and [7]. However, the reader who consults these papers will see that it is not at all clear how to go about extracting this information. I think that the proof in this paper, based on my conversations with Michel Talagrand, should be quite useful. I am very grateful to him for helping me straighten this out. Besides the intrinsic interest of this theorem itself, it is also used in a significant way in Theorem 1.5, [3].

References

[1] M. Arcones and E. Giné. On decoupling, series expansion, and tail behavior of chaos processes. *Jour. Theoret. Prob.*, 6:101–122, 1993.

[2] M. Ledoux and M. Talagrand. *Probability in Banach Spaces*. Springer-Verlag, New York, 1991.

[3] M.B. Marcus and J. Rosen. Gaussian chaos and sample path properties of additive functionals of symmetric Markov processes. Annals of Probability, (1996), to appear.

[4] M.B. Marcus and J. Rosen. Renormalized self-intersection local times and Wick power chaos processes. preprint.

[5] M.B. Marcus and M. Talagrand. Continuity conditions for a class of Gaussian chaos processes related to continuous additive functionals of Lévy processes. Jour. Theoret. Prob., to appear.

[6] M. Talagrand. Regularity of infinitely divisible processes. *Ann. Probab.*, 21:362–432, 1993.

[7] M. Talagrand. The supremum of some cannonical processes. *Amer. Journ. of Math.*, 116:283–325, 1994.

Michael Marcus
Department of Mathematics
The City College of CUNY
New York, NY 10031
mbmcc@cunyvm.cuny.edu

Progress in Probability, Vol. 43
© 1998 Birkhäuser Verlag Basel/Switzerland

On Wald's Equation and First Exit Times for Randomly Stopped Processes with Independent Increments

VICTOR H. DE LA PEÑA[*]

ABSTRACT. The general theme of this paper is the study of the properties of randomly stopped processes with independent increments with CADLAG (right continuous with left limits) paths with values in general Banach spaces. First we present a general version of de la Peña and Eisenbaum (1997)[(1)] which extends the Burkholder-Gundy inequality for randomly stopped Brownian motion to general Banach spaces and processes with CADLAG paths and independent increments. We continue by providing a proof (from first principles) of the upper bound of (1) in the case the process has continuous paths. We apply our results to extend Wald's equation to this type of process and to obtain information on first exit times of the processes involved. An example involving the joint behavior of several stocks in a market underlines the importance of the fact that the results hold in general Banach spaces. This fact is also used to obtain bounds for the L_p-norms of randomly stopped Bessel and related processes. The issue of constants in exponential bounds is also addressed.

Throughout our presentation, we will be referring to the following class of functions.

For fixed $\alpha > 0$, let $\Phi : [0, \infty) \to R$, $\Phi \epsilon \mathcal{F}_\alpha$ where,

$$\mathcal{F}_\alpha = \{\Phi : \Phi(0) = 0 \quad and \quad \Phi(cx) \leq c^\alpha \Phi(x) \quad for \quad all \quad c \geq 2, \quad x \geq 0\}, \qquad (1.0)$$

and the functions Φ are non-decreasing and continuous.

We begin with Theorem 1.2 of de la Peña and Eisenbaum (1997) which is an extension of Klass (1988, 1990). Due to an oversight at the time of publication Theorem 1.2 was stated for processes with continuous paths. Nontheless, both the theorem and its proof are trivially valid for CADLAG processes. To see this just use right continuity instead of continuity in its proof. After we state this result in its full generality (Theorem 1) we present an alternative proof in the case of processes with continuous paths.

THEOREM 1. (de la Peña and Eisenbaum). *Fix any $\alpha > 0$. Let $\Phi \epsilon \mathcal{F}_\alpha$ as in (1.0). Let $N_t, t \geq 0$, be a continuous time process with independent increments and CADLAG paths taking values in a Banach space $(B, || \cdot ||)$. Let T be a stopping time adapted to $\sigma(\{N_s, s \leq t\})$. Then*

$$c_\alpha E a_T^* \leq E \sup_{0 \leq s \leq T} \Phi(||N_s||) \leq C_\alpha E a_T^*, \qquad (1.1)$$

*)   Partially supported by NSF grant DMS-96-26175.

where $a_t^* = E \sup_{0 \le s \le t} \Phi(||N_s||)$, *and* c_α, C_α *are constants depending on* α *only (indeed,* $C_\alpha \le 20(18)^\alpha$).

Equivalently if $\tilde{N}_t, t \ge 0$ is an independent copy of $N_t, t \ge 0$ and independent of T, then

$$c_\alpha E \sup_{0 \le s \le T} \Phi(||\tilde{N}_s||) \le E \sup_{0 \le s \le T} \Phi(||N_s||) \le C_\alpha E \sup_{0 \le s \le T} \Phi(||\tilde{N}_s||). \qquad (1.2)$$

REMARK 1. The shape of the constant C_α allows the extension of this result to exponential functions as discussed in de la Peña and Eisenbaum (1997). In that paper it is shown that under the assumptions of Theorem 1, for all $\lambda, p > 0$, $E \exp\{\lambda ||N_T^*||^p\} \le 20 E \exp\{\lambda 18^p ||\tilde{N}_T^*||^p\}$.

REMARK 2. In the special case where $N_t = (B_t^{(1)}, \cdots, B_t^{(d)})$ and $||N_t|| = \sqrt{(B_t^{(1)})^2 + \cdots + (B_t^{(d)})^2}$ with $B_t^{(i)}$, i.i.d. standard Brownian motions then $||N_t||$ is known as a Bessel Process. This type of process has been studied by many researchers. Among others, we can cite the works of Burkholder (1977), Davis (1978), De Blassie (1987) (which contains a nice survey) and Graversen and Peskir (1994) were they obtain related approximations to $E\Phi(N_T^*)$. However their results do not extend directly to the generality given in Theorem 1 in particular since in our result the components of N_t need not be independent from one another. Nonetheless, each component should have independent increments. Part of our motivation for looking at Bessel processes came from discussions with Larry Shepp concerning Dubins, Shepp and Shiryaev (1993).

REMARK 3. We would like to stress the fact that if $N_t = (X_t^{(1)}, \ldots, X_t^{(d)})$ then the coordinate processes forming N_t can have different distributions. For example, $X_t^{(1)}$ could be a Poisson process while $X_t^{(2)}$ a Gaussian process, etc. etc.

Before proving Theorem 1 we will state a key lemma of Burkholder and Gundy.

LEMMA 1. Let U, $V \ge 0$ be non-negative variables. Assume that for all $y > 0$, there exist $\beta, \delta > 0$ and $0 < \gamma < 1$ such that

$$P(U > \beta y, V \le \delta y) \le \gamma P(U > y), \qquad (1.3)$$

with $\frac{1}{\beta} - \gamma > 0$ then,

$$EU \le \frac{1}{\delta} \frac{1}{\frac{1}{\beta} - \gamma} EV. \qquad (1.4)$$

Proof of Lemma 1: From (1.3) we have,

$$P(U > \beta y) \le P(U > \beta y, V \le \delta y) + P(V > \delta y) \le \gamma P(U > y) + P(V > \delta y). \quad (1.5)$$

Integrating over y we have,

$$\frac{EU}{\beta} = \int P(U > \beta y) dy \le \gamma EU + \frac{EV}{\delta}. \qquad (1.6)$$

Solving for EU gives,

$$EU \leq \frac{1}{\delta} \frac{1}{\frac{1}{\beta} - \gamma} EV. \tag{1.7}$$

Proof of Theorem 1: We remark that the statement of Theorem 1 as presented in de la Peña and Eisenbaum (1997) includes only the special case of pth moments. However, the limiting nature of the proof makes it obvious that the proof is valid for the case of functions as in (1.0) as obtained by Klass (1988, 1990). We will present here an alternative proof of the right hand of (1.1) for processes with continuous paths and functions Φ belonging to the class introduced in (1.0). The proof uses some of the ideas of Klass (1988) adapted to the case of continuous time processes.

Let $U = \Phi(\sup_{s \leq T} ||N_s||)$, $V = a_T^* = E[\Phi(\sup_{s \leq T} ||\tilde{N}_s||)|T]$. Observe that $Ea_T^* = E(E[\Phi(\sup_{s \leq T} ||\tilde{N}_s||)|T]) = E\Phi(\sup_{s \leq T} ||\tilde{N}_s||)$. Then, if we get (1.3) for U and V, this will give us the right hand side of (1.1). To this end let

$$T_y = \inf\{t > 0 : \Phi(||N_t||) > y\}, \quad r^* = \sup\{s > 0 : a_s^* \leq \delta y\},$$

and $N_t^* = \sup_{s \leq t} ||N_s||$. We also observe that if $\mathcal{F}_{T_y} = \sigma\{N_s, s \leq T_y\}$, then conditioning on $\mathcal{F}_{T_y}$, for all fixed $s > T_y$, the joint distribution of $||N_s - N_{T_y}||$ is equal to that of $||\tilde{N}_s - \tilde{N}_{T_y}||$. The growth properties of Φ give that $\Phi(x + y) \leq \Phi(2\max\{x, y\}) \leq 2^\alpha[\Phi(x) + \Phi(y)]$. Then, for any fixed $\delta > 0, \beta > 2^\alpha$ and all $y > 0$,

$$P(\Phi(N_T^*) > \beta y, a_T^* \leq \delta y) \leq \tag{1.8}$$

$$P(2^\alpha\{\Phi(\sup_{T_y < s \leq T} ||N_s - N_{T_y}||) + \Phi(||N_{T_y}||)\} > \beta y, a_T^* \leq \delta y, T \geq T_y) \leq$$

$$P(2^\alpha\Phi(\sup_{T_y < s \leq T} ||N_s - N_{T_y}||) > (\beta - 2^\alpha)y, T \leq r^*, T \geq T_y) \leq$$

(by the definition of r^*)

$$\frac{2^\alpha}{(\beta - 2^\alpha)y} E\Phi(\sup_{T_y < s \leq T} ||N_s - N_{T_y}||)1(T \leq r^*, T \geq T_y) \leq$$

(by Markov's inequality)

$$\frac{2^\alpha}{(\beta - 2^\alpha)y} E\Phi(\sup_{T_y < s \leq r^*} ||N_s - N_{T_y}||)1(T \geq T_y) =$$

$$\frac{2^\alpha}{(\beta - 2^\alpha)y} E(E[\Phi(\sup_{T_y < s \leq r^*} ||N_s - N_{T_y}||)|\mathcal{F}_{T_y} 1(T \geq T_y)]) =$$

$$\frac{2^\alpha}{(\beta - 2^\alpha)y} E(E[\Phi(\sup_{T_y < s \leq r^*} ||\tilde{N}_s - \tilde{N}_{T_y}||)|\mathcal{F}_{T_y}]1(T \geq T_y)) \leq \tag{1.9}$$

(by equality of the joint distributions)

$$\frac{2^\alpha}{(\beta - 2^\alpha)y} 2^\alpha E\Phi(\sup_{s \leq r^*} ||\tilde{N}_s||)P(T \geq T_y) \leq$$

(by independence between $\{\tilde{N}_t\}$ and $\{N_t\}, T$)

$$\frac{4^\alpha \delta y}{(\beta - 2^\alpha)y} P(T \geq T_y) = \tag{1.10}$$

(by the definition of r^*)

$$\frac{4^\alpha \delta}{(\beta - 2^\alpha)} P(\Phi(\sup_{s \leq T} ||N_s||) > y),$$

where the last line follows by the definition of T_y. Using the above bound in Lemma 1 with $U = \Phi(N_T^*)$ and $V = a_T^*$ we get,

$$E\Phi(\sup_{s \leq T} ||N_s||) \leq \frac{1}{\delta} \frac{1}{\frac{1}{\beta} - \gamma} E(a_T^*) = \frac{1}{\delta} \frac{1}{\frac{1}{\beta} - \gamma} E\Phi(\sup_{s \leq T} ||\tilde{N}_s||), \tag{1.11}$$

for $\beta > \frac{1}{\gamma}$ and $\gamma = \frac{4^\alpha \delta}{(\beta - 2^\alpha)}$. A use of (1.4) gives that for all Φ as in (1.0),

$$E\Phi(N_T^*) \leq C_\alpha E\Phi(\sup_{s \leq T} ||\tilde{N}_s||). \tag{1.12}$$

As a tool for obtaining a useful variant to Theorem 1, we state an extension of Levy's inequality to the case of all i.i.d. variables due to Montgomery-Smith (1993).

THEOREM 2. *Let $\{X_i\}$ be a sequence of i.i.d. random elements with values in a Banach space $(B, || \cdot ||)$. Then, for all $y \geq 0$, there exists a universal constant $0 < C < \infty$ such that*

$$P(\max_{j \leq n} ||\sum_{i=1}^{j} X_i|| \geq t) \leq C P(C||\sum_{i=1}^{n} X_i|| \geq t). \tag{1.13}$$

We extend the above result to the case of continuous time processes with stationary and independent increments. The proof is very straightforward. Throughout we will say a process $\{N_t\}$ has stationary increments, if the distribution of $N_t - N_s$, $t > s$ is the same as that of $N_{t'} - N_{s'}$, $s' > t'$ whenever $t - s = t' - s'$.

COROLLARY 1. *Let N_t be a continuous time process with increments that are stationary and independent. We also assume that the process has CADLAG paths taking values in a Banach space, $(B, || \cdot ||)$. Then, there exists a universal constant $0 < C < \infty$ such that for all $y \geq 0$,*

$$P(\sup_{0 \leq s \leq t} ||N_s|| \geq y) \leq C P(C||N_t|| \geq y). \tag{1.14}$$

Proof. For a fixed t, consider the subdivision τ_n of $[0, t]$, $\tau_n = (s_i^n)_{0 \leq i \leq n}$, with $0 = s_0^n < s_1^n < \ldots < s_{n-1}^n < s_n^n = t$, and for $i = 0, \ldots, 2^n$, $s_i^n = \frac{i}{2^n}t$. Then $\lim_{n \to +\infty} |\tau_n| = 0$ and τ_{n+1} contains τ_n. Moreover, $N_{s_j^n} = \sum_{i=1}^{j}(N_{s_i^n} - N_{s_{i-1}^n}) + N_{s_0^n}$.

and the restriction of N to τ_n is a discrete time process with i.i.d. increments. Then, by Theorem 2 we have for each fixed $n \geq 1$, and all $y > 0$,

$$P(\sup_{s \in \tau_n, s \leq t} ||N_s|| \geq y) \leq CP(C||N_t|| \geq y).$$

The result follows using the Dominated Convergence Theorem, letting n tend to infinity.

The following result contains as a special case the Burkholder-Gundy inequalities for randomly stopped Brownian motion as well as results for Bessel processes and homogeneous Poisson processes.   □

THEOREM 3. *Let N_t be a continuous time process with increments that are stationary and independent. We also assume that the process has CADLAG paths taking values in a Banach space, $(B, ||\cdot||)$. Let $\tilde{N}_t$ be an independent copy of N_t. Fix $\alpha > 0$. Then, for any $\Phi \epsilon \mathcal{F}_\alpha$ as in (1.0), there exist constants c_α and C_α with $0 < c_\alpha, C_\alpha < \infty$ depending on α only such that for any stopping time T, adapted to $\sigma(\{N_s, s \leq t\})$,*

$$c_\alpha E\Phi(||\tilde{N}_T||) \leq E \sup_{0 \leq t \leq T} \Phi(||N_t||) \leq C_\alpha E\Phi(||\tilde{N}_T||). \qquad (1.15)$$

Proof. The lower bound follows trivially from Theorem 1. As for the bound from above, using Theorem 1 followed by Corollary 1 and (1.0) we get

$$E\sup_{t \leq T} \Phi(||N_t||) \leq C_\alpha E \sup_{t \leq T} \Phi(||\tilde{N}_t||) = C_\alpha E[E(\sup_{t \leq T} \Phi(||\tilde{N}_t||)|T)] \leq$$

$$C_\alpha E[CE(\Phi(C||\tilde{N}_T||)|T)] \leq C'_\alpha E[E(\Phi(||\tilde{N}_T||)|T)] = C'_\alpha E\Phi(||\tilde{N}_T||),$$

which concludes the proof.   □

Using the assumed independence between $\tilde{N}_t$ and N_t, if we let $N_t = B_t$, a standard Brownian motion, and apply the scaling properties of this process we get the Burkholder-Gundy inequality.

Our next theorem extends Wald's equation to the case of a wide class of processes with independent increments.

COROLLARY 2. *Under the assumptions of Theorem 1 (resp. Theorem 3), let N_t be real valued. Set $EN_t = b_t$, $E|N_t - b_t| = d_t$ and $d_t^* = E\sup_{s \leq t} |N_s - b_t|$. Then,*

$$EN_T = Eb_T = E\tilde{N}_T, \qquad (1.16)$$

whenever $Ed_T^* < \infty$ (resp. whenever $Ed_T < \infty$).

Proof. Let $M_t = N_t - b_t$. Then, M_t is a zero-mean martingale with independent increments. By the optional sampling theorem, for each fixed $t < \infty$, $EM_{T \wedge t} = 0$. Using Theorem 1 (Theorem 3 resp.), taking $\tilde{M}_t$ to be an independent copy of M_t, independent of T as well, we have that

$$E \sup_{s \leq T} |M_s| \leq CE \sup_{s \leq T} |\tilde{M}_s| = CEd_T^*, \qquad (1.17)$$

where the bound is CEd_T if we assume stationarity of the process. To complete the proof, use the finiteness of the rhs of (1.17) and the DCT to obtain,

$$0 = \lim_{t \to \infty} EM_{T \wedge t} = E \lim_{t \to \infty} M_{T \wedge t} = EM_T. \qquad (1.18)$$

In the case of sums of i.i.d. random variables $\{X_i\}$ with mean μ, Corollary 2 gives $E \sum_{i=1}^{T} X_i = \mu ET$ whenever $Ed_T < \infty$ where $d_n = E|\sum_{i=1}^{n}(X_i - \mu)|$, for an independent copy $\{\tilde{X}_i\}$ of $\{X_i\}$. If the variables have a finite variance then $d_n \approx n^{1/2}$. In the case N_t is a homogeneous Poisson process with rate λ, then $EN_T = \lambda ET$ whenever $Ed_T < \infty$ where $d_t = E|N_t - \lambda t|$. □

The following corollary to Theorem 1 from de la Peña and Eisenbaum (1997) reminds us that we actually have an extension of the Burkholder Gundy inequality for randomly stopped Brownian motion.

COROLLARY 3. *Let $B_t, t \geq 0$ be a standard Brownian motion and T a stopping time adapted to $\sigma(B_s, s > 0)$ then,*

$$c_p E|B_1|^p ET^{p/2} \leq E \sup_{t \leq T} |B_t|^p \leq C_p E|B_1|^p ET^{p/2}. \qquad (1.19)$$

EXAMPLE 1. Consider a vector consisting of d processes some of which are continuous and others have jumps. This system could represent e.g., the earnings of several stock brokers in a market (where future earnings are independent of past earnings). We impose the following conditions on the processes $N_t = (X_t^{(1)}, \ldots, X_t^{(d)})$: for each i, $X_t^{(i)}$ has independent increments and paths which are right continuous and with left limits (CADLAG). However as i varies over $1, \ldots, d$, the $X_t^{(i)}$'s might depend on one another and could come from different distributions. The following random variables and associated stopping times are of great importance.

$$U_t = |X_t^{(1)}|, \quad V_t = \sup_{i \leq d} |X_t^{(i)}|, \quad W_t = \sqrt{(X_t^{(1)})^2 + \cdots + (X_t^{(d)})^2}, \qquad (1.20)$$

and for a positive constant a,

$$T_1 = \{\inf_{t>0} : U_t \geq a\}, \quad T_2 = \{\inf_{t>0} : V_t \geq a\}, \quad T_3 = \{\inf_{t>0} : W_t \geq a\}. \qquad (1.21)$$

Here U_t involves a study of the extremes of an individual process while V_t contains information on the extremal (among d) process. The stopping time T_2

records the first time one of the processes is bigger than a or less than $-a$. This value could stand for a pre-assigned boundary and the information on the hitting time of such a barrier could be important in preventing drastic losses or in motivating the introduction of a more agressive investment strategy. The pair (W_t, T_3) provides information on the first time the vector $(X_t^{(1)}, \ldots, X_t^{(d)})$ exits from the d-dimensional sphere of radius a.

Concerning T_1, if we take $\{Y_i\}$ to be a sequence of independent Bernoulli random variables with $P(Y_i = 1) = P(Y_i = -1) = \frac{1}{2}$ and set $X_t^{(1)} = \sum_{i=1}^{n} Y_i$, for $n \leq t < n+1$, $n = 1, \ldots$, Wald's second equation gives $ET_1 = a^2$

In the case $N_t = (X_t^{(1)}, \ldots, X_t^{(d)})$ we can get information on T_2 and T_3 by using Theorem 3 with $\Phi(x) = |x|^p$ as we describe next.

Letting $||N_t|| = \sup_{i \leq d} |X_t^{(i)}|$ we get that for each $p > 0$,

$$c_p E \sup_{i \leq d} |\tilde{X}_T^{(i)}|^p \leq E \sup_{0 \leq t \leq T} \sup_{i \leq d} |X_t^{(i)}|^p \leq C_p E \sup_{i \leq d} |\tilde{X}_T^{(i)}|^p. \tag{1.22}$$

If we take $N_t = (B_t^{(1)}, \ldots, B_t^{(d)})$ where the $B_t^{(i)}$ are standard Brownian motions, the independence between T and $\tilde{N}_t = (\tilde{B}_t^{(1)}, \ldots, \tilde{B}_t^{(d)})$ gives

$$E \sup_{i \leq d} |\tilde{B}_T^{(i)}|^p = E \sup_{i \leq d} |\sqrt{T} \tilde{B}_1^{(i)}|^p = ET^{p/2} E \sup_{i \leq d} |B_1^{(i)}|^p, \tag{1.23}$$

from which we get the following approximation on the average values of T_2^p for all fixed $p > 0$,

$$ET_2^p \approx_p \frac{a^{2p}}{E \sup_{i \leq d} |B_1^{(i)}|^{2p}}, \tag{1.24}$$

where $A \approx_p B$ is used to denote that there exist positive finite constants c_p, C_p depending on p only and for which $c_p \leq \frac{A}{B} \leq C_p$.

Letting $||N_t|| = \sqrt{(X_t^{(1)})^2 + \cdots + (X_t^{(d)})^2}$ in Theorem 3, we get for all $p > 0$,

$$c_p E (\sum_{i=1}^{d} (\tilde{X}_T^{(i)})^2)^{p/2} \leq E \sup_{t \leq T} (\sum_{i=1}^{d} (X_t^{(i)})^2)^{p/2} \leq C_p E (\sum_{i=1}^{d} (\tilde{X}_T^{(i)})^2)^{p/2}. \tag{1.25}$$

If in addition we assume that $\{B_t^{(i)}\}$ is a sequence of independent standard Brownian motions then conditioning one obtains that,

$$E (\sum_{i=1}^{d} (\tilde{B}_T^{(i)})^2)^{p/2} = ET^{p/2} (\sum_{i=1}^{d} (\tilde{B}_1^{(i)})^2)^{p/2} = ET^{p/2} E (\mathcal{X}_d^2)^{p/2}, \tag{1.26}$$

where $\mathcal{X}_d^2$ is a Chi-squared random variable with d degrees of freedom. Applying the above bound one gets that $ET_3^p \approx_p \frac{a^{2p}}{E(\mathcal{X}_d^2)^p}$.

We continue by providing an approach for developing exponential decoupling inequalities in addition to the one introduced in Remark 1. Some of the constants obtained are improvements over the ones presented earlier.

THEOREM 4. *Let $\{X_t\}$ be a real valued sub-martingale with independent increments and moment generating function $M_{X_t}(\lambda) = Ee^{\lambda X_t} = g(t, \lambda)$, a function of the pair (t, λ) only, $-\infty < \lambda < \infty$.*

Let T be a stopping time adapted to $\sigma\{X_t\}$. Consider an independent copy $\{\tilde{X}_t\}$ of $\{X_t\}$ with $\{\tilde{X}_t\}$ independent of T as well. Then,

$$Ee^{\lambda \sup_{t \le T} X_t} \le e\sqrt{Ee^{2\lambda \sup_{t \le T} \tilde{X}_t}}. \tag{1.27}$$

Proof. By assumption, $E\frac{e^{\lambda X_t}}{g(\lambda, t)} = 1$ for all t. Moreover since X_t has independent increments, $\frac{e^{\lambda X_t}}{g(\lambda, t)}$ is a non-negative martingale. Let k, $0 < k < \infty$ be a constant. Then, $T_k = T \wedge k$ is a stopping time and hence $E\frac{e^{\lambda X_{T_k}}}{g(\lambda, T_k)} = 1$. Let $X = e^{\lambda X_{T_k}}$ and $Y = g(\lambda, T_k)$, then, $E\frac{X}{Y} = 1$. Using Hölder's inequality one obtains,

$$E\sqrt{X} = E\sqrt{\frac{X}{Y}} \times \sqrt{Y} \le \sqrt{E\frac{X}{Y} \times EY} = \sqrt{EY}. \tag{1.28}$$

Therefore

$$Ee^{\frac{\lambda}{2} X_{T_k}} \le \sqrt{Eg(\lambda, T_k)} = \sqrt{E(E(e^{\lambda \tilde{X}_{T_k}} | T_k))} = \sqrt{Ee^{\lambda \tilde{X}_{T_k}}}, \tag{1.29}$$

since the moment generating function of $\{\tilde{X}_t\}$ is $g(\lambda, t)$ and T is independent of $\{\tilde{X}_t\}$. The convexity of the exponential function gives that $e^{\lambda X_{T_k}}$ is a sub-martingale. From Doob's inequality it follows that for any p, $1 < p < \infty$

$$Ee^{\lambda \sup_{t \le T_k} X_t} = E(e^{\frac{\lambda}{p} \sup_{t \le T_k} X_t})^p \le (\frac{p}{p-1})^p E(e^{\frac{\lambda}{p} X_{T_k}})^p = (\frac{p}{p-1})^p Ee^{\lambda X_{T_k}}. \tag{1.30}$$

Taking the infimum over $p > 1$ and using the decoupling inequality proved above we have

$$Ee^{\frac{\lambda}{2} \sup_{t \le T_k} X_t} \le eEe^{\frac{\lambda}{2} X_{T_k}} \le e\sqrt{Ee^{\lambda \tilde{X}_{T_k}}} \le e\sqrt{Ee^{\lambda \sup_{t \le T_k} \tilde{X}_t}}. \tag{1.31}$$

Letting $k \to \infty$, a use of the monotone convergence theorem gives the desired result. $\qquad\square$

The following result should be compared with the result stated in Remark 1.

COROLLARY 4. *Under the assumptions of Theorem 4, for all $\lambda \ge 0$,*

$$Ee^{\lambda \sup_{t \le T} |X_t|} \le 2e^{5/2}\sqrt{Ee^{2\lambda |\tilde{X}_T|}}. \tag{1.32}$$

Proof. From Theorem 4 and an application of Doob's inequality, we have that for all $\lambda \ge 0$,

$$Ee^{\lambda X_T} \le e\sqrt{eEe^{2\lambda \tilde{X}_T}},$$

with the same result valid with λ replaced by $-\lambda$. Therefore,

$$Ee^{\lambda |X_T|} \le Ee^{\lambda X_T} + Ee^{-\lambda X_T} \le e\sqrt{eEe^{2\lambda \tilde{X}_T}} + e\sqrt{eEe^{-2\lambda \tilde{X}_T}} \le 2e\sqrt{eEe^{2\lambda |\tilde{X}_T|}}.$$

A further use of Doob's inequality gives the result. $\qquad\square$

Applying Corollary 4 to a standard Brownian motion $\{B_t, t \geq 0\}$ we obtain

COROLLARY 5.   *Let $\{B_t, t \geq 0\}$ be a standard Brownian motion. Let T be a stopping time adapted to its filtration. Then, for all $\lambda \geq 0$,*

$$Ee^{\lambda \sup_{t \leq T} |B_t|} \leq 2e^{5/2}\sqrt{Ee^{2\lambda^2 T}}. \tag{1.33}$$

Proof.   Apply the proof of Corollary 4 observing that $\tilde{B}_T$ has the same distribution as $\tilde{B}_1 \sqrt{T}$ and bound the moment generating function of this variable.   □

REMARK 5.   Corollary 5 provides improved constants to the ones presented in de la Peña and Eisenbaum (1997) that dealt with the general case of bounding $Ee^{\lambda \sup_{t \leq T} |B_t|^p}$ for all $0 \leq p \leq 2$ by using the exponential inequality stated in Remark 1 of this paper. For related results see Peskir (1996).

REMARK 6.   Theorem 1.2 of de la Peña and Eisenbaum (1997) and some of its applications to exponential inequalities were announced in de la Peña and Eisenbaum (1994) and presented at the "IV Simposio de Probabilidad y Procesos Estocasticos" in Guanajuato, México the Spring of 1996.

Acknowledgements. We thank Karl Sigmund for helpful comments and remarks.

References

[1]   Burkholder, D.L. (1973). Distribution function inequalities for martingales. *Ann. Probab.* **1** (1), 19–42.

[2]   Burkholder, D.L. (1977). Exit times of Brownian motion, harmonic majorization and Hardy spaces. *Advances in Math.* **26**, 182–205.

[3]   Davis, B. (1978). On Stopping times for n dimensional Brownian motion. *Ann. Probab.* **6** (4), 651–659.

[4]   De Blassie, D. R. (1987). Stopping times of Bessel Processes. *Ann. Probab.* **15** (3), 1044–1051.

[5]   de la Peña, V. H. (1994). A bound on the moment generating function of a sum of dependent variables with an application to simple random sampling without replacement. *Annales de L'Institute Henry Poincaré. Probabilités et Statistiques,* **30** (2), 197–211.

[6]   de la Peña, V. H., and Eisenbaum, N. (1994). Decoupling inequalities for the local times of a linear Brownian motion. Preprint.

[7]   de la Peña, V. H., and Eisenbaum, N. (1997). Exponential Burkholder Davis Gundy inequalities. *Bull. London Math. Soc.* **29**, 239–242.

[8]   Dubins, L. E., Shepp, L. and Shiryaev, A. N. (1993). Optimal stopping rules and maximal inequalities for Bessel processes. *Theory Probab. Appl.* **38**, 226–261.

[9]   Graversen, S. E. and Peskir, G. (1994). Maximal Inequalities for Bessel processes. Math. Inst. Aarhus, Preprint Ser. No. 23. To appear in J. Inequal. Appl.

[10]   Klass, M. J. (1988). A best possible improvement of Wald's equation. *Ann. Probab.* **15** (2), 840–863.

[11]   Klass, M. J. (1990). Uniform lower bounds for randomly stopped Banach space valued random sums. *Ann. Probab.* **18** (2), 780–809.

[12] Montgomery-Smith, S. J. (1993). Comparison of sums of independent identically distributed random variables. *Probability and Mathematical Statistics*, **14** (2), 281–285.

[13] Peskir, G. (1996). On the exponential Orlicz norms of stopped Brownian motion. *Studia Mathematica* **117** (3), 253–273.

Victor H. de la Peña
Department of Statistics
Columbia University
New York, N. Y. 10027
vp@wald.stat.columbia.edu

Progress in Probability, Vol. 43
© 1998 Birkhäuser Verlag Basel/Switzerland

The Best Doob-Type Bounds for the Maximum of Brownian Paths

GORAN PEŠKIR

ABSTRACT. Let $B = (B_t)_{t \geq 0}$ be standard Brownian motion started at zero. Then the following inequality is shown to be satisfied:

$$E\left(\max_{0 \leq t \leq \tau} |B_t|^p \right) \leq \gamma^*_{p,q} \left(E \int_0^\tau |B_t|^{q-1} \, dt \right)^{p/(q+1)}$$

for all stopping times τ for B, all $0 < p < 1+q$, and all $q > 0$, with the best possible value for the constant being equal:

$$\gamma^*_{p,q} = (1+\kappa) \left(\frac{s_*}{\kappa^\kappa} \right)^{1/(1+\kappa)}$$

where $\kappa = p/(q-p+1)$, and s_* is the zero point of the (unique) maximal solution $s \mapsto g_*(s)$ of the differential equation:

$$g^\alpha(s) \left(s^\beta - g^\beta(s) \right) \frac{dg}{ds}(s) = K$$

satisfying $0 < g_*(s) < s$ for all $s > s_*$, where $\alpha = q/p - 1$, $\beta = 1/p$ and $K = p/2$. This solution is also characterized by $g_*(s)/s \to 1$ for $s \to \infty$. The equality above is attained at the stopping time:

$$\tau_* = \inf \{ t > 0 \mid X_t = g_*(S_t) \}$$

where $X_t = |B_t|^p$ and $S_t = \max_{0 \leq r \leq t} |B_r|^p$. In the case $p = 1$ the closed form for $s \mapsto g_*(s)$ is found. This yields $\gamma^*_{1,q} = (q(q+1)/2)^{1/(q+1)} (\Gamma(1+(q+1)/q))^{q/(q+1)}$ for all $q > 0$. In the case $p \neq 1$ no closed form for $s \mapsto g_*(s)$ seems to exist. The inequality above holds also in the case $p = q+1$ (Doob's maximal inequality). In this case the equation above (with $K = p/2c$) admits $g_*(s) = \lambda s$ as the maximal solution, and the equality is attained only in the limit through the stopping times $\tau_* = \tau^*(c)$ when c tends to the best value $\gamma^*_{q+1,q} = (q+1)^{q+2}/2q^q$ from above. The method of proof relies upon the principle of smooth fit of Kolmogorov and the maximality principle. The results obtained extend to the case when B starts at any given point, as well as to all non-negative submartingales.

1. Introduction

Let $B = (B_t)_{t \geq 0}$ be standard Brownian motion started at zero. Then the following comparison inequalities are known to be valid:

$$A_p E\left(\int_0^\tau |B_t|^{p-2} dt \right) \leq E\left(\max_{0 \leq t \leq \tau} |B_t|^p \right) \leq B_p E\left(\int_0^\tau |B_t|^{p-2} dt \right) \qquad (1.1)$$

for all stopping times τ for B, and all $p > 1$, where A_p and B_p are some universal constants. In this paper we shall address the case $0 < p \leq 1$ when these inequalities fail to hold (note that $E(\int_0^\tau |B_t|^{-1} dt) = \infty$ whenever $B_\tau \neq 0$ P-a.s.) More precisely, for $0 < p \leq 1 + q$ and $q > 0$ given and fixed, we shall answer the question on sharpness of the following inequality:

$$E\left(\max_{0 \leq t \leq \tau} |B_t|^p \right) \leq \gamma_{p,q} E\left(\int_0^\tau |B_t|^{q-1} dt \right)^{p/q+1} \tag{1.2}$$

where τ is any stopping time for B, and $\gamma_{p,q}$ is a universal constant. In our main result below (Theorem 2.1) we derive (1.2) with the best possible value $\gamma_{p,q}^*$ for the constant $\gamma_{p,q}$, and we find the stopping time τ_* at which the equality in (1.2) is attained (in the limit for $p = 1 + q$).

In order to give a more familiar form to the inequality (1.2), note that by Itô formula and the optional sampling theorem we have:

$$E\left(\int_0^\tau |B_t|^{q-1} dt \right) = \frac{2}{q(q+1)} E|B_\tau|^{q+1} \tag{1.3}$$

whenever τ is a stopping time for B satisfying $E(\tau^{(q+1)/2}) < \infty$ for $q > 0$. Hence the right-hand inequality in (1.1) is the well-known Doob's maximal inequality for non-negative submartingales being applied to $|B| = (|B_t|)_{t \geq 0}$ (see [3]). The advantage of the formulation (1.1) lies in its validity for all stopping times. It is well-known that in the case $p = 1$ the analogue of Doob's inequality fails. In this case the $L \log L$-inequality of Hardy and Littlewood is the adequate substitute (see [7] for the best bounds and [11] for a new probabilistic proof which exhibits the optimal stopping times too). Instead of introducing a log-term as in the Hardy-Littlewood inequality, in the inequality (1.2) we use Doob's bound (1.3) on the right-hand side.

While the inequality (1.2) (with some constant $\gamma_{p,q} > 0$) follows easily from (1.1) and (1.3) by Jensen's inequality, the question of its sharpness is far from being trivial and has gained some interest. The case $p = 1$ was treated independently by Jacka [14] and Gilat [8] both who found the best possible value $\gamma_{1,q}^*$ for $q > 0$. This in particular yields $\gamma_{1,1}^* = \sqrt{2}$ which was obtained independently earlier by Dubins and Schwarz [5], and later again by Dubins, Shepp and Shiryaev [6] who studied a more general case of Bessel processes. (A simple proof for $\gamma_{1,1}^* = \sqrt{2}$ is given in [9]). In the case $p = 1 + q$ with $q > 0$, the inequality (1.2) is Doob's maximal inequality (it follows by (1.3) above). The best constants in Doob's maximal inequality and the corresponding optimal stopping times are well-known (see [16]). That the equality in Doob's maximal inequality (for any $p > 1$) cannot be attained by a non-zero (sub)martingale was observed by Cox [2]. The reader should note that this fact also follows from the method and results below (the equality in (1.2) is attained only in the limit when $p = 1 + q$).

In this paper we present a proof which gives the best values $\gamma_{p,q}^*$ in (1.2) and the corresponding optimal stopping times τ^* for all $0 < p \leq 1 + q$ and all $q > 0$. Our method relies upon the principle of smooth fit of Kolmogorov [6] and the maximality principle [10] (which is the main novelty in this context). It should be noted that the results extend to the case when Brownian motion B starts at any given point (Remark 2.2). Finally, due to its extreme properties, it is clear that the results obtained for reflected Brownian motion $|B|$ extend to all non-negative submartingales. This can be done by using the maximal embedding result of Jacka [13]. For reader's convenience we state this extension and outline the proof (Corollary 2.3).

2. The results and proofs

In this section we present the main results and proofs. In view of the well-known extreme properties in such a context we study the case of Brownian motion. The results obtained are then extended to all non-negative submartingales. The principal result of the paper is contained in the following theorem. It extends the result of Jacka [14] by a different method.

THEOREM 2.1 Let $B = (B_t)_{t \geq 0}$ be standard Brownian motion started at zero. Then the following inequality is shown to be satisfied:

$$E\left(\max_{0 \leq t \leq \tau} |B_t|^p \right) \leq \gamma^*_{p,q} \left(E \int_0^\tau |B_t|^{q-1} \, dt \right)^{p/(q+1)} \tag{2.1}$$

for all stopping times τ for B, all $0 < p < 1 + q$, and all $q > 0$, with the best possible value for the constant $\gamma^*_{p,q}$ being equal:

$$\gamma^*_{p,q} = (1+\kappa)\left(\frac{s_*}{\kappa^\kappa} \right)^{1/(1+\kappa)} \tag{2.2}$$

where $\kappa = p/(q-p+1)$, and s_* is the zero point of the (unique) maximal solution $s \mapsto g_*(s)$ of the differential equation:

$$g^\alpha(s)\left(s^\beta - g^\beta(s) \right)\frac{dg}{ds}(s) = K \tag{2.3}$$

satisfying $0 < g_*(s) < s$ for all $s > s_*$, where $\alpha = q/p - 1$, $\beta = 1/p$ and $K = p/2$. This solution is also characterized by $g_*(s)/s \to 1$ for $s \to \infty$.

The equality in (2.1) is attained at the stopping time:

$$\tau_* = \inf \{ t > 0 \mid X_t = g_*(S_t) \} \tag{2.4}$$

where $X_t = |B_t|^p$ and $S_t = \max_{0 \leq r \leq t} |B_r|^p$. In the case $p = 1$ the closed form for $s \mapsto g_*(s)$ is found:

$$s \exp\left(-\frac{g^q_*(s)}{Kq} \right) + \int_0^{g_*(s)} \frac{t^q}{K} \exp\left(-\frac{t^q}{Kq} \right) \, dt = q^{1/q}\,\Gamma\big((q+1)/q\big)\,K^{1/q} \tag{2.5}$$

for $s \geq s_*$. This, in particular, yields:

$$\gamma^*_{1,q} = \big(q(q+1)/2 \big)^{1/(q+1)} \big(\Gamma\big(1+(q+1)/q\big) \big)^{q/(q+1)} \tag{2.6}$$

for all $q > 0$. In the case $p \neq 1$ no closed form for $s \mapsto g_*(s)$ seem to exist.

The inequality (2.1) holds also in the case $p = q+1$. In this case the equation (2.3) (with $K = p/2c$) admits a linear solution $g_*(s) = \lambda s$ as the maximal solution satisfying $0 < g_*(s) < s$ for $s > 0$, where λ is the maximal root of the equation:

$$\lambda^{q/(q+1)} - \lambda = (q+1)/2c \qquad (2.7)$$

satisfying $0 < \lambda < 1$. The equality in (2.1) is attained in the limit through the stopping times $\tau_* = \tau^*(c)$ of the form (2.4) when c tends to the best value:

$$\gamma^*_{q+1,q} = (q+1)^{q+2}/2q^q \qquad (2.8)$$

from above.

Proof. 1. Given $0 \le x \le s$ and $c > 0$, consider the optimal stopping problem:

$$V_c(x, s) = \sup_\tau E_{x,s}\big(S_\tau - cI_\tau\big) \qquad (2.9)$$

where the supremum is taken over all stopping times τ for B, and the maximum process $S = (S_t)_{t\ge0}$ and the integral process $I = (I_t)_{t\ge0}$ associated with the process $X = (X_t)_{t\ge0}$ are respectively given by:

$$S_t = \Big(\max_{0\le r\le t} X_r\Big) \vee s \qquad (2.10)$$

$$I_t = \int_0^t (X_r)^{(q-1)/p}\, dr \qquad (2.11)$$

$$X_t = |B_t + x^{1/p}|^p \qquad (2.12)$$

with $0 < p < 1 + q$ and $q > 0$ given and fixed. Note that under $P_{x,s} := P$ the process (X, S) starts at (x, s).

2. By the scaling property of Brownian motion it is easily verified:

$$V_c(0, 0) = c^{-\kappa} V_1(0, 0) \qquad (2.13)$$

where $\kappa = p/(q-p+1)$. This yields:

$$E(S_\tau) \le \inf_{c>0} \Big(cE(\tau) + c^{-\kappa} V_1(0, 0)\Big) \qquad (2.14)$$

for all stopping times τ for B. We shall show below that:

$$V_1(0, 0) = s_* \qquad (2.15)$$

where s_* is the zero point of the (unique) maximal solution $s \mapsto g_*(s)$ of (2.3) satisfying $0 < g_*(s) < s$ for all $s > s_*$. Then by computing the infimum in (2.14) we get (2.1) with (2.2). Moreover, we shall see below that the equality in (2.15) is attained at τ_* from (2.4). Thus by (2.14) the same holds for (2.1). Finally, we shall prove that:

$$\lim_{s\to\infty} \frac{g_*(s)}{s} = 1 \qquad (2.16)$$

which leads to the closed form (2.5) when $p = 1$. In this case the equation (2.3) admits an integrating factor $(\mu(x, y) = \exp(-y^q/Kq))$, and the unknown constant from the closed form of its solution is then easily specified by using (2.16). From (2.5) we find the closed expression (2.6). Thus to complete the proof in the case $0 < p < 1 + q$ with $q > 0$, it is necessary and sufficient to prove (2.15) with (2.16).

3. We begin by proving (2.15). For this note that (X, S) is a two-dimensional diffusion which changes (increases) in the second coordinate only (instantly) after the process (X, S) hits the diagonal $x = s$. Thus the infinitesimal operator of (X, S) outside the diagonal equals the infinitesimal operator of X, which is easily verified on $]0, \infty[$ to be equal:

$$\mathbf{L_X} = \frac{p(p-1)}{2} x^{1-2/p} \frac{\partial}{\partial x} + \frac{p^2}{2} x^{2-2/p} \frac{\partial^2}{\partial x^2}. \tag{2.17}$$

Assuming now that the supremum in (2.9) is attained at the exit time by (X, S) from an open set, by the general Markov processes theory we know that $x \mapsto V_c(x, s)$ is to satisfy:

$$(\mathbf{L_X} V_c)(x, s) = cx^{(q-1)/p} \tag{2.18}$$

for $g_*(s) < x < s$, where $s > 0$ is given and fixed, while $s \mapsto g_*(s)$ is the optimal stopping boundary to be found. The following boundary conditions seem evident:

$$V_c(x, s)\Big|_{x=g_*(s)+} = s \qquad \text{(instantaneous stopping)} \tag{2.19}$$

$$\frac{\partial V_c}{\partial x}(x, s)\Big|_{x=g_*(s)+} = 0 \qquad \text{(smooth fit)} \tag{2.20}$$

$$\frac{\partial V_c}{\partial s}(x, s)\Big|_{x=s-} = 0 \qquad \text{(normal reflection)}. \tag{2.21}$$

Note that (2.18)–(2.21) is a Stephan problem with moving (free) boundary. The condition (2.20) is the principle of smooth fit of Kolmogorov (see [6]). The equation (2.18) is of Cauchy type. Its general solution is given by:

$$V_c(x, s) = A(s) x^{1/p} + B(s) + \frac{2c}{q(q+1)} x^{(q+1)/p} \tag{2.22}$$

where $s \mapsto A(s)$ and $s \mapsto B(s)$ are unknown functions. By (2.19) and (2.20) we find:

$$A(s) = -\frac{2c}{q} g_*^{q/p}(s) \tag{2.23}$$

$$B(s) = s + \frac{2c}{q+1} g_*^{(q+1)/p}(s). \tag{2.24}$$

Inserting this into (2.22) and using (2.21) we see that $s \mapsto g_*(s)$ is to satisfy (2.3) with $K = p/2c$. Motivated by the maximality principle [10] we let $s \mapsto g_*(s)$ be the maximal solution of (2.3) satisfying $0 < g_*(s) < s$ for $s > s_*$, where s_* is the (unique) zero point of $s \mapsto g_*(s)$. Thus (2.22) with (2.23) and (2.24) gives $V_c(x, s)$ for $g_*(s) \le x \le s$ only when $s \ge s_*$. Clearly $V_c(x, s) = s$ for $0 \le x \le g_*(s)$ with $s \ge s_*$.

To get $V_c(x, s)$ for $0 \le x \le s < s_*$, note by the strong Markov property that:

$$V_c(x, s) = V_c(s_*, s_*) - cE_{x,s}(I_{\sigma_*}) \tag{2.25}$$

for all $0 \le x \le s < s_*$, where $\sigma_* = \inf\{ t > 0 \mid X_t = s_* \}$. By Itô formula and the optional sampling theorem we find:

$$E_{x,s}(I_{\sigma_*}) = \frac{2}{q(q+1)} \left(s_*^{(q+1)/p} - x^{(q+1)/p} \right) \tag{2.26}$$

for all $0 \le x \le s < s_*$. Inserting this into (2.25), and using (2.22) with (2.23) and (2.24), we get:

$$V_c(x, s) = s_* + \frac{2c}{q(q+1)} x^{(q+1)/p} \tag{2.27}$$

for all $0 \le x \le s < s_*$. This formula in particular gives (2.15) when $x = s = 0$ and $c = 1$.

4. To deduce (2.16) note (since $g_*(s) \le s \le V_c(s, s)$) from (2.22) with (2.23) and (2.24) that:

$$0 \le \limsup_{s \to \infty} \frac{V_c(s, s)}{s^{(q+1)/p}} \le \frac{2c}{q(q+1)} \tag{2.28}$$

for all $c > 0$. The "limsup" in (2.28) is decreasing in c, thus after letting $c \downarrow 0$ we see that the "limsup" must be zero for all $c > 0$. Using this fact and going back to (2.22) with (2.23) and (2.24) we easily obtain (2.16). (Note that this can be proved in a similar way by looking at the definition of $V_c(x, s)$ in (2.9).) By the standard arguments based on Picard's method of successive approximations one can verify that the equation (2.3) admits a (unique) solution satisfying (2.16). This ends the first part of the proof (guess). In the next step we verify its validity by using Itô-Tanaka's formula.

5. To verify that the candidate (2.22) with (2.23) and (2.24) is indeed the payoff (2.9) with the optimal stopping time given by:

$$\tau_* = \inf \{ t > 0 \mid X_t \le g_*(S_t) \} \tag{2.29}$$

denote this candidate by $V_*(x, s)$, and apply Itô-Tanaka's formula (two-dimensionally) to the process $V_*(X_t, S_t)$. For this note by (2.20) that $x \mapsto V_*(x, s)$ is C^2 on $[0, s\,[$, except at $g_*(s)$ where it is C^1, while $(x, s) \mapsto V_*(x, s)$ is C^2 away from $\{ (g_*(s), s) \mid s > 0 \}$, so that clearly Itô-Tanaka's formula can be applied. In this way by (2.17) and (2.21) we get:

$$
\begin{aligned}
V_*(X_t, S_t) &= V_*(x, s) + \int_0^t \frac{\partial V_*}{\partial x}(X_r, S_r)\, dX_r \\
&\quad + \int_0^t \frac{\partial V_*}{\partial s}(X_r, S_r)\, dS_r + \frac{1}{2}\int_0^t \frac{\partial^2 V_*}{\partial x^2}(X_r, S_r)\, d\langle X, X \rangle_r \\
&= V_*(x, s) + \int_0^t \mathbf{L_X}(V_*)(X_r, S_r)\, dr + \int_0^t \frac{\partial V_*}{\partial x}(X_r, S_r)\, \sigma(X_r)\, dB_r \\
&= V_*(x, s) + \int_0^t \mathbf{L_X}(V_*)(X_r, S_r)\, dr + M_t
\end{aligned}
\tag{2.30}
$$

where $\sigma(x) = px^{1-1/p}$ and $M = (M_t)_{t\geq 0}$ is a continuous local martingale. Hence by (2.18):

$$V_*(X_\tau, S_\tau) \leq V_*(x, s) + cI_\tau + M_\tau \qquad (2.31)$$

for all stopping times τ for B, with the equality if $\tau \leq \tau_*$. By the optional sampling theorem and Burkholder-Davis-Gundy's inequality for continuous local martingales it is easily verified that:

$$E_{x,s}(M_\tau) = 0 \qquad (2.32)$$

whenever τ is a stopping time satisfying $E_{x,s}(\tau^{(q+1)/2}) < \infty$. It is well-known that the stopping time τ_* satisfies such an integrability condition (this can be obtained by the methods used in [1] and presented in [15] p. 258–264). Since $s \leq V_*(x, s)$ thus from (2.31) then it follows:

$$E_{x,s}(S_\tau) \leq E_{x,s}(V_*(X_\tau, S_\tau)) \leq V_*(x, s) + cE_{x,s}(I_\tau) \qquad (2.33)$$

for all stopping times τ for B, with the equalities if $\tau = \tau_*$. This completes the proof in the case $0 < p < 1 + q$ with $q > 0$.

6. The proof just presented extends to the case $p = 1+q$ with $q > 0$ with some minor modifications. In this case the maximal solution of (2.3) with $K = p/2c$ is given by:

$$g_*(s) = \lambda s \qquad (2.34)$$

where λ is the maximal root (out of two possible ones) of the equation:

$$\lambda^{q/(q+1)} - \lambda = \frac{q+1}{2c} \qquad (2.35)$$

satisfying $0 < \lambda < 1$. By the standard argument one can verify that this happens if and only if $c \geq \gamma^*_{q+1,q}$ with $\gamma^*_{q+1,q}$ from (2.8). The essential difference from the case $p < 1+q$ is that (2.32) fails for $c \leq \gamma^*_{q+1,q}$. Note as above that $E_{x,s}(\tau^{(q+1)/2}) < \infty$ if and only if $c > \gamma^*_{q+1,q}$. Thus the inequality (2.1) for $p = 1+q$ (with the equality) can be obtained in the limit when $c \downarrow \gamma^*_{q+1,q}$. Note that $V_c(0,0) = 0$ for all $c \geq \gamma^*_{q+1,q}$ (compare this with (2.13) above), as well as that (2.22) with (2.23) and (2.24) reads:

$$V_c(x, s) = s + \frac{2c}{(q+1)} g_*(s) - \frac{2c}{q} g_*^{q/(q+1)}(s) x^{1/(q+1)} + \frac{2c}{q(q+1)} x \qquad (2.36)$$

for all $0 \leq x \leq s$, where $s \mapsto g_*(s)$ is given by (2.34). The proof is complete. $\square$

REMARK 2.2 The reader should note that the results stated in Theorem 2.1 above extend to the case when the Brownian motion starts at any given point. The proof above shows that whenever $0 < p \leq 1 + q$ and $q > 0$ are given and fixed, the following inequality is satisfied:

$$E\left(\max_{0\leq t\leq \tau} |B_t + x|^p\right) \leq cE\left(\int_0^\tau |B_t + x|^{q-1}\, dt\right) + V_c(x^p, x^p) \qquad (2.37)$$

for all stopping times τ for B and all $c > 0$ (all $c \geq \gamma^*_{q+1,q}$ if $p = 1+q$), where in the case $0 < p < 1 + q$ the function $V_c(x, x)$ is given by (2.22) with (2.23) and (2.24) if $x > s_*$ and by (2.27) if $0 \leq x \leq s_*$ (with $s_* > 0$ as in Theorem 2.1 above), while in the case $p = 1+q$ the function $V_c(x, x)$ is given by (2.36) for $x \geq 0$. Note, moreover, that by Itô formula and the optional sampling theorem, the right-hand side in (2.37) can be modified to read as follows:

$$E\left(\max_{0 \leq t \leq \tau} |B_t + x|^p \right) \leq \frac{2c}{q(q+1)} E|B_\tau + x|^{q+1} - \frac{2c}{q(q+1)} x^{q+1} + V_c(x^p, x^p) \quad (2.38)$$

whenever $E(\tau^{(q+1)/2}) < \infty$. The inequalities (2.37) and (2.38) are sharp for each fixed c and x (the equality is attained (in the limit if $p = 1 + q$) at $\tau_* = \tau_*(c; x)$ from (2.4) with $X_t = |B_t + x|^p$ and $S_t = \max_{0 \leq r \leq t} |B_r + x|^p$ where $s \mapsto g_*(s)$ is the maximal solution of (2.3) with $K = p/2c$.)

Moreover, in the case $0 < p < 1 + q$ by taking the infimum over all $c > 0$ on the right-hand side in either (2.37) or (2.38) we get a sharp inequality for each fixed x (the equality will be attained at each $\tau_* = \tau_*(c; x)$ for all $c > 0$). For simplicity we omit the explicit expressions. Note, however, that in the case $p = 1 + q > 1$ such a procedure (when $c \downarrow \gamma^*_{q+1,q}$) gives:

$$E\left(\max_{0 \leq t \leq \tau} |B_t + x|^p \right) \leq \left(\frac{p}{p-1} \right)^p E|B_\tau + x|^p - \left(\frac{p}{p-1} \right) x^p \quad (2.39)$$

for all $x \geq 0$ and all stopping times τ for B satisfying $E(\tau^{p/2}) < \infty$.

Observe that Cox [2] derived inequality (2.39) for discrete non-negative submartingales by a different method. Our proof above shows that this inequality is sharp for each fixed x (the equality is attained at $\tau_* = \tau_*(c; x)$ from (2.4) with $X_t = |B_t + x|^p$ and $S_t = \max_{0 \leq r \leq t} |B_r + x|^p$ where $s \mapsto g_*(s)$ is from (2.34) above). $\square$

Due to the extreme properties of Brownian motion, the inequalities (2.38) and (2.39) extend to all non-negative submartingales. This can be obtained by using the maximal embedding result of Jacka [13]. For reader's convenience, we state the result and outline the proof.

COROLLARY 2.3 *Let* $X = (X_t)_{t \geq 0}$ *be a non-negative cadlag (right continuous with left limits) uniformly integrable submartingale started at* $x \geq 0$ *under* P. *Let* X_∞ *denote the* P-*a.s. limit of* X *for* $t \to \infty$. *Then the following inequality is satisfied:*

$$E\left(\sup_{t > 0} X_t^p \right) \leq \frac{2c}{q(q+1)} E(X_\infty^{q+1}) - \frac{2c}{q(q+1)} x^{q+1} + V_c(x^p, x^p) \quad (2.40)$$

for all $0 < p < 1 + q$ *and all* $q > 0$, *where* $V_c(x, x)$ *is given by (2.22) with (2.23) and (2.24) if* $x > s_*$ *and by (2.27) if* $0 \leq x \leq s_*$ *(with* $s_* > 0$ *as in Theorem 2.1 above). This inequality is sharp.*

Similarly, the following inequality is satisfied:

$$E\left(\sup_{t>0} X_t^p\right) \leq \left(\frac{p}{p-1}\right)^p E\left(X_\infty^p\right) - \left(\frac{p}{p-1}\right) x^p \qquad (2.41)$$

for all $p > 1$. This inequality is sharp.

Proof. Given such a submartingale $X = (X_t)_{t\geq 0}$ satisfying $E(X_\infty) < \infty$, and a Brownian motion $B = (B_t)_{t\geq 0}$ started at $X_0 = x$ under P_x, by the result of Jacka [13] we know that there exists a stopping time τ for B, such that $|B_\tau| \sim X_\infty$ and $P\{\sup_{t\geq 0} X_t \geq \lambda\} \leq P_x\{\max_{0\leq t\leq \tau} |B_t| \geq \lambda\}$ for all $\lambda > 0$, with $(B_{t\wedge\tau})_{t\geq 0}$ being uniformly integrable. The result then easily follows from the proof of Theorem 2.1 as indicated in Remark 2.2 by using the integration by parts formula. $\qquad\square$

References

[1] Azema, J. *and* Yor, M. (1979). Une solution simple au probleme de Skorokhod. *Lecture Notes in Math.* 721, Springer (90–115).

[2] Cox, D. C. (1984). Some sharp martingale inequalities related to Doob's inequality. *IMS Lecture Notes Monograph Ser.* 5 (78–83).

[3] Doob, J. L. (1953). *Stochastic Processes.* John Wiley & Sons.

[4] Dubins, L. E. *and* Gilat, D. (1978). On the distribution of maxima of martingales. *Proc. Amer. Math. Soc.* 68 (337–338).

[5] Dubins, L. E. *and* Schwarz, G. (1988). A sharp inequality for sub-martingales and stopping times. *Astérisque* 157–158 (129–145).

[6] Dubins, L. E. Shepp, L. A. *and* Shiryaev, A. N. (1993). Optimal stopping rules and maximal inequalities for Bessel processes. *Teor. Veroyatnost. i Primenen.* 38 (288–330) (Russian); (226–261) (English translation).

[7] Gilat, D. (1986). The best bound in the $L \log L$-inequality of Hardy and Littlewood and its martingale counterpart. *Proc. Amer. Math. Soc.* 97 (429–436).

[8] Gilat, D. (1988). On the ratio of the expected maximum of a martingale and the L_p-norm of its last term. *Israel J. Math.* 63 (270–280).

[9] Graversen, S. E. *and* Peškir, G. (1994). Solution to a Wald's type optional stopping problem for Brownian motion. *Math. Inst. Aarhus, Preprint Ser.* No. 10 (15 pp). To appear in *J. Appl. Probab.*

[10] Graversen. S. E. *and* Peškir, G. (1995). Optimal stopping and maximal inequalities for linear diffusions. *Research Report* No. 335, *Dept. Theoret. Statist. Aarhus* (18 pp). To appear in *J. Theoret. Probab.*

[11] Graversen, S. E. *and* Peškir, G. (1996). Optimal stopping in the $L \log L$-inequality of Hardy and Littlewood. *Research Report* No. 360, *Dept. Theoret. Statist. Aarhus* (12 pp). To appear in *J. London Math. Soc.*

[12] Hardy, G. H. *and* Littlewood, J. E. (1930). A maximal theorem with function-theoretic applications. *Acta Math.* 54 (81–116).

[13] Jacka, S. D. (1988). Doob's inequalities revisited: A maximal H^1-embedding. *Stochastic Process. Appl.* 29 (281–290).

[14] Jacka, S. D. (1991). Optimal stopping and best constants for Doob-like inequalities I: The case $p = 1$. *Ann. Probab.* 19 (1798–1821).

[15] Revuz, D. *and* Yor, M. (1994). *Continuous Martingales and Brownian Motion.* Springer-Verlag.

[16] Wang, G. (1991). Sharp maximal inequalities for conditionally symmetric martingales and Brownian motion. *Proc. Amer. Math. Soc.* 112 (579–586).

Goran Peškir
Institute of Mathematics,
University of Aarhus
Ny Munkegade,
8000 Aarhus, Denmark
goran@mi.aau.dk

(Department of Mathematics
University of Zagreb
Bijenička 30,
41000 Zagreb, Croatia)

Progress in Probability, Vol. 43
© 1998 Birkhäuser Verlag Basel/Switzerland

Optimal Tail Comparison Based on Comparison of Moments

Iosif Pinelis

1. Introduction

Many of extremal problems of probability and statistics may be put in the following general form. Suppose that one has a moment comparison inequality

$$\mathbf{E}\varphi(\xi) \le \mathbf{E}\varphi(\eta) \quad \text{for all } \varphi \in \mathcal{F}, \tag{1}$$

where ξ and η are random elements of a measurable space (K, Σ) and $\mathcal{F}$ is a non-empty set of measurable real functions on K. Given the class $\mathcal{F}$, the random element η, and a function ψ on K, one is asked to

$$\text{maximize } \mathbf{E}\psi(\xi) \tag{2}$$

over all random elements ξ satisfying (1). This kind of problem goes back to Chebyshev and Markov; see, e.g., Karlin and Studden (1966) [12], Kemperman (1968, 1983) [15, 14], Karr (1983) [13], and Pinelis and Utev (1989) [22].

We are especially interested in the situations when (1) implies a probability comparison inequality of the form

$$\mathbf{P}\left(\xi \ge x\right) \le c\,\mathbf{P}\left(\eta \ge x\right) \quad \text{for all } x. \tag{3}$$

This corresponds to the problem (1)-(2) with $\psi = I_{[x,\infty)}$.

Without loss of generality (w.l.o.g.), we can assume that $\mathcal{F}$ is a convex cone containing all the constants.

Let us write $\xi \preceq \eta$ or $\mu \preceq \nu$, where μ and ν are the distributions of ξ and η respectively, adding $(\mathrm{mod}\,\mathcal{F})$ when necessary, if (1) takes place. Thus, (1) is seen as a generalized majorization condition, the usual majorization corresponding to the important case when $\mathcal{F}$ is the class of all convex functions; cf. Marshall and Olkin (1979) [18].

If $\varphi \in \mathcal{F}$ and $\varphi \ge \psi$, then $\int \psi d\mu \le \int \varphi d\mu \le \int \varphi d\nu$ by (1), so that $\int \varphi d\nu$ is an upper bound on $\int \psi d\mu$; this simple idea goes back again to Chebyshev; what is important is that such upper bounds with appropriate chosen functions φ are "normally" exact or almost exact, giving thus a solution to the above problem; this is in fact a very general duality phenomenon: given ν, $\mathcal{F}$, and ψ,

$$\sup\left\{\int \psi d\mu\colon \mu \preceq \nu \ (\mathrm{mod}\mathcal{F})\right\} = \inf\left\{\int \varphi d\nu\colon \varphi \ge \psi,\ \varphi \in \mathcal{F}\right\}. \tag{4}$$

This follows under certain very general sufficient conditions from Theorem 6 of Kemperman [15]; note that the above restrictions $\mathbf{E}\varphi(\xi) \le \mathbf{E}\varphi(\eta)$ may be replaced by more general inequalities of the form $\mathbf{E}\varphi(\xi) \leq c(\varphi)$; since $\mathbf{E}\varphi(\xi)$ is linear in φ, the function c may be assumed w.l.o.g. to satisfy the conditions (i) c is convex and (ii) $c(t\varphi) = tc(\varphi)$ for all $t \geq 0$ [otherwise, $c(\varphi)$ may be replaced by $\inf\{\sum t_i c(\varphi_i): t_i \geq 0, \varphi_i \in \mathcal{F} \ \forall i, \sum t_i \varphi_i = \varphi\}$]. Then a necessary and sufficient condition for

$$\sup\left\{\int \psi d\mu: \int \varphi d\mu \leq c(\varphi) \ \forall \varphi \in \mathcal{F}\right\} = \inf\{c(\varphi): \varphi \geq \psi, \ \varphi \in \mathcal{F}\} \qquad (5)$$

is that the "perturbation" functional

$$\delta \mapsto \sup\left\{\int \psi d\mu: \int \varphi d\mu \leq c(\varphi) + \delta(\varphi) \ \forall \varphi \in \mathcal{F}\right\}$$

is upper-semicontinuous with respect to the "perturbation" δ at 0 in the topology of pointwise convergence in the space of, say, all real functions on $\mathcal{F}$; this is immediate from Pinelis (1991) [20] upon consideration of the Lagrangian

$$L(\mu, \varphi) := \int \psi d\mu - \int \varphi d\mu + c(\varphi)$$

[conditions concerning topological features, measurability, and integrability do not matter here as long as (i) $L(\mu, \varphi)$ is defined and real for all μ belonging to a convex set containing all delta-measures on K and for all $\varphi \in \mathcal{F}$ and (ii) $L(\mu, \varphi)$ is affine (or, more generally, concave) in μ and convex in φ]. However, the mentioned sufficient conditions of Kemperman [15] are easier to check and will suffice for our purposes within this article.

In the special case when K is compact and the cone $\mathcal{F}$ is closed under $\bigwedge$ (for instance, when $\mathcal{F}$ is the cone of all concave functions on K), the duality (4) is a statement of the measure balayage theory (see, e.g., XI **T20** in Meyer (1966) [19]).

Thus, if the duality (4) or (5) takes place, the problem (1)-(2) is reduced to that of evaluation of the R.H.S. of (4) or (5). Even when (4) or (5) does not hold, the corresponding inequality L.H.S. $\leq$ R.H.S. is always true, so that the R.H.S. of (4) or (5) is still an upper bound on $\mathbf{E}\psi(\xi)$, although not the least one.

In this paper, we evaluate the R.H.S. of (4) in the case when $\mathcal{F}$ is the set of the α-convex functions, defined below, and for a certain class of functions ψ. Next, we provide exact upper bounds on and asymptotics of the R.H.S. of (4) provided certain regularity of ν. Finally, we give some applications, including exact bounds for the distributions of Gaussian random vectors, multivariate stochastic integrals, and (semi)martingales; in particular, we give a stronger version of the Hoeffding-Azuma inequality.

DEFINITION 1.1 Let $\alpha > 0$. A real function φ on $\mathbf{R} \cup \{-\infty\}$ is α-convex if it is non-decreasing on $\mathbf{R} \cup \{-\infty\}$ and $(\varphi - \varphi(-\infty))^{1/\alpha}$ is convex on $\mathbf{R}$. Let $\mathcal{F}_\alpha$ stand for the set of all α-convex functions.

REMARK 1.2 If $\alpha \geq 1$, then $\mathcal{F}_\alpha$ is a convex cone. This follows because for $\alpha \geq 1$, the function $\mathbf{R}_+^2 \ni (g_1, g_2) \mapsto (g_1^\alpha + g_2^\alpha)^{1/\alpha}$ is convex and non-decreasing in either argument.

Obviously, $0 < \alpha < \beta$ implies $\mathcal{F}_\alpha \supseteq \mathcal{F}_\beta$.

2. Evaluation of $\mathcal{I}(\nu, \psi, \alpha) := \inf \left\{ \int \varphi d\nu \colon \varphi \geq \psi, \ \varphi \in \mathcal{F}_\alpha \right\}$

Because of the duality (4), the evaluation mentioned in the title of this section will provide the value of (2) subject to (1) in the case $\mathcal{F} = \mathcal{F}_\alpha$.

The following lemma is needed to prove Theorem 2.2 below.

LEMMA 2.1 *Suppose that $\alpha > 0$, $x \in \mathbf{R}$, and ψ is a nonnegative function on $\mathbf{R}$ such that $\psi^{1/\alpha}$ is concave on $[x, \infty)$, and $\psi(u) = 0$ for $u < x$. Suppose also that $\varphi \geq \psi$ and $\varphi \in \mathcal{F}_\alpha$. Then there exist $t \leq x$ and $k > 0$ such that*

$$\varphi(u) \geq k(u - t)_+^\alpha \geq \psi(u) \quad \forall u \in \mathbf{R},$$

or $\varphi \geq c \geq \psi$ for some real constant c.

Proof. W.l.o.g., $\alpha = 1$. Then $\{(u, y) \colon u \in \mathbf{R}, \ y > \varphi(u)\}$ and $\{(u, y) \colon u \geq x, \ y \leq \psi(u)\}$ are non-empty disjoint convex subsets of $\mathbf{R}^2$, and the former set is open. Hence, by a version of the Hahn-Banach theorem, there exists $(c_1, c_2, c_3) \in \mathbf{R}^3$ such that inequalities $u \geq x$ and $y \leq \psi(u)$ imply $c_1 u + c_2 y \leq c_3$ and inequality $y > \varphi(u)$ implies $c_1 u + c_2 y > c_3$, for any real u and y. It is easy to see that here, with necessity, $c_2 > 0$. One thus has $\varphi(u) \geq ku + c_3/c_2 \ \forall u \in \mathbf{R}$ and $ku + c_3/c_2 \geq \psi(u) \ \forall u \geq x$, where $k := -c_1/c_2$. Note that, with necessity, $k \geq 0$. If $k = 0$, then $\varphi \geq c \geq \psi$ for $c := c_3/c_2$. Otherwise, for $t := -(c_3/c_2)/k$, one obtains $\varphi(u) \geq k(u - t)_+ \ \forall u \in \mathbf{R}$, since $\varphi \geq \psi \geq 0$; also, $t \leq x$ [because $0 \leq \psi(x) \leq k(x - t)$], whence $\psi(u) = 0$ for $u < t$, and so, $k(u - t)_+ \geq \psi(u) \ \forall u \in \mathbf{R}$. $\qquad\square$

The following theorem provides a useful infimum-convolution expression for $\mathcal{I}(\nu, \psi, \alpha)$.

THEOREM 2.2 *Suppose that $\alpha > 0$, $x \in \mathbf{R}$, and ψ is a nonnegative function on $\mathbf{R} \cup \{-\infty\}$ such that $\psi^{1/\alpha}$ is concave on $[x, \infty)$ and $\psi(u) = 0$ for $u \in [-\infty, x)$. Let η be any real-valued random variable (r.v.) and let ν be its distribution. Then*

$$\mathcal{I}(\nu, \psi, \alpha) = (\sup \psi) \wedge \inf_{t \in (-\infty, x)} \kappa_{\psi,\alpha}(t) \mathbf{E}(\eta - t)_+^\alpha, \ \text{where} \ \kappa_{\psi,\alpha}(t) := \sup_{u \in (t, \infty)} \frac{\psi(u)}{(u - t)^\alpha}.$$

Proof. It follows from Lemma 2.1 that

$$\mathcal{I}(\nu, \psi, \alpha) = (\sup \psi) \wedge \inf\{k \mathbf{E}(\eta - t)_+^\alpha \colon k > 0, \ t \leq x, \ k(u - t)_+^\alpha \geq \psi(u) \ \forall u \in \mathbf{R}\}.$$

Now the theorem is immediate. $\qquad\square$

COROLLARY 2.3 *Suppose that $\alpha > \beta \geq 0$, $x \in \mathbf{R}$, and let $\psi_{\beta,x}(u) := (u - x)_+^\beta$ if $\beta > 0$ and $\psi_{0,x}(u) := I\{u > x\}$. Let η be any real-valued r.v. such that $\mathbf{E}\eta_+^\alpha < \infty$ and let ν be its distribution. Then*

$$\mathcal{I}(\nu, \psi_{\beta,x}, \alpha) = C_{\alpha,\beta} \inf_{t \in (-\infty, x)} \frac{\mathbf{E}(\eta - t)_+^\alpha}{(x - t)^{\alpha - \beta}}, \tag{6}$$

where

$$C_{\alpha,\beta} := (\alpha - \beta)^{\alpha - \beta} \beta^\beta \alpha^{-\alpha}, \ 0^0 := 1. \tag{7}$$

Proof. This follows because for $t < x$ and $\alpha > \beta \geq 0$, $\kappa_{\psi_{\beta,x},\alpha}(t) := C_{\alpha,\beta}/(x-t)^{\alpha-\beta}$, and because, by the Lebesgue theorem, $C_{\alpha,\alpha}(x-t)^{-\alpha}\mathbf{E}(\eta - t)_+^\alpha \to C_{\alpha,\alpha} = 1$ as $t \to -\infty$. $\qquad\square$

REMARK 2.4 $C_{\alpha,\beta} < 1 = C_{\alpha,0} = C_{\alpha,\alpha}$ if $\alpha > \beta > 0$, because $u \ln u$ is convex in $u > 0$.

The following theorem details the evaluation of the R.H.S. of (6).

THEOREM 2.5 *Suppose that* $\alpha \geq 1$, $\alpha > \beta \geq 0$, $x \in \mathbf{R}$, *and let* $\psi_{\beta,x}(u) := (u-x)_+^\beta$ *if* $\beta > 0$ *and* $\psi_{0,x}(u) := I\{u > x\}$. *Let* η *be any real-valued r.v. such that* $\mathbf{E}\eta_+^\alpha < \infty$, *and let* ν *be its distribution; for* $t \in \mathbf{R}$, *let*

$$\gamma(t) := \gamma_\alpha(t) := \gamma_{\alpha,\eta}(t) := \mathbf{E}(\eta - t)_+^\alpha \tag{8}$$

and

$$m(t) := m_{\alpha,\beta}(t) := t - (\alpha - \beta)\frac{\gamma(t)}{\gamma'(t)} \quad \text{if } \gamma'(t) \neq 0 \text{ (i.e., if } \mathbf{P}(\eta > t) > 0)), \tag{9}$$

where γ' *is the [right-hand-side if* $\alpha = 1$*] derivative of* γ. *Then*

(i) $m_{\alpha,\beta}$ *is monotone in the sense that* $s < t$ *implies* $m_{\alpha,\beta}(s) \leq m_{\alpha,\beta}(t)$ *and, moreover,* $m_{\alpha,\beta}(s) < m_{\alpha,\beta}(t)$ *if* $(\alpha - 1)(-1 + \mathrm{card}((t,\infty) \cap \mathrm{supp}\,\nu)) > 0$ *or* $\beta > 0$ *or* $\mathbf{P}(s < \eta \leq t)\mathbf{P}(\eta > t) > 0$;

(ii) $m_{\alpha,\beta}(-\infty) := \lim_{t\to-\infty} m_{\alpha,\beta}(t) = -\infty$ *if* $\beta > 0$, *and* $m_{\alpha,0}(-\infty) = \mathbf{E}\eta \in [-\infty,\infty)$;

(iii) for every $x > m(-\infty)$, *there exists* $t_x \in \mathbf{R}$ *such that*

$$m(t_x - 0) \leq x \leq m(t_x);$$

moreover, t_x *is uniquely determined if* $\alpha > 1$ *and* $\sup \mathrm{supp}\,\nu = \infty$, *or* $\beta > 0$, *or* $\mathrm{supp}\,\nu \supset (m(-\infty),\infty)$;

(iv) for every $x > m(-\infty)$,

$$\mathcal{I}(\nu,\psi_{\beta,x},\alpha) = C_{\alpha,\beta}\frac{\gamma(t_x)}{(x - t_x)^{\alpha-\beta}},$$

where $C_{\alpha,\beta}$ *is defined by (7); in particular, for every* $x > m(-\infty)$,

$$\mathbf{P}(\eta > t_x) \leq \mathcal{I}(\nu,\psi_{0,x},1) \leq \mathbf{P}(\eta \geq t_x); \tag{10}$$

finally, if $x \leq m(-\infty)$ *(in which case necessarily* $\beta = 0$ *and* $m(-\infty) = \mathbf{E}\eta$*), then* $\mathcal{I}(\nu,\psi_{\beta,x},\alpha) = 1$.

Proof. (i) Let $c := \alpha - 1$ and $s < t$. Note that

$$\gamma'(t) = -\alpha \int_{t+0}^\infty (u - t)^c \nu(du), \tag{11}$$

and so,

$$m_{\alpha,0}(t) = \frac{\int_{t+0}^{\infty} u(u-t)^c \nu(du)}{\int_{t+0}^{\infty} (u-t)^c \nu(du)}. \tag{12}$$

Using simple algebra and the identity

$$\int\int g(x,u)\nu(dx)\nu(du) = \frac{1}{2}\int\int [g(x,u)+g(u,x)]\nu(dx)\nu(du),$$

one obtains

$$[m_{\alpha,0}(t) - m_{\alpha,0}(s)] \int_{t+0}^{\infty} (x-t)^c \nu(dx) \int_{s+0}^{\infty} (u-s)^c \nu(du) = I_1 + \frac{1}{2} I_2,$$

where

$$I_1 := \int_{t+0}^{\infty} (x-t)^c \nu(dx) \int_{s+0}^{t+0} (u-s)^c (x-u)\nu(du)$$

and

$$I_2 := \int_{t+0}^{\infty} \int_{t+0}^{\infty} (x-s)^c (u-s)^c (x-u) \left[\left(\frac{x-t}{x-s} \right)^c - \left(\frac{u-t}{u-s} \right)^c \right] \nu(dx)\nu(du).$$

Obviously, $I_1 \geq 0$; moreover, $I_1 > 0$ if $\nu((s,t])\nu((t,\infty)) > 0$. Next, $I_2 \geq 0$ [$I_2 > 0$ if $c \int_{t+0}^{\infty} \int_{t+0}^{\infty} I\{x \neq u\}\nu(dx)\nu(du) > 0$], because $(t,\infty) \ni u \mapsto (u-t)/(u-s)$ is increasing. Thus, (i) is proved for $\beta = 0$; the case $\beta > 0$ is now obvious from

$$m_{\alpha,\beta}(t) = \frac{\alpha - \beta}{\alpha} m_{\alpha,0}(t) + \frac{\beta}{\alpha} t.$$

(ii) Here, again, it suffices to consider the case $\beta = 0$. Let $t \to -\infty$ and $A \to \infty$ in such a way that $A = o(|t|)$. If $\int_{-\infty}^{0} |u|\nu(du) < \infty$, then

$$|t|^{-c} J_1 := |t|^{-c} \int_{(t,-A]} u(u-t)^c \nu(du) = O\left(\int_{(t,-A]} |u|\nu(du) \right) = o(1);$$

otherwise, it will suffice to notice that eventually $J_1 \leq 0$. Next,

$$|t|^{-c} J_2 := |t|^{-c} \int_{(-A,A]} u(u-t)^c \nu(du) \to \int u\nu(du)$$

and

$$|t|^{-c} J_3 := |t|^{-c} \int_{(A,\infty)} u(u-t)^c \nu(du) = O\left(\int_{(A,\infty)} (u + u^{c+1})\nu(du) \right) = o(1).$$

Hence, $|t|^{-c} \int_{(t,\infty)} u(u-t)^c \nu(du) \to \int u\nu(du)$. Similarly, $|t|^{-c} \int_{(t,\infty)} (u-t)^c \nu(du) \to 1$. Thus, in view of (12), (ii) is also proved.

(iii) This is immediate from (i) and (ii), because $m(t) \geq t$, and so, $m(t) \to \infty$ as $t \to \infty$.

(iv) Let

$$F(t) := F(t, x) := \frac{\gamma(t)}{(x - t)^{\alpha - \beta}}. \tag{13}$$

Then

$$F'(t) := F'(t + 0) = \frac{\gamma'(t)}{(x - t)^{\alpha - \beta + 1}} [x - m(t)].$$

Hence, for every $x > m(-\infty)$, $F'(t) \leq 0$ for $t < t_x$ and $F'(t) \geq 0$ for $t \geq t_x$. Now, in view of Corollary 2.3, (iv) follows for $x > m(-\infty)$. If $x \leq m(-\infty)$ [and so, by (ii), $\beta = 0$], then $F'(t) \geq 0$ for all $t \in \mathbf{R}$; hence,

$$\inf_{t \in \mathbf{R}} F(t) = \lim_{t \to -\infty} \mathbf{E} \frac{(\eta - t)_+^\alpha}{(x - t)^\alpha} = 1,$$

as in the proof of Corollary 2.3. Thus, (iv) is completely proved. □

REMARK 2.6 Condition $\alpha \geq 1$ is essential in Theorem 2.5. Indeed, if $0 < \alpha < 1$ and ν is e.g. a two-point distribution on $\mathbf{R}$, then $m_{\alpha,0}(t)$ is not monotone. However, if $0 < \alpha < 1$ and ν is absolutely continuous on (b, ∞) with a density $f(t)$, which is nonzero and non-increasing for large enough t, then $m_{\alpha,0}$ is increasing on (b, ∞), because for $t > b$, $\gamma'_\alpha(t) = -\int (u - t)_+^\alpha df(u)$ is non-increasing, whence $\gamma''_a(t) := \liminf_{\delta \downarrow 0} [\gamma'_a(t + \delta) - \gamma'_a(t)]/\delta \geq 0$, and so,

$$m'_{\alpha,0}(t) := \liminf_{\delta \downarrow 0} \frac{m_{\alpha,0}(t + \delta) - m_{\alpha,0}(t)}{\delta} = 1 - \alpha + \frac{\alpha \gamma_a(t) \gamma''_a(t)}{\gamma'_a(t)^2} > 0.$$

3.   Exact upper bounds on $\mathcal{I}(\nu, \psi_{\beta,x}, \alpha)$ for r-concave tails of ν

DEFINITION 3.1 Suppose that $r > 0$ and K is a convex subset of a linear space. A positive real function q on K is r-concave if $q^{-1/r}$ is convex.

REMARK 3.2 An important characterization of r-concavity in terms of a generalized Brunn-Minkowski inequality follows from Theorem 3.2 of Borell (1975) [3].

The following proposition presents a simple idea, which will be useful in the proof of Theorem 3.11, one of the main results in this paper.

PROPOSITION 3.3 (i) If q is r-concave and $\beta \in (0, r)$, then q is β-concave.
    (ii) q is r-concave for all $r > 0$ iff q is log-concave.
    (iii) Suppose that q is a positive real function on $\mathbf{R}$. Then q is r-concave for some $r > 0$ iff for every $x \in \mathbf{R}$ there exist $a \in \mathbf{R}$ and $b \in \mathbf{R}$ such that

$$(ax + b)_+^{-r} = q(x) \quad \text{and} \quad (at + b)_+^{-r} \geq q(t) \; \forall t \in \mathbf{R}; \tag{14}$$

q is log-concave iff for every $x \in \mathbf{R}$ there exist $a \in \mathbf{R}$ and $b \in \mathbf{R}$ such that

$$\exp[-(ax + b)] = q(x) \quad \text{and} \quad \exp[-(at + b)] \geq q(t) \; \forall t \in \mathbf{R}; \tag{15}$$

here, $0^{-r} := \infty$. Thus, q is a lower envelope of a family of power or exponential functions.

Proof. (i) This follows because $q^{-1/\beta} = (q^{-1/r})^{r/\beta}$ and the function $(0, \infty) \ni x \mapsto x^c$ is increasing and convex if $c \geq 1$.

(ii) Similarly, if q is log-concave, then it is r-concave for all $r > 0$, because $q^{-1/r} = \exp[-(1/r)\ln q]$ and the function $x \mapsto e^x$ is increasing and convex. Conversely, if q is r-concave for all $r > 0$, then it is log-concave, because $-\ln q = \lim_{r \to \infty} r(q^{-1/r} - 1)$ is convex.

(iii) This follows because for any convex real function f on $\mathbf{R}$ and any $x \in \mathbf{R}$, $f(x) = \max\{f(t) + f'(t+0)(x-t): t \in \mathbf{R}\}$ and, on the other hand, if $f(t) = \max\{a(x)t + b(x): x \in \mathbf{R}\}$ for some real functions a and b on $\mathbf{R}$ and for all $t \in \mathbf{R}$, then f is a convex real function. $\square$

REMARK 3.4 In view of Part (ii) of Proposition 3.3, one can refer to the log-concavity as to the ∞-concavity.

DEFINITION 3.5 Suppose that $r > 0$. A positive real function q on $\mathbf{R}$ is asymptotically r-concave if for every $\beta \in (0, r)$, there exists $x \in \mathbf{R}$ such that $q^{-1/\beta}$ is convex on $[x, \infty)$; q is asymptotically log-concave (or, to say the same, asymptotically ∞-concave) if for every $\beta > 0$, there exists $x \in \mathbf{R}$ such that $q^{-1/\beta}$ is convex on $[x, \infty)$.

Closely related with the class $\mathcal{F}_\alpha$ is the following well-known notion of the semigroup of the operators of fractional integration.

DEFINITION 3.6 Suppose that q is a nonnegative Borel function on $\mathbf{R}$. Let

$$(T^\alpha q)(t) := \frac{1}{\Gamma(\alpha)} \int_t^\infty (u-t)^{\alpha-1} q(u) du, \quad \text{for } \alpha > 0 \text{ and } t \in \mathbf{R},$$

and $T^0 q := q$.

REMARK 3.7 (i) If $\alpha \geq 0$, $\beta \geq 0$, and q is a nonnegative Borel function on $\mathbf{R}$, then $T^\alpha T^\beta q = T^{\alpha+\beta} q$.

(ii) Suppose that $\alpha \geq 0$ and $q(x) = \mathbf{P}(\eta \geq x)$, the tail of a real r.v. η with the distribution ν [or, alternatively, the probability density function (p.d.f.) of η]. Then $T^1 q$ is the integrated tail of η [respectively, the tail of η]; thus, $T^\alpha q$ interpolates the sequence $T^0 q = q, T^1 q, T^2 q, \dots$ of the iteratively integrated tails. Integrated tails arise naturally in boundary problems for random walks and hence in the theory of queue systems. Note also that for $\alpha > 0$,

$$(T^\alpha q)(t) := \frac{1}{\Gamma(\alpha+1)} \int_t^\infty (u-t)^\alpha \nu(du) = \frac{\mathbf{E}(\eta-t)_+^\alpha}{\Gamma(\alpha+1)} = \frac{\gamma_\alpha(t)}{\Gamma(\alpha+1)}, \quad t \in \mathbf{R} \quad (16)$$

[cf. (8)]. It follows that for all $t \in \mathbf{R}$, $(T^\alpha q)(t) \to q(t+0)$ as $\alpha \downarrow 0$, for all $t \in \mathbf{R}$, provided that $(T^\alpha q)(t) < \infty$ for some $\alpha > 0$.

(iii) $\xi \preceq \eta \bmod \mathcal{F}_\alpha$ iff $T^\alpha p \leq T^\alpha q$, where $p(x) := \mathbf{P}(\xi \geq x)$ and $q(x) := \mathbf{P}(\eta \geq x)$.

The next proposition is needed in the proof of Theorem 3.11 below.

PROPOSITION 3.8 *If $\alpha > \beta \geq 0$, $a \in \mathbf{R}$, and q is an r-concave function on (a, ∞), then $T^\beta q$ is $(r - \beta)$-concave on (a, ∞). In particular, if q is a log-concave function on (a, ∞), then $T^\beta q$ is so for all $\beta \geq 0$ [the latter statement is well known for $\beta = 1$].*

Proof. will be given elsewhere.   □

PROPOSITION 3.9 *If $\infty \geq r > 0$ and q is a positive twice differentiable function such that*

$$qq'' \leq \left(1 + \frac{1}{r}\right) q'^2, \quad \text{where } \frac{1}{\infty} := 0, \tag{17}$$

then q is r-concave; if

$$\limsup_{u \to \infty} \frac{q(u)q''(u)}{q'(u)^2} \leq 1 + \frac{1}{r}, \tag{18}$$

then q is asymptotically r-concave.

Proof. Straightforward.   □

EXAMPLE 3.10 The p.d.f. of the Student t distribution with k degrees of freedom is $(k+1)$-concave, and so, its tail is k-concave. In particular, the p.d.f. of the normal distribution is log-concave, and hence so is its tail; moreover, the same holds for the distribution of $\chi_d := \sqrt{\chi_d^2}$, where χ_d^2 is a r.v. having the χ^2 distribution with $k \geq 2$ degrees of freedom.

   If $k \geq 2$, then the p.d.f. of the F distribution with k and m degrees of freedom is $(m/2 + 1)$-concave, and so, its tail is $(m/2)$-concave. In particular, the p.d.f. of the χ^2 distribution with $k \geq 2$ degrees of freedom is log-concave, and hence so is its tail.

   More examples can be found in Borell (1975) [3].

   The following theorem is one of the main results in this paper.

THEOREM 3.11 *Suppose that $\infty \geq r > \alpha > \beta \geq 0$, $\xi \preceq \eta \pmod{\mathcal{F}_\alpha}$, and the tail function $q(x) := \mathbf{P}(\eta \geq x)$ is r-concave. Let $p(x) := \mathbf{P}(\xi \geq x)$. Then*

$$\mathbf{E}(\xi - x)_+^\beta \leq c(r; \alpha, \beta)\mathbf{E}(\eta - x)_+^\beta \quad \forall x \in \mathbf{R}, \tag{19}$$

where

$$c(r; \alpha, \beta) := \frac{\alpha^{-\alpha}\Gamma(\alpha + 1)\Gamma(r - \alpha)(r - \alpha)^{\alpha - r}}{\beta^{-\beta}\Gamma(\beta + 1)\Gamma(r - \beta)(r - \beta)^{\beta - r}} \tag{20}$$

if $r < \infty$, and

$$c(\infty; \alpha, \beta) := \lim_{r \to \infty} c(r; \alpha, \beta) = \frac{\alpha^{-\alpha}\Gamma(\alpha + 1)e^\alpha}{\beta^{-\beta}\Gamma(\beta + 1)e^\beta} \tag{21}$$

[recall that the ∞-concavity is the same as the log-concavity].

In particular,

$$\mathbf{P}(\xi \geq x) \leq c(r; \alpha, 0)\mathbf{P}(\eta \geq x) \quad \forall x \in \mathbf{R}, \tag{22}$$

and

$$c(r; \alpha, 0) := \frac{\alpha^{-\alpha}\Gamma(\alpha + 1)\Gamma(r - \alpha)(r - \alpha)^{\alpha - r}}{\Gamma(r)r^{-r}}$$

if $r < \infty$ and

$$c(\infty; \alpha, 0) = \alpha^{-\alpha}\Gamma(\alpha + 1)e^{\alpha}.$$

The constant $c(r; \alpha, \beta)$ is the best possible in the conditions of the theorem if $\alpha \geq 1$.

Proof. is given here only for the case $r < \infty$; case $r = \infty$ is similar. Fix any $x \in \mathbf{R}$. By Proposition 3.8, the function $T^{\beta}q$ is $(r - \beta)$-concave; besides, $T^{\beta}q(t) \downarrow 0$ as $t \uparrow \infty$. Therefore, $\forall \varepsilon > 0 \ \exists a > 0 \ \exists b \in \mathbf{R}$

$$\tilde{q}_{x,\beta,\varepsilon}(t) := (at + b)_{+}^{-r+\beta} \geq (T^{\beta}q)(t) \ \forall t \quad \text{and} \quad \tilde{q}_{x,\beta,\varepsilon}(x) \leq (T^{\beta}q)(x) + \varepsilon \tag{23}$$

[cf. (14); here, the ε takes part since we want $a > 0$]. Using now the obvious inequality $\mathbf{E}(\xi - x)_{+}^{\alpha} \leq \mathcal{I}(\nu, \psi_{\beta,x}, \alpha)$, Corollary 2.3, and (16), one has

$$\begin{aligned} (T^{\beta}p)(x) \ &\leq \ C_{\alpha,\beta}\frac{\Gamma(\alpha + 1)}{\Gamma(\beta + 1)}\inf_{t \in (-\infty, x)}\frac{(T^{\alpha}q)(t)}{(x - t)^{\alpha - \beta}} \\ &\leq \ C_{\alpha,\beta}\frac{\Gamma(\alpha + 1)}{\Gamma(\beta + 1)}\inf_{t \in (-\infty, x)}\frac{(T^{\alpha - \beta}\tilde{q}_{x,\beta,\varepsilon})(t)}{(x - t)^{\alpha - \beta}}. \end{aligned} \tag{24}$$

Note that

$$(T^{\alpha - \beta}\tilde{q}_{x,\beta,\varepsilon})(t) = \frac{\Gamma(r - \alpha)}{a^{\alpha - \beta}\Gamma(r - \beta)}(at + b)_{+}^{\alpha - r} \tag{25}$$

and

$$\inf_{t \in (-\infty, x)}\frac{(at + b)_{+}^{\alpha - r}}{(x - t)^{\alpha - \beta}} = \frac{a^{\alpha - \beta}}{(\alpha - \beta)^{\alpha - \beta}}\frac{(r - \beta)^{r - \beta}}{(r - \alpha)^{r - \alpha}}(ax + b)_{+}^{\beta - r}. \tag{26}$$

The result follows from (7), (16), (24), (25), (26), the second part of (23), and, for the optimality, Theorem 6 of [15]. □

REMARK 3.12 Condition $\alpha \geq 1$ is needed in the latter proof in order for $\mathcal{F}_{\alpha}$ to be convex. As the author has been kindly advised by Jon Wellner, a similar result for the case when $\alpha = 1$ and $\beta = 0$ is due to Kemperman and is contained in the book by Shorack and Wellner [24], pages 797–799; note that $c(r; 1, 0) = (r/(r - 1))^r$ and $c(\infty; 1, 0) = e$.

REMARK 3.13 A useful point is that the requirement of the r-concavity of $q(x) := \mathbf{P}(\eta \geq x)$ in Theorem 3.11 can be relaxed by replacing $q(x) = \mathbf{P}(\eta \geq x)$ in (19) and (22) by any [e.g., the least] r-concave majorant of q. However, then the optimality of $c(r; \alpha, \beta)$ is not guaranteed. Related to this remark is Theorem 3.14 below.

THEOREM 3.14 *Suppose that $\infty \geq r > \alpha > \beta \geq 0$, $\xi \preceq \eta$ (mod $\mathcal{F}_\alpha$), and the tail function $q(x) := \mathbf{P}(\eta \geq x)$ is asymptotically r-concave. Let $p(x) := \mathbf{P}(\xi \geq x)$. Then*

$$\limsup_{x \to \infty} \frac{(T^\beta p)(x)}{(T^\beta q)(x)} \leq c(r; \alpha, \beta) \tag{27}$$

where $c(r; \alpha, \beta)$ is given by (20) and (21).

In particular,

$$\limsup_{x \to \infty} \frac{\mathbf{P}(\xi \geq x)}{\mathbf{P}(\eta \geq x)} \leq c(r; \alpha, 0) \quad \forall x \in \mathbf{R}. \tag{28}$$

The constant $c(r; \alpha, \beta)$ is the best possible in the conditions of the theorem if $\alpha \geq 1$.

Proof. Similar to that of Theorem 3.11. □

REMARK 3.15 Inequalities (22) and (28) may be compared with results found by de la Peña, Montgomery-Smith, and Szulga (1994) [5], where relations like

$$\limsup_{u \to \infty} \Psi(u) \mathbf{P}(\xi \geq u) \leq \limsup_{u \to \infty} \Psi(u) \mathbf{P}(\eta \geq u/c_\alpha) \tag{29}$$

for $\Psi(u)$ increasing faster than u^α, $\alpha > 1$, and some $c_\alpha > 0$ were obtained – see Proposition 2.4 in de la Peña, Montgomery-Smith, and Szulga (1994); see also Theorem 3.5 therein. The assumptions on moment comparison between ξ and η in de la Peña, Montgomery-Smith, and Szulga (1994) are more stringent than here in some respects and more relaxed in others; the assumptions on the regularity of the tail of the distribution of η are less stringent; however, their method seemingly does not provide for best constants; note also that c_α in the R.H.S. of (29) is under the probability sign, which is of concern especially when the tail $\mathbf{P}(\eta \geq u)$ decreases faster than any power function.

4. Exact asymptotics

DEFINITION 4.1 Suppose that $\infty \geq r > 0$. A positive twice differentiable function q on $\mathbf{R}$ is *like x^{-r}* if

$$\lim_{x \to \infty} \frac{q(x) q''(x)}{q'(x)^2} = 1 + \frac{1}{r}; \tag{30}$$

cf. Proposition 3.9.

E.g., if $q(x) = C x^{-r} \ln^\alpha x$ for all $x \geq x_0$ and some $r \in (0, \infty)$, $\alpha \in \mathbf{R}$, $C > 0$, and $x_0 \in (1, \infty)$, then q is like x^{-r}. On the other hand, if q is like $x^{-\infty}$, then $\forall r$ $q(x) = o(x^{-r})$ as $x \to \infty$.

Let us write $f \sim g$ if $f/g \to 1$.

THEOREM 4.2 *Suppose that $\infty \geq r > \alpha > \beta \geq 0$, $x \in \mathbf{R}$, η is any real-valued r.v. such that $\mathbf{E}\eta_+^\alpha < \infty$, and $q(x) := \mathbf{P}(\eta \geq x)$ is like x^{-r}. Then*

$$\sup\{\mathbf{E}(\xi - x)_+^\beta : \xi \preceq \eta \ (\text{mod } \mathcal{F}_\alpha)\} \sim c(r; \alpha, \beta) \mathbf{E}(\eta - x)_+^\beta, \quad x \to \infty,$$

where $c(r; \alpha, \beta)$ is given by (20) and (21). In particular,

$$\sup\{\mathbf{P}\,(\xi \geq x)\colon \xi \preceq \eta \pmod{\mathcal{F}_\alpha}\} \sim c(r; \alpha, 0)\mathbf{P}\,(\eta \geq x), \quad x \to \infty,$$

REMARK 4.3 Theorem 4.2 remains true if $q(x)/q_0(x) \to 1$ as $x \to \infty$ for some q_0 which is like x^{-r}.

Proof of Theorem 4.2.

LEMMA 4.4 *Suppose that a real function $h(u)$ on $\mathbf{R}$ is positive and differentiable for all large enough u and that $h'(u) \to c \geq 0$ as $u \to \infty$. Then for every $b > 0$,*

$$\int_t^{t+bh(t)} \frac{du}{h(u)} \to \begin{cases} \dfrac{1}{c}\ln(1 + bc) & \text{if } c \neq 0 \\ b & \text{if } c = 0 \end{cases}$$

Proof. One has

$$h(t + sh(t)) = h(t) + \int_t^{t+sh(t)} h'(u)du \sim h(t)(1 + sc), \quad t \to \infty, \qquad (31)$$

unformly in $s \in [0, b]$. Hence,

$$\int_t^{t+bh(t)} \frac{du}{h(u)} = \int_0^b \frac{h(t)ds}{h(t+sh(t))} \sim \int_0^b \frac{ds}{1+sc}, \quad t \to \infty,$$

which implies the lemma. $\qquad\square$

LEMMA 4.5 *Suppose that for all large enough u, a real function $h(u)$ on $\mathbf{R}$ is positive and differentiable and that $h'(u) \leq c \geq 0$. Then $\exists t_0 \in \mathbf{R} \; \forall t \geq t_0 \; \forall b \geq 0$*

$$\int_t^{t+bh(t)} \frac{du}{h(u)} \geq \begin{cases} \dfrac{1}{c}\ln(1 + bc) & \text{if } c \neq 0 \\ b & \text{if } c = 0 \end{cases}$$

Proof. Similar to that of Lemma 4.4. $\qquad\square$

Let us return to the proof of Theorem 4.2. Introduce $h := -\dfrac{q}{q'}$. Observing that

$$h'(u) = \frac{q(u)q''(u)}{q'(u)^2} - 1 \to \frac{1}{r} \quad \text{as } u \to \infty \qquad (32)$$

and applying Lemmas 4.4 and 4.5, one has

$$\frac{q(t)}{q(t + bh(t))} = \exp\left\{\int_t^{t+bh(t)} \frac{du}{h(u)}\right\} \to \left(1 + \frac{b}{r}\right)^r \quad \text{as } t \to \infty \qquad (33)$$

for every $b > 0$, and $\forall p \in (\alpha, r) \; \exists t_0 \in \mathbf{R} \; \forall t \geq t_0 \; \forall b \geq 0$

$$\frac{q(t)}{q(t + bh(t))} \geq \left(1 + \frac{b}{p}\right)^p.$$

Here and elsewhere the values of the expressions at $r = \infty$ are understood as the limits as $r \to \infty$. Using now the Lebesgue theorem and recalling (8), one obtains

$$\gamma_\alpha(t) = \alpha h(t)^\alpha q(t) \int_0^\infty b^{\alpha-1} \frac{q(t + bh(t))}{q(t)} db \sim \frac{r^\alpha \Gamma(\alpha + 1)\Gamma(r - \alpha)}{\Gamma(r)} h(t)^\alpha q(t)$$

$$\text{as } t \to \infty. \qquad (34)$$

Hence,

$$\gamma_\alpha'(t) = -\alpha \gamma_{\alpha-1}(t) \sim -\frac{r^{\alpha-1}\Gamma(\alpha+1)\Gamma(r-\alpha+1)}{\Gamma(r)} h(t)^{\alpha-1} q(t) \quad \text{as } t \to \infty,$$

and so,

$$m(t) - t \sim (\alpha - \beta)\frac{r}{r - \alpha} h(t) \quad \text{as } t \to \infty, \qquad (35)$$

where $m(t)$ is defined by (9). By (31),

$$h(m(t)) \sim \frac{r - \beta}{r - \alpha} h(t) \quad \text{as } t \to \infty, \qquad (36)$$

and by (33),

$$q(m(t)) \sim \left(\frac{r - \alpha}{r - \beta}\right)^r q(t) \quad \text{as } t \to \infty. \qquad (37)$$

It follows from (34), (36), and (37) that

$$\gamma_\beta(m(t)) \sim \frac{r^\beta \Gamma(\beta + 1)\Gamma(r - \beta)}{\Gamma(r)} \left(\frac{r - \beta}{r - \alpha}\right)^{\beta - r} h(t)^\beta q(t) \quad \text{as } t \to \infty. \qquad (38)$$

Now Theorem 4.2 follows from Theorem 2.5 together with Remark 2.6, (34), (35), (38), and (7). $\qquad \square$

REMARK 4.6 If $r = \infty$ in Theorem 4.2 or Remark 4.3, that is, roughly speaking, if tail q decreases more rapidly than any power, then $\xi \preceq \eta \pmod{\mathcal{F}_\alpha}$ implies

$$\mathbf{P}(\xi \geq x) \leq \mathbf{P}(\eta \geq t_x),$$

and

$$t_x \sim x$$

as $x \to \infty$, so that if $\xi \preceq \eta$, then the quantiles of ξ asymptotically cannot exceed the corresponding quantiles of η. This follows from (10), (32) [by which $h(u) = o(u)$ as $u \to \infty$], and (35).

5. Applications

Let X and X_d be, respectively, a zero-mean Gaussian random vector in a separable Banach space $(\mathcal{X}, \|\cdot\|)$ and a canonical zero-mean Gaussian random vector in $\mathbf{R}^d$. For a function $F : \mathcal{X} \to E$, where E is a normed space, let us write $F \in$

Lip $(\mathcal{X}, E, \sigma)$, where $\sigma > 0$, if $\|F(x) - F(y)\| \leq \sigma\|x - y\|$ for all $x \in \mathcal{X}$, $y \in \mathcal{X}$; let Lip $(\mathcal{X}, \sigma) := $ Lip $(\mathcal{X}, \mathbf{R}, \sigma)$.

For any $F \in $ Lip $(\mathbf{R}^d, \sigma)$, the isoperimetric inequality due to Borell (1975) [2] implies

$$\mathbf{P}\left(F(X_d) - \text{Median } F(X_d) \geq u\right) \leq \mathbf{P}\left(\sigma X_1 \geq u\right) \quad u > 0. \tag{39}$$

One can prove that for any $F \in $ Lip $(\mathbf{R}^d, \sigma)$,

$$F(X_d) - \mathbf{E}F(X_d) \preceq \sigma X_1 \pmod{\mathcal{F}_1}. \tag{40}$$

Inequality (40) is a generalization of the Maurey-Pisier bound [Pisier (1985, pages 180–181) [23]]

$$\mathbf{E}\exp\left\{F(X_d) - \mathbf{E}F(X_d)\right\} \leq e^{\sigma^2/2}. \tag{41}$$

One possible proof of (40) is to a certain extent similar to that of (41) – see the more general inequality (52) below. Obviously, (40) cannot hold in general if $\mathbf{E}F(X_d)$ is replaced by Median $F(X_d)$.

Nevertheless, the following may be deduced from above statements, by means of Theorem 3.11.

For any $F \in $ Lip $(\mathbf{R}^d, \sigma)$, one has [since $c(\infty; 1, 0) = e$]

$$\mathbf{P}\left(F(X_d) - \mathbf{E}F(X_d) \geq u\right) \leq e\,\mathbf{P}\left(\sigma X_1 \geq u\right), \quad \forall u \in \mathbf{R}. \tag{42}$$

Ibragimov, Sudakov, and Tsirel'son (1976) [11] obtained a better version of (42), with the constant factor 2 in place of e, using a random change of time argument, which also could give (40). Whether 2 is the best constant here is not known; on the other hand, the optimal constant must obviously be strictly greater than 1.

Recall now another result by Pisier (1985) [23]. Suppose that $\mathcal{X}$ and E are finite-dimensional Banach spaces; X is, as above, a Gaussian random vector in $\mathcal{X}$; Y is an independent copy of X; and $f\colon \mathcal{X} \to E$ is a locally Lipshitz function. Then for any convex function $\Psi\colon E \to \mathbf{R}$, one has

$$\mathbf{E}\Psi(f(X) - \mathbf{E}f(X)) \leq \mathbf{E}\Psi\left(\frac{\pi}{2}f'(X) \cdot Y\right), \tag{43}$$

where $f'(x) \cdot y$ stands for the directional derivative of f at x in the direction y, which exists almost everywhere; this may be considered a vector counterpart of (40). It follows that if $f \in $ Lip$(\mathbf{R}^d, E, \sigma)$, then

$$\mathbf{P}\left(\|f(X_d) - \mathbf{E}f(X_d))\|_E \geq u\right) \leq e\,\mathbf{P}\left(\frac{\pi}{2}\sigma|X_d| \geq u\right), \quad \forall u \in \mathbf{R}. \tag{44}$$

In the case $E = \mathbf{R}^m$, one also has

$$\mathbf{E}\Psi(f(X_d) - \mathbf{E}f(X_d)) \leq \mathbf{E}\Psi\left(\frac{\pi}{2}\sigma X_m\right), \tag{45}$$

and so,

$$\mathbf{P}\left(|f(X_d) - \mathbf{E}f(X_d)| \geq u\right) \leq e\,\mathbf{P}\left(\frac{\pi}{2}\sigma|X_m| \geq u\right), \quad \forall u \in \mathbf{R}; \tag{46}$$

the latter bound is better than that in (44) in the typical case $m < d$. Inequality (45) follows from (43), because (i) for any matrix A, one has

$$\|A\| \leq \sigma \iff AA^T \leq \sigma^2 I \tag{47}$$

and (ii) by the Jensen inequality,

$$\mathrm{Cov}\, X \leq \mathrm{Cov}\, Y \implies \mathbf{E}F(X) \leq \mathbf{E}F(Y) \tag{48}$$

whenever X and Y are centered Gaussian random vectors and F is a convex function.

Moreover, using a modification of mentioned the Maurey-Pisier method, one can remove the factor $\frac{\pi}{2}$ from the bounds in (45) and (46). This seems to be impossible to achieve via the random change of time argument but is implied by the following statement for vector Itô integrals.

THEOREM 5.1 *Suppose that* $V\colon [0,1] \to \mathbf{R}^{m \times d}$ *is a random matrix function, adapted to a standard Brownian motion* $B_d(\cdot)$ *in* $\mathbf{R}^d$ *and such that* $\|V(t)\| \leq v(t)\ \forall t \in [0,1]$ *almost surely for some non-random real function* v. *Then*

$$\mathbf{E}F\left(\int_0^1 V(t)dB_d(t)\right) \leq \mathbf{E}F\left(\sigma X_m\right), \tag{49}$$

for any convex $F\colon \mathbf{R}^m \to \mathbf{R}$ *and* $\sigma := (\int_0^1 v^2(t)dt)^{1/2}$. *Hence,*

$$\mathbf{P}\left(\left|\int_0^1 V(t)dB_d(t)\right| \geq u\right) \leq e\,\mathbf{P}\left(\sigma|X_m| \geq u\right), \quad \forall u \in \mathbf{R}. \tag{50}$$

This theorem is immediate from Theorem 3.11 and the following general lemma for certain random variables including possibly multivariate martingale transforms.

LEMMA 5.2 *Let* $S := \sum_{k=1}^n V_{k-1}Y_k$, *where* (V_k) *and* (Y_k) *are random sequences in* $\mathbf{R}^{m \times d}$ *and* $\mathbf{R}^d$, *respectively, adapted to a filtration* $(\mathcal{G}_k)$, *and such that for all* k, $v_k := \mathrm{ess\,sup}\|V_k\| < \infty$ *and* $\mathcal{L}(Y_k|\mathcal{G}_k) \preceq N(\mathbf{0}, \sigma_k^2 I_d)$ *a.s. (mod* $\mathcal{F}$); *here,* $\|\cdot\|$ *is the Euclidian operator norm,* $\sigma_k \in \mathbf{R}$, *and* $\mathcal{F}$ *is any class of convex real functions on* $\mathbf{R}^d$. *Let* F *be any real function on* $\mathbf{R}^m$ *such that for every choice of* $A \in \mathbf{R}^{m \times d}$ *and* $b \in \mathbf{R}^m$, *the function* $\mathbf{R}^d \ni x \mapsto F(Ax+b)$ *belongs to* $\mathcal{F}$. *Then*

$$\mathbf{E}F(S) \leq \mathbf{E}F(\sigma X_m), \quad \text{where } \sigma := \left(\sum_k v_{k-1}^2 \sigma_k^2\right)^{1/2}.$$

Proof. Let $S_i := \sum_{k=1}^i V_{k-1}Y_k$, $R_i := \sum_{k=i+1}^n v_{k-1}\sigma_k X_m^{(k)}$, where, for every natural p, $X_p, X_p^{(1)}, X_p^{(2)}, \ldots$ are iid and also independent of $\mathcal{G}_n$. Also let $T_i := S_i + R_i$, $\forall i$. Then $\forall i$

$$\begin{aligned}
\mathbf{E}F(T_i) &= \mathbf{E}\mathbf{E}(F(S_{i-1} + V_{i-1}Y_i + R_i)|\mathcal{G}_{i-1}, R_i) \\
&\leq \mathbf{E}\mathbf{E}(F(S_{i-1} + \sigma_i V_{i-1}X_d^{(i)} + R_i)|\mathcal{G}_{i-1}, R_i) \\
&\leq \mathbf{E}\mathbf{E}(F(S_{i-1} + \sigma_i v_{i-1}X_m^{(i)} + R_i)|\mathcal{G}_{i-1}, R_i) \\
&= \mathbf{E}F(T_{i-1}).
\end{aligned} \tag{51}$$

The first inequality here follows because $\mathcal{L}(Y_k|\mathcal{G}_k) \preceq N(\mathbf{0}, \sigma_k^2 I_d)$ a.s. (mod $\mathcal{F}$) and the function $\mathbf{R}^d \ni x \mapsto F(Ax + b)$ belongs to $\mathcal{F}$; the second, because $V_{i-1}X_d^{(i)} \preceq v_{i-1}X_m^{(i)}$ (mod $\mathcal{F}$); in turn, the latter inequality follows in view of (47) and (48), since every function in $\mathcal{F}$ is convex. Thus,

$$\mathbf{E}F(S) = \mathbf{E}F(T_n) \leq \mathbf{E}F(T_0) = \mathbf{E}F(\sigma X_m).$$

□

Using the mentioned Maurey-Pisier argument and Theorem 5.1, one removes the factor $\frac{\pi}{2}$ from (45) and (46):

THEOREM 5.3 *For Ψ and f as in (45) and (46),*

$$\mathbf{E}\Psi(f(X_d) - \mathbf{E}f(X_d)) \leq \mathbf{E}\Psi(\sigma X_m), \tag{52}$$

and so,

$$\mathbf{P}\left(|f(X_d) - \mathbf{E}f(X_d)| \geq u\right) \leq e\,\mathbf{P}\left(\sigma|X_m| \geq u\right), \quad \forall u \in \mathbf{R}. \tag{53}$$

Proof. Modulo Theorem 5.1, the proof is very similar to the mentioned Maurey-Pisier argument and given here for the reader's convenience. Let $\xi(t) := \mathbf{E}(f(B_d(1)|B_d(t)) = g(t, B_d(t))$, where $g(t, x) := \mathbf{E}f(x + B_d(1 - t))$, $0 \leq t \leq 1$. W.l.o.g., f is sufficiently smooth with $\|f'(x)\| \leq \sigma \ \forall x \in \mathbf{R}^d$. Hence, by the Îto formula and the heat equation $g_t' + \frac{1}{2}\Delta g = 0$, one has

$$f(B_d(1)) - \mathbf{E}f(B_d(1)) = \xi(1) - \xi(0) = \int_0^1 g_x'(t, B_d(t))dB_d(t);$$

note also that a.s. for all t, $\|g_x'(t, B_d(t))\| \leq \sup_x \|f'(x)\| \leq \sigma$. It remains to apply Theorem 5.1.

□

The next inequality is a slight improvement of Theorem 2.3 of Pinelis (1994) [21]. One has

$$|\eta_1 x_1 + \cdots + \eta_n x_n| \preceq |\xi_1 x_1 + \cdots + \xi_n x_n| \pmod{\mathcal{F}_3}, \tag{54}$$

where $\eta_1, \ldots, \eta_n, \xi_1, \ldots, \xi_n$ are any independent real valued random variables (no two of them are necessarily identically distributed), $\mathbf{E}\eta_i = 0$, $|\eta_i| \leq 1$ almost surely (a.s.), the ξ_i's are symmetrically distributed, $\mathbf{E}\xi_i^2 = 1$ for all i, and the x_i's are nonrandom (or independent of the η_i's and ξ_i's) vectors in a Hilbert space $(\mathcal{H}, |\cdot|)$.

The proof of (54) is similar to that of Theorem 2.3 of Pinelis (1994) [21]; in order to eliminate the assumption of the evenness of the moment functions in [21], it suffices to note that if in the proof of Lemma 3.1 therein one takes $t < 0$, then the expression in (ii) is valid for all $u \in \mathbf{R}$.

In fact, (54) is a duality type result, showing that both the minimum of the R.H.S. of (54) and the maximum of its L.H.S. are attained when both the η_i's and ξ_i's are Rademacher. Eaton (1970, 1974) [6, 7] used a different method to obtain a similar result for $\mathcal{H} = \mathbf{R}$.

[Cf. Ledoux and Talagrand (1991, page 97) [17]: for all convex functions $F \geq 0$, x_i in a separable Banach space $(\mathcal{X}, \|\cdot\|)$, g_i iid $N(0,1)$,

$$\mathbf{E}F\left(\left\|\sum \varepsilon_i x_i\right\|\right) \leq \mathbf{E}F\left(\left(\frac{\pi}{2}\right)^{1/2} \left\|\sum \gamma_i x_i\right\|\right).] \tag{55}$$

The case, which is particularly interesting from the point of view of certain statistical applications (as in Efron (1969) [9], Eaton and Efron (1970) [8], Pinelis (1994) [21]), is the one when the (random) Gram matrix of the tuple $(x_1, \ldots, x_n)$ of random vectors, independent of the η_i's and ξ_i's, is that of an orthoprojector whose rank does not exceed a given natural number, say d, almost surely. Taking then, besides, the ξ_i's to be normally distributed, one obtains from (54):

$$|\eta_1 x_1 + \cdots + \eta_n x_n| \preceq |X_d| \pmod{\mathcal{F}_3}. \tag{56}$$

This, together with (22), implies

$$\mathbf{P}\left(|\eta_1 x_1 + \cdots + \eta_n x_n| \geq u\right) \leq c\,\mathbf{P}\left(|X_d| \geq u\right), \ u \in \mathbf{R}, \tag{57}$$

where $c = c(\infty; 3, 0) = 2e^3/9$ is the best constant factor obtainable based on (56). Inequality (57) is the main result in [21], where the proof is based on specific properties of the χ_d distribution.

To appreciate this result, note that even in the much simpler special case, namely, $d = 1$, the weaker version of (57) – with $\sqrt{2/\pi}\exp(-u^2/2)/u$ in place of $\mathbf{P}(\chi_1 \geq u)$ and only for $u > \sqrt{2}$ – had long been an unproved conjecture by Eaton (1974) [7].

Note that (54) also implies

$$\mathbf{P}\left(|\eta_1 x_1 + \cdots + \eta_n x_n| \geq u\right) \leq \left(\frac{e}{d\sigma^2}\right)^{d/2}\left(\prod_{i=d+1}^{\infty}\frac{1}{\sqrt{1 - \sigma_i^2/\sigma^2}}\right)u^d \exp\left\{-\frac{u^2}{2\sigma^2}\right\}, \tag{58}$$

where $u > \sigma\sqrt{d}$ and $\sigma^2 := \sigma_1^2 = \ldots = \sigma_d^2 > \sigma_{d+1}^2 \geq \sigma_{d+2}^2 \ldots$ are the eigenvalues of G; this easily follows from (54). The R.H.S. of (58) is much smaller for large u than an expression of the type $1 - \Phi((u - m)/\sigma)$, where Φ is the distribution function of $N(0, 1)$.

Yet another application of Theorem 3.11 and Lemma 5.2 [with $d = m = 1$, $V_k = 1$, $Y_k = d_k$, and $\mathcal{F} = \mathcal{F}_3$], taking into account also (54), is the following improved version of the Hoeffding-Azuma inequality.

THEOREM 5.4 Let $f_0 = 0$, $f_n = \sum_{i=1}^n d_i$ be a real-valued martingale. One has

$$f_n \preceq \sigma X_1 (\mathrm{mod}\ \mathcal{F}_3), \quad where \ \sigma := \left(\sum \|d_i\|_\infty^2\right)^{1/2}, \tag{59}$$

and so,

$$\mathbf{P}\left(f_n \geq u\right) \leq \frac{2e^3}{9}\left(1 - \Phi\left(\frac{u}{\sigma}\right)\right), \ \forall u \in \mathbf{R}. \tag{60}$$

The latter bound is much smaller than the Hoeffding-Azuma upper bound $e^{-u^2/(2\sigma^2)}$ whenever $e^{-u^2/(2\sigma^2)}$ is much smaller than 1.

Let us also list some applications to nonnegative sub- and super-martingales.

If $f := (f_n)$ is a nonnegative submartingale bounded from above by 1, and g is strongly subordinate to f [see Burkholder (1994) [4] for the definition of being strongly subordinate], then for all n,

$$\mathbf{P}\left(|g_n| > 2u\right) < \frac{e^2}{3} e^{-u}, \; u > 2.$$

If $X := (X_t)$ is a nonnegative submartingale bounded from above by 1, and Y is the integral of H with respect to X, where H is a predictable process such that $|H| \leq 1$, then for all t,

$$\mathbf{P}\left(|Y_t| > 2u\right) \leq \frac{e^2}{3} e^{-u}, \; u > 2.$$

If $f := (f_n)$ is a nonnegative supermartingale bounded from above by 1, and g is strongly subordinate to f, then for all n,

$$\mathbf{P}\left(|g_n| > 2u\right) < \frac{e^2}{2} \mathbf{E} f_0 e^{-u}, \; u > 2.$$

If $X := (X_t)$ is a nonnegative supermartingale bounded from above by 1, and Y is the integral of H with respect to X, where H is a predictable process such that $|H| \leq 1$, then for all t,

$$\mathbf{P}\left(|Y_t| > 2u\right) \leq \frac{e^2}{2} \mathbf{E} X_0 e^{-u}, \; u > 2.$$

These results are obtained using Theorems 6.1, 7.1, 10.1, and 11.1 of Burkholder (1994) [4] by making the choice $\Phi(t) = (t-u+2)_+^2/4$, $u > 2$, suggested by Theorem 2.5; note that $c(\infty; 2, 0) = e^2/2$.

References

[1] Billingsley, P. (1968) *Convergence of Probability Measures* Wiley, New York.

[2] Borell, C. (1975). The Brunn-Minkowski inequality in Gauss spaces. *Invent. Math.* **6** 387–400.

[3] Borell, C. (1975). Convex set functions in d-space. *Period. Math. Hungar.* **6** 111-136.

[4] Burkholder, D.L. (1994). Strong differential subordination and stochastic integration. *Ann. Probab* **22** 995–1025.

[5] De la Peña, V. H.; Montgomery-Smith, S. J.; and Szulga, J. (1994). Decoupling inequalities for tail probabilities of multilinear forms of symmetric and hypercontractive variables. *Ann. Prob* **22** 1745–1765.

[6] Eaton, M. L. (1970). A note on symmetric Bernoulli random variables. *Ann. Math. Statist.* **41** 1223–1226.

[7] Eaton, M. L. (1974). A probability inequality for linear combinations of bounded random variables. *Ann. Statist.* **2** 609–614.

[8] Eaton, M. L. and Efron, B. (1970). Hotelling's T^2 test under symmetry conditions. *J. Amer. Statist. Assoc.* **65** 702–711.

[9] Efron, B. (1969). Student's t-test under symmetry condition. *J. Amer. Statist. Assoc.* **64** 1278–1302.

[10] Fan, Ky (1953). Minimax theorems. *Proc. Nat. Acad. Sci. USA* **39** 42–47.

[11] Ibragimov, I. A., Sudakov, V. N., and Tsirel'son, B. S. Norms of Gaussian sample functions. Proceedings of the third Japan-USSR Symposium on Probability Theory. Lecture Notes in Math. 550, 20–41 (1976). Springer-Verlag.

[12] Karlin, S. and Studden, W. (1966). Tchebycheff Systems: with Applications in Analysis and Statistics. Wiley, New York.

[13] Karr, A. F. (1983) Extreme points of certain sets of probability measures, with applications. *Math. Oper. Res.* **8** 74–85.

[14] Kemperman, J. H. B. (1968). The general moment problem, a geometric approach. *Ann. Math. Statist.* **39** 93–122.

[15] Kemperman, J. H. B. (1983). On the role of duality in the theory of moments. In: *Semi-Infinite Programming and Applications. Internat. Symp. Austin, Texas, 1981; Lect. Notes Econ. Math. Systems.* bf 215 Springer, Berlin – Heidelberg.

[16] Ledoux, M. (1993) Inégalités isopérimétriques en analyse et probabilités. Séminaire Bourbaki, 45ème année, n° 773, 1–27.

[17] Ledoux, M. and Talagrand, M. (1991). Probability in Banach Spaces. Springer, New York.

[18] Marshall, A. W. and Olkin, I. (1979). *Inequalities: theory of majorization and its applications.* Academic, New York.

[19] Meyer, P.A. (1966). *Probability and Potentials.* Blaisdell Publishing Co.

[20] Pinelis, I.F. (1991). Criterion for complete determinacy for concave-convexlike games. *Math. Notes* **49** 277-279.

[21] Pinelis, I. (1994). Extremal probabilistic problems and Hotelling's T^2 test under a symmetry condition. *Annals of Statistics* **22** 357-368.

[22] Pinelis, I.F. and Utev, S.A. (1989). Exact exponential estimates for sums of independent random variables. *Theory Probab. Appl.* **34** 340-346.

[23] Pisier, G. (1985). Probability methods in the geometry of Banach spaces. *Probability and Analysis. Lect. Notes. Math.* **1206** 168-241.

[24] Shorack, G. R. and Wellner, J. A. (1986). Empirical Processes with Applications to Statistics. Wiley, New York

Iosif Pinelis
Department of Mathematical Sciences
Michigan Technological University
Houghton, Michigan 49931, USA

Progress in Probability, Vol. 43
© 1998 Birkhäuser Verlag Basel/Switzerland

The Bootstrap of Empirical Processes
for α-Mixing Sequences

DRAGAN RADULOVIĆ*

ABSTRACT. It is shown that the blockwise bootstrap of the empirical process for stationary α-mixing sequences, indexed by indicators of the half lines, converges weakly to the appropriate Gaussian process, conditionally in probability. The conditions imposed are weaker than the weakest known sufficient conditions for the regular CLT for these processes.

1. Introduction and statement of the result

Efron [8] introduced the "bootstrap", a resampling method for approximating the distribution functions of statistics $H_n(X_1, \ldots, X_n; P)$, where the random variables X_i are independent, identically distributed with common law P [i.i.d. (P)]. This method has been validated with limit theorems for many statistics H_n, and in particular Bickel and Freedman [2] proved that it works for empirical processes in the i.i.d. case.

Künsch [11] and Liu and Singh [12] have independently introduced a modification of Efron's bootstrap that applies to weakly dependent stationary sequences, namely the Moving Blocks Bootstrap (MBB). Recently, Radulović [16] showed that for the concrete problem of bootstrapping the mean the situation in the dependent case is similar to the independent case. Namely, a strong mixing stationary sequence satisfying the Central Limit Theorem (CLT) for the mean automatically satisfies the MBB CLT in probability.

Several recent papers have studied the blockwise bootstrap for empirical processes under the assumption of weak dependence. Shao and Yu [19] established the bootstrap version of the CLT for the classical empirical processes under ρ-mixing type of dependence, while Radulović [17] showed the same for a larger class of empirical processes (indexed by VC-subgraph classes of functions) under β-mixing type of dependence. The results for the least restrictive α-mixing, in the classical empirical processes case, were obtained by Naik-Nimbalkar and Rajarshi [13] and Bühlmann [5]. However, the rates of the mixing coefficient assumed by Naik-Nimbalkar and Rajarshi and Bühlmann are much stronger than needed for the original CLT. A recent preprint of Peligrad [14] claims that the bootstrap a.s. for α-mixing sequences holds under weaker conditions than assumed in [13] and [5]. The proof of tightness in her paper had a gap, and the result in the present paper constitutes what I can prove in the same direction.

In this paper (Theorem 1) we offer the bootstrap CLT for empirical processes with α-mixing type of dependence under conditions weaker than the weakest known sufficient conditions for the original CLT (see Yoshihara [22] and Shao

---

*) Research partly supported by NSF Grant No DMS-93-00725

and Yu [20]), and hence significantly improve the results of [13] and [5]. Before formally stating our result let us introduce notation and several definitions.

As a measure of dependence we will use the α-mixing coefficient, as defined in Rosenblatt [18]. Let $\mathcal{A}$ and $\mathcal{B}$ be two σ-fields in $(\Omega, T, \mathbf{P})$. Then

$$\alpha(\mathcal{A}, \mathcal{B}) = \sup_{(A,B) \in \mathcal{A} \times \mathcal{B}} |\mathbf{P}(A \cap B) - \mathbf{P}(A)\mathbf{P}(B)|.$$

Let $\{X_i\}_{i=-\infty}^{\infty}$ be a strictly stationary sequence of random variables (r.v.'s), that is, for each $i_1, \ldots, i_n, n, k \in \mathbf{Z}$, the law of the vector $(X_{i_1+k}, \ldots, X_{i_n+k})$ does not depend on k. Because of stationarity, the mixing coefficients α_n associated to $\{X_i\}_{i=-\infty}^{\infty}$, can be defined as $\alpha_n = \alpha(\mathcal{F}_1, \mathcal{G}_n)$, where $\mathcal{F}_1 = \sigma(X_i : i \leq 1)$ and $\mathcal{G}_n = \sigma(X_i : i \geq n+1)$. Then we say $\{X_i\}_{i=-\infty}^{\infty}$ is a strong (alpha) mixing sequence if $\lim_{n \to \infty} \alpha_n = 0$.

Let us describe the bootstrap procedure: Given the sample $X_1, \ldots, X_n$ and $b \in \mathbf{N}$, $b < n$ we first set $X_{n+i} := X_i$ for $i \in \{1, \ldots, b\}$. Then the MBB sample with block size b is defined as follows: let $B_{i,b} = \{X_i, \ldots, X_{i+b-1}\}$, $i \leq n$, be the block of b observations starting from X_i and let $I_1, \ldots, I_k$, $k := [\frac{n}{b}]$, be i.i.d. uniform on $\{1, \ldots, n\}$. The MBB sample consists of all the data points X_i that belong to the blocks $B_{I_1,b}, \ldots, B_{I_k,b}$, i.e., $X_1^* = X_{I_1}, \ldots, X_b^* = X_{I_1+b-1}, X_{b+1}^* = X_{I_2}, \ldots, X_l^* = X_{I_k+b-1}$ where $l = kb$. Without loss of generality we can assume $l = n$. As of now P^*, E^* and Var^* will stand for conditional P, E, and Var given the sample. It is obvious that X_i^* and therefore P^*, E^*, and Var^* depend on n but, in order to ease notation, we will not make this dependence explicit.

For a given stationary sequence of real valued r.v.'s $\{X_i\}_{i=-\infty}^{\infty}$, and class of functions $\mathcal{F} = \{f_t(x) = 1_{x \leq t}(x) : t \in \mathbf{R}\}$, we define the empirical process indexed by the class $\mathcal{F}$ as

$$Z_n(f)_{f \in \mathcal{F}} := \sqrt{n}(P_n - P)(f) = \frac{1}{\sqrt{n}} \sum_{i=1}^{n} (f(X_i) - P(f)), \tag{1}$$

where $P_n(\omega) = \frac{1}{n} \sum_{i=1}^{n} \delta_{X_i(\omega)}$. The bootstrap version of this process is defined as

$$Z_n^*(f, \omega)_{f \in \mathcal{F}} := \sqrt{n}\left(\widehat{P}_n(\omega) - P_n(\omega)\right)(f) = \frac{1}{\sqrt{n}} \sum_{i=1}^{n} (f(X_i^*(\omega)) - P_n(f, \omega)), \tag{2}$$

where $\{X_i^*\}_{i=1}^{n}$ is obtained by the MBB procedure with block size $b(n)$ and $\widehat{P}_n(\omega) = \frac{1}{n} \sum_{i=1}^{n} \delta_{X_i^*(\omega)}$. Obviously the bootstrap process (2) is defined given the sample (given ω). It is also clear that $Z_n(f)$ can be viewed as a random element in $l^\infty(\mathcal{F})$, and therefore we can study convergence of

$$Z_n(f)_{f \in \mathcal{F}} \overset{W}{\longmapsto} G_P(f)_{f \in \mathcal{F}} \text{ in } l^\infty(\mathcal{F}) \tag{3}$$

in the sense of Hoffmann-Jørgensen [10]. The Gaussian process $G_P(f)_{f \in \mathcal{F}}$ from statement (3) is determined by its covariance structure, that is

$$Cov(G_P(f), G_P(g)) = \lim_{n \to \infty} Cov(Z_n(f), Z_n(g)), \tag{4}$$

if the above limit exists. Analogously, we can consider weak convergence in $l^\infty(\mathcal{F})$ of the bootstrap version of the process

$$Z_n^*(f,\omega)_{f\in\mathcal{F}} \overset{W}{\longmapsto} G(f)_{f\in\mathcal{F}} \text{ in probability or a.s.} \qquad (5)$$

for the centered Gaussian process G, independent of ω. We say that the bootstrap works if the limiting process $G(f)$ coincides with $G_P(f)$. In this paper we consider only the bootstrap "in probability" as defined in Giné and Zinn [9], chapter 3. Namely, we say that statement 5 holds in probability if

$$d_{BL}[\mathcal{L}(Z_n^*(f) \mid X_1,\ldots X_n), \mathcal{L}(G(f))] \underset{n\to\infty}{\longmapsto} 0 \text{ in outer probability,} \qquad (6)$$

where the distance $d_{BL}(\mu,\nu) := \sup\{|\int f d\mu - \int f d\nu| : \|f\|_{BL} \le 1\}$, and $\| . \|_{BL}$ stands for the bounded Lipschitz norm (for more details see [1], chapter 2). Since most of the standard techniques do not apply for outer probability, we are temporarily assuming that the quantity defined in (6) is measurable. At the end of Chapter 3 we address this problem more carefully . It is well known that the above defined bootstrap is sufficient for most of the applications since it allows us to construct asymptotic confidence regions for P. Now we state our result.

Theorem 1 *Let $\{X_i\}_{i=-\infty}^{\infty}$ be a strictly stationary α-mixing sequence of real random variables. Suppose also that $\alpha_n \le n^{-\nu}$ for some $\nu > 1$ and that $\{X_i^*\}_{i=1}^{n}$ is generated by the MBB procedure with block size $b_n \to \infty$ and $b_n = O(n^\mu)$ for some $0 < \mu < \min(\frac{1}{3}, \frac{\nu}{4} - \frac{1}{6})$. Then*

$$Z_n^*(f)_{f\in\mathcal{F}} \overset{W}{\to} G_P(f)_{f\in\mathcal{F}} \text{ in probability,} \qquad (7)$$

where $G_P(f)$ is a centered Gaussian process defined by the covariance structure in (4).

The weakest known sufficient conditions for the original CLT are given by Yoshihara [22]. His assumption on the α-mixing coefficients is

$$\alpha_n = O(n^{-\nu}) \text{ for some } \nu > 3. \qquad (8)$$

Naik-Nimbalkar's and Rajarshi's conditions for the bootstrap version of the empirical CLT are

$$\sum_{i=1}^{\infty}(i+1)^7 \alpha_i^{\frac{1}{2}-\tau} < \infty \text{ and } b_n = O(n^{\frac{1}{2}-\epsilon}) \text{ for some } \tau, \epsilon \in (0, \frac{1}{2}), \qquad (9)$$

while Buhlmann's are

$$\sum_{i=1}^{\infty}(i+1)^{32} \alpha_i^{\frac{1}{2}} < \infty \text{ and } b_n = O(n^{\frac{1}{2}-\epsilon}) \text{ for some } \epsilon \in (0, \frac{1}{2}). \qquad (10)$$

It should be noted that Buhlmann's result also applies to the a.s. bootstrap as well as for vector-valued X_1, with different conditions on α-mixing. It is clear that our conditions are much weaker than (9) and (10) and they are also (surprisingly) weaker than (8). This is true even if we consider the size of the blocks (b_n) to be ϵ close to the optimal $b_n = Cn^{1/3}$. (For a discussion of optimality see [11] Remark 3.3). Moreover, Bradley [4] showed that $\alpha_n = O(1/n)$ is not sufficient for the original CLT to hold.

The reason for such nice bootstrap behavior lies in the fact that with independent resampling the MBB introduces some extra regularity into the system. In particular, we do not encounter the dependence structure of the data while working on the CLT part of the theorem but only when we "open" the MBB sample and deal with the LLN (Lemmas 1-3), which is much easier to handle.

Here is one of the possible applications of Theorem 1. By definition (3) and Yoshihara's result, it is clear that if $F(t)$ is a continuous distribution function and if the stationary sequence $\{X_i\}_{i=-\infty}^{\infty}$ satisfies the appropriate α-mixing condition, then the original empirical CLT implies

$$G_n = \sup_{t\epsilon\mathbf{R}} |\sqrt{n}(F_n(t) - F(t))| \overset{W}{\longmapsto} \sup_{t\epsilon\mathbf{R}} |G(t)|.$$

This formally enables us to do hypothesis testing on the distribution function of the given sample. Unfortunately, unlike in the i.i.d. case, $G(t)$ is not the Brownian bridge but a centered Gaussian process with covariance structure

$$Cov(G(s), G(t)) = \sum_{k=-\infty}^{\infty} Cov(1_{X_0 \leq s}, 1_{X_{|k|} \leq t}).$$

Therefore, it is very hard, if not impossible, to compute the appropriate confidence interval. Since by Theorem 1 we have that Yoshihara's conditions for the original CLT imply the blockwise bootstrap CLT, we have the following:

$$G_n^* = \sup_{t\epsilon\mathbf{R}} |\sqrt{n}(F_n^*(t) - F_n(t))| \overset{W}{\longmapsto} \sup_{t\epsilon\mathbf{R}} |G(t)| \text{in probability.}$$

This in turns implies that we can estimate the distribution of $\sup_{t\epsilon\mathbf{R}} |G(t)|$ using Monte Carlo simulation, and hence the desired confidence interval.

2. Preliminary results

The proof of Theorem 1 will rely heavily on the following results. First we recall Davydov's [6] covariance inequality

$$|Cov(X,Y)| \leq Const.\alpha(X,Y)^r \|X\|_p \|Y\|_q, \tag{11}$$

where r, p, q are positive real numbers such that $r + \frac{1}{p} + \frac{1}{q} = 1$. As a consequence of this inequality Yokoyama [21] has proved the following:

Theorem A *Let $\{X_i\}_{i\in\mathbf{Z}}$ be a strictly stationary strong mixing sequence with $EX_i = 0$ and $\|X_i\|_\infty \leq C < \infty$. Suppose also that for some $r > 2$, $\sum_{i=0}^{\infty}(i+1)^{r/2-1}\alpha_i < \infty$. Then there exist a constant M such that*

$$E \left| \frac{1}{\sqrt{n}} \sum_{i=1}^{n} X_i \right|^r \leq M < \infty. \tag{12}$$

Finally, before proving Theorem 1 we need to establish the following three lemmas.

Lemma 1 Let $\{Z_i^n, i \in \{1, \ldots, n\}, n \in \mathbf{N}\}$ be a triangular array of centered r.v.'s such that for every fixed n the 'row' $\{Z_i^n, i = 1, \ldots, n\}$ is strictly stationary. Let us also assume that $\alpha(Z_1^n, Z_i^n) = a_{0 \vee (i-b_n)}$ where $\{a_i\}_{i \geq 0}$ is some decreasing sequence of positive real numbers and $\{b_n\}_{n > 0}$ is some unbounded non-decreasing sequence of integers. Finally if $\sup_{i, n \in \mathbf{N}} E|Z_i^n|^s = K < \infty$ for some $s \leq 4$ and if r and p are positive numbers such that $r + \frac{4}{p} = 1$ then

$$E\left|\frac{1}{n}\sum_{i=1}^{n} Z_i^n\right|^4 \leq C\left(\frac{b_n^2}{n^2}\|Z_1^n\|_\infty^{4-s} + \|Z_1^n\|_\infty^{\frac{4(p-s)}{p}}\frac{1}{n^2}\sum_{i=1}^{n} ia_i^r\right),$$

where constant C depend on K.

Proof of Lemma 1. (In the following computation the constant C might change from line to line) In order to ease notation we will occasionally write Z_i, and b instead Z_i^n and b_n. It is easy to see that there exists constant C such that

$$E\left|\frac{1}{n}\sum_{i=1}^{n} Z_i\right|^4 \leq \frac{C}{n^4}\{\sum_{i=1}^{n} E(Z_i)^4 + \sum_{1 \leq i < j \leq n} E(Z_i^2 Z_j^2) + \sum_{1 \leq i < j \leq n} |E(Z_i Z_j^3)| \tag{13}$$
$$+ \sum_{1 \leq i < j < k \leq n} |E(Z_i Z_j Z_k^2)| + \sum_{1 \leq i < j < k < l \leq n} |E(Z_i Z_j Z_k Z_l)|\}.$$

The first three terms can be easily estimated by Hölder's inequality. They are bounded by

$$C\frac{n^2}{n^4}E(Z_1)^4 \leq C\frac{1}{n^2}\|Z_1^n\|_\infty^{4-s}. \tag{14}$$

In order to estimate the fourth term we first observe that by the stationarity

$$\frac{C}{n^4}\sum_{1 \leq i < j < k \leq n} |E(Z_i Z_j Z_k^2)| \leq \frac{C}{n^3}\sum_{1 < j < k \leq n} |E(Z_1 Z_j Z_k^2)|$$

(By Davydov's inequality (11) and Hölder's inequality)

$$\leq \frac{Cn}{n^3}\sum_{j=1}^{n} \alpha(Z_1, Z_j)^r (E|Z_1|^p)^{\frac{1}{p}} (E|Z_1|^{3q})^{\frac{1}{q}} \text{ where } r + \frac{1}{p} + \frac{1}{q} = 1$$

(Letting $r = 0$ for $j = 1, \ldots, b$ and taking $p = 3q$)

$$\leq \frac{C}{n^2}\sum_{j=1}^{b} E(Z_1)^4 + \frac{C}{n^2}\sum_{j=1}^{n} a_j^r (E|Z_1|^p)^{\frac{4}{p}} \text{ for } r + \frac{4}{p} = 1$$

(By the assumptions of Lemma 1)

$$\leq C\left(\frac{b_n}{n^2}\|Z_1^n\|_\infty^{4-s} + \frac{1}{n^2}\|Z_1^n\|_\infty^{\frac{4(p-s)}{p}}\sum_{j=1}^{n} a_j^r\right). \tag{15}$$

Finally, in order to estimate the last expression in (13) we proceed as follows. Consider four-tuples (i, j, k, l), $1 \leq i < j < k < l \leq n$. It is easy to see that there are at most $6dn^2$ four-tuples such that $\max(j - i, l - k) = d$. For each such four-tuple, using Davidov's and Hölder's inequality we have the following estimate:

$$|E(Z_i Z_j Z_k Z_l)| \leq \alpha(Z_1 Z_d)^r (E|Z_1|^p)^{\frac{4}{p}} \text{ for } r + \frac{4}{p} = 1.$$

Therefore by stationarity

$$\frac{C_1}{n^4} \sum_{1 \leq i < j < k < l \leq n} |E(Z_i Z_j Z_k Z_l)| \leq \frac{C_2}{n^2} \sum_{i=1}^{n} i\alpha(Z_1 Z_i)^r (E|Z_1|^p)^{\frac{4}{p}}$$

(Letting $r = 0$ for $i = 1, \ldots, b$ and using the assumptions of Lemma 1)

$$\leq C_1 \left(\frac{b_n^2}{n^2} \|Z_1^n\|_{\infty}^{4-s} + \frac{1}{n^2} \|Z_1^n\|_{\infty}^{\frac{4(p-s)}{p}} \sum_{i=1}^{n} i a_i^r \right). \tag{16}$$

Combining (14),(15) and (16) we have proved Lemma 1.

Lemma 2 *Let $\{X_i\}_{i=-\infty}^{\infty}$ be a strictly stationary sequence of real r.v.'s such that X_i is uniformly distributed on $[0, 1]$ and let α_n, b_n, μ and ν be the same as in Theorem 1. Let H_n be some sequence of subsets of $[0, 1]^2$ such that $Card(H_n) \leq Cn^{2/3}$ and let $P_n(s, t) = n^{-1} \sum_{i=1}^{n} 1_{s < X_i \leq t}$. Finally, if $\{X_i^*\}_{i=1}^{n}$ is generated by the MBB procedure with block size b_n, then there exists $\rho > 0$ such that*

$$n^\rho \sup_{(s,t) \in H_n} \left| E^* \left(b_n^{-1/2} \sum_{i=1}^{b_n} \left(1_{\left(s < X_i^* \leq t\right)} - P_n(s, t) \right) \right)^2 \right.$$

$$\left. - E \left(b_n^{-1/2} \sum_{i=1}^{b_n} \left(1_{\left(s < X_i \leq t\right)} - (t - s) \right) \right)^2 \right| \to 0$$

in probability.

Proof of Lemma 2. In order to ease notation we will write b instead of b_n. First we show that we can replace $P_n(s, t)$ with $(t - s)$. Adding and subtracting $(t - s)$ and using the notation $A_b^* := b^{-1/2} \sum_{i=1}^{b} 1_{\left(s < X_i^* \leq t\right)}$ and $A_b := b^{-1/2} \sum_{i=1}^{b} 1_{\left(s < X_i \leq t\right)}$, the above expression is bounded by

$$n^\rho \sup_{(s,t) \in H_n} \left| E^* \left(A_b^* - \sqrt{b}(t - s) \right)^2 - E \left(A_b - \sqrt{b}(t - s) \right)^2 \right| \tag{17}$$

$$+ n^\rho b \sup_{(s,t) \in H_n} |P_n(s, t) - (t - s)|^2$$

$$+ 2n^\rho \sqrt{b} \sup_{(s,t) \in H_n} |P_n(s, t) - (t - s)| \left| E^* \left(A_b^* - \sqrt{b}(t - s) \right) \right| = I_n + II_n + III_n.$$

Since $E^*\left(A_b^* - \sqrt{b}(t-s)\right) = \sqrt{b}(P_n(s,t) - (t-s))$, in order to show that II_n and III_n converge to 0 in probability it is sufficient to show that II_n converges to 0 in probability. To prove this we proceed as follows. (In the following computation the constants C, which depends on ϵ, might change from line to line). For every $\epsilon > 0$

$$P(n^\rho II_n > \epsilon) \le Card(H_n)n^{2\rho}b^2 C \sup_{(s,t)\in H_n} E\left|\frac{1}{n}\sum_{i=1}^n 1_{(s<X_i\le t)} - (t-s)\right|^4$$

$$\le Cn^{2\rho}\frac{n^{2/3}b^2}{n^2} \sup_{(s,t)\in H_n} E\left|\frac{1}{\sqrt{n}}\sum_{i=1}^n [1_{(s<X_i\le t)} - (t-s)]\right|^4 \qquad (18)$$

$$\le Cn^{2\rho}\frac{b^2}{n^{4/3}}n^{(4-r)/2} \sup_{(s,t)\in H_n} E\left|\frac{1}{\sqrt{n}}\sum_{i=1}^n [1_{(s<X_i\le t)} - (t-s)]\right|^r,$$

$$\text{for any } 2 < r \le \min(4, 2\nu). \qquad (19)$$

For $\nu > 2$, we can take $r = 4$ and by Theorem A the above is bounded by $Cn^{2\rho}b^2n^{-4/3}$, and we can obviously choose $\rho > 0$ for which this converges to 0. For $1 < \nu \le 2$ we let $2 < r < 2\nu$. Theorem A implies that (19) is bounded by

$$Cn^{2\rho}\frac{b^2}{n^{4/3}}n^{(4-r)/2}.$$

Since in this case $b^2 \le n^{(\nu/2-1/3-\tau)}$ for some $\tau > 0$, we can obviously choose $\rho > 0$ and $\gamma > 0$ such that $r = 2\nu - \gamma$ and such that the above expression converges to 0. Finally in order to finish the proof of Lemma 2 we proceed as follows. Let $Y_i^n := \left(b^{-1/2}\sum_{j=i}^{i+b}\left(1_{(s<X_i\le t)} - (t-s)\right)\right)^2$. Then for every $\epsilon > 0$

$$P\left(I_n > \epsilon\right) = P\left(\sup_{(s,t)\in H_n} n^\rho\left|\frac{1}{n}\sum_{i=1}^n Y_i^n - EY_1^n\right| > \epsilon\right). \qquad (20)$$

Because of the 'wrapped' data in the definition of the MBB we do not have stationarity of $\{Y_i^n\}_{i=1}^n$, namely the last b of them are not identically distributed. However, without loss of generality we can assume this, since

$$\left\|\frac{1}{n}\sum_{i=n-b+1}^n\left(b^{-1/2}\sum_{j=i}^{i+b}\left(1_{(s<X_i\le t)} - (t-s)\right)\right)^2\right\|_\infty \le C\frac{b^2}{n}, \qquad (21)$$

which converges to 0 by the assumptions on b_n. Therefore (20) is trivially bounded by

$$CCard(H_n)\sup_{(s,t)\in H_n} P\left(n^\rho\left|\frac{1}{n}\sum_{i=1}^n(Y_i^n - EY_i^n)\right| > \epsilon\right)$$

$$\le Cn^{2/3}n^{4\rho}\sup_{(s,t)\in H_n} E\left|\frac{1}{n}\sum_{i=1}^n(Y_i^n - EY_i^n)\right|^4. \qquad (22)$$

For $Z_i^n := Y_i^n - EY_i^n$, Theorem A implies $\sup_{i,n\in\mathbf{N}} E\,|Z_i^n|^s < \infty$ for any fixed $s < \nu$, therefore we can apply Lemma 1 and bound expression (22) with

$$Cn^{4\rho}n^{2/3}\left(\frac{b^2}{n^2}\,\|Z_1^n\|_\infty^{4-s} + \frac{1}{n^2}\,\|Z_1^n\|_\infty^{\frac{4(p-s)}{p}}\sum_{i=1}^n i\alpha_i^r\right) \tag{23}$$

(letting $p = \infty$ and by definition of Z_i^n)

$$\leq Cn^{4\rho}\frac{1}{n^{4/3}}\left(b^{6-s} + b^4\sum_{i=1}^n i\alpha_i\right)$$

If $\nu > 2$, we can take $s = 2$ and the above expression is dominated by

$$Cn^{4\rho}\frac{b^4}{n^{4/3}}.$$

Obviously we can choose $\rho > 0$ such that the above converges to 0 as n tends to infinity. If $\nu \in (1,2]$ then the same expression can be bounded by

$$Cn^{4\rho}\ln(n)\frac{1}{n^{4/3}}\left(b^{6-s} + b^4n^{2-\nu}\right).$$

($\ln(n)$ is here to take care of the case $\nu = 2$). By the assumption on the block size b_n there exists $\tau > 0$ such that $b \leq n^{(\nu/4-1/6-\tau)}$. Since we can choose s arbitrarily close to ν, there exist $\rho > 0$ and $\gamma > 0$ such that $s = \nu - \gamma$ and such that the above expression converges to 0. This proves Lemma 2.

Before stating the next result we will need several definitions. Let $\mathcal{F}$ be a class of functions and let d be a pseudo metric on $\mathcal{F}$. Then for every $\epsilon > 0$, the covering number $N(\epsilon, \mathcal{F}, d)$ is defined by

$$N(\epsilon, \mathcal{F}, d) = \min\{m : \text{there are } f_1, \ldots, f_m \in \mathcal{F} \text{ such that } \sup_{f\in\mathcal{F}}\min_{1\leq j\leq m} d(f, f_j) \leq \epsilon\}.$$

The collection $\mathcal{F}_\epsilon = \{f_1, \ldots f_m\}$ is called an ϵ-net in $\mathcal{F}$. Finally we set

$$\sigma_b^2(f) := Var(b^{-1/2}\sum_{i=1}^b f(X_i)) \text{ and } \widehat{\sigma}_b^2(f) := Var^*(b^{-1/2}\sum_{i=1}^b f(X_i^*)).$$

Now we are in position to formally state the result.

Lemma 3 *Let $\{X_i\}_{i\in\mathbf{Z}}$, $\{X_i^*\}_{i=1}^n$, b_n, ρ and α_n be the same as in Lemma 2. Let $\mathcal{F} = \{f_t(x) = 1_{0<x\leq t}, t \in [0,1]\}$, and $\mathcal{F}_n$ be an $n^{-1/3}$-net in $\mathcal{F}$ under the pseudo distance $d(f_t, f_s) := |t - s|$ such that $Card(\mathcal{F}_n) \leq Const.n^{1/3}$. Finally, let $\mathcal{F}_n' := \{f - g, f, g \in \mathcal{F}_n\}$. Then*

$$A_n = n^\rho \sup_{h\in\mathcal{F}_n'}|\widehat{\sigma}_b^2(h) - \sigma_b^2(h)| \longmapsto 0 \;. \text{ in probability} \tag{24}$$

and

$$B_n = (\ln n)^2 \sup_{f,g\in\mathcal{F};d(f,g)\leq n^{-1/3}} \widehat{\sigma}_b^2(f-g) \longmapsto 0 \;. \text{ in probability.} \tag{25}$$

Proof of Lemma 3. Let us establish (24). First we observe that $Card(\mathcal{F}'_n) \leq Const.n^{2/3}$. By an easy application of Lemma 2 and by the definitions of $\hat{\sigma}_b^2(h)$ and $\sigma_b^2(h)$ we have that $A_n \longmapsto 0$ in probability. In order to show the second part of Lemma 3, we proceed as follows. Since $d(f,g) \leq n^{-1/3}$, (11) implies that $\sigma_b^2(f-g) \leq Const.n^{-\delta}$ for some $\delta > 0$, and it is sufficient to estimate the following, for every $\epsilon > 0$:

$$P((\ln n)^2 \sup_{f,g\in\mathcal{F};d(f,g)<n^{-1/3}} |\hat{\sigma}_b^2(f-g) - \sigma_b^2(f-g)| > \epsilon).$$

Let

$$f_i^b(s,t) := (b^{-1/2} \sum_{j=i}^{i+b} [1_{s<X_j\leq t} - (t-s)])^2$$

and

$$\hat{f}_i^b(s,t) := (b^{-1/2} \sum_{j=i}^{i+b} [1_{s<X_j\leq t} - P_n(s,t)])^2.$$

Using the definition of $\hat{\sigma}_b^2$ and σ_b^2, splitting the above supremum into the cases $s < t$ and $s > t$, and considering an argument similar to that used in (21) it suffices to estimate the following:

$$P(\sup_{\substack{s<t\in[0,1]\\|s-t|\leq n^{-1/3}}} (\ln n)^2 |\frac{1}{n}\sum_{i=1}^n [\hat{f}_i^b(s,t) - E\hat{f}_i^b(s,t)]| > \epsilon). \tag{26}$$

First we would like to replace $P_n(s,t)$, (which appears in the definition of $\hat{f}_i^b(s,t)$) with $(t-s)$. In order to do so, it is sufficient to prove that

$$\sup_{\substack{s<t\in[0,1]\\|s-t|\leq n^{-1/3}}} b(\ln n)^2 |P_n(s,t) - (t-s)| \to 0 \text{ in probability.} \tag{27}$$

The above follows by an easy argument: Consider the grid $\{\frac{p}{n^{1/3}} : p = 1,\ldots,n^{1/3}\}$ of the unit interval $[0,1]$. Using the simple inequality (where $F_n(t)$ is empirical cumulative distribution)

$$F_n(\frac{r-1}{n^{1/3}}) - \frac{r-1}{n^{1/3}} - \frac{1}{n^{1/3}} \leq F_n(t) - t \leq F_n(\frac{r}{n^{1/3}}) - \frac{r}{n^{1/3}} + \frac{1}{n^{1/3}}, \tag{28}$$

for every $t \in [\frac{r-1}{n^{1/3}}, \frac{r}{n^{1/3}}]$, and the trivial fact that $A \leq B \leq C \Rightarrow |B| \leq \max(|A|,|C|)$ it easily follows that in order to prove (27) it suffices to show that

$$\max_{\substack{0\leq p\leq r\leq n^{1/3}\\|p-r|\leq 2}} b(\ln n)^2 |P_n(\frac{p}{n^{1/3}}, \frac{r}{n^{1/3}}) - (\frac{r}{n^{1/3}} - \frac{p}{n^{1/3}})| + \frac{2b}{n^{1/3}} = I_n + II_n,$$

converges to 0 in probability. In the following computation the constant C might change from line to line. Since II_n converges to 0 it is sufficient to show that for every $\epsilon > 0$

$$P(I_n > \epsilon) \leq C(\ln n)^8 n^{1/3} b^4 \sup_{s,t\in[0,1]} E(\frac{1}{n}\sum_{i=1}^n [1_{s<X_i\leq t} - (t-s)])^4$$

(by Theorem A)

$$\leq C \frac{b^4}{n^{4/3}} n^{(4-r)/2}, \text{ for any } 2 < r < \min(4, 2\nu),$$

which converges to 0 by the assumptions on the size of b_n and argument similar to one used in Lemma 2. Therefore we can proceed with estimating (26) using $f_i^b(s,t)$ instead of $\widehat{f}_i^b(s,t)$. Using a grid $\{(\frac{p}{\sqrt{n}}, \frac{r}{\sqrt{n}}) : p, r = 1, \ldots, \sqrt{n}\}$ of the unit rectangle and the inequality (28), we would like to replace $\sup \{s < t \in [0,1]\}$ with $\max\{0 \leq p \leq r \leq \sqrt{n}\}$. In order to do so, we define r_t and $p_s \in \{1, \ldots, \sqrt{n}\}$ such that $p_s^n = p_s n^{-1/2} \leq s \leq (p_s + 1)n^{-1/2} = p_{s+1}^n$ and $r_{t-1}^n = (r_t - 1)n^{-1/2} \leq t \leq r_t n^{-1/2} = r_t^n$. Adding and subtracting the appropriate term, in order to prove the lemma, it is sufficient to estimate the following:

$$P(\sup_{\substack{s<t\in[0,1] \\ |s-t|\leq n^{-1/3}}} (\ln n)^2 |\frac{1}{n} \sum_{i=1}^n [f_i^b(s,t) - Ef_i^b(s,t) - f_i^b(p_s^n, r_t^n) + Ef_i^b(p_s^n, r_t^n)]| > \epsilon)$$

$$+ P(\max_{\substack{0\leq p\leq r\leq\sqrt{n} \\ |p-r|\leq 2n^{1/6}}} (\ln n)^2 |\frac{1}{n} \sum_{i=1}^n [f_i^b(\frac{p}{n^{1/2}}, \frac{r}{n^{1/2}}) - Ef_i^b(\frac{p}{n^{1/2}}, \frac{r}{n^{1/2}})]| > \epsilon) \qquad (29)$$

In order to estimate the first term we define the following:

$$A_i := b^{-1/2} \sum_{j=i}^{i+b} 1_{s<X_j\leq t}, \qquad B := b^{1/2}(t - s)$$

$$\widetilde{A}_i := b^{-1/2} \sum_{j=i}^{i+b} 1_{p_s^n<X_j\leq r_t^n}, \qquad \widetilde{B} := b^{1/2}(r_t^n - p_s^n)$$

and $\delta_\iota := \widetilde{A}_i - A_i$, $\beta := \widetilde{B} - B$. (Each of these terms obviously depend on n but in order to ease the notation we will not make this dependence explicit) First we observe that:

$$f_i^b(s,t) - f_i^b(p_s^n, r_t^n) = -\delta_i^2 - \beta^2 + 2A_i\beta - 2A_i\delta_i + 2\delta_i B - 2B\beta.$$

Trivially, $\beta \leq 2b^{1/2}n^{-1/2}$. This implies $\sup_{s,t\in[0,1]} |\beta^2 + 2A_i\beta - 2B\beta| \leq 10bn^{-1/2}$ which converges to 0. Also, we observe that

$$\sup_{s,t\in[0,1]} E(\delta_i^2 + 2A_i\delta_i + 2\delta_i B)$$

$$\leq \sup_{s,t\in[0,1]} 6b^{1/2}E(\delta_1) \leq 12b \sup_{s\in[0,1]} E1_{p_s^n<X_1\leq p_{s+1}^n} \leq 12bn^{-1/2},$$

which converges to 0. Therefore we only need to estimate

$$P(\sup_{s<t\in[0,1]} (\ln n)^2 |\frac{1}{n} \sum_{i=1}^n [\delta_i^2 + 2A_i\delta_i + 2\delta_i B]| > \epsilon)$$

$$\leq P(\sup_{s<t\in[0,1]} (\ln n)^2|\frac{6b^{1/2}}{n}\sum_{i=1}^{n}\delta_i| > \epsilon)$$

$$\leq 2P(\sup_{s\in[0,1]} (\ln n)^2|\frac{6b^{1/2}}{n}\sum_{i=1}^{n}b^{-1/2}\sum_{j=i}^{i+b}1_{p_s^n<X_j\leq p_{s+1}^n}| > \epsilon/2)$$

(data are wrapped, i.e. $X_{n+i} := X_i$ for $i = 1,\ldots,b$)

$$= 2P(\max_{p\in\{0,\ldots,\sqrt{n}\}} (\ln n)^2|\frac{6b}{n}\sum_{i=1}^{n}1_{p_s^n<X_j\leq p_{s+1}^n}| > \epsilon/2)$$

(for n large enough)

$$\leq 2P(\max_{p\in\{0,\ldots,\sqrt{n}\}} (\ln n)^2|\frac{6b}{n}\sum_{i=1}^{n}(1_{p_s^n<X_j\leq p_{s+1}^n} - n^{-1/2})| > \epsilon/3)$$

(constant C depends on ϵ)

$$\leq \frac{C(\ln n)^8 n^{1/2}b^4}{n^2} \max_{p\in\{0,\ldots,\sqrt{n}\}} E|\frac{1}{n^{1/2}}\sum_{i=1}^{n}(1_{p_s^n<X_j\leq p_{s+1}^n} - n^{-1/2})|^4$$

$$\leq \frac{C(\ln n)^8 b^4}{n^{3/2}} n^{(4-k)/2} \max_{p\in\{0,\ldots,\sqrt{n}\}} E|\frac{1}{n^{1/2}}\sum_{i=1}^{n}(1_{p_s^n<X_j\leq p_{s+1}^n} - n^{-1/2})|^k$$

for any $2 < k < \min(4, 2\nu)$

(by Theorem A and with different constant C)

$$\leq \frac{C(\ln n)^8 b^4}{n^{3/2}} n^{(4-k)/2}.$$

Using the argument similar to the one in Lemma 2 the above converges to 0 by the choice of b. Finally in order to finish the proof of Lemma 3, we observe that the cardinality of the set $H_n = \{0 \leq p \leq r \leq n^{1/2} : |p - r| \leq 2n^{1/6}\}$ is smaller than $Const.n^{2/3}$. Therefore Lemma 2 implies that (29) converges to 0. This proves Lemma 3.

3. Proof of the main result

We will first assume that the sequence X_i is uniformly distributed on $[0, 1]$. It is well known that in order to prove Theorem 1 we need to establish finite dimensional convergence and stochastic equicontinuity of the process (2), conditionally in probability. To establish this, in a view of definition (6) it is sufficient to show that for every subsequence n_k there exists a further subsequence n_{k_l} and a set $C \subset \Omega$, $P(C) = 1$, such that:

a) for every $w \in C$ and for every finite collection $t_{j_1},\ldots,t_{j_d} \in [0,1]$

$$\left(Z_{n_{k_l}}^*(f_{t_{j_1}}),\ldots,Z_{n_{k_l}}^*(f_{t_{j_d}})\right) \overset{W}{\longmapsto} (G_P(f_{t_{j_1}}),\ldots,G_P(f_{t_{j_d}})) \tag{30}$$

Dragan Radulović

b) for every $\omega \in C$ and for every $\tau > 0$

$$\lim_{\delta \to 0} \limsup_{l \to \infty} P^*(\sup_{d(s,t)<\delta} |Z^*_{n_{k_l}}(f_s, f_t)| > \tau) = 0. \tag{31}$$

for some (any) pseudo distance $d(.)$ for which $\mathcal{F}$ is totally bounded. In order to establish (30) and (31) we will proceed in the same way as in the Radulović [17] paper.

First we observe that, as in the mentioned paper, (30) easily follows by the MBB CLT for the mean (see [16]). Next, we notice that by Davydov's inequality (11), there exists $\phi' > 0$ and a constant C such that

$$\sigma_b^2(f_s, f_t) \leq C \|f_t - f_s\|_{\phi'} = C(E1_{s<X\leq t})^{1/\phi'} = C|t - s|^{1/\phi'}. \tag{32}$$

Now we let $\phi = \min(1/\phi', 1/6, \rho)$ and define $d(s,t) = |t - s|^\phi$ (ρ is from Lemma 2).We divide the proof into two steps.

Claim 1. Let $\beta(n) = n^{-\phi/3}$ and let $f_{s,\beta(n)}$ be the closest point in $\mathcal{F}_{\beta(n)}$ (the $\beta(n)$ net under pseudo distance d) to f_s. Then for every subsequence n_k there exists a further subsequence n_{k_l} such that for every $\tau > 0$ and for every $\delta > 0$ sufficiently small

$$P^*(\sup_{d(s,t)<\delta} |Z^*_{n_{k_l}}(f_s) - Z^*_{n_{k_l}}(f_t)| > 3\tau)$$

$$\leq 2\Psi(\delta, n_{k_l}) + 2P^*(\sup_{s\in[0,1]} |Z^*_{n_{k_l}}(f_s) - Z^*_{n_{k_l}}(f_{s,\beta(n_{k_l})})| > \tau),$$

eventually a.s., where $\Psi(\delta, n)$ is a real function such that $\lim_{\delta \to 0} \lim_{n \to \infty} \Psi(\delta, n) = 0$.

Proof of Claim 1. The argument is based on the restricted chaining technique as presented in Pollard [15] (Theorem 26 p. 160). To mach the conditions of the mentioned theorem we first observe that since our class $\mathcal{F}$ is obviously VC-subgraph, the covering number increases polynomialy and covering integral ($J(\delta, d, T)$ in [15]) is bounded. Second, we have to establish the following: (i.e. expression (24) in [15]) For every subsequence n_k there exists a further subsequence n_{k_l} (call it n') such that

$$d(s,t) < \delta \Rightarrow P^*(|Z^*_{n'}(f_s) - Z^*_{n'}(f_t)| > \eta) \leq 2\exp(-\frac{1}{2}\eta^2/D^2\delta^2), \text{ eventually a.s.} \tag{33}$$

for every $\delta \geq \beta(n')\eta^{1/2}$ and uniformly in s and t such that $d(s,t) \geq \beta(n')$. Using Bernstein's inequality (see [15] p. 194) we have

$$P^*(|Z^*_{n'}(f_s) - Z^*_{n'}(f_t)| > \eta) \leq 2\exp\left(-\frac{1}{2}\frac{\eta^2}{\widehat{\sigma}_{b'}^2(f_s - f_t) + \frac{b'}{3n'^{1/2}}\eta}\right),$$

where $b' = b(n')$. We would like to replace the bootstrap variance $\widehat{\sigma}_{b'}^2$ with $\sigma_{b'}^2$. A careful look at the proof of the mentioned theorem in [15] reviles that in order to use restricted chaining we do not really need (33) to hold for all s, t such that $d(s,t) \geq \beta(n')$ but only for the members of $\beta(n')$-net. (i.e. for $f \in \mathcal{F}_{\beta(n')}$). This is an important observations which allows us to bypass the uniform law of large numbers and use the first part of Lemma 3 instead. Namely, the cardinality of $\mathcal{F}_{\beta(n')}$ is obviously bounded by $Const.(n')^{2/3}$. Therefore Lemma 3, expression (32) and subsequences characterizations of convergence "in probability" imply that there exits a constant D such that

$$\hat{\sigma}_{b'}^2(f_s - f_t) + \frac{b'}{3(n')^{1/2}}\eta \leq \frac{1}{(n')^\phi} + |t - s|^\phi + \frac{1}{3(n')^{1/6}}$$

$$\leq D^2|t - s|^\phi \text{ evenually a.s.}$$

and uniformly in $\{s, t : d(s,t) \geq \beta(n'), f_s, f_t \in \mathcal{F}_{\beta(n')}\}$. Now Claim 1 follows by an easy application of mentioned theorem from [15].

Claim 2. Let $n', b', \beta(n')$ and $f_{s,\beta(n')}$ be the same as in Claim 2. Then for every $\tau > 0$

$$P^*(\sup_{s\in[0,1]} |Z_{n'}^*(f_s) - Z_{n'}^*(f_{s,\beta(n')})| > \tau) \longmapsto 0 \text{ eventually a.s.}$$

Proof of Claim 2. Let $\mathcal{H} := \{f_s - f_{s,\beta(n')} : f \in \mathcal{F}\}$, and let $\mathcal{H}_{n'}$ be a $\frac{1}{(n')^2}$-net for a $\mathcal{H}$ under the pseudo-distance $\tilde{d}(f,g) := P_{n'}(|f - g|)$. It is trivial to see that since $\mathcal{F}$ is a VC-subgraph the cardinality of $\mathcal{H}_{n'}$ is bounded by $C(n')^\lambda$ for some constants $C, \lambda > 0$. Let $\tilde{h}_{n'}(h) \in \mathcal{H}_{n'}$ be the closest element to h under semi-distance $\tilde{d}(f,g)$. To ease the notation we will write $\tilde{h}$ instead of $\tilde{h}_{n'}(h)$. (the constant C in the following computations might change from line to line). Trivially

$$\sup_{h\in\mathcal{H}} |Z_{n'}^*(h - \tilde{h})|$$

$(k' = \frac{n'}{b'})$

$$\leq \sup_{h\in\mathcal{H}} \frac{1}{\sqrt{k'}} \sum_{i=1}^{k'} \frac{1}{\sqrt{b'}} \sum_{j=(i-1)b'+1}^{ib'} |[(h - \tilde{h})(X_j^*) - P_{n'}(h - \tilde{h})]|$$

$$\leq \sup_{h\in\mathcal{H}} \left(\frac{k'}{\sqrt{n'}} \sum_{i=1}^{n'} \sum_{j=i}^{i+b'-1} |[(h - \tilde{h})(X_j)| + \sqrt{n'}P_{n'}|(h - \tilde{h})|\right)$$

$$\leq \sup_{h\in\mathcal{H}} \left(\frac{k'b'}{\sqrt{n'}} \sum_{i=1}^{n'} |[(h - \tilde{h})(X_j)| + \sqrt{n'}P_{n'}|(h - \tilde{h})|\right)$$

$$\leq \sup_{h\in\mathcal{H}} (2n'^{\frac{3}{2}} P_{n'}|(h - \tilde{h})|)$$

(by the definition of $\tilde{h}$)

$$\leq 2n'^{\frac{3}{2}} \frac{1}{n'^2} \leq \frac{2}{\sqrt{n'}}.$$

Finally, we are able to estimate

$$P^*(\sup_{s\in[0,1]} |Z_{n'}^*(f_s) - Z_{n'}^*(f_{s,\beta(n)})| > \tau) = P^*(\sup_{h\in\mathcal{H}} |Z_{n'}^*(h)| > \tau)$$

$$= P^*(\sup_{h \in \mathcal{H}} |Z_{n'}^*(h) - Z_{n'}^*(\tilde{h}) + Z_{n'}^*(\tilde{h})| > \tau)$$

(for $n' > \frac{16}{\tau^2}$)

$$\leq P^*(\sup_{h \in \mathcal{H}} |Z_{n'}^*(\tilde{h})| > \frac{\tau}{2})$$

$$\leq Card(\mathcal{H}_{n'}) \sup_{h \in \mathcal{H}} P^*(|Z_{n'}^*(h)| > \frac{\tau}{2})$$

(for n' large enough)

$$\leq C(n')^\lambda \sup_{h \in \mathcal{H}} P^*(|Z_{n'}^*(h)| > \frac{\tau}{2})$$

(letting $M = \left\| \frac{1}{\sqrt{n'}} \sum_{j=1}^{b'} [h(X_j^*) - P_{n'}(h)] \right\|_\infty$, using definition of MBB and by Bernstein's inequality)

$$\leq C(n')^\lambda \sup_{h \in \mathcal{H}} \exp\left(\frac{-\tau^2}{2\hat{\sigma}_{b'}^2(h) + \frac{2}{3}\tau M} \right)$$

($h \in \mathcal{H}$ implies $h \in \mathcal{F}' = \{f_s - f_t : s, t \in [0,1]\}$ and $d(s,t) \leq \beta(n')$)

$$\leq C(n')^\lambda \sup_{s,t \in [0,1]; d(s,t) \leq \beta(n')} \exp\left(\frac{-\tau^2}{2\hat{\sigma}_{b'}^2(f_t - f_s) + \frac{2}{3}\tau M} \right) \tag{34}$$

In order to replace $\hat{\sigma}_{b'}^2$, we observe that Lemma 3, the fact that $d(s,t) \leq \beta(n') \Leftrightarrow |t - s| \leq n^{-1/3}$, and the characterization of convergence in probability imply that for every subsequence n_k there exists a further subsequence n_{k_r} (call it n'') such that for n'' large enough

$$\hat{\sigma}_{b''}^2(f_t - f_s) \leq \frac{1}{(\ln n'')^2} \text{ uniformly for } d(s,t) \leq \beta(n').$$

Since A_n and B_n from Lemma 3 are both positive and since obviously $A_n + B_n$ converges to 0 in probability we can find a subsequence n' which will work for both A_n and B_n. Therefore the above is true if we replace n'' with n' Also, for n' large enough

$$M \leq b'/n'^{1/2} \leq \frac{3}{2} \ln(n')^{-2}.$$

Therefore (34) is bounded by

$$C(n')^\lambda \exp(-\frac{\tau^2}{2 \ln(n')^{-3} + \tau(\ln n')^{-2}}),$$

which obviously converges to 0 as n' tends to infinity. (C, λ and τ are fixed real numbers greater then zero). This concludes the proof of Claim 2.

Combining Claim 1 and Claim 2 we have proven Theorem 1 for uniform random variables. In order to relax this assumption we proceed as in Billingsley [3] p. 197. Namely, if X_1 has continuous distribution function F_X then $Y_1 := F_X^{-1}(X_1)$

is obviously uniformly distributed. Therefore, by the continuous mapping theorem we can easily extend our proof. In the case when X_1 does not have a continuous distribution function we define a generalized inverse $\widehat{F}_X$ (see [3] p. 142).

Finally, we address the measurability problem. In Chapter 1 we assume the measurability of expression (6) in order to justify the use of the subsequence characterization of convergence in probability (SCCP). It is clear (see [17], Appendix) that in order to use the SCCP we only need measurability of the sets $A_{\delta,\tau} = \{\omega : \sup_{|s-t|<\delta;s,t\in[0,1]} |Z_n(s,t)| > \tau\}$. Since the class of functions $\mathcal{F} = \{f_t = 1_{x\leq t} : t \in [0,1]\}$ is obviously permissible, this follows by the same argument as in Yu [23] (see Appendix; Permissible class, Measurable Suprema). This proves Theorem 1.

Acknowledgements. I would like to thank Prof. Evarist Giné for his valuable comments, discussions and advice. My gratitude also extends to Prof. Hans R. Künsch who pointed out this problem to me.

References

[1] Araujo, A. and Giné, E., *The Central Limit Theorem for Real and Banach Valued Random Variables*, Wiley, New York, 1980.

[2] Bickel, P.J. and Freedman, D., Some asymptotic theory for the bootstrap. *Ann. Statist.* 1981 pp. 1196–1216.

[3] Billingsley, P., *Convergence of Probability Measures*, Wiley, 1968.

[4] Bradley, R.C., A stationary, pairwise independent, absolutely regular sequences for which the central limit theorem fails. *Probab. Th. Rel. Fields* 1989. 81, 1–10.

[5] Bühlmann, P., Blockwise bootstrapped empirical process for stationary sequences. *Ann. Statist.* 1994. 22, 995–1012.

[6] Davydov, Y.A., Convergence of distributions generated by stationary stochastic processes. *Theor. Probab. Appl.* 1968 pp 691–696.

[7] Dudley, R.M., Central limit theorem for empirical processes. *Ann. Prob.* 1978 pp. 899–929.

[8] Efron, B., Bootstrap Methods: Another Look at the Jackknife. *Ann. Statist.*, 1979. 7 pp. 1–26.

[9] Giné, E. and Zinn, J., Bootstrapping general empirical measures. *Ann. Probab.* 1990. 18 pp. 851–869.

[10] Hoffmann-Jørgensen, J., Convergence of stochastic processes on Polish spaces. Unpublished manuscript. 1984.

[11] Künsch, H.R., The Jackknife and the Bootstrap for General Stationary Observations, *Ann. Statist.* 1989. 17 pp. 1217–1241.

[12] Liu, R.Y. and Singh, K. Moving Blocks Jackknife and Bootstrap Capture Weak Dependence, in *Exploring the Limits of Bootstrap*, ed. R. LePage and L.Billard, John Wiley, New York, 1992, pp. 225–248.

[13] Naik-Nimbalkar, U.V. and Rajarshi, M.B., Validity of blockwise bootstrap for empirical processes with stationary observations. *Ann. Statist.* 1994. 22 pp. 980–994.

[14] Peligrad, M., On the blockwise bootstrap for empirical processes for stationary sequences, *preprint* 1995.

[15] Pollard, D., *Convergence of stochastic processes*. Springer, New York. 1984.

[16] Radulović, D., The bootstrap of the mean for strong mixing sequences under minimal conditions. *Stat. and Prob. Letters* 1996. 28. pp 65–72.

[17] Radulović, D., The bootstrap for empirical processes based on stationary observations. *Stochastic Processes and their Applications* 1996. 65. pp 259–279.

[18] Rosenblatt, M., A Central Limit Theorem and a Strongly Mixing Condition. *Proc. Nat. Acad. Sci. U. S. A.* 1956 42. pp. 43–47.

[19] Shao, Q. and Yu, H., Bootstrapping empirical process for stationary ρ-mixing sequences (preprint).

[20] Shao, Q. and Yu, H., Weak convergence for weighted empirical processes of dependent sequences. *Ann. Probab.* 1996, 24. pp. 098–2127.

[21] Yokoyama, R., Moment bounds for stationary mixing sequences. Z. Wahrsch. verw. Gebiete 1980. 52. pp. 45–57.

[22] Yoshihara, K., Note on an almost sure invariance principle for some empirical processes, *Yokoyama Math. J.* 1979. 27. pp. 105–110.

[23] Yu, B., Rates of Convergence for Empirical Processes of Stationary Mixing Sequences. *Ann. Probab.* 1994, 22. pp. 94–116

Dragan Radulović
Department of Mathematics
University of Connecticut
Storrs, CT 06269

PP - Progress in Probability

Edited by
Th. M. Liggett / Ch. Newman / L. Pitt

Progress in Probability is designed for the publication of workshops, seminars and conference proceedings on all aspects of probability theory and stochastic processes, as well as their connections with and applications to other areas such as mathematical statistics and statistical physics.

PA - Probability and its Applications

Edited by
Th. M. Liggett / Ch. Newman / L. Pitt

Probability and its Applications publishes research-level monographs and advanced graduate texts dealing with all aspects of probability theory and stochastic processes, as well as their connections with and applications to other areas such as mathematical statistics and statistical physics.

STATISTICS • FINANCE • INSURANCE

R.-D. Reiss / M. Thomas,
Universität-Gesamthochschule Siegen, Germany

Statistical Analysis
of Extreme Values

1997. 336 pages. Softcover
ISBN 3-7643-5768-1

The statistical analysis of extreme data is important for various disciplines, including insurance, finance, hydrology, engineering and environmental sciences. This book provides a self-contained introduction to the parametric modeling, exploratory analysis and statistical inference for extreme events.

Besides numerous data-based examples, the book contains special chapters about insurance (coauthored by M. Radtke), returns in asset prices (coauthored by C.G. de Vries) and flood frequency analysis. In addition, five longer case studies are included.

The assessment of the adequacy of the parametric modeling and the statistical inference is facilitated by the included statistical software XTREMES, an interactive menu-driven system which runs under Windows 3.1 (16 bit version) and Windows 95, NT (32 bit version). The applicability of the system is enhanced by the Pascal-like integrated programming language XPL.

For orders originating from all over
the world except USA and Canada:
Birkhäuser Verlag AG
P.O Box 133
CH-4010 Basel/Switzerland
Fax: +41/61/205 07 92
e-mail: farnik@birkhauser.ch

For orders originating in the
USA and Canada:
Birkhäuser
333 Meadowland Parkway
USA-Secaurus, NJ 07094-2491
Fax: +1 201 348 4033
e-mail: orders@birkhauser.com

Birkhäuser

Birkhäuser Verlag AG
Basel · Boston · Berlin

http://www.birkhauser.ch